Springer Series in Statistics

For further volumes:
http://www.springer.com/series/692

Lajos Horváth • Piotr Kokoszka

Inference for Functional Data with Applications

Springer

Lajos Horváth
Department of Mathematics
University of Utah
UT, USA

Piotr Kokoszka
Department of Statistics
Colorado State University
CO, USA

ISSN 0172-7397
ISBN 978-1-4614-3654-6 ISBN 978-1-4614-3655-3 (eBook)
DOI 10.1007/978-1-4614-3655-3
Springer New York Heidelberg Dordrecht London

Library of Congress Control Number: 2012937071

Printed on acid-free paper

Springer is part of Springer Science+Business Media (www.springer.com)

To our families

Preface

Aims, scope and audience

Over the last two decades, functional data analysis has established itself as an important and dynamic area of statistics. It offers effective new tools and has stimulated new methodological and theoretical developments. The field has become very broad, with many specialized directions of research. This book focuses on inferential methods grounded in the Hilbert space formalism. It is primarily concerned with statistical hypothesis tests in various functional data analytic settings. Special attention is devoted to methods based on the functional principal components and model specification tests, including change point tests. While most procedures presented in this book are carefully justified by asymptotic arguments, the emphasis is on practically applicable methodology, rather than theoretical insights into the structure of the relevant statistical models. The methodology we present is motivated by questions and data arising in several fields, most notably space physics and finance, but we solve important general problems of inference for functional data, and so the methodology explained in this book has a much broader applicability. Detailed derivations are presented, so readers will be able to modify and extend the specific procedures we describe to other inferential problems.

The book can be read at two levels. Readers interested primarily in methodology will find detailed descriptions of the testing algorithms, together with the assessment of their performance by means of simulations studies, followed by, typically, two examples of application to real data. Researchers interested in mathematical foundations of these procedures will find carefully developed asymptotic theory. We provide both the introduction to the requisite Hilbert space theory, which many graduate students or advanced undergraduate students may find useful, and some novel asymptotic arguments which may be of interest to advanced researchers. A more detailed description of the contents is given below.

As noted above, functional data analysis has become a broad research area, and this book does not aim at giving a comprehensive account of all recent developments. It focuses on the construction of test statistics and the relevant asymptotic theory, with emphasis on models for dependent functional data. Many areas of functional data analysis that have seen a rapid development over the last decade are not covered. These include dynamical systems, sparse logitudinal data and nonparametric methods. The collection of Ferraty and Romain (2011) covers many of the topics which are the focus of this book, including functional regression models and the functional principal components, but it also contains excellent review papers on important topics not discussed in this book, including resampling, curve registration, classification, analysis of sparse data, with special emphasis on nonparametric methods and mathematical theory. The books of Ferraty and Vieu (2006) and Ramsay *et al.* (2009) are complementary to our work, as they cover, respectively, nonparametric methods and computational issues. The monograph of Ramsay and Silverman (2005) offers an excellent and accessible introduction to many topics mentioned above, while Bosq (2000) and Bosq and Blanke (2007) study mathematical foundations. Our list of references is comprehensive, but it is no longer possible to refer even to a majority of important and influential papers on functional data analysis. We cite only the papers most closely related to the research presented in this book.

Outline of the contents

Chapters 1, 2 and 3 introduce the prerequisites, and should be read before any other chapters. Readers not interested in the asymptotic theory, may merely go over Chapter 2 to become familiar with the concepts and definitions central to the whole book. The remaining chapters can essentially be read independently of each other. There are some connections between them, but appropriate references can be followed only if desired. Many chapters end with bibliographical notes that direct the reader to related research papers. The book consists of three parts. Part I is concerned with the independent and identically distributed functions, a functional random sample. Part II studies the functional regression model. Part III focuses on functional data structures that exhibit dependence, in time or in space.

Chapter 1 sets the stage for the remainder of the book by discussing several examples of functional data and their analyses. Some of the data sets introduced in Chapter 1 are revisited in the following chapters. Section 1.5 provides a brief introduction to software packages and the fundamental ideas used in the numerical implementation of the procedures discussed in the book. Part I begins with Chapter 2 which introduces the central mathematical ideas of the book, the covariance operator and its eigenfunctions. Chapter 3 follows with the definition of the functional principal components, which are the most important ingredient of the methodology we study. Chapters 4, 5 and 6 focus, respectively, on functional counterparts of the multivariate canonical correlation analysis, the two sample problem and the change point problem. Part I concludes with Chapter 7 which discusses a test designed to verify if a sample of functional data can be viewed as a collection of independent identically distributed functions. Part II begins with Chapter 8 which offers an overview of the various functional linear models and of the related inference. The remaining three chapters, 9, 10 and 11, focus on three inferential problems of increasing complexity, testing for the nullity of the autoregressive kernel, testing for the equality of two regression kernels, and testing for the error correlation. The methodology explained in these chapters is illustrated by many data examples. Part III considers functional time series, Chapters 13–16, and modeling spatially distributed curves, Chapters 17 and 18. The initial sections of Chapter 13 review the central ideas of functional autoregression, but the focus is on the methodology and theory developed after the publication of the monograph of Bosq (2000). Spatially distributed functional data have not been discussed in any other book, as far as we know.

Acknowledgements

To a large extent, this book is an account of our research conducted with superb collaborators: Alexander Aue, István Berkes, Siegfried Hörmann, Marie Hušková, Jan Sojka, Josef Steinebach and Lie Zhu. Our graduate students: Devin Didericksen, Stefan Fremdt, Robertas Gabrys, Sasha Gromenko, Inga Maslova, Ron Reeder, Matt Reimherr, Greg Rice and Xi Zhang have been coauthors on almost all papers in which the research reported in this book had been initially published. They prepared all figures and tables, and performed the numerical calculations. We are grateful to many researchers for stimulating comments and discussions, in particular, to John Aston, Pedro Delicado, Diogo Dubart Norinho, Pedro Galeano, Peter Hall, Mevin Hooten, Claudia Kirch, Ping Ma, Hans–Georg Müller, Debashis Paul, Dimitris Politis, Mohsen Pourahmadi, James Ramsay, Philip Reiss, Xiaofeng Shao, John Stevens, James Stock, Jane–Ling Wang and Fang Yao. We acknowledge stimulating research environments at the University of Utah and Utah State University, and the generous funding from the National Science Foundation.

We thank Springer editors John Kimmel and Marc Strauss, and the editorial assistants Hannah Bracken and Kathryn Schell the for the advice they offered at the various stages of the preparation of this book. Our special thanks go to Rajiv Monsurate of Springer who prepared the final LaTeX document. The following publishers kindly granted licences to reproduce figures: John Wiley and Sons for Figure 1.4, Oxford University Press for Figure 1.9, and Statistica Sinica for Figure 8.1.

Lajos Horváth
Piotr Kokoszka

Contents

Chapter 1
Functional data structures

Statistics is concerned with obtaining information from observations X_1, $X_2, \ldots, X_N$. The X_n can be scalars, vectors or other objects. For example, each X_n can be a satellite image, in some spectral bandwidth, of a particular region of the Earth taken at time n. Functional Data Analysis (FDA) is concerned with observations which are viewed as functions defined over some set T. A satellite image processed to show surface temperature can be viewed as a function X defined on a subset T of a sphere, $X(t)$ being the temperature at location t. The value $X_n(t)$ is then the temperature at location t at time n. Clearly, due to finite resolution, the values of X_n are available only at a finite grid of points, but the temperature does exist at every location, so it is natural to view X_n as a function defined over the whole set T.

Some functional data belong to the class of high dimensional data in the sense that every data object consists of a large number of scalar values, and the number of measurements per objects may be larger than the sample size N. If there are m measurements per object, such data falls into the "large m small N" paradigm. However, for functional data, the values within one functional object (a curve or surface) for neighboring arguments are similar. Typical functional objects are thus smooth curves or surfaces that can be approximated by smooth functions. Thus, to perform computations, a functional object can be replaced by a few smooth standard building blocks. The central idea of this book is to study the approximations of functional objects consisting of large number of measurements by objects that can be described using only p coefficients of the standard building blocks, with p being much smaller than N. Such approximations give rise to many interesting and challenging questions not encountered in statistical inference for scalars or vectors.

1.1 Examples of functional data

The data that motivated the research presented in this book is of the form $X_n(t)$, $t \in [a, b]$, where $[a, b]$ is an interval on the line. Each observation is thus a curve. Such curves can arise in many ways. Figure 1.1 shows a reading of a magnetometer over a period of one week. A magnetometer is an instrument that measures the three components of the magnetic field at a location where it is placed. There are over 100 magnetic observatories located on the surface of the Earth, and most of them have digital magnetometers. These magnetometers record the strength and direction of the field every five seconds, but the magnetic field exists at any moment of time, so it is natural to think of a magnetogram as an approximation to a continuous record. The raw magnetometer data are cleaned and reported as averages over one minute intervals. Such averages were used to produce Figure 1.1. Thus $7\times24\times60 = 10,080$ values (of one component of the field) were used to draw Figure 1.1. The dotted vertical lines separate days in Universal Time (UT). It is natural to view a curve defined over one UT day as a single observation because one of the main sources influencing the shape of the record is the daily rotation of the Earth. When an observatory faces the Sun, it records the magnetic field generated by wind currents flowing in the ionosphere which are driven mostly by solar heating. Thus, Figure 1.1 shows seven consecutive functional observations.

Many important examples of data that can be naturally treated as functional come from financial records. Figure 1.2 shows two consecutive weeks of Microsoft stock prices in one minute resolution. In contrast to the magnetic field, the price of an asset exists only when the asset is traded. A great deal of financial research has been done using the closing daily price, i.e. the price in the last transaction of a trading day. However many assets are traded so frequently that one can practically

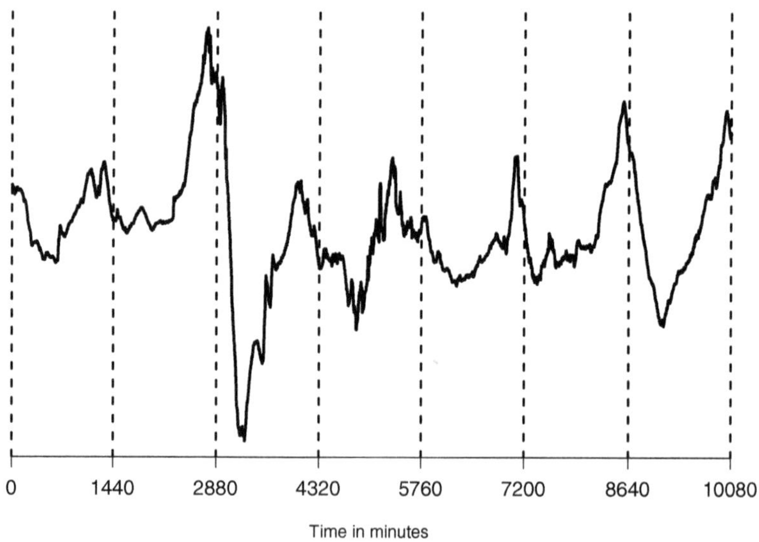

Fig. 1.1 The horizontal component of the magnetic field measured in one minute resolution at Honolulu magnetic observatory from 1/1/2001 00:00 UT to 1/7/2001 24:00 UT.

think of a price curve that is defined at any moment of time. The Microsoft stock is traded several hundred times per minute. The values used to draw the graph in Figure 1.2 are the closing prices in one-minute intervals. It is natural to choose one trading day as the underlying time interval. If we do so, Figure 1.2 shows 10 consecutive functional observations. From these functional observations, various statistics can be computed. For example, the top panels of Figure 1.3 show the mean functions for the two weeks computed as $\hat{\mu}(t) = 5^{-1} \sum_{i=1}^{5} X_i(t)$, where $X_i(t)$ is the price at time t on the ith day of the week. We see that the mean functions have roughly the same shape (even though they have different ranges), and we may ask if it is reasonable to assume that after adjusting for the ranges, the differences in these curves can be explained by chance, or these curves are really different. This is clearly a setting for a statistical hypothesis test which requires the usual steps of model building and inference. Most chapters of this book focus on inferential procedures in models for functional data. The bottom panels of Figure 1.3 show the five curves $X_i(t) - \hat{\mu}(t)$ for each week. We will often work with functional data centered in this way, and will exhibit the curves using the graphs as those in the bottom panels of Figure 1.3.

Functional data arise not only from finely spaced measurements. For example, when measurements on human subjects are made, it is often difficult to ensure that they are made at the same time in the life of a subject, and there may be different numbers of measurements for different subjects. A typical example are growth curves, i.e. $X_n(t)$ is the height of subject n at time t after birth. Even though every

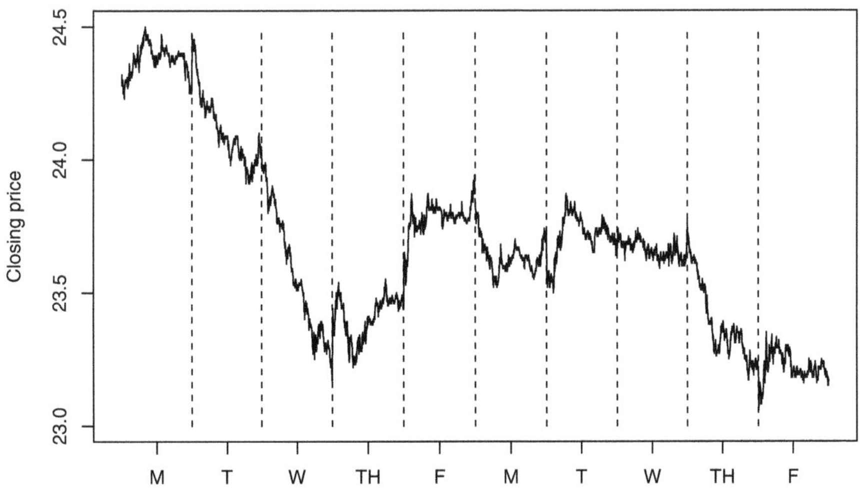

Fig. 1.2 Microsoft stock prices in one-minute resolution, May 1-5, 8-12, 2006

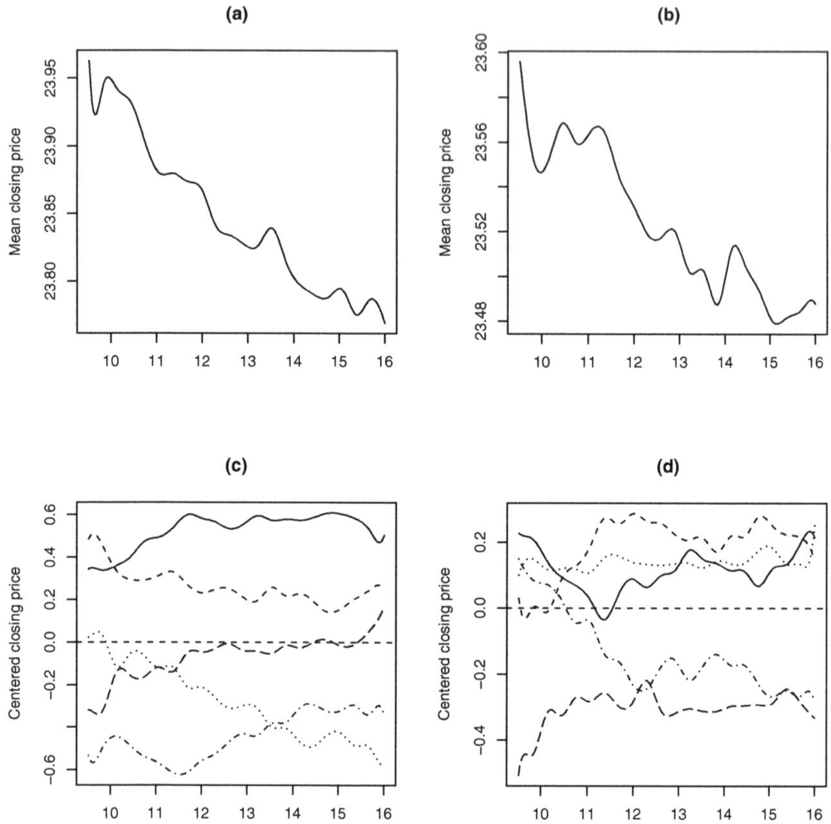

Fig. 1.3 (a) Mean function of Microsoft stock prices, May 1-5, 2006; (b) Mean function of Microsoft stock prices, May 8-12, 2006; (c) Centered prices of Microsoft stock, May 1-5, 2006; (d) Centered prices of Microsoft stock, May 8-12, 2006.

individual has a height at any time t, it is measured only relatively rarely. Thus it has been necessary to develop methods of estimating growth curves from such sparse unequally spaced data, in which smoothing and regularization play a crucial role. Examples and methodology of this type are discussed in the monographs of Ramsay and Silverman (2002, 2005).

It is often useful to treat as functional data measurements that are neither sparse nor dense. Figure 1.4, shows the concentration of nitrogen oxide pollutants, referred to as NO_x, measured at Barcelona's neighborhood of Poblenou. The NO_x concentration is measured every hour, so we have only 24 measurements per day. It is nevertheless informative to treat these data as a collection of daily curves because the pattern of pollution becomes immediately apparent. The pollution peaks in morning hours, declines in the afternoon, and then increases again in the evening. This

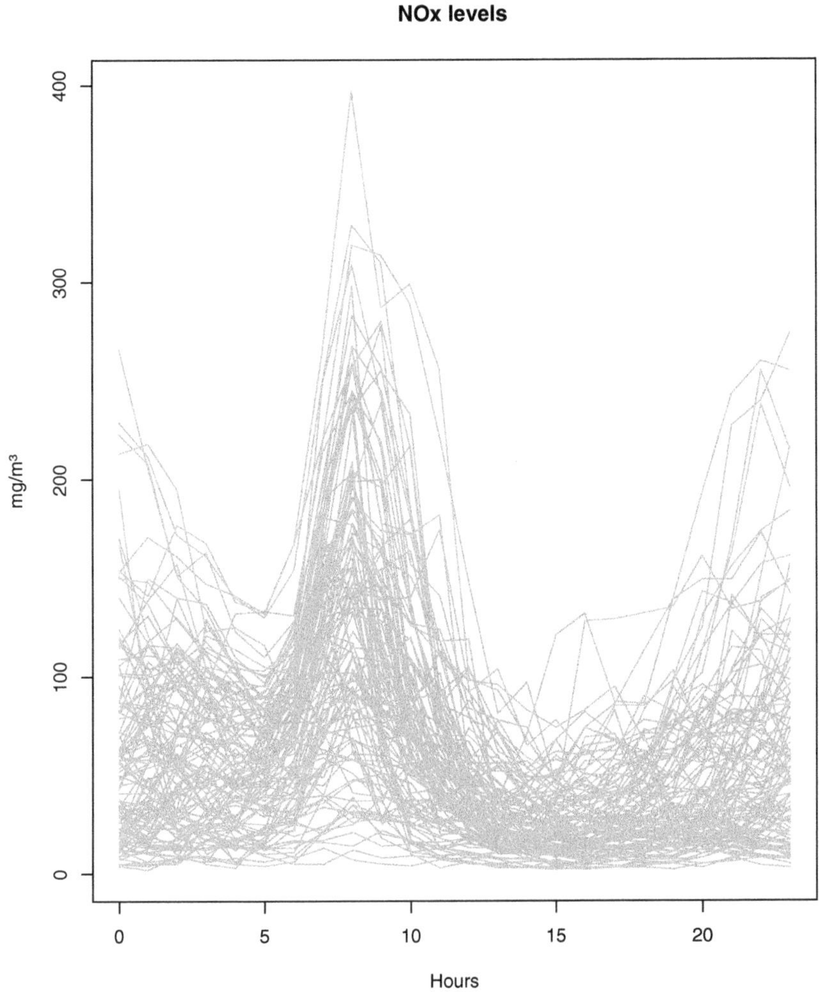

Fig. 1.4 Hourly levels of NO_x pollutants measured in Poblenou, Spain. Each curve represents one day.

pattern is easy to explain because the monitoring station is in a city center, and road traffic is a major source of NO_x pollution. Broadly speaking, for functional data the information contained in the *shape* of the curves matters a great deal. The above data set was studied by Febrero *et al.* (2008), Jones and Rice (1992) study ozone levels In Upland, California.

The usefulness of the functional approach has been recognized in many other fields of science. Borggaard and Thodberg (1992) provide interesting applications of the functional principal component analysis to chemistry. A spectrum is a sampling of a continuous function at a set of fixed wavelengths or energies. Borggaard and Thodberg (1992) point out that multivariate linear regression often fails because the number of input variables is very large. Their simulations and examples show that functional regression provides much better results. Spectra are studied in detail in Ferraty and Vieu (2006). Starting with Kirkpatrick and Heckman (1989), it has been recognized that evolutionary important traits are better modeled as infinite–dimensional data. The examples in Griswold *et al.* (2008) are organismal growth trajectories, thermal performance curves and morphological shapes. Griswold *et al.* (2008) argue that the functional approach provides a more convenient framework than the classical multivariate methods. Many recent applications of functional data analysis are discussed in Ferraty (2011).

In the remaining sections of this chapter, we present a few analyses which illustrate some ideas of FDA. The discussion in this chapter is informal, in the following chapters the exposition will be more detailed and rigorous. In Section 1.2, we discuss in the functional context the concepts of the center of a distribution and outliers. Section 1.3 show how temporal dependence in functional data can be modeled. Finally, Section 1.4 focuses on modeling the dependence between two samples.

1.2 Detection of abnormal NO_x pollution levels

In this section, based on the work of Febrero *et al.* (2008), we show how the fundamental statistical concepts of the center of a distribution and of an outlier can be defined in the functional context. A center of a sample of scalars can be defined by the median, the mean, the trimmed mean, or other similar measures. The definition of an outlier is less clear, but for relatively small samples even visual inspection may reveal suspect observations. For a collection of curves, like those shown in Figure 1.4, it is not clear how to define central curves or outlying curves. The value of a function at every point t may not be an outlier, but the curve itself may be a functional outlier. Generally speaking, once incorrectly recorded curves have been removed, a curve is an outlier if it comes from a populations with a different distribution in a function space than the majority of the curves. An outlier may be far away from the other curves, or may have a different shape. The concept of depth of functional data offers a possible framework for identifying central and outlying observations; those with maximal depth are central, and those with minimal depth are potential outliers.

The depth of a scalar data point can be defined in many ways, see Zuo and Serfling (2000). To illustrate, suppose $X_1, X_2, \dots X_N$ are scalar observations, and

$$F_N(x) = \frac{1}{N} \sum_{n=1}^{N} \mathbf{I}\{X_n \le x\}$$

is their empirical distribution function. The halfspace depth of the observation X_i is defined as

$$HSD_N(X_i) = \min\{F_N(X_i), 1 - F_N(X_i)\}.$$

If X_i is the median, then $F_N(X_i) = 1/2$, and so $HSD_N(X_i) = 1/2$, the largest possible depth. If X_i is the largest point, then $F_N(X_i) = 1$, and so $HSD_N(X_i) = 0$, the least possible depth. Another way of measuring depth is to define

$$D_N(X_i) = 1 - \left| \frac{1}{2} - F_N(X_i) \right|.$$

The largest possible depth is now 1, and the smallest 1/2.

Suppose now that we have a sample of functions $\{X_n(t),\ t \in [a,b],\ n = 1, 2, \ldots, N\}$. We define the empirical distribution function at point t by

$$F_{N,t}(x) = \frac{1}{N} \sum_{n=1}^{N} \mathbf{I}\{X_n(t) \le x\},$$

and we can define a functional depth by integrating one of the univariate depths. For example, Fraiman and Muniz (2001) define the functional depth of the curve X_i as

$$FD_N(X_i) = \int_a^b \left[1 - \left| \frac{1}{2} - F_{N,t}(X_i(t)) \right| \right] dt.$$

There are also other approaches to defining functional depth, an interested reader is referred to Febrero *et al.* (2008) and López-Pintado and Romo (2009).

Once a measure of a functional depth, denote it generically by FD_N, has been chosen, we can use the following algorithm to identify outliers:

1. Calculate $FD_N(X_1), FD_N(X_2), \ldots, FD_N(X_N)$.
2. Remove curves with depth smaller than a threshold C from the sample and classify them as outliers. If there are no such curves, the procedure ends here.
3. Go back to step 1 and apply it to the sample without outliers removed in step 2.

The critical element of this procedure is determining the value of C which should be so small that only a small fraction, say 1%, of the curves are classified as outliers, if there are in fact no outliers. The value of C can then be computed from the sample using some form of bootstrap, two approaches are described in Febrero *et al.* (2008). Step 3 is introduced to avoid masking, which takes place when "large" outliers mask the presence of other outliers.

Febrero *et al.* (2008) applied this procedure with three measures FD_N to the data shown in Figure 1.4, but split into working and non-working days. The two samples containing 76 working and 39 nonworking days between February 23 and June 26, 2005 are shown in Figure 1.5, with outliers identified by black lines. For working

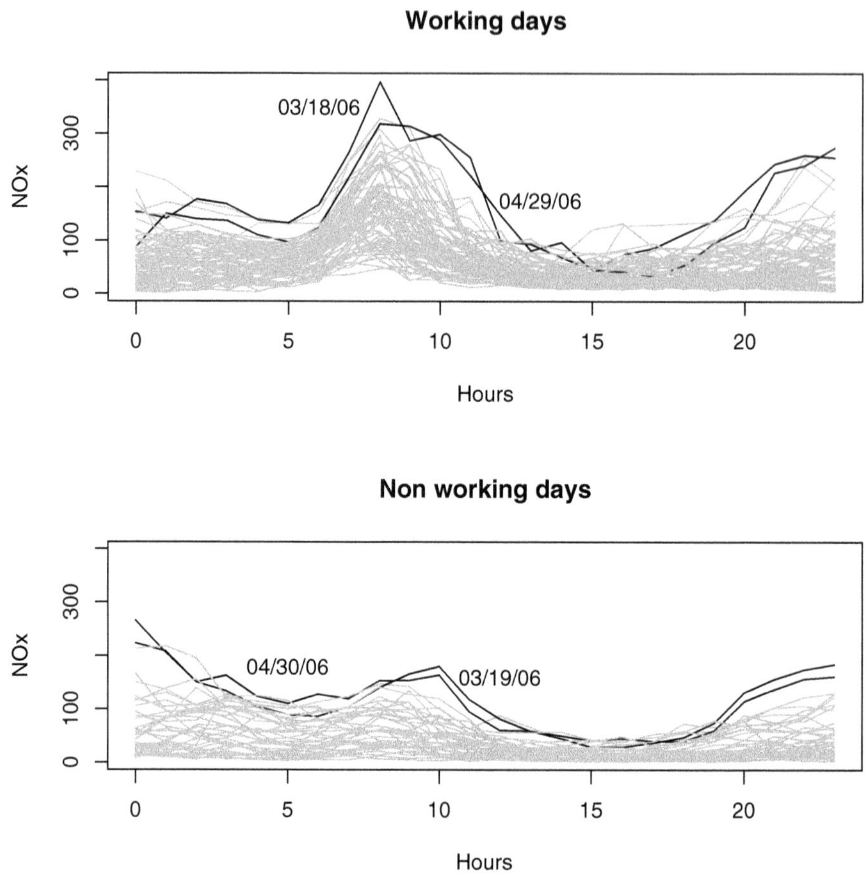

Fig. 1.5 Outliers in NO_x concentration curves in the samples of working and nonworking days.

days, these are Friday, March 18, and Friday, April 29. For non-working days, the outliers are the following Saturdays, March 19 and April 30. These days are the beginning of long weekend holidays in Spain. This validates the identification of the NO_x curves on these days as outliers, as the traffic pattern can be expected to be different on holidays.

Febrero *et al.* (2008) did not attempt to develop an asymptotic justification for the procedure described in this section. Its performance is assessed by application to a real data set. Such an approach is common. In this book, however, we focus on statistical procedures whose asymptotic validity can be established. Resampling procedures for functional data taking values in a general measurable space are reviewed by McMurry and Politis (2010).

1.3 Prediction of the volume of credit card transactions

In this section, based on the work of Laukaitis and Račkauskas (2002), we describe the prediction of the volume of credit card transactions using the functional autoregressive process, which will be studied in detail in Chapter 13.

The data available for this analysis consists of all transactions completed using credit cards issued by Vilnius Bank, Lithuania. Details of every transaction are documented, but here we are interested only in predicting the daily pattern of the volume of transactions. For our exposition, we simplified the analysis of Laukaitis and Račkauskas (2002), and denote by $D_n(t_i)$ the number of credit card transactions on day n, $n = 1, \ldots, 200$, (03/11/2000 – 10/02/2001) between times t_{i-1} and t_i, where $t_i - t_{i-1} = 8$ min, $i = 1, \ldots, 128$. We thus have $N = 200$ daily curves, which we view as individual observations. The grid of 8 minutes was chosen for ease of exposition, Laukaitis and Račkauskas (2002) divide each day into 1024 intervals of equal length. The transactions are normalized to have time stamps in the interval $[0, 1]$, which thus corresponds to one day. The left most panel of Figure 1.6 shows the $D_n(t_i)$ for two randomly chosen days. The center and right panels show smoothed functional versions $D_n(t)$ obtained, respectively, with 40 and 80 Fourier basis functions as follows. Each vector $[D_n(t_1), D_n(t_2), \ldots, D_n(t_{128})]$ is approximated using sine and cosine functions $B_m(t)$, $t \in [0, 1]$, whose frequencies increase with m. We write this approximation as

$$D_n(t_i) \approx \sum_{m=1}^{M} c_{nm} B_m(t_i), \quad n = 1, 2, \ldots, N.$$

The trigonometric functions are defined on the whole interval $[0, 1]$, not just at the points t_i, so we can continue to work with truly functional data

$$Y_n(t) = \sum_{m=1}^{M} c_{nm} B_m(t), \quad n = 1, 2, \ldots, N.$$

In this step, we reduced the the number of scalars needed to represent each curve from 128 to M (40 or 80). If the original data are reported on a scale finer than 8 minutes, the computational gain is even greater. The step of expanding the data with respect to a fixed basis is however often only a preliminary step to further dimension reduction. The number M is still too large for many matrix operations, and the choice of the trigonometric basis is somewhat arbitrary, a spline or a wavelet basis could be used as well. The next step will attempt to construct an "optimal" basis.

Before we move on to the next step, we remove the weekly periodicity by computing the differences $X_n(t) = Y_n(t) - Y_{n-7}(t)$, $n = 8, 9, \ldots, 200$. Figure 1.7 displays the first three weeks of these data. The most important steps of the analysis are performed on the curves $X_n(t), n = 8, 9, \ldots, 200$. which we view as a stationary functional time series. Thus, while each X_n is assumed to have the same distribution in a function space, they are dependent. We assume that each X_n is an

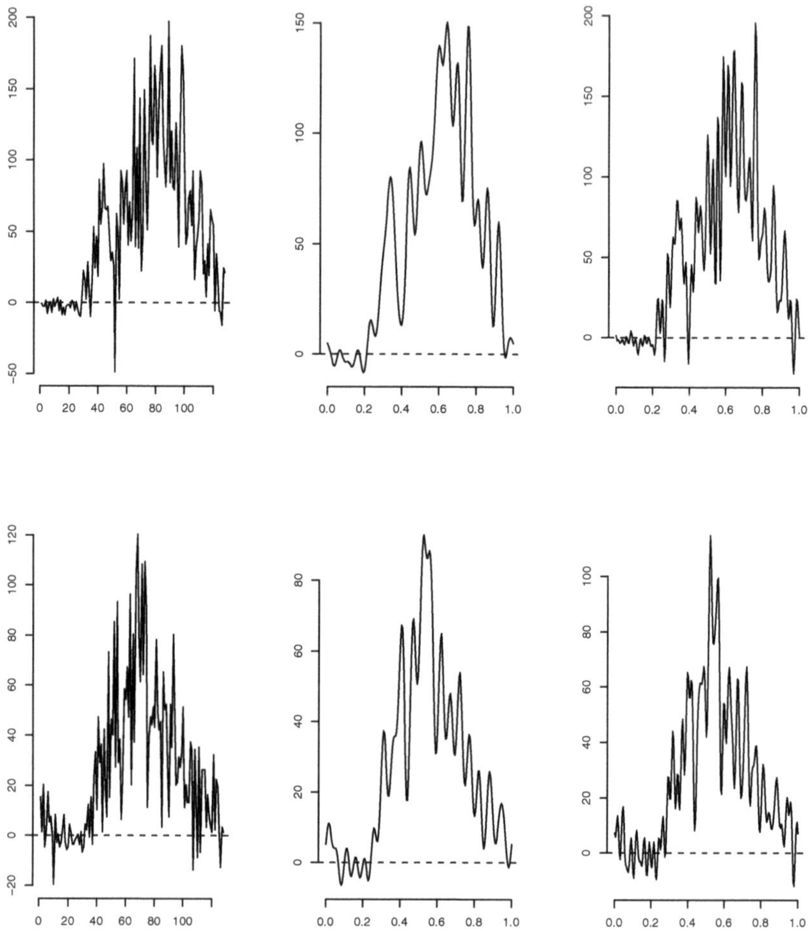

Fig. 1.6 Two functional observations X_n derived from the credit card transactions (left–most panel) together with smooths obtained by projection on 40 and 80 Fourier basis functions.

element of the space $L^2 = L^2([0,1])$ of square integrable functions on $[0,1]$, and that there is a function $\psi(t,s)$, $t \in [0,1], s \in [0,1]$, such that

$$X_n(t) = \int_0^1 \psi(t,s) X_{n-1}(s) ds + \varepsilon_n(t),$$

where the errors ε_n are iid elements of L^2. The above equation extends to the functional setting the most popular model of time series analysis, the AR(1) model, in which the scalar observations X_i are assumed to satisfy $X_i = \psi X_{i-1} + \varepsilon_i$, see e.g.

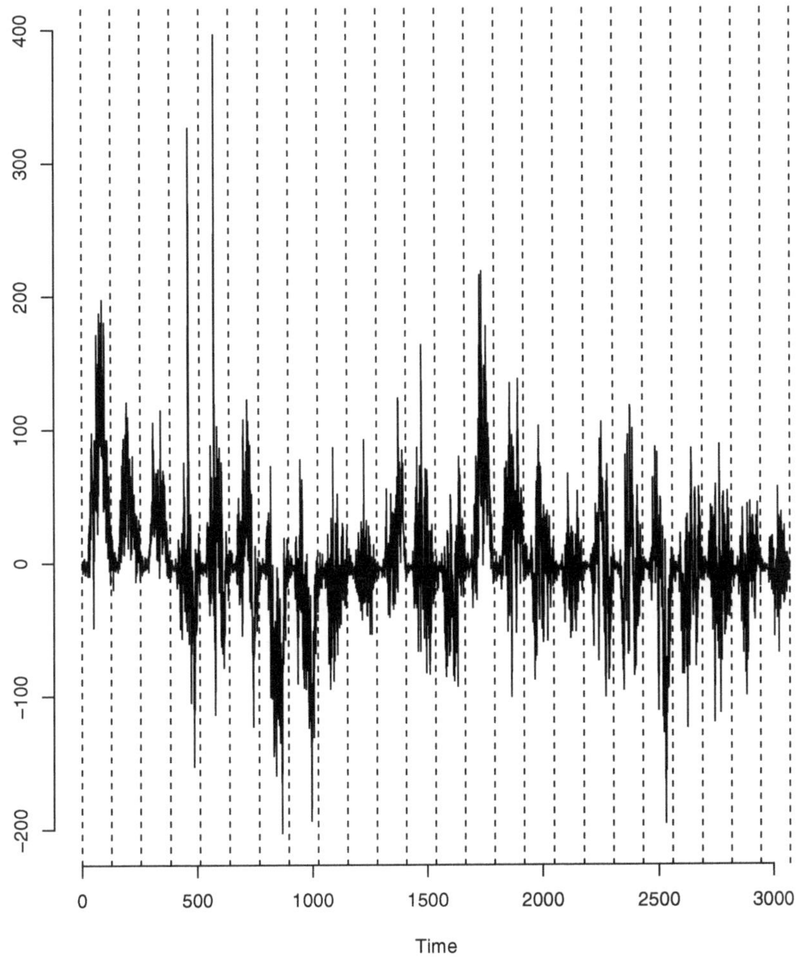

Fig. 1.7 Three weeks of centered time series of $\{X_n(t_i)\}$ derived from credit card transaction data. The vertical dotted lines separate days.

Chapter 3 of Box *et al.* (1994). To compute an estimate of the kernel $\psi(t,s)$, the curves X_n are approximated by an expansion of the form

$$X_n(t) \approx \sum_{k=1}^{p} \xi_{kn} v_k(t),$$

where the v_k are the functional principal components (FPC's) of the X_n, $n = 8, 9, \ldots, 200$. The idea of expansion with respect to FPC's will be taken up in

Chapter 3. Here we note that p is generally much smaller than the number of the points at which the curves are evaluated (128 in this example) or the number M of basis functions (40 or 80 in this example). The v_k are orthonormal, and form an "optimal" system for expressing the observations. Laukaitis and Račkauskas (2002) recommend using $p = 4$ FPC's. Once an estimator $\hat{\psi}$ has been constructed, we can predict X_n via $\hat{X}_n = \int_0^1 \hat{\psi}(t,s)X_{n-1}(s)ds$ and the transaction volume curves via

$$\hat{Y}_{n+1}(t) = Y_{n-6}(t) + \int_0^1 \hat{\psi}(t,s)[Y_n(s) - Y_{n-7}(s)]ds.$$

Figure 1.8 shows examples of two curves Y_n ($n = 150$ and $n = 190$) and their predictions $\hat{Y}_n$. In general, the predictions tend to underestimate the transaction volume. This is because even for the scalar AR(1) process, the series of prediction $\hat{X}_n = \hat{\phi} X_{n-1}$ has a smaller range than the observations $X_n = \phi X_{n-1} + \varepsilon_n$. The problem of prediction of functional time series is studied in detail in Chapter 13.

1.4 Classification of temporal gene expression data

This section, based on the work of Leng and Müller (2006), introduces one of many formulations of the functional linear model. We introduce such models in Chapter 8, and study them in Chapters 9, 11 and 10. Our presentation focuses only on the central idea and omits many details, which can be found in Leng and Müller (2006) and Müller and Stadtmüller (2005).

Figure 1.9 shows expression time courses of 90 genes. The expressions are measured at 18 time points t_i with $t_i - t_{i-1} = 7$ minutes. The genes can be classified as G1 phase and non–G1 phase. A classification performed using traditional methods yielded 44 G1 and 46 non–G1 genes. Leng and Müller (2006) proposed a statistical method of classifying genes based exclusively on their expression trajectories. Their approach can be summarized as follows.

After rescaling time, each trajectory is viewed as a smooth curve $X_n(t)$, $t \in [0,1]$, observed, with some error, at discrete time points t_i. It is assumed that the curves are independent and identically distributed with the mean function $\mu(t) = EX_n(t)$ and the FPC's v_k, so that they admit a representation

$$X_n(t) = \mu(t) + \sum_{k=1}^{\infty} \xi_{kn} v_k,$$

with

$$\xi_{kn} = \int_0^1 (X_n(t) - \mu(t)) v_k(t) dt.$$

The unknown curves X_n must be estimated, as outlined below, but the idea is that the scalars

$$\eta_n = \alpha + \int_0^1 \beta(t)\,(X_n(t) - \mu(t))\,dt,$$

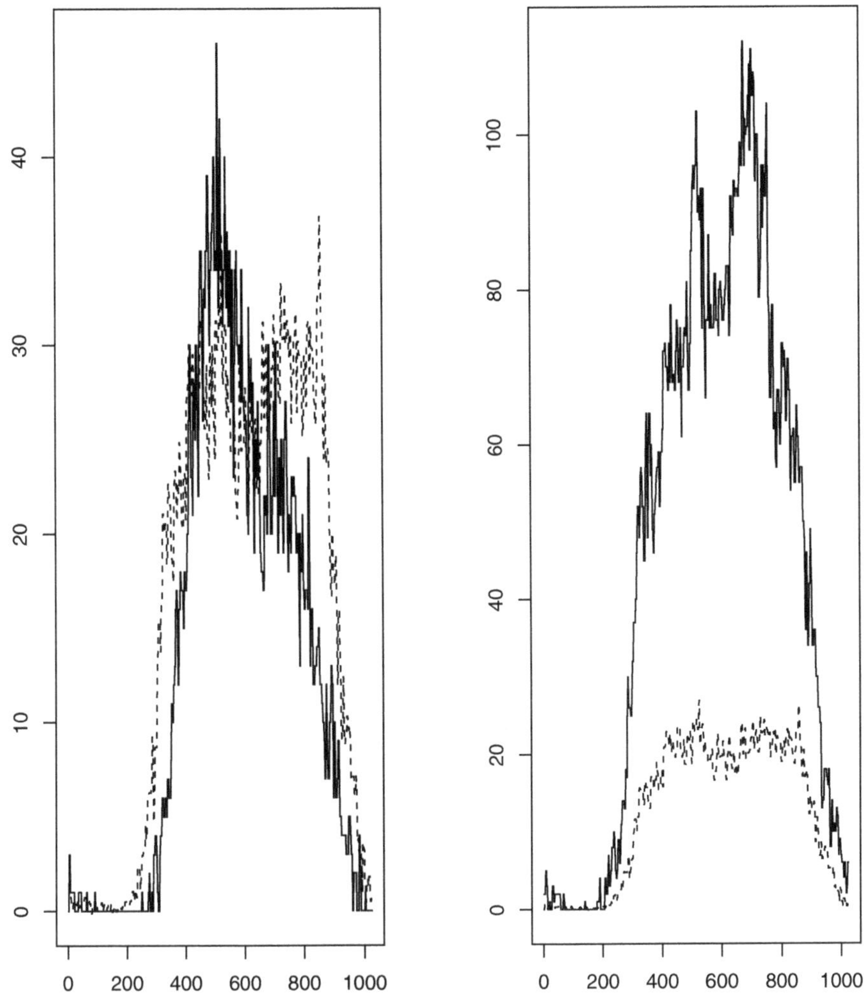

Fig. 1.8 Two credit card transaction volume curves Y_n (solid lines) and their predictions $\hat{Y}_n$ (dotted lines)

for some parameters α and $\beta(t), t \in [0, 1]$, can be used to classify the genes as G1 or non–G1. Note that the parameter β is a smooth curve. The idea of classification is that we set a cut–off probability p_1, and classify a gene as G1 phase if $h(\eta_n) > p_1$, where

$$h(\eta) = \frac{e^\eta}{1 + e^\eta}.$$

The central issue is thus to approximate the linear predictors η_n, and this involves the estimation of the curves X_n and the parameters α and β.

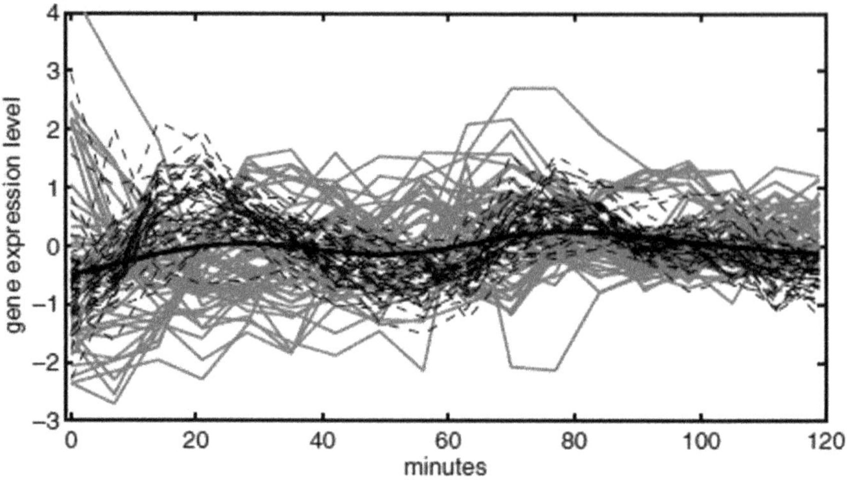

Fig. 1.9 Temporal gene expression profiles of yeast cell cycle. Dashed lines: G1 phase; Gray solid lines: non-G1 phases; Black solid line: overall mean curve.

The curves X_n are estimated by smooth curves

$$X_n^{(p)}(t) = \hat{\mu}(t) + \sum_{k=1}^{p} \hat{\xi}_{kn} \hat{v}_k(t).$$

For the curves shown in Figure 1.9, using $p = 5$ is appropriate. Estimation of the FPC's v_k involves obtaining a smooth estimate of the covariance surface

$$c(t,s) = E\left\{(X_n(t) - \mu(t))(X_n(s) - \mu(s))\right\}, \quad t, s \in [0,1].$$

Inserting $X_n^{(p)}$ into the equation defining η_n yields

$$\eta_n^{(p)}(\alpha, \beta) = \alpha + \int_0^1 \beta(t) \left(\sum_{k=1}^{p} \hat{\xi}_{kn} \hat{v}_k(t) \right) dt,$$

i.e

$$\eta_n^{(p)}(\alpha, \beta) = \alpha + \sum_{k=1}^{p} \beta_k \hat{\xi}_{kn},$$

where

$$\beta_k = \int_0^1 \beta(t) \hat{v}_k(t) dt, \quad k = 1, 2, \ldots, p.$$

The parameters $\alpha, \beta_1, \ldots, \beta_p$ are estimated using the generalized linear model

$$Y_n = h\left(\alpha + \sum_{k=1}^{p} \beta_k \hat{\xi}_{kn} \right) + e_n,$$

where $Y_n = 1$ if a gene is classified as G1 using traditional methods, and $Y_n = 0$ otherwise. This is done by solving an appropriate score equation. Denoting the estimates by $\hat{\alpha}, \hat{\beta}_1, \ldots, \hat{\beta}_p$, we can compute the linear predictor

$$\hat{\eta}_n = \hat{\alpha} + \sum_{k=1}^{p} \hat{\beta}_k \hat{\xi}_{kn}$$

for any trajectory, and classify the gene as G1 phase if $h(\hat{\eta}) > p_1$.

Leng and Müller (2006) applied this method to the time courses of 6,178 genes in the yeast cell cycle, and found that their method compares favorably with an earlier method. In the training sample of the 90 trajectories, they found 5 genes which their method classified as non–G1, but the traditional method as G1. They argued that the traditional method may have classified some of these 5 genes incorrectly.

1.5 Statistical packages, bases, and functional objects

All procedures described in this book can be implemented in readily available statistical software without writing additional code in FORTRAN or C++. We have implemented them using the `R` package `fda`. When applied to a single data sets, these procedures are reasonably fast, and never take more then a few minutes on a single processor laptop or desktop. Some simulations, which require running the same procedure thousands of times can however take hours, or days if bootstrap is involved.

Ramsay *et al.* (2009) provide a solid introduction to computational issues for functional data, and numerous examples. Their book describes not only the `R` package `fda`, but contains many examples implemented in `Matlab`. Clarkson *et al.* (2005) describes the implementation in `S+`.

Throughout this book we often refer the choice of a *basis* and the number of basis functions. This is an important step in the analysis of functional data, which is often not addressed in detail in the subsequent chapters, so we explain it here. This is followed by brief comments on sparsely observed data.

We assume that the collected raw data are already cleaned and organized. Let t be the one-dimensional argument. Functions of t are observed at discrete sampling values t_j, $j = 1, \ldots, J$, which may or may not be equally spaced. We work with N functions with indexes $i = 1, \ldots, N$; these are our functional data. These data are converted to the functional form, i.e. a functional object is created. In order to do this, we need to specify a basis. A basis is a system of basis functions, a linear combination of which defines the functional objects. The elements of a basis may or may not be orthogonal. We express a functional observation X_i as

$$X_i(t) \approx \sum_{k=1}^{K} c_{ik} \phi_k(t),$$

where the ϕ_k, $k = 1, \ldots, K$, are the basis functions. One of the advantages of this approach is that instead of storing all the data points, one stores the coefficients of the expansion, i.e. the c_{ik}. As indicated in Section 1.3, this step thus involves an initial dimension reduction and some smoothing. It is also critical for all subsequent computations which are performed on the matrices built from the coefficients c_{ik}. The number K of the basis functions impacts the performance of some procedures, but other are fairly insensitive to its choice. We discuss this issue in subsequent chapters on a case by case basis. We generally choose K so that the plotted functional objects resemble original data with some smoothing that eliminates the most obvious noise. If the performance of a test depends on K, we indicate what values of K give correct size. The choice of the basis is typically important. We work in this book with two systems: the *Fourier basis* and the *B–spline* basis. The Fourier basis is usually used for periodic, or nearly periodic, data. Fourier series are useful for expanding functions with no strong local features and a roughly constant curvature. They are inappropriate for data with discontinuities in the function itself or in low order derivatives. The B-spline basis is typically used for non-periodic locally smooth data. Spline coefficients are fast to compute and B–splines form a very flexible system, so a good approximation can be achieved with a relatively small K.

In R, bases are created with calls like:

```
minutebasis<-create.fourier.basis(rangeval=c(0,1440),nbasis=49)
```

```
minutebasis<-create.bspline.basis(rangeval=c(0,1440),nbasis=49)
```

The parameter `rangeval` is a vector containing the initial and final values of the argument t. The bases created above will be used for magnetometer data, which consist of 1440 data points per day. These data are in one minute resolution, and there are 1440 minutes in a day. The argument `nbasis` is the number of basis functions.

Once a basis is created, the data are converted into functional objects. This is needed to reduce the computational burden; only the coefficients c_{ik} are used after this conversion. In our example, the reduction is from 1440 to 49 numbers. In order to convert raw data into a functional object the function `data2fd` is used. The code below produces Figure 1.10. The data are the daily records of the magnetic intensity stored in the matrix `data`.

```
minutetime<-seq(from = 1, to = 1440, by = 1)
minutebasis<-create.bspline.basis(rangeval=c(0,1440),nbasis=69)
data.fd<-data2fd(data, minutetime, basisobj=minutebasis)
plot.fd(data.fd, col="black")
title("Functional data, March -- April, 2001")
mean.function<-mean.fd(data.fd)
lines(mean, lw=7)
```

The `fda` package contains a variety of display functions and summary statistics such as `plot.fd`, `mean.fd`, `var.fd`, `sd.fd`, `center.fd`, etc. All these functions use functional objects as input.

The data we work with in this book are available at very densely spaced (typically equispaced) and numerous points t_j (often over a thousand per curve). We

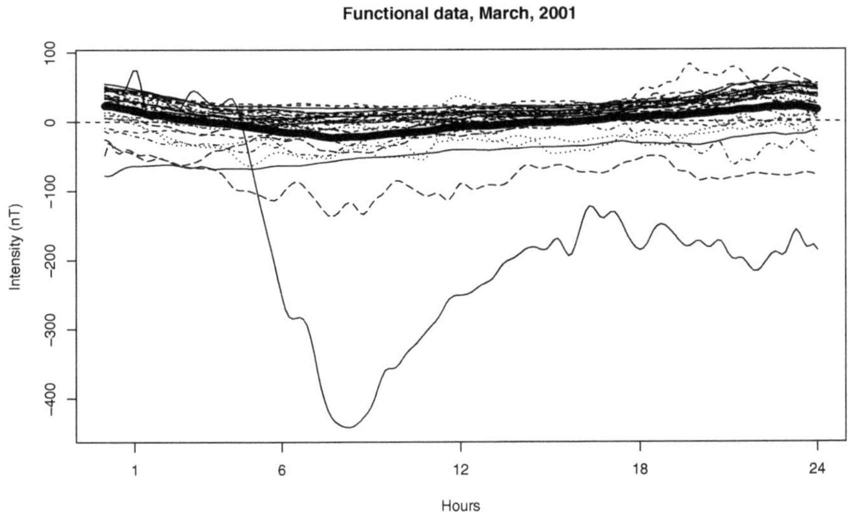

Fig. 1.10 31 magnetic intensity functions with the mean function (thick line)

need smoothing with a basis expansion as a means to make further calculations feasible. The measurement errors for our data are typically very small relative to the magnitude of the curves, and so are negligible. In many application, the data are available only at a few sparsely distributed points t_j, which may be different for different curves, and the data are available with non–negligible measurement errors, Yao *et al.* (2005a) introduce such data structures. For such data, smoothing with a basis expansion is at best inappropriate, and often not feasible. Different smoothing techniques are required to produce smooth curves which can be used as input data for the procedures described in this book. These techniques are implemented in the `Matlab` package `PACE` developed at the University of California at Davis, available at `http://anson.ucdavis.edu/$\sim$mueller/data/software.html`, at the time of writing.

After the data have been represented as functional objects, we often construct a more sparse representation by expanding them with respect to the orthonormal system formed by the functional principal components v_k, as illustrated in Section 1.3 and 1.4. In `R`, this is done by using the function `pca.fd`. For the procedures described in this book, the argument `centerfns` must be set to `TRUE`. This means that the sample mean function $\bar{X}_N(t) = N^{-1} \sum_{i=1}^{N} X_i(t)$ is subtracted from each $X_i(t)$ before the v_k are estimated. The FPC's are thus computed for the *centered* data. Further details are presented in Section 3.4.

Part I
Independent functional observations

Chapter 2
Hilbert space model for functional data

In this Chapter we introduce some fundamental concepts of the theory of operators in a Hilbert space, and then focus of the properties of random samples in the space L^2 of square integrable functions. The space L^2 is sufficient to handle most procedures considered in this book. We also present a few technical results that fit into the framework considered in this chapter, and are used in subsequent chapters.

2.1 Operators in a Hilbert space

In this section we follow closely the exposition in Bosq (2000). Good references on Hilbert spaces are Riesz and Sz.-Nagy (1990), Akhiezier and Glazman (1993) and Debnath and Mikusinski (2005). An in–depth theory of operators in a Hilbert space is developed in Gohberg *et al.* (1990), where the proofs of all results stated in this section can be found.

We consider a separable Hilbert space H with inner product $\langle \cdot, \cdot \rangle$ which generates the norm $\| \cdot \|$, and denote by $\mathcal{L}$ the space of bounded (continuous) linear operators on H with the norm

$$\|\Psi\|_{\mathcal{L}} = \sup\{\|\Psi(x)\| : \|x\| \le 1\}.$$

An operator $\Psi \in \mathcal{L}$ is said to be *compact* if there exist two orthonormal bases $\{v_j\}$ and $\{f_j\}$, and a real sequence $\{\lambda_j\}$ converging to zero, such that

$$\Psi(x) = \sum_{j=1}^{\infty} \lambda_j \langle x, v_j \rangle f_j, \quad x \in H, \tag{2.1}$$

The λ_j may be assumed positive because one can replace f_j by $-f_j$, if needed.

The existence of representation (2.1) is equivalent to the condition: Ψ maps every bounded set into a compact set. Another equivalent condition is the following: the convergence $\langle y, x_n \rangle \to \langle y, x \rangle$ for every $y \in H$ implies that $\|\Psi(x_n) - \Psi(x)\| \to 0$.

Compact operators are also called *completely continuous* operators. Representation (2.1) is called the *singular value decomposition.*

A compact operator admitting representation (2.1) is said to be a *Hilbert–Schmidt* operator if $\sum_{j=1}^{\infty} \lambda_j^2 < \infty$. The space $\mathcal{S}$ of Hilbert–Schmidt operators is a separable Hilbert space with the scalar product

$$\langle \Psi_1, \Psi_2 \rangle_{\mathcal{S}} = \sum_{i=1}^{\infty} \langle \Psi_1(e_i), \Psi_2(e_i) \rangle, \tag{2.2}$$

where $\{e_i\}$ is an arbitrary orthonormal basis, the value of (2.2) does not depend on it. One can show that $\|\Psi\|_{\mathcal{S}}^2 = \sum_{j \geq 1} \lambda_j^2$ and

$$\|\Psi\|_{\mathcal{L}} \leq \|\Psi\|_{\mathcal{S}}. \tag{2.3}$$

An operator $\Psi \in \mathcal{L}$ is said to be *symmetric* if

$$\langle \Psi(x), y \rangle = \langle x, \Psi(y) \rangle, \quad x, y \in H,$$

and positive–definite if

$$\langle \Psi(x), x \rangle \geq 0, \quad x \in H.$$

(An operator with the last property is sometimes called positive semidefinite, and the term positive–definite is used when $\langle \Psi(x), x \rangle > 0$ for $x \neq 0$.)

A symmetric positive–definite Hilbert–Schmidt operator Ψ admits the decomposition

$$\Psi(x) = \sum_{j=1}^{\infty} \lambda_j \langle x, v_j \rangle v_j, \quad x \in H, \tag{2.4}$$

with orthonormal v_j which are the eigenfunctions of Ψ, i.e. $\Psi(v_j) = \lambda_j v_j$. The v_j can be extended to a basis by adding a complete orthonormal system in the orthogonal complement of the subspace spanned by the original v_j. The v_j in (2.4) can thus be assumed to form a basis, but some λ_j may be zero.

2.2 The space L^2

The space $L^2 = L^2([0, 1])$ is the set of measurable real–valued functions x defined on $[0, 1]$ satisfying $\int_0^1 x^2(t)dt < \infty$. The space L^2 is a separable Hilbert space with the inner product

$$\langle x, y \rangle = \int x(t)y(t)dt.$$

An integral sign without the limits of integration is meant to denote the integral over the whole interval $[0, 1]$. If $x, y \in L^2$, the equality $x = y$ always means $\int [x(t) - y(t)]^2 dt = 0$.

An important class of operators in L^2 are the integral operators defined by

$$\Psi(x)(t) = \int \psi(t,s)x(s)ds, \quad x \in L^2,$$

with the real kernel $\psi(\cdot,\cdot)$. Such operators are Hilbert–Schmidt if and only if

$$\iint \psi^2(t,s)dtds < \infty,$$

in which case

$$\|\Psi\|_{\mathcal{S}}^2 = \iint \psi^2(t,s)dtds. \tag{2.5}$$

If $\psi(s,t) = \psi(t,s)$ and $\iint \psi(t,s)x(t)x(s)dt\,ds \geq 0$, the integral operator Ψ is symmetric and positive–definite, and it follows from (2.4) that

$$\psi(t,s) = \sum_{j=1}^{\infty} \lambda_j v_j(t)v_j(s) \quad \text{in } L^2([0,1]\times[0,1]). \tag{2.6}$$

If ψ is continuous, the above expansions holds for all $s,t \in [0,1]$, and the series converges uniformly. This result is known as Mercer's theorem, see e.g. Riesz and Sz.-Nagy (1990).

2.3 Random elements in L^2 and the covariance operator

We view a random curve $X = \{X(t),\ t \in [0,1]\}$ as a random element of L^2 equipped with the Borel σ–algebra. We say that X is integrable if $E\|X\| = E[\int X^2(t)dt]^{1/2} < \infty$. If X is integrable, there is a unique function $\mu \in L^2$ such that $E\langle y, X\rangle = \langle y, \mu\rangle$ for any $y \in L^2$. It follows that $\mu(t) = E[X(t)]$ for almost all $t \in [0,1]$. The expectation commutes with bounded operators, i.e. if $\Psi \in \mathcal{L}$ and X is integrable, then $E\Psi(X) = \Psi(EX)$.

If X is square integrable, i.e.

$$E\|X\|^2 = E\int X^2(t)dt < \infty,$$

and $EX = 0$, the covariance operator of X is defined by

$$C(y) = E[\langle X, y\rangle X], \quad y \in L^2.$$

It is easy to see that

$$C(y)(t) = \int c(t,s)y(s)ds, \quad \text{where } c(t,s) = E[X(t)X(s)].$$

Clearly, $c(t,s) = c(s,t)$ and

$$\iint c(t,s)y(t)y(s)dt\,ds = \iint E[X(t)X(s)]y(t)y(s)dt\,ds$$
$$= E\left[\left(\int X(t)y(t)dt\right)^2\right] \geq 0.$$

Thus C is symmetric and positive–definite, so it has nonnegative eigenvalues.

However, not every symmetric positive–definite operator in L^2 is a covariance operator. To explain, let $v_j, \lambda_j, j \geq 1$, be the eigenfunctions and the eigenvalues of the covariance operator C. The relation $C(v_j) = \lambda_j v_j$ implies that

$$\lambda_j = \langle C v_j, v_j \rangle = \langle E[\langle X, v_j \rangle X], v_j \rangle = E\left[\langle X, v_j \rangle^2\right].$$

The eigenfunctions v_j are orthogonal, and they can be normalized to have unit norm, so that $\{v_j\}$ forms a basis in L^2. Consequently, by Parseval's equality,

$$\sum_{j=1}^{\infty} \lambda_j = \sum_{j=1}^{\infty} E\left[\langle X, v_j \rangle^2\right] = E\|X\|^2 < \infty. \tag{2.7}$$

One can show that, in fact, $C \in \mathcal{L}(L^2)$ is a covariance operator if and only if it is symmetric positive–definite and its eigenvalues satisfy $\sum_{j=1}^{\infty} \lambda_j < \infty$.

To give a specific example of a bounded, symmetric, positive–definite operator which is not a covariance operator, consider an arbitrary orthonormal basis $\{e_j, j \geq 1\}$, so that every $x \in L^2$ can be expanded as $x = \sum_j \langle x, e_j \rangle e_j$. Define

$$\Psi(x) = \sum_j \langle x, e_j \rangle j^{-1} e_j.$$

The operator Ψ is bounded because

$$\|\Psi(x)\|^2 = \sum_j \langle x, e_j \rangle^2 j^{-2} \leq \sum_j \langle x, e_j \rangle^2 = \|x\|^2.$$

Thus, in fact, $\|\Psi\|_{\mathcal{L}} \leq 1$. To see that this operator is symmetric, observe that

$$\langle \Psi(x), y \rangle = \left\langle \sum_j \langle x, e_j \rangle j^{-1} e_j, \sum_k \langle y, e_k \rangle e_k \right\rangle$$
$$= \sum_{j,k} \langle x, e_j \rangle \langle y, e_k \rangle \langle e_j, e_k \rangle j^{-1}$$
$$= \sum_j \langle x, e_j \rangle \langle y, e_j \rangle j^{-1} = \langle x, \Psi(y) \rangle.$$

Since

$$\langle \Psi(x), x\rangle = \sum_j \langle x, e_j\rangle^2 j^{-1} \geq 0,$$

the operator Ψ is positive–definite.

The eigenvalues of Ψ are equal to j^{-1} because $\Psi(e_j) = j^{-1}e_j,\ j \geq 1$. Since $\sum_j j^{-1} = \infty$, Ψ is not a covariance operator.

Throughout the book, we will often use the following central limit theorem , which is stated (and proven) as Theorem 2.7 in Bosq (2000).

A more general version is stated and proven as Theorem 6.2, and an extension to dependent summands is given in Theorem 16.10.

Theorem 2.1. *Suppose $\{X_n,\ n \geq 1\}$ is a sequence of iid mean zero random elements in a separable Hilbert space such that $E\|X_i\|^2 < \infty$. Then*

$$N^{-1/2}\sum_{n=1}^{N} X_n \xrightarrow{d} Z,$$

where Z is a Gaussian random element with the covariance operator

$$C(x) = E[\langle Z, x\rangle Z] = E[\langle X_1, x\rangle X_1].$$

Notice that a normally distributed function Z with a covariance operator C admits the expansion

$$Z \stackrel{d}{=} \sum_{j=1}^{\infty} \sqrt{\lambda_j} N_j v_j, \tag{2.8}$$

with independent standard normal N_j. This follows because C is the covariance operator of the right–hand side of (2.8), and it determines the distributions, as both sides are normally distributed.

For ease of reference, we also recall the law of large numbers , which is stated and proved as Theorem 2.4 in Bosq (2000).

Theorem 2.2. *Suppose $\{X_n,\ n \geq 1\}$ is a sequence of iid random elements in a separable Hilbert space such that $E\|X_i\|^2 < \infty$. Then $\mu = EX_i$ is uniquely defined by $\langle \mu, x\rangle = E\langle X, x\rangle$, and*

$$N^{-1}\sum_{n=1}^{N} X_n \xrightarrow{a.s.} \mu.$$

2.4 Estimation of mean and covariance functions

In applications, we observe a sample consisting of N curves $X_1, X_2, \ldots X_N$. We view each curve as a realization of a random function X, or as a random element of L^2 with the same distribution as X. We can often assume that the X_i are independent, especially if these curves arise from measurements on subjects randomly selected from a large population.

Assumption 2.1. *The observations $X_1, X_2, \dots X_N$ are iid in L^2, and have the same distribution as X, which is assumed to be square integrable.*

We define the following parameters:

$$\begin{aligned} \mu(t) &= E[X(t)] \quad \text{(mean function)}; \\ c(t,s) &= E[(X(t)-\mu(t))(X(s)-\mu(s))] \quad \text{(covariance function)}; \\ C &= E[\langle (X-\mu), \cdot \rangle (X-\mu)] \quad \text{(covariance operator)}. \end{aligned}$$

The mean function μ is estimated by the sample mean function

$$\hat{\mu}(t) = N^{-1} \sum_{i=1}^{N} X_i(t)$$

and the covariance function by its sample counterpart

$$\hat{c}(t,s) = N^{-1} \sum_{i=1}^{N} (X_i(t) - \hat{\mu}(t))(X_i(s) - \hat{\mu}(s)).$$

The sample covariance operator is defined by

$$\hat{C}(x) = N^{-1} \sum_{i=1}^{N} \langle X_i - \hat{\mu}, x \rangle (X_i - \hat{\mu}), \quad x \in L^2.$$

Note that $\hat{C}$ maps L^2 into a finite dimensional subspace spanned by X_1, $X_2, \dots, X_N$. This illustrates the limitations of statistical inference for functional observations; a finite sample can recover an infinite dimensional object only with limited precision.

The first theorem states that $\hat{\mu}$ is an unbiased MSE consistent estimator of μ, and implies that it is consistent, in a sense that $\|\hat{\mu} - \mu\| \xrightarrow{P} 0$. The theorem, and its proof, parallel analogous results for the average of scalar observations.

Theorem 2.3. *If Assumption 2.1 holds, then $E\hat{\mu} = \mu$ and $E\|\hat{\mu} - \mu\|^2 = O(N^{-1})$.*

Proof. For every i, for almost all $t \in [0,1]$, $EX_i(t) = \mu(t)$, so it follows that $E\hat{\mu} = \mu$ in L^2. Observe that

$$\begin{aligned} E\|\hat{\mu} - \mu\|^2 &= N^{-2} \sum_{i,j=1}^{N} E[\langle (X_i - \mu), (X_j - \mu) \rangle] \\ &= N^{-2} \sum_{i=1}^{N} E\|X_i - \mu\|^2 = N^{-1} E\|X - \mu\|^2. \end{aligned}$$ □

In the proof we used the following lemma which follows from conditioning on X_2 and the definition of expectation in a Hilbert space, see Section 2.3.

Lemma 2.1. *If $X_1, X_2 \in L^2$ are independent, square integrable and $EX_1 = 0$, then $E[\langle X_1, X_2 \rangle] = 0$.*

To study the properties of $\hat{c}(t,s)$, we must first choose an appropriate norm for measuring the distance between $\hat{c}(t,s)$ and $c(t,s)$. Observe that $c(\cdot,\cdot) \in L^2([0,1] \times [0,1])$ because

$$\begin{aligned}
&\iint c^2(t,s)dt\,ds \\
&= \iint E[(X(t)-\mu(t))(X(s)-\mu(s))]^2 dt\,ds \\
&\le \iint E[(X(t)-\mu(t))^2]E[(X(s)-\mu(s))^2]dt\,ds \\
&= \left(\int E[(X(t)-\mu(t))^2]dt \right)^2 \\
&= \left(E \int (X(t)-\mu(t))^2 dt \right)^2 \\
&= \left(E\|X-\mu\|^2 \right)^2 .
\end{aligned}$$

It follows that the covariance operator C is Hilbert–Schmidt. Just as in the scalar case, $\hat{c}(t,s)$ is a biased estimator of $c(t,s)$. An elementary verification shows that

$$E[\hat{c}(t,s)] = \frac{N}{N-1}\,c(t,s) \quad (\text{in } L^2([0,1]\times[0,1])).$$

The bias of $\hat{c}$ is asymptotically negligible, and is introduced by the estimation of the mean function μ. Replacing μ by $\hat{\mu}$ in general has a negligible effect, and in theoretical work, it is convenient to assume that μ is known and equal to zero. This simplifies many formulas. When applying such results to real data, it is important to remember to first subtract the sample mean function $\hat{\mu}$ from functional observations.

From now on, except when explicitly stated, we thus assume that the observations have mean zero. We therefore have

$$\hat{c}(t,s) = N^{-1}\sum_{i=1}^{N} X_i(t)X_i(s); \quad \hat{C}(x) = N^{-1}\sum_{i=1}^{N} \langle X_n, x\rangle X_n$$

and

$$\hat{C}(x)(t) = \int \hat{c}(t,s)x(s)ds, \quad x \in L^2. \tag{2.9}$$

We will see in Theorem 2.4 that $E\|X\|^4 < \infty$ implies $E\|\hat{C}\|_{\mathcal{S}}^2 < \infty$, where $\|\cdot\|_{\mathcal{S}}$ is the Hilbert–Schmidt norm. By (2.5) and (2.9), this implies that with probability one $\hat{c}(\cdot,\cdot) \in L^2([0,1]\times[0,1])$ because then $E\iint \hat{c}^2(t,s)dt\,ds < \infty$. The assumption $E\|X\|^4 < \infty$ is however only a sufficient condition. A direct verification shows that if for each $1 \le n \le N$, $X_n(\cdot) \in L^2([0,1])$ a.s., then $\hat{c}(\cdot,\cdot) \in L^2([0,1]\times[0,1])$ a.s..

Theorem 2.4. *If* $E\|X\|^4 < \infty,\ EX = 0$, *and Assumption 2.1 holds, then*

$$E\|\hat{C}\|_{\mathcal{S}}^2 \le E\|X\|^4.$$

Proof. By the triangle inequality

$$E\|\hat{C}\|_{\mathcal{S}} = E\left\| N^{-1}\sum_{n=1}^{N} \langle X_n, \cdot\rangle X_n \right\|_{\mathcal{S}} \le E\| \langle X, \cdot\rangle X\|_{\mathcal{S}}.$$

By (2.2), for any orthonormal basis $\{e_j,\ j \ge 1\}$,

$$\| \langle X, \cdot\rangle X\|_{\mathcal{S}}^2 = \sum_{j=1}^{\infty} \| \langle X, e_j\rangle X\|^2 = \|X\|^2 \sum_{j=1}^{\infty} |\langle X, e_j\rangle|^2 = \|X\|^4. \qquad \square$$

Theorem 2.5. *If* $E\|X\|^4 < \infty,\ EX = 0$, *and Assumption 2.1 holds, then*

$$E\|\hat{C} - C\|_{\mathcal{S}}^2 \le N^{-1}E\|X\|^4.$$

Proof. By (2.2), for any orthonormal basis $\{e_j,\ j \ge 1\}$,

$$
\begin{aligned}
&\|\hat{C} - C\|_{\mathcal{S}}^2 \\
&= \sum_{j=1}^{\infty} \left\| \frac{1}{N}\sum_{n=1}^{N} \langle X_n, e_j\rangle X_n - E[\langle X, e_j\rangle X] \right\|^2 \\
&= \sum_{j=1}^{\infty} \left\langle \frac{1}{N}\sum_{n=1}^{N} \{\langle X_n, e_j\rangle X_n - E[\langle X_n, e_j\rangle X_n]\}, \right. \\
&\qquad\qquad \left. \frac{1}{N}\sum_{m=1}^{N} \{\langle X_m, e_j\rangle X_m - E[\langle X_m, e_j\rangle X_m]\} \right\rangle \\
&= \frac{1}{N^2}\sum_{j=1}^{\infty}\sum_{n=1}^{N}\sum_{m=1}^{N} \langle \{\langle X_n, e_j\rangle X_n - E[\langle X_n, e_j\rangle X_n]\} \\
&\qquad\qquad\qquad \times \{\langle X_m, e_j\rangle X_m - E[\langle X_m, e_j\rangle X_m]\}\rangle.
\end{aligned}
$$

By Lemma 2.1, for $n \ne m$,

$$E\langle \{\langle X_n, e_j\rangle X_n - E[\langle X_n, e_j\rangle X_n]\}, \{\langle X_m, e_j\rangle X_m - E[\langle X_m, e_j\rangle X_m]\}\rangle = 0.$$

Therefore,

$$
\begin{aligned}
&E\|\hat{C}-C\|_{\mathcal{S}}^2\\
&=\frac{1}{N^2}\sum_{j=1}^{\infty}\sum_{n=1}^{N}E\left\|\langle X_n,e_j\rangle X_n-E[\langle X_n,e_j\rangle X_n]\right\|^2\\
&=\frac{1}{N}\sum_{j=1}^{\infty}E\left\|\langle X,e_j\rangle X-E[\langle X,e_j\rangle X]\right\|^2\\
&\le\frac{1}{N}\sum_{j=1}^{\infty}E\left\|\langle X,e_j\rangle X\right\|^2\\
&=N^{-1}E\sum_{j=1}^{\infty}\left\|\langle X,e_j\rangle X\right\|^2\\
&=N^{-1}E\left[\|X\|^2\sum_{j=1}^{\infty}|\langle X,e_j\rangle|^2\right]\\
&=N^{-1}E\|X\|^4.
\end{aligned}
$$

By (2.5), the conclusion of Theorem 2.5 can be equivalently stated as

$$E\iint[\hat{c}(t,s)-c(t,s)]^2dt\,ds\le N^{-1}E\|X\|^4.$$

This implies that $\hat{c}(t,s)$ is a mean squared consistent estimator of the covariance function $c(t,s)$.

The following application of Theorem 2.5 will be used in subsequent chapters dealing with change point analysis. We first formulate the assumptions and introduce some additional notation.

Assumption 2.2. *The $X, Y, X_i, Y_i,\ i \ge 1$, are random elements of L^2 which satisfy the following conditions:*

C1: $X, X_i,\ i \ge 1$ are independent and identically distributed,
C2: $Y, Y_i,\ i \ge 1$ are independent and identically distributed,
C3: $\{X, X_i,\ i \ge 1\}$ and $\{Y, Y_i,\ i \ge 1\}$ are independent,
C4: $EX = 0,\ E\|X\|^4 < \infty,\ \ EY = 0,\ E\|Y\|^4 < \infty.$

Let $k^* = k_N^*$ be a sequence of integers satisfying $1 \le k_N^* \le N$ and

$$\lim_{N\to\infty}\frac{k_N^*}{N}=\theta,\quad \text{for some } 0\le\theta\le 1. \tag{2.10}$$

Define

$$\hat{c}_N^*(t,s) = \frac{1}{N}\left(\sum_{1\le i\le k^*} X_i(t)X_i(s) + \sum_{k^*<i\le N} Y_i(t)Y_i(s)\right)$$

and

$$c_\theta(t,s) = \theta E[X(t)X(s)] + (1-\theta)E[Y(t)Y(s)].$$

Introduce the corresponding operators $\hat{C}_N^*$ and C_θ defined by

$$\hat{C}_N^*(x)(t) = \int \hat{c}_N^*(t,s)x(s)ds, \quad x \in L^2,$$
$$C_\theta(x)(t) = \int c_\theta(t,s)x(s)ds, \quad x \in L^2.$$

Theorem 2.6. *If Assumption 2.2 and condition* (2.10) *hold, then*

$$E\|\hat{C}_N^* - C_\theta\|_{\mathcal{S}}^2 \to 0.$$

Proof. To avoid introducing additional operators, we identify the operators with their kernels, and, somewhat abusing the notation, use the arguments t and s, when we actually mean the corresponding operators.

Since $\|\cdot\|_{\mathcal{S}}$ is a norm, we get

$$\begin{aligned}
&\left\|\hat{C}_N^* - \left(\frac{k^*}{N}E[X(t)X(s)] + \frac{N-k^*}{N}E[Y(t)Y(s)]\right)\right\|_{\mathcal{S}} \\
&\le \frac{k^*}{N}\left\|\frac{1}{k^*}\sum_{1\le i\le k^*} X_i(t)X_i(s) - E[X(t)X(s)]\right\|_{\mathcal{S}} \\
&\quad + \frac{N-k^*}{N}\left\|\frac{1}{N-k^*}\sum_{k^*<i\le N} Y_i(t)Y_i(s) - E[Y(t)Y(s)]\right\|_{\mathcal{S}}.
\end{aligned}$$

Hence

$$\begin{aligned}
&\left\|\hat{C}_N^* - \left(\frac{k^*}{N}E[X(t)X(s)] + \frac{N-k^*}{N}E[Y(t)Y(s)]\right)\right\|_{\mathcal{S}}^2 \\
&\le 2\left\{\left(\frac{k^*}{N}\right)^2\left\|\frac{1}{k^*}\sum_{1\le i\le k^*} X_i(t)X_i(s) - E[X(t)X(s)]\right\|_{\mathcal{S}}^2\right. \\
&\quad \left. + \left(\frac{N-k^*}{N}\right)^2\left\|\frac{1}{N-k^*}\sum_{k^*<i\le N} Y_i(t)Y_i(s) - E[Y(t)Y(s)]\right\|_{\mathcal{S}}^2\right\}.
\end{aligned}$$

Therefore, by Theorem 2.5, we have

$$E\left\|\hat{C}_N^* - \left(\frac{k^*}{N}E[X(t)X(s)] + \frac{N-k^*}{N}E[Y(t)Y(s)]\right)\right\|_{\mathcal{S}}^2$$
$$\leq 2\left\{\frac{k^*}{N^2}E\|X\|^4 + \frac{N-k^*}{N^2}E\|Y\|^4\right\}.$$

On the other hand,

$$\left\|\left(\frac{k^*}{N} - \theta\right)E[X(t)X(s)] + \left(\frac{N-k^*}{N} - (1-\theta)\right)E[Y(t)Y(s)]\right\|_{\mathcal{S}} \to 0,$$

on account of (2.10). □

2.5 Estimation of the eigenvalues and the eigenfunctions

We often must estimate the eigenvalues and eigenfunctions of C, but the interpretation of these quantities as parameters, and their estimation, must be approached with care. The eigenvalues must be identifiable, so we must assume that $\lambda_1 > \lambda_2 > \cdots$. In practice, we can estimate only the p largest eigenvalues, and assume that $\lambda_1 > \lambda_2 > \cdots > \lambda_p > \lambda_{p+1}$, which implies that the first p eigenvalues are nonzero. The eigenfunctions v_j are defined by $Cv_j = \lambda_j v_j$, so if v_j is an eigenfunction, then so is av_j, for any nonzero scalar a (by definition, eigenfunctions are nonzero). The v_j are typically normalized, so that $\|v_j\| = 1$, but this does not determine the sign of v_j. Thus if $\hat{v}_j$ is an estimate computed from the data, we can only hope that $\hat{c}_j\hat{v}_j$ is close to v_j, where

$$\hat{c}_j = \text{sign}(\langle\hat{v}_j, v_j\rangle).$$

Note that $\hat{c}_j$ cannot be computed form the data, so it must be ensured that the statistics we want to work with do not depend on $\hat{c}_j$.

With these preliminaries in mind, we define the estimated eigenelements by

$$\int \hat{c}(t,s)\hat{v}_j(s)ds = \hat{\lambda}_j\hat{v}_j(t), \quad j = 1, 2, \ldots N. \tag{2.11}$$

We will often use the following result established in Dauxois *et al.* (1982) and Bosq (2000). Its proof is presented in Section 2.7.

Theorem 2.7. *Suppose* $E\|X\|^4 < \infty$, $EX = 0$, *Assumption 2.1 holds, and*

$$\lambda_1 > \lambda_2 > \cdots > \lambda_p > \lambda_{p+1}. \tag{2.12}$$

Then, for each $1 \leq j \leq p$,

$$\limsup_{N\to\infty} NE\left[\|\hat{c}_j\hat{v}_j - v_j\|^2\right] < \infty, \quad \limsup_{N\to\infty} NE\left[|\lambda_j - \hat{\lambda}_j|^2\right] < \infty. \tag{2.13}$$

If Assumption (2.12) is replaced by $\lambda_j > \lambda_{j+1} > 0$ for every $j \geq 1$, then (2.13) holds for every $j \geq 1$.

Theorem 2.7 implies that, under regularity conditions, the population eigenfunctions can be consistently estimated by the empirical eigenfunctions. If the assumptions do not hold, the direction of the $\hat{v}_k$ may not be close to the v_k. Examples of this type, with many references, are discussed in Johnstone and Lu (2009).

The study of several change point procedures to be introduced in subsequent chapters requires a version of Theorem 2.7. By Theorem 2.6, the empirical covariance operator $\hat{C}_N^*$ converges to C_θ. Hence the empirical eigenvalues and eigenfunctions should be compared to $\lambda_{j,\theta}$ and $v_{j,\theta}$, the eigenvalues and eigenfunctions of C_θ defined by

$$\int c_\theta(t,s) v_{j,\theta}(s) ds = \lambda_{j,\theta} v_{j,\theta}(t), \quad j \geq 1.$$

We also define

$$\hat{c}_{j,\theta} = \operatorname{sign}(\langle \hat{v}_j, v_{j,\theta} \rangle)$$

and state the following theorem.

Theorem 2.8. *Suppose Assumption 2.2 and condition* (2.10) *hold, and*

$$\lambda_{1,\theta} > \lambda_{2,\theta} > \cdots > \lambda_{p,\theta} > \lambda_{p+1,\theta}.$$

Then, for each $1 \leq j \leq p$,

$$E\left[\|\hat{c}_{j,\theta}\hat{v}_j - v_{j,\theta}\|^2\right] \to 0 \quad \text{and} \quad E\left[|\hat{\lambda}_j - \lambda_{j,\theta}|^2\right] \to 0.$$

Proof. The result follows from Theorem 2.6 and Lemmas 2.2 and 2.3 because the kernel $c_\theta(\cdot,\cdot)$ is symmetric. □

2.6 Asymptotic normality of the eigenfunctions

In most applications considered in this book, the L^2 and in probability bounds implied by (2.13) are sufficient. These bounds are optimal, as the random functions $N^{1/2}(\hat{c}_j\hat{v}_j - v_j)$ and the random variables $N^{1/2}(\lambda_j - \hat{\lambda}_j)$ have nondegenerate limits. It can be shown that they are asymptotically normal, see Mas (2002). Theorem 2.10 is a simplified version of an approximation result obtained by Hall and Hosseini-Nasab (2006) which implies the asymptotic normality under stronger assumptions on the distribution of the observations X_n.

First, we introduce the random functions

$$Z_N(t,s) = N^{1/2}(\hat{c}(t,s) - c(t,s)),$$

where $\hat{c}(t,s)$, $c(t,s)$ can be either centered with the (sample) mean function, or uncentered if the mean function is assumed zero, see Section 2.4. The following theorem establishes the weak convergence of $Z_N(\cdot,\cdot)$ in the space $L^2([0,1]\times[0,1])$ assuming mean zero observations. A proof in the centered case is a simple modification.

Theorem 2.9. *If Assumption 2.1 holds with* $EX(t) = 0$ *and* $E\|X\|^4 < \infty$, *then* $Z_N(t,s)$ *converges weakly in* $L^2([0,1] \times [0,1])$ *to a Gaussian process* $\Gamma(t,s)$ *with* $E\Gamma(t,s) = 0$ *and*

$$E[\Gamma(t,s)\Gamma(t',s')] = E[X(t)X(s)X(t')X(s')] - c(t,s)c(t',s').$$

Proof. Writing

$$Z_N(t,s) = N^{-1/2}\sum_{n=1}^{N}[X_n(t)X_n(s) - c(t,s)],$$

we observe that $Z_N(t,s)$ is a normalized partial sums process of independent, identically distributed random processes taking values in $L^2([0,1] \times [0,1])$. Hence the CLT in a Hilbert space, holds if $E\iint (X(t)X(s))^2 dt\, ds < \infty$. This condition holds, because

$$E\iint (X(t)X(s))^2 dt\, ds = E\int X^2(t)dt \int X^2(s)ds$$
$$\leq \left\{E\left[\int X^2(t)dt\right]^2\left[\int X^2(s)ds\right]^2\right\}^{1/2} = \left(E\|X\|^4\right)^2 < \infty. \qquad \square$$

We note that if the X_n are strongly mixing random functions, then the functions $X_n(t)X_n(s)$ are also strongly mixing with the same rate. Hence, assuming some moment conditions, for example those in Theorem 2.17 of Bosq (2000), the weak convergence of the sequence Z_N can also be established in the dependent case.

Since $Z_N(\cdot,\cdot)$ converges weakly in the space $L^2([0,1] \times [0,1])$, the asymptotic normality of $\hat{\lambda}_j - \lambda_j$ and $\hat{c}_j\hat{v}_j - v_j$ is an immediate consequence of Theorem 2.10 which follows for the results of Hall and Hosseini-Nasab (2006), see also Hall and Hosseini-Nasab (2007). First we state the required conditions.

Assumption 2.3. *The random function* $X \in L^2$ *which has the same distribution as the* X_n *satisfies the following conditions:*

C1: For all $\kappa > 0$, $\sup_{0\leq t\leq 1} E|X(t)|^\kappa < \infty$,
C2: there is $\gamma > 0$ *such that for all* $\kappa > 0$

$$\sup_{0\leq t,s\leq 1} E\left[|t-s|^{-\gamma}|X(t) - X(s)|\right]^\kappa < \infty,$$

C3: for each integer $r \geq 0$, *the sequence* $\left\{\lambda_j^{-1}E\langle X, v_j\rangle^{2r}\right\}$ *is bounded.*

Theorem 2.10. *If Assumptions 2.1 and 2.3 and condition* (2.12) *hold, then for* $1 \leq j \leq p$,

$$N^{1/2}(\hat{\lambda}_j - \lambda_j) = \iint Z_N(t,s)v_j(t)v_j(s)dt\, ds + o_P(1)$$

and

$$\sup_{0\leq t\leq 1} |N^{1/2}(\hat{c}_j\hat{v}_j(t) - v_j(t)) - \hat{T}_j(t)| = o_P(1),$$

where

$$\hat{T}_j(t) = \sum_{k \neq j} (\lambda_j - \lambda_k)^{-1} v_k(t) \iint Z_n(t,s) v_j(t) v_j(s) dt\, ds.$$

2.7 Proof of Theorem 2.7

The proof of Theorem 2.7 is based on Lemmas 2.2 and 2.3. These lemmas have wider applicability, and will be used in subsequent chapters. They state that if the operators are close, then their eigenvalues and eigenfunctions (adjusted for the sign) are also close.

Consider two compact operators $C, K \in \mathcal{L}$ with singular value decompositions

$$C(x) = \sum_{j=1}^{\infty} \lambda_j \langle x, v_j \rangle f_j, \quad K(x) = \sum_{j=1}^{\infty} \gamma_j \langle x, u_j \rangle g_j. \tag{2.14}$$

The following Lemma is proven in Section VI.1 of Gohberg *et al.* (1990), see their Corollary 1.6 on p. 99.

Lemma 2.2. *Suppose $C, K \in \mathcal{L}$ are two compact operators with singular value decompositions* (2.14). *Then, for each $j \geq 1$, $|\gamma_j - \lambda_j| \leq \|K - C\|_{\mathcal{L}}$.*

We now tighten the conditions on the operator C by assuming that it is symmetric and $C(v_j) = \lambda_j v_j$, i.e. $f_j = v_j$ in (2.14). Notice that any covariance operator C satisfies these conditions. We also define

$$v'_j = c_j v_j, \quad c_j = \text{sign}(\langle u_j, v_j \rangle).$$

Lemma 2.3. *Suppose $C, K \in \mathcal{L}$ are two compact operators with singular value decompositions* (2.14). *If C is symmetric, $f_j = v_j$ in* (2.14), *and its eigenvalues satisfy* (2.12), *then*

$$\|u_j - v'_j\| \leq \frac{2\sqrt{2}}{\alpha_j} \|K - C\|_{\mathcal{L}}, \quad 1 \leq j \leq p,$$

where $\alpha_1 = \lambda_1 - \lambda_2$ and $\alpha_j = \min(\lambda_{j-1} - \lambda_j, \lambda_j - \lambda_{j+1})$, $2 \leq j \leq p$.

Proof. For a fixed $1 \leq j \leq p$, introduce the following quantities

$$D_j = \|C(u_j) - \lambda_j u_j\|, \quad S_j = \sum_{k \neq j} \langle u_j, v_k \rangle^2.$$

The claim will follow, once we have established that

$$\|u_j - v'_j\|^2 \leq 2S_j, \tag{2.15}$$

$$\alpha_j^2 S_j \leq D_j^2, \tag{2.16}$$

and

$$D_j \leq 2\|K - C\|_{\mathcal{L}}. \tag{2.17}$$

Verification of (2.15): By the Parseval identity

$$\|u_j - v'_j\|^2 = \sum_{k=1}^{\infty} (\langle u_j, v_k \rangle - c_j \langle v_j, v_k \rangle)^2 = (\langle u_j, v_j \rangle - c_j)^2 + S_j.$$

If $c_j = 0$, then (2.15) clearly holds, since in this case $\|u_j - v'_j\| = \|u_j\| = 1$ and $S_j = \|u_j\| - \langle u_j, v_j \rangle = 1$.

If $|c_j| = 1$, then $(\langle u_j, v_j \rangle - c_j)^2 = (1 - |\langle u_j, v_j \rangle|)^2$, and using the identity $\sum_k \langle u_j, v_k \rangle^2 = 1$, we obtain

$$(1 - |\langle u_j, v_j \rangle|)^2 = \sum_{k=1}^{\infty} \langle u_j, v_k \rangle^2 - 2|\langle u_j, v_j \rangle| + \langle u_j, v_j \rangle^2.$$

Thus, if $|c_j| = 1$,

$$(\langle u_j, v_j \rangle - c_j)^2 = S_j + 2(\langle u_j, v_j \rangle^2 - |\langle u_j, v_j \rangle| \leq S_j.$$

Verification of (2.16): By the Parseval identity

$$D_j^2 = \sum_{k=1}^{\infty} (\langle C(u_j), v_k \rangle - \lambda_j \langle u_j, v_k \rangle)^2.$$

Since C is *symmetric* and $C(v_j) = \lambda_j v_j$, $\langle C(u_j), v_k \rangle = \lambda_k \langle u_j, v_k \rangle$. Therefore

$$D_j^2 = \sum_{k \neq j} (\lambda_k - \lambda_j)^2 \langle u_j, v_k \rangle^2 \geq S_j \min_{k \neq j} (\lambda_k - \lambda_j)^2.$$

Verification of (2.15): Observe that

$$C(u_j) - \lambda_j u_j = (C - K)(u_j) + (\gamma_j - \lambda_j) u_j.$$

Therefore, by Lemma 2.2,

$$D_j \leq \|C - K\|_{\mathcal{L}} \|u_j\| + |\gamma_j - \lambda_j| \|u_j\| \leq 2\|K - C\|_{\mathcal{L}}.$$ □

Proof of Theorem 2.7.. By (2.3), $\|Z_N\|_{\mathcal{L}} \leq \|Z_N\|_{\mathcal{S}}$, where the kernel Z_N is defined in Section 2.6. By Theorem 2.5, $E\|Z_N\|_{\mathcal{S}}^2 = O(N)$, so the result follows from Lemmas 2.2 and 2.3. □

2.8 Bibliographical notes

In the `R` package `fda`, and in most other numerical implementations, the curves X_i are smoothed before the estimates $\hat{\mu}(t)$ and $\hat{c}(t,s)$ introduced in Section 2.4 are computed. Smoothing is done by approximating the X_i by finite linear combinations of smooth basis functions,as explained in Section 1.5. This step ensures that $\hat{\mu}(t)$ and $\hat{c}(t,s)$, and the estimated eigenfunctions $\hat{v}_j(t)$ defined in Section 2.5 are also smooth. If the smoothing is not too severe, i.e. if it eliminates only noise and the smoothed curves retain the main features of the original curves, then the estimates resulting by smoothing the X_i first are approximately equivalent to the estimates considered in this chapter. To formulate this property precisely, a statistical model with a noise and a smooth component needs to be formulated. An interested reader is referred to Boente and Fraiman (2000) and Zhang and Chen (2007) who consider, respectively, kernel–based and local polynomial smoothing.

Just like the average of scalar data, the estimates $\hat{\mu}(t)$ and $\hat{c}(t,s)$ are not robust to outlying curves. It is possible to define the functional median, which is a more robust measure of central tendency. If $E\|X\| < \infty$, the population median m is defined as the minimizer of $E\|X - m\|$, but it is also possible to define it and its sample counterpart without assuming the finite first moment, see Gervini (2008), who also proposes a method of robust estimation of the eigenfunctions v_k. The idea is that instead of using the eigenfunctions of the kernel $\hat{c}(t,s)$, cf. (2.11), the eigenfunctions of a weighted version of $\hat{c}(t,s)$ should be used.

Delaigle and Hall (2010) propose a definition of the mode of the distribution a random function. The definition, and the estimation procedure, involve the functional principal component expansions discussed in Chapter 3. This chapter focused on moments of the distribution of a random function and on their estimation. Delaigle and Hall (2010) argue that a density function cannot be meaningfully defined.

Chapter 3
Functional principal components

This chapter introduces one of the most fundamental concepts of FDA, that of the functional principal components (FPC's). FPC's allow us to reduce the dimension of infinitely dimensional functional data to a small finite dimension in an optimal way. In Sections 3.1 and 3.2, we introduce the FPC's from two angles, as coordinates maximizing variability, and as an optimal orthonormal basis. In Section 3.3, we identify the FPC's with the eigenfunctions of the covariance operator, and show how its eigenvalues decompose the variance of the functional data. We conclude with Section 3.4 which explains how to compute the FPC's in the `R` package `fda`.

3.1 A maximization problem

In this section we present some preliminary results which are fundamental for the remainder of this chapter. To motivate, we begin with a vector case, and then move on to the space L^2.

We first state the following well–known result, see e.g. Chapter 6 of Leon (2006).

Theorem 3.1. *(Principal axis theorem). Suppose* $\mathbf{A}$ *is a symmetric* $p \times p$ *matrix. Then, there is an orthonormal matrix* $\mathbf{U} = [\mathbf{u}_1, \ldots, \mathbf{u}_p]$ *whose columns are the eigenvectors of* $\mathbf{A}$*, i.e.*

$$\mathbf{U}^T\mathbf{U} = \mathbf{I} \quad \text{and} \quad \mathbf{A}\mathbf{u}_j = \lambda_j \mathbf{u}_j.$$

Moreover,

$$\mathbf{U}^T\mathbf{A}\mathbf{U} = \boldsymbol{\Lambda} = \text{diag}[\lambda_1, \lambda_2, \ldots, \lambda_p].$$

The orthonormality of $\mathbf{U}$ is equivalent to the assertion that the vectors $\mathbf{u}_1, \ldots, \mathbf{u}_p$ form an orthonormal basis in the Euclidean space R^p. Theorem 3.1 implies that $\mathbf{A} = \mathbf{U}\boldsymbol{\Lambda}\mathbf{U}^T$, a representation known as the spectral decomposition of $\mathbf{A}$. It can be used to solve the following maximization problem. Suppose $\mathbf{A}$ is symmetric and positive–definite with distinct eigenvalues arranged in decreasing order:

$\lambda_1 > \lambda_2 > \cdots > \lambda_p$. We want to find a unit length vector $\mathbf{x}$ such that $\mathbf{x}^T\mathbf{A}\mathbf{x}$ is maximum. By the spectral decomposition, $\mathbf{x}^T\mathbf{A}\mathbf{x} = \mathbf{y}^T\boldsymbol{\Lambda}\mathbf{y}$, where $\mathbf{y} = \mathbf{U}^T\mathbf{x}$. Since $\mathbf{U}$ is orthonormal, $\|\mathbf{y}\| = \|\mathbf{x}\|$, so it is enough to find a unit length vector $\mathbf{y}$ such that $\mathbf{y}\boldsymbol{\Lambda}\mathbf{y}^T$ is maximum, and then set $\mathbf{x} = \mathbf{U}\mathbf{y}$. Since $\mathbf{y}\boldsymbol{\Lambda}\mathbf{y}^T = \sum_{j=1}^p \lambda_j y_j^2$, clearly $\mathbf{y} = [1, 0, \ldots, 0]^T$, and $\mathbf{x} = \mathbf{u}_1$ with the maximum being λ_1.

The above ideas can be easily extended to a separable Hilbert space, where they become even more transparent. Suppose Ψ is a symmetric positive–definite Hilbert–Schmidt operator in L^2. We have seen in Section 2.4 that the covariance operator C and its sample counterpart $\hat{C}$ are in this class, provided $E\|X\|^4 < \infty$. The operator Ψ then admits the spectral decomposition (2.4), and the problem of maximizing $\langle \Psi(x), x\rangle$ subject to $\|x\| = 1$ becomes trivial because

$$\langle \Psi(x), x\rangle = \left\langle \sum_{j=1}^{\infty} \lambda_j \langle x, v_j\rangle v_j,\ x \right\rangle = \sum_{j=1}^{\infty} \lambda_j \langle x, v_j\rangle^2 .$$

By Parseval's equality, we must maximize the above, subject to $\sum_{j=1}^{\infty} \langle x, v_j\rangle^2 = 1$. To ensure uniqueness, suppose $\lambda_1 > \lambda_2 > \cdots$, so we take $\langle x, v_1\rangle^2 = 1$ and $\langle x, v_j\rangle = 0$ for $j > 1$. Thus, $\langle \Psi(x), x\rangle$ is maximized at v_1 (or $-v_1$), and the maximum is λ_1. Suppose now that we want to maximize $\langle \Psi(x), x\rangle$ subject not only to the condition $\|x\| = 1$, but also to $\langle x, v_1\rangle = 0$. Thus we want to find another unit norm function which is orthogonal to the function found in the first step. Such a function, clearly satisfies $\langle \Psi(x), x\rangle = \sum_{j=2}^{\infty} \lambda_j \langle x, v_j\rangle^2$ and $\sum_{j=2}^{\infty} \langle x, v_j\rangle^2 = 1$, so $x = v_2$, and the maximum now is λ_2. Repeating this procedure, we arrive at the following theorem.

Theorem 3.2. *Suppose Ψ is a symmetric, positive definite Hilbert–Schmidt operator with eigenfunctions v_j and eigenvalues λ_j satisfying* (2.12). *Then,*

$$\sup\left\{\langle \Psi(x), x\rangle :\ \|x\| = 1,\ \langle x, v_j\rangle = 0,\ 1 \le j \le i-1,\ i < p\right\} = \lambda_i,$$

and the supremum is reached if $x = v_i$. The maximizing function is unique up to a sign.

3.2 Optimal empirical orthonormal basis

The approach developed in the previous section can be applied to the following important problem. Suppose we observe functions $x_1, x_2, \ldots, x_N$. In this section it is not necessary to view these functions as random, but we can think of them as the observed realizations of random functions in L^2. Fix an integer $p < N$. We think of p as being much smaller than N, typically a single digit number. We want to find an orthonormal basis $u_1, u_2, \ldots, u_p$ such that

$$\hat{S}^2 = \sum_{i=1}^{N} \left\| x_i - \sum_{k=1}^{p} \langle x_i, u_k\rangle u_k \right\|^2$$

is minimum. Once such a basis is found, we can replace each curve x_i by $\sum_{k=1}^{p} \langle x_i, u_k \rangle u_k$, to a good approximation. For the p we have chosen, this approximation is uniformly optimal, in the sense of minimizing $\hat{S}^2$. This means that instead of working with infinitely dimensional curves x_i, we can work with p–dimensional vectors

$$\mathbf{x}_i = [\langle x_i, u_1 \rangle, \langle x_i, u_2 \rangle, \ldots, \langle x_i, u_p \rangle]^T.$$

This is a central idea of functional data analysis, as to perform any practical calculations we must reduce the dimension from infinity to a finite number. The functions u_j are called collectively the *optimal empirical orthonormal basis* or *natural orthonormal components*, the words "empirical" and "natural" emphasizing that they are computed directly from the functional data.

The functions $u_1, u_2, \ldots, u_p$ minimizing $\hat{S}^2$ are equal (up to a sign) to the normalized eigenfunctions of the sample covariance operator, see (2.11). To see this, suppose first that $p = 1$, i.e. we want to find u with $\|u\| = 1$ which minimizes

$$\begin{aligned}\sum_{i=1}^{N} \|x_i - \langle x_i, u \rangle u\|^2 &= \sum_{i=1}^{N} \|x_i\|^2 - 2\sum_{i=1}^{N} \langle x_i, u \rangle^2 + \sum_{i=1}^{N} \langle x_i, u \rangle^2 \|u\|^2 \\ &= \sum_{i=1}^{N} \|x_i\|^2 - \sum_{i=1}^{N} \langle x_i, u \rangle^2,\end{aligned}$$

i.e. maximizes $\sum_{i=1}^{N} \langle x_i, u \rangle^2 = \left\langle \hat{C} u, u \right\rangle$. By Theorem 3.2, we conclude that $u = \hat{v}_1$.

The general case is treated analogously. Since

$$\hat{S}^2 = \sum_{i=1}^{N} \|x_i\|^2 - \sum_{i=1}^{N} \sum_{k=1}^{p} \langle x_i, u_k \rangle^2,$$

we need to maximize

$$\begin{aligned}\sum_{k=1}^{p} \sum_{i=1}^{N} \langle x_i, u_k \rangle^2 &= \sum_{k=1}^{p} \left\langle \hat{C}(u_k), u_k \right\rangle \\ &= \sum_{j=1}^{\infty} \hat{\lambda}_j \left\langle u_1, \hat{v}_j \right\rangle^2 + \sum_{j=1}^{\infty} \hat{\lambda}_j \left\langle u_2, \hat{v}_j \right\rangle^2 + \cdots + \sum_{j=1}^{\infty} \hat{\lambda}_j \left\langle u_p, \hat{v}_j \right\rangle^2.\end{aligned}$$

By Theorem 3.2, the sum cannot exceed $\sum_{k=1}^{p} \hat{\lambda}_k$, and this maximum is attained if $u_1 = \hat{v}_1, u_2 = \hat{v}_2, \ldots, u_p = \hat{v}_p$.

3.3 Functional principal components

Suppose $X_1, X_2, \ldots, X_N$ are functional observations. The eigenfunctions of the sample covariance operator $\hat{C}$ are called the empirical functional principal components (EFPC's) of the data $X_1, X_2, \ldots, X_N$. If these observations have the same distribution as a square integrable L^2–valued random function X, we define the functional principal components (FPC's) as the eigenfunctions of the covariance operator C. We have seen in Section 2.5 that under regularity conditions the EFPC estimate the FPC's (up to a sign).

Section 3.2 explains that the EFPC's can be interpreted as an optimal orthonormal basis with respect to which we can expand the data. The inner product $\langle X_i, \hat{v}_j \rangle = \int X_i(t)\hat{v}_j(t)dt$ is called the jth score of X_i. It can be interpreted as the weight of the contribution of the FPC $\hat{v}_j$ to the curve X_i.

Another interpretation of EFPC's follows from Section 3.1. Observe that under the assumption $EX_i = 0$, the statistic

$$\frac{1}{N}\sum_{i=1}^{N} \langle X_i, x \rangle^2 = \left\langle \hat{C}(x), x \right\rangle$$

can be viewed as the sample variance of the data "in the direction" of the function x. If we are interested in finding the function x which is "most correlated" with the variability of the data (away from the mean if the data are not centered), we must thus find x which maximizes $\langle \hat{C}(x), x \rangle$. Clearly, we must impose a restriction on the norm of x, so if we require that $\|x\| = 1$, we see from Theorem 3.2 that $x = \hat{v}_1$, the first EFPC. Next, we want to find a second direction, orthogonal to $\hat{v}_1$, which is "most correlated" with the variability of the data. By Theorem 3.2, this direction is $\hat{v}_2$. Observe that since the $\hat{v}_j$, $i = 1, \ldots, N$, form a basis in R^N,

$$\frac{1}{N}\sum_{i=1}^{N} \|X_i\|^2 = \frac{1}{N}\sum_{i=1}^{N}\sum_{j=1}^{N} \left\langle X_i, \hat{v}_j \right\rangle^2 = \sum_{j=1}^{N} \frac{1}{N}\sum_{i=1}^{N} \left\langle X_i, \hat{v}_j \right\rangle^2 = \sum_{j=1}^{N} \hat{\lambda}_j.$$

Thus, we may say that the variance in the direction $\hat{v}_j$ is $\hat{\lambda}_j$, or that $\hat{v}_j$ explains the fraction of the total sample variance equal to $\hat{\lambda}_j / (\sum_{k=1}^{N} \hat{\lambda}_k)$. We also have the corresponding population analysis of variance:

$$E\|X\|^2 = \sum_{j=1}^{\infty} E[\langle X, v_j \rangle^2] = \sum_{j=1}^{\infty} \langle C v_j, v_j \rangle = \sum_{j=1}^{\infty} \lambda_j.$$

We now present an example that describes how functional data with specified FPC's can be generated. Set

$$X_n(t) = \sum_{j=1}^{p} a_j Z_{jn} e_j(t), \tag{3.1}$$

where a_j are real numbers, for every n, the Z_{jn} are iid mean zero random variables with unit variance, and the e_j are orthogonal functions with unit norm. To compute the covariance operator, we do not have to specify the dependence between the sequences $\{Z_{jn}, j \geq 1\}$. This is needed to claim the convergence of the EFPC's to the FPC's, see Section 2.5. Denote by X a random function with the same distribution as each X_n, i.e. $X(t) = \sum_{j=1}^{p} a_j Z_j e_j(t)$. Then the covariance operator of X acting on x is equal to

$$C(x)(t) = E\left[\left(\int X(s)x(s)ds\right) X(t)\right] = \int E[X(t)X(s)]x(s)ds.$$

By the independence of the Z_j, the covariance function is equal to

$$E[X(t)X(s)] = E\left[\sum_{j=1}^{p} a_j Z_j e_j(t) \sum_{i=1}^{p} a_i Z_i e_i(s)\right] = \sum_{j=1}^{p} a_j^2 e_j(t) e_j(s).$$

Therefore,

$$C(x)(t) = \sum_{j=1}^{p} a_j^2 \left(\int e_j(s)x(s)ds\right) e_j(t).$$

It follows that the EPC's of the X_n are the e_j, and the eigenvalues are $\lambda_j = a_j^2$.

Methods of functional data analysis which use EFPC's assume that the observations are well approximated by an expansion like (3.1) with a small p and relatively smooth functions e_j.

In most applications, it is important to determine a value of p such that the actual data can be replaced by the approximation $\sum_{i=1}^{p} \langle \hat{v}_j, X_n \rangle \hat{v}_j$. A popular method is the *scree plot*. This is a graphical method proposed, in a different context, by Cattell (1966). To apply it, one plots the successive eigenvalues $\hat{\lambda}_j$ against j (see Figure 9.6). The method suggests to find j where the decrease of the eigenvalues appears to level off. This point is used as the selected value of p. To the right of it, one finds only the "factorial scree" ("scree" is a geological term referring to the debris which collects on the lower part of a rocky slope). The method that works best for the applications discussed in this book is the *CPV method* defined as follows. The cumulative percentage of total variance (CPV) explained by the first p EFPC's is

$$CPV(p) = \frac{\sum_{k=1}^{p} \hat{\lambda}_k}{\sum_{k=1}^{N} \hat{\lambda}_k}.$$

We choose p for which $CPV(p)$ exceeds a desired level, 85% is the recommended value. Other methods, known as *pseudo–AIC* and *cross–validation* have also been proposed. All these methods are described and implemented in the MATLAB package PACE developed at the University of California at Davis.

This section has merely set out the fundamental definitions and properties. Interpretation and estimation of the functional principal components has been a subject of extensive research, in which concepts of smoothing and regularization play a major role, see Chapters 8, 9, 10 of Ramsay and Silverman (2005).

3.4 Computation of functional principal components

The R function pca.fd computes the EFPC's $\hat{v}_j$, the corresponding eigenvalues $\hat{\lambda}_j$, and the scores $\langle X_i - \bar{X}_N, \hat{v}_j \rangle$. Its argument must be a functional object, see Section 1.5. A typical call is

```
pca<-pca.fd(data.fd, nharm = 3, centerfns = TRUE)
```

The functional object data.fd contains the 31 magnetic intensity functions introduced in Section 1.5. The argument nharm specifies the number p of the EFPC's (also called harmonics) to be estimated. As explained at the end of Section 1.5, centerfns = TRUE means that the EFPC's and the scores are computed for the centered functions $X_i - \bar{X}_N$.

Once the object pca has been created, $\hat{v}_1, \hat{v}_2, \ldots \hat{v}_p$ can be extracted as pca$harmonics, $\hat{\lambda}_1, \hat{\lambda}_2, \ldots \hat{\lambda}_p$ as pca$values. The scores $\langle X_i - \bar{X}_N, \hat{v}_j \rangle$, $i = 1, 2, \ldots, N$, $j = 1, 2, \ldots p$ are in the $N \times p$ matrix pca$scores.

To illustrate, the left–most panel of Figure 3.1 shows the scatter plot of the pairs $(\langle X_i - \bar{X}_N, \hat{v}_1 \rangle, \langle X_i - \bar{X}_N, \hat{v}_2 \rangle)$. The other two panels show the scatter plots for the remaining two combinations of pairs. Plots of this type are often used to detect outliers, verify normality of the observations, or the validity of a model. Figure 3.1 was created with the following code

```
par(mfrow=c(1,3))
plot(pca$scores[,1], pca$scores[,2], xlab="1st PC scores",
      ylab="2nd PC scores")
plot(pca$scores[,1], pca$scores[,3], xlab="1st PC scores",
      ylab="3rd PC scores")
plot(pca$scores[,2], pca$scores[,3], xlab="2nd PC scores",
      ylab="3rd PC scores")
```

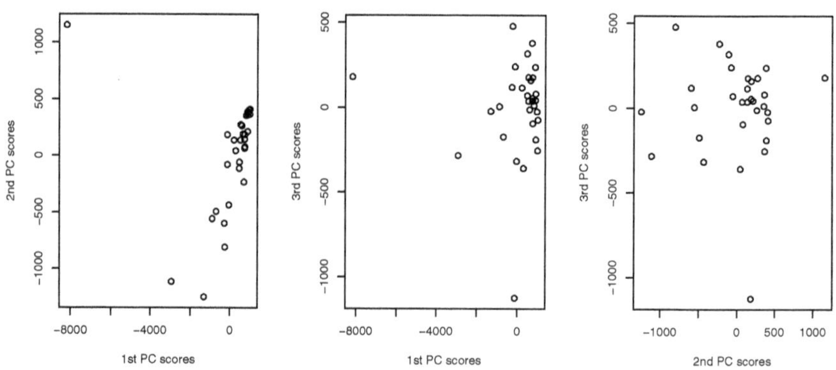

Fig. 3.1 Scatter plots of the scores of the magnetic intensity data.

3.5 Bibliographical notes

A comprehensive modern treatment of principal component analysis is given in the monograph of Jolliffe (2002), who also gives a brief account of the history of the subject. Following Jolliffe (2002), we note that the mathematical ideas behind the PCA are related to those of the singular value decomposition of matrices, which was obtained independently by Beltrami and Jordan in 1870's. Most authors cite the papers of Pearson (1901) and Hotelling (1933) as giving rise to the statistical idea of the PCA form two different angles. The PCA was not widely used until late 1960's, but recent decades have seen its very extensive use in practically all fields where statistics is applied. Ruppert (2011), Chapter 17, gives illustrative examples of PCA applied to yield curves, similar to Eurodollar futures studied later in this book, and to daily returns. Like many other authors, he uses the terminology in which the vectors of scores are called the principal components, while for what we call the principal components the more direct term eigenvectors is used.

Related to the subject of this book is recent interest in the properties of sample functional principal components in the "small n large p" setting, see Jung and Marron (2009) among several contributions. This setting is different from the FDA framework because functional data consist of scalar observations which are naturally organized, say, in time, and can be assumed to be generated by underlying curves with some degree of smoothness. If the data lack smoothness, different approaches are required. The difficulties arising for non-smooth data, and possible solutions, are discussed in Johnstone and Lu (2009).

The approaches described in this book are suitable for data which can be viewed as curves observed at fine time grids with a measurement error which is negligible relative to the size of the data or the purpose of analysis. This approach is not suitable for data that are available only at sparse time points, possibly with a large measurement error . A group of researchers at UC Davis developed a new approach to deal with such data. Its essence is explained in Müller (2009). The idea is that smoothing must be applied to all observations collectively, not to individual trajectories (which basically do not exist for sparse data). This methodology is motivated by data arising in medical longitudinal studies in which the individuals can be regarded as independent cases. The focus is on surface smoothing and measurement error evaluation.

The EFPC's represent the data as linear combinations of functions estimated from the data, and so this technique is data driven. This is of great value if little is known a priori about the structure of the data, and a general purpose linear dimension reduction is sought. In many applications however, linear decompositions with respect to fixed, not necessarily orthonormal, bases are useful, or even nonlinear decompositions, as discussed in Izem and Marron (2007).

Chapter 4
Canonical correlation analysis

Canonical correlation analysis (CCA) is one of the most important tools of multivariate statistical analysis. Its extension to the functional context is not trivial, and in many ways illustrates the differences between multivariate and functional data. One of the most influential contributions has been made by Leurgans *et al.* (1993) who showed that smoothing is necessary in order to define the functional canonical correlations meaningfully.

This chapter is organized as follows. Section 4.1 reviews multivariate population and sample canonical correlation analysis (CCA). In Section 4.2, we explain how functional population CCA should be defined, but postpone the difficult question of its existence to Section 4.6. First, in Section 4.3, we discuss two ways in which its sample version has been defined, and then, in Section 4.4, we show the usefulness of the functional sample CCA by applying it the analysis of space physics data. After this numerical example, we return, in Section 4.6, to the theoretical question of the existence of the population functional CCA. Section 4.6 uses some properties of the square root of the covariance operator, so we first review the relevant concepts in Section 4.5.

4.1 Multivariate canonical components

In this section we review the definition and some properties of the multivariate CCA. Proofs of the results stated in this section are presented e.g. in Johnson and Wichern (2002).

Suppose $\mathbf{X}$ and $\mathbf{Y}$ are two random vectors, respectively, in R^p and R^q. For deterministic vectors $\mathbf{a} \in R^p$ and $\mathbf{b} \in R^q$ define the random variables

$$A = \mathbf{a}^T \mathbf{X}, \quad B = \mathbf{b}^T \mathbf{Y}.$$

We want to find $\mathbf{a}$ and $\mathbf{b}$ which maximize

$$\operatorname{Corr}(A, B) = \frac{\operatorname{Cov}(A, B)}{\sqrt{\operatorname{Var}[A]\operatorname{Var}[B]}}. \tag{4.1}$$

Clearly, if $\mathbf{a}$ and $\mathbf{b}$ maximize (4.1), then so do $c\mathbf{a}$ and $d\mathbf{b}$ for any $c, d > 0$. Therefore, we impose a normalizing condition

$$\text{Var}[A] = 1, \quad \text{Var}[B] = 1. \tag{4.2}$$

If such $\mathbf{a}$ and $\mathbf{b}$ exist, we denote them $\mathbf{a}_1$ and $\mathbf{b}_1$, and set $A_1 = \mathbf{a}_1^T\mathbf{X}$, $B_1 = \mathbf{b}_1^T\mathbf{Y}$. We call (A_1, B_1) the *first pair of canonical variables* and

$$\rho_1 = \text{Cov}(A_1, B_1) = \max\left\{\text{Cov}(\mathbf{a}^T\mathbf{X}, \mathbf{b}^T\mathbf{Y}) : \text{Var}[\mathbf{a}^T\mathbf{X}] = \text{Var}[\mathbf{b}^T\mathbf{Y}] = 1\right\} \tag{4.3}$$

the *first canonical correlation.*

Once $\mathbf{a}_1$ and $\mathbf{b}_1$ have been found, we want to find another pair $(\mathbf{a}, \mathbf{b})$ which maximizes (4.1) subject to (4.2), but also satisfies

$$\text{Cov}(A, A_1) = \text{Cov}(A, B_1) = \text{Cov}(B, B_1) = \text{Cov}(B, A_1) = 0. \tag{4.4}$$

If such $\mathbf{a}$ and $\mathbf{b}$ exist, we denote them $\mathbf{a}_2$ and $\mathbf{b}_2$ and call $A_2 = \mathbf{a}_2^T\mathbf{X}$, $B_2 = \mathbf{b}_2^T\mathbf{Y}$ the *second pair of canonical variables* and the resulting value ρ_2 of (4.1) the *second canonical correlation*. Notice that $\rho_2 \leq \rho_1$ because ρ_2 is a maximum over a smaller subspace (condition (4.4) is added).

We can continue in this way to find *kth canonical components* $(\rho_k, \mathbf{a}_k, \mathbf{b}_k, A_k, B_k)$ by requiring that the pair (A_k, B_k) maximizes (4.1) subject to (4.2) and

$$\text{Cov}(A_k, A_j) = \text{Cov}(A_k, B_j) = \text{Cov}(B_k, B_j) = \text{Cov}(B_k, A_j) = 0, \quad j < k. \tag{4.5}$$

One can show that,under mild assumptions, the canonical components exist for $k \leq \min(p, q)$, and can be computed as follows. Assume, to lighten the notation, that

$$E\mathbf{X} = \mathbf{0} \quad \text{and} \quad E\mathbf{Y} = \mathbf{0}$$

and define the covariance matrices

$$\mathbf{C}_{11} = E[\mathbf{X}\mathbf{X}^T], \quad \mathbf{C}_{22} = E[\mathbf{Y}\mathbf{Y}^T], \quad \mathbf{C}_{12} = E[\mathbf{X}\mathbf{Y}^T], \quad \mathbf{C}_{21} = E[\mathbf{Y}\mathbf{X}^T].$$

Assume that $\mathbf{C}_{11}$ and $\mathbf{C}_{22}$ are nonsingular and introduce the correlation matrices

$$\mathbf{R} = \mathbf{C}_{11}^{-1/2}\mathbf{C}_{12}\mathbf{C}_{22}^{-1/2}, \quad \mathbf{R}^T = \mathbf{C}_{22}^{-1/2}\mathbf{C}_{21}\mathbf{C}_{11}^{-1/2}.$$

Setting $m = \min(p, q)$, it can be shown that the first m eigenvalues of the matrices

$$\mathbf{M}_X = \mathbf{R}\mathbf{R}^T = \mathbf{C}_{11}^{-1/2}\mathbf{C}_{12}\mathbf{C}_{22}^{-1}\mathbf{C}_{21}\mathbf{C}_{11}^{-1/2}$$

and

$$\mathbf{M}_Y = \mathbf{R}^T\mathbf{R} = \mathbf{C}_{22}^{-1/2}\mathbf{C}_{21}\mathbf{C}_{11}^{-1}\mathbf{C}_{12}\mathbf{C}_{22}^{-1/2}$$

are the same and positive, and are equal to

$$\rho_1^2 \geq \rho_2^2 \geq \cdots \rho_m^2 > 0.$$

Define the corresponding eigenvectors by

$$\mathbf{M}_X \mathbf{e}_k = \rho_k^2 \mathbf{e}_k, \quad \mathbf{M}_Y \mathbf{f}_k = \rho_k^2 \mathbf{f}_k, \quad k = 1, 2, \ldots m.$$

Then

$$\mathbf{a}_k = \mathbf{C}_{11}^{-1/2} \mathbf{e}_k, \quad \mathbf{b}_k = \mathbf{C}_{22}^{-1/2} \mathbf{f}_k$$

are the weights of the kth pair of canonical variables, and ρ_k is the kth canonical correlation. It is easy to check the the vectors $\mathbf{e}_k$ and $\mathbf{f}_k$ have unit norm and are related via

$$\mathbf{e}_k = \rho_k^{-1} \mathbf{R} \mathbf{f}_k, \quad \mathbf{f}_k = \rho_k^{-1} \mathbf{R}^T \mathbf{e}_k.$$

For the development in the subsequent sections, it is convenient to summarize the above using the inner product notation. Observe that

$$\begin{aligned} \mathrm{Cov}(\mathbf{a}^T \mathbf{X}, \mathbf{b}^T \mathbf{Y}) &= E\left[\mathbf{a}^T \mathbf{X} \mathbf{b}^T \mathbf{Y}\right] \\ &= E\left[\mathbf{a}^T \mathbf{X} \mathbf{Y}^T \mathbf{b}\right] = \mathbf{a}^T E\left[\mathbf{X} \mathbf{Y}^T\right] \mathbf{b} = \langle \mathbf{a}, \mathbf{C}_{12} \mathbf{b} \rangle \end{aligned}$$

and

$$\mathrm{Var}\left[\mathbf{a}^T \mathbf{X}\right] = \langle \mathbf{a}, \mathbf{C}_{11} \mathbf{a} \rangle, \quad \mathrm{Var}\left[\mathbf{b}^T \mathbf{Y}\right] = \langle \mathbf{b}, \mathbf{C}_{22} \mathbf{b} \rangle.$$

Thus

$$\begin{aligned} \rho_k &= \langle \mathbf{a}_k, \mathbf{C}_{12} \mathbf{b}_k \rangle \\ &= \max \left\{ \langle \mathbf{a}, \mathbf{C}_{12} \mathbf{b} \rangle : \mathbf{a} \in R^p, \mathbf{b} \in R^q, \langle \mathbf{a}, \mathbf{C}_{11} \mathbf{a} \rangle = 1, \langle \mathbf{b}, \mathbf{C}_{22} \mathbf{b} \rangle = 1 \right\} \end{aligned} \tag{4.6}$$

subject to the conditions

$$\langle A_k, A_j \rangle = \langle A_k, B_j \rangle = \langle B_k, B_j \rangle = \langle B_k, A_j \rangle = 0, \quad j < k, \tag{4.7}$$

where $A_j = \langle \mathbf{a}_j, \mathbf{X} \rangle$, $B_j = \langle \mathbf{b}_j, \mathbf{Y} \rangle$.

We conclude this section by describing the multivariate CCA for a sample

$$(\mathbf{x}_1, \mathbf{y}_1), \ (\mathbf{x}_2, \mathbf{y}_2), \ldots, (\mathbf{x}_N, \mathbf{y}_N), \tag{4.8}$$

in which each $\mathbf{x}_j$ is an observed vector of dimension p and $\mathbf{y}_j$ is of dimension q. The goal of the CCA is to find vectors $\hat{\mathbf{a}} \in R^p$, $\hat{\mathbf{b}} \in R^q$ such that the sample correlation between the $N \times 1$ vectors

$$\hat{\mathbf{A}} = [\hat{\mathbf{a}}^T \mathbf{x}_1, \ \hat{\mathbf{a}}^T \mathbf{x}_2, \ldots, \hat{\mathbf{a}}^T \mathbf{x}_N]^T$$

and

$$\hat{\mathbf{B}} = [\hat{\mathbf{b}}^T \mathbf{y}_1, \ \hat{\mathbf{b}}^T \mathbf{y}_2, \ldots, \hat{\mathbf{b}}^T \mathbf{y}_N]^T$$

is maximum, provided the vectors $\hat{\mathbf{A}}$ and $\hat{\mathbf{B}}$ have unit sample variance. Once the weight vectors $\hat{\mathbf{a}}$ and $\hat{\mathbf{b}}$ have been found, they are denoted $\hat{\mathbf{a}}_1$ and $\hat{\mathbf{b}}_1$, and the corresponding *first pair of sample canonical variates* by $\hat{\mathbf{A}}_1$ and $\hat{\mathbf{B}}_1$. We then search for another pair $(\hat{\mathbf{a}}_2, \hat{\mathbf{b}}_2)$ such that the sample correlation between analogously defined $\hat{\mathbf{A}}_2$ and $\hat{\mathbf{B}}_2$ is maximum subject to the conditions of unit sample variances and the lack of sample correlation with the $\hat{\mathbf{A}}_1$ and $\hat{\mathbf{B}}_1$. Conditions for the existence of sample multivariate canonical components are fully analogous to those stated for the population CCA. We define the matrices $\hat{\mathbf{C}}_{ij},\ i,j = 1,2$, by analogy with the definition of the matrices $\mathbf{C}_{ij}$. For example, the $p \times q$ matrix $\hat{\mathbf{C}}_{12}$ is defined as

$$\hat{\mathbf{C}}_{12} = \frac{1}{N-1}\sum_{j=1}^{N}\left[\mathbf{x}_j - \frac{1}{N}\sum_{j=1}^{N}\mathbf{x}_j\right]\left[\mathbf{y}_j - \frac{1}{N}\sum_{j=1}^{N}\mathbf{y}_j\right]^T.$$

If the matrices $\hat{\mathbf{C}}_{11}$ and $\hat{\mathbf{C}}_{22}$ are nonsingular, then the sample multivariate canonical components $(\hat{\rho}_k, \hat{\mathbf{a}}_k, \hat{\mathbf{b}}_k, \hat{\mathbf{A}}_k, \hat{\mathbf{B}}_k)$ exist for $k \le \min(p,q)$, and are calculated by replacing the matrices $\mathbf{C}_{ij}$ by the $\hat{\mathbf{C}}_{ij}$.

4.2 Functional canonical components

We now define the functional canonical components (FCC) by analogy to the multivariate setting. Their existence will be investigated in Section 4.6. We work with two L^2 spaces $\mathcal{H}_1 = L^2(T_1)$ and $\mathcal{H}_2 = L^2(T_2)$, where T_1 and T_2 are, possibly different, subsets of a Euclidean space. We consider square integrable random functions $X \in \mathcal{H}_1,\ Y \in \mathcal{H}_2$ and, to simplify the notation and some formulas, we continue to assume that they have mean zero. The canonical components are determined solely by the covariance structure and do not depend on the means. Thus, we define the covariance functions

$$\begin{aligned} c_{11}(t,s) &= E[X(t)X(s)], \\ c_{12}(t,s) &= E[X(t)Y(s)], \\ c_{21}(t,s) &= E[Y(t)X(s)], \\ c_{22}(t,s) &= E[Y(t)Y(s)]. \end{aligned}$$

Next, we define the operators

$$\begin{aligned} C_{11} &: \mathcal{H}_1 \to \mathcal{H}_1, \quad & C_{12} &: \mathcal{H}_2 \to \mathcal{H}_1, \\ C_{21} &: \mathcal{H}_1 \to \mathcal{H}_2, \quad & C_{22} &: \mathcal{H}_2 \to \mathcal{H}_2 \end{aligned}$$

via

$$\begin{aligned} C_{11}(x)(t) &= \int_{T_1} c_{11}(t,s)x(s)ds = E[\langle X, x\rangle\, X(t)], \\ C_{12}(y)(t) &= \int_{T_2} c_{12}(t,s)y(s)ds = E[\langle Y, y\rangle\, X(t)], \\ C_{21}(x)(t) &= \int_{T1} c_{21}(t,s)x(s)ds = E[\langle X, x\rangle\, Y(t)], \\ C_{22}(y)(t) &= \int_{T_2} c_{22}(t,s)y(s)ds = E[\langle Y, Y\rangle\, Y(t)]. \end{aligned}$$

The operators C_{11} and C_{22} are just the covariance operators introduced in Chapter 2, so they are symmetric, positive–definite and Hilbert–Schmidt. It is easy to extend the definition of a Hilbert–Schmidt operator to the space $\mathcal{L}(\mathcal{H}_2, \mathcal{H}_1)$ of bounded operators from $\mathcal{H}_2$ to $\mathcal{H}_1$, see Section 4.5. It is then seen that C_{12} is Hilbert–Schmidt because by the Cauchy–Schwartz inequality

$$\int_{T_1}\int_{T_2} c_{12}^2(t,s)dt\,ds \le E\|X\|^2 E\|Y\|^2.$$

Analogous statements are true for C_{21}.

We define the kth canonical correlation ρ_k and the associated weight functions a_k and b_k, if they exist, by

$$\rho_k = \operatorname{Cov}(\langle a_k, X\rangle, \langle b_k, Y\rangle) = \sup\{\operatorname{Cov}(\langle a, X\rangle, \langle b, Y\rangle) : a \in \mathcal{H}_1, b \in \mathcal{H}_2\}, \tag{4.9}$$

where a and b are subject to

$$\operatorname{Var}[\langle a, X\rangle] = 1, \quad \operatorname{Var}[\langle b, Y\rangle] = 1, \tag{4.10}$$

and for $k > 1$ also to (4.7) with $A_j = \langle a_j, X\rangle$, $B_j = \langle b_j, Y\rangle$. We call $(\rho_k, a_k, b_k, A_k, B_k)$ the kth canonical components.

Notice that

$$\begin{aligned}\operatorname{Cov}(\langle a, X\rangle, \langle b, Y\rangle) &= E[\langle a, X\rangle \langle b, Y\rangle] = E[\langle a, \langle Y, b\rangle X\rangle] \\ &= \langle a, E[\langle Y, b\rangle X]\rangle = \langle a, C_{12}(b)\rangle .\end{aligned}$$

Similarly,

$$\operatorname{Var}[\langle a, X\rangle] = \langle a, C_{11}(a)\rangle, \quad \operatorname{Var}[\langle a, Y\rangle] = \langle b, C_{22}(b)\rangle .$$

Therefore, just as in the multivariate setting, conditions (4.9) and (4.10) can be, respectively, rewritten as

$$\rho_k = \langle a_k, C_{12}(b_k)\rangle = \sup\{\langle a, C_{12}(b)\rangle : a \in \mathcal{H}_1, b \in \mathcal{H}_2\}, \tag{4.11}$$

$$\langle a, C_{11}(a)\rangle = 1, \quad \langle b, C_{22}(b)\rangle = 1. \tag{4.12}$$

4.3 Sample functional canonical components

Suppose we observe a sample of pairs of functions

$$(x_1, y_1),\ (x_2, y_2), \ldots, (x_N, y_N),$$

and we would like to obtain sample FCC analogously to the multivariate sample canonical components described in Section 4.1. Replacing the inner product in a Euclidean space by the inner product in L^2, we would thus like to maximize the sample correlation between the $N \times 1$ vectors

$$\hat{\mathbf{A}} = [\langle a, x_1\rangle, \langle a, x_2\rangle, \ldots, \langle a, x_N\rangle]^T \tag{4.13}$$

and

$$\hat{\mathbf{B}} = [\langle b, y_1 \rangle, \langle b, y_2 \rangle, \ldots, \langle b, y_N \rangle]^T. \tag{4.14}$$

As discovered by Leurgans *et al.* (1993), see also Chapter 11 of Ramsay and Silverman (2005), this is not a meaningful approach because it is possible to find functions a and b such that the sample correlation of $\hat{\mathbf{A}}$ and $\hat{\mathbf{B}}$ is arbitrarily close to 1. This can be intuitively explained as follows: in the multivariate case, the vectors **a** and **b** lie in spaces of finite dimension ($p-1$ and $q-1$), whereas the functions a and b are in an infinitely dimensional subspace. This gives too much flexibility. To illustrate, consider two samples of curves shown in Figure 4.1. These curves are obtained in an intermediate step leading to the construction of a global index of magnetic activity which is described in Section 4.4. The sample in the left panel reflects measurements obtained at Honolulu, the sample on the right those obtained at Kakioka, Japan. The curves a and b obtained by numerically maximizing the correlation between the vectors (4.13) and (4.14) are shown in the bottom row of Figure 4.2, (c) for Honolulu, (d) for Kakioka; they do not convey any meaningful information, and produce the first canonical correlation of 0.995. The weight functions showed in Panels (a) and (b) are more informative and reflect the presence of a very large storm which, as a global event, leaves almost the same signature at both Honolulu and Kakioka. These curves produce the first canonical correlation of 0.98. This correlation is so high because the storm event leaves almost identical signatures at Honolulu and Kakioka, and the remaining curves appear almost as noise relative

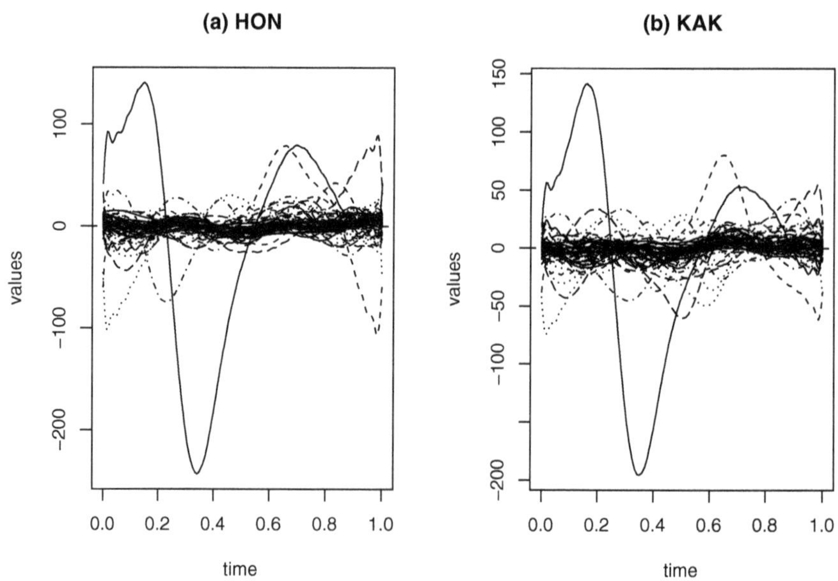

Fig. 4.1 Daily curves reflecting geomagnetic activity.

to this storm. We explain toward the end of this section how the curves in Figure 4.2 were obtained.

Several ways of defining sample FCC's have been put forward. All of them involve some form of smoothing and/or dimension reduction. We describe here a method recommended by He *et al.* (2004), which is closely related to the theory introduced in Section 4.6. Then we turn to the method of Leurgans *et al.* (1993), which emphasizes smoothing the weight functions a and b. It is implemented in the R package `fda`.

Denote by $\hat{\lambda}_i$ and $\hat{v}_i$ the eigenvalues and the eigenfunctions of the sample covariance operator of the functions x_i, and define analogously $\hat{\gamma}_j$ and $\hat{u}_j$ for the y_j. Determine the numbers p and q such that $\sum_{i\le p}\hat{\lambda}_i$ and $\sum_{j\le q}\hat{\gamma}_j$ explain the required proportion of the variance, see Section 3.3. Methods of selecting p and q which involve cross–validation are described in Section 2.5 of He *et al.* (2004). Next, compute the scores

$$\hat{\xi}_{in} = \langle \hat{v}_i, X_n \rangle, \; i = 1, 2, \ldots, p, \qquad \hat{\zeta}_{jn} = \langle \hat{u}_j, Y_n \rangle, \; j = 1, 2, \ldots, q.$$

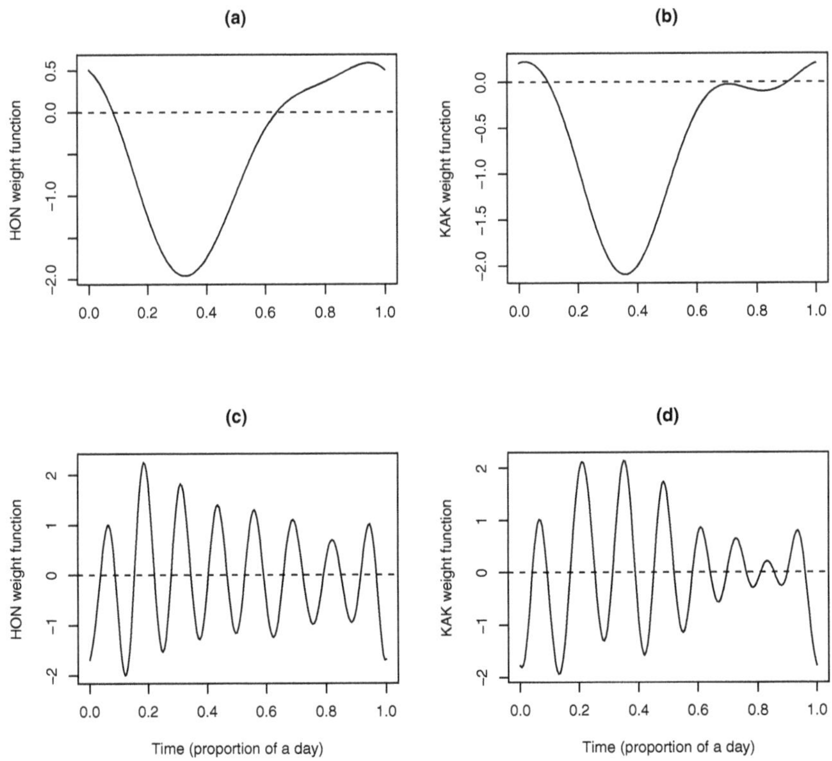

Fig. 4.2 Weight functions for first canonical correlations of the curves displayed in Figure 4.1. Top row with penalty, bottom row without penalty.

Now, we can work with the finite expansions

$$\hat{X}_n = \sum_{i=1}^{p} \hat{\xi}_{in} \hat{v}_i, \qquad \hat{Y}_n = \sum_{j=1}^{q} \hat{\zeta}_{jn} \hat{u}_j.$$

The curves $\hat{X}_n$ and $\hat{Y}_n$ are smoothed versions of the original observations X_n and Y_n, while the vectors

$$\hat{\boldsymbol{\xi}}_n = [\langle X_n, \hat{v}_1\rangle, \langle X_n, \hat{v}_2\rangle, \ldots \langle X_n, \hat{v}_p\rangle]^T,$$
$$\hat{\boldsymbol{\zeta}}_n = [\langle Y_n, \hat{u}_1\rangle, \langle Y_n, \hat{u}_2\rangle, \ldots \langle Y_n, \hat{u}_q\rangle]^T$$

allow us to reduce the the problem to the multivariate sample CCA described in Section 4.1. The collection of pairs

$$(\hat{\boldsymbol{\xi}}_1, \hat{\boldsymbol{\zeta}}_1), (\hat{\boldsymbol{\xi}}_2, \hat{\boldsymbol{\zeta}}_2), \ldots, (\hat{\boldsymbol{\xi}}_N, \hat{\boldsymbol{\zeta}}_N)$$

now plays the role of the multivariate sample (4.8), and allows us to find the multivariate sample canonical components $(\rho_k, \hat{\mathbf{a}}_k, \hat{\mathbf{b}}_k, \hat{\mathbf{A}}_k, \hat{\mathbf{B}}_k)$. In analogy to (4.25), we define the functional canonical components as $(\rho_k, \hat{a}_k, \hat{b}_k, \hat{\mathbf{A}}_k, \hat{\mathbf{B}}_k)$, where

$$\hat{a}_k = \hat{\mathbf{a}}_k^T [\hat{v}_1, \ldots \hat{v}_p]^T, \qquad \hat{b}_k = \hat{\mathbf{b}}_k^T [\hat{u}_1, \ldots \hat{u}_q]^T.$$

He *et al.* (2004) recommend an additional smoothing step. After the $\hat{v}_i$ and the $\hat{u}_j$ have been computed, they can be smoothed in some way, for example using polynomial smoothing. Denote the smoothed FPC's by $\tilde{v}_i$ and $\tilde{u}_j$, and construct the vectors

$$\tilde{\boldsymbol{\xi}}_n = [\langle X_n, \tilde{v}_1\rangle, \langle X_n, \tilde{v}_2\rangle, \ldots \langle X_n, \tilde{v}_p\rangle]^T,$$
$$\tilde{\boldsymbol{\zeta}}_n = [\langle Y_n, \tilde{u}_1\rangle, \langle Y_n, \tilde{u}_2\rangle, \ldots \langle Y_n, \tilde{u}_q\rangle]^T.$$

The pairs $(\tilde{\boldsymbol{\xi}}_n, \tilde{\boldsymbol{\zeta}}_n)$, $n = 1, 2, \ldots, N$, lead to multivariate sample canonical components, which again yield functional sample canonical components.

To explain the approach of Leurgans *et al.* (1993), denote by $C_N(\mathbf{A}, \mathbf{B})$ the sample covariance of the N–dimensional vectors $\mathbf{A}$ and $\mathbf{B}$. The naive approach to finding sample functional canonical correlations is to maximize $C_N(\hat{\mathbf{A}}, \hat{\mathbf{B}})$ subject to $C_N(\hat{\mathbf{A}}, \hat{\mathbf{A}}) = 1$, $C_N(\hat{\mathbf{B}}, \hat{\mathbf{B}}) = 1$ (and orthogonality conditions), where $\hat{\mathbf{A}}, \hat{\mathbf{B}}$ are defined by (4.13) (4.14). We have seen that it is then always possible to find functions a and b such that $C_N(\hat{\mathbf{A}}, \hat{\mathbf{B}}) = 1$. In order to restrict the set of function over which the maximum is found, we assume that that $T_1 = T_2 = [0, 1]$, and consider only functions a, b such that a'', b'' (second derivatives) exist and are elements of L^2. We then maximize $C_N(\hat{\mathbf{A}}, \hat{\mathbf{B}})$ subject to

$$C_N(\hat{\mathbf{A}}, \hat{\mathbf{A}}) + \lambda \|a''\|^2 = 1, \quad \text{and} \quad C_N(\hat{\mathbf{B}}, \hat{\mathbf{B}}) + \lambda \|b''\|^2 = 1.$$

The number $\lambda > 0$ is a smoothing parameter which penalizes for using functions a, b which are highly irregular. It can be chosen by cross-validation, or subjectively, in order to obtain informative weight functions a and b. The weight functions in Panels (a) and (b) of Figure 4.2 were obtained with $\lambda = 10^{-11.5}$.

The following R code implements the method of Leurgans *et al.* (1993) and produces Figure 4.2.

```
harmaccelLfd <- vec2Lfd(c(0, 0, (2*pi)\^{} 2, 0))
lambda=10^(-11.5)
```

Define a functional parameter object:

```
fdPar <- fdPar(fdobj = basis, Lfdobj = harmaccelLfd,
           lambda = lambda)
```

Create a functional object that smooths the data using specified roughness penalty:

```
fd.s.1 <- smooth.basis(t, t(z1.new), fdPar)\$fd
fd.s.2 <- smooth.basis(t, t(z2.new), fdPar)\$fd
ccafdPar1 <- fdPar(fd.s.1, harmaccelLfd, lambda = 1e-8)
ccafdPar2 <- fdPar(fd.s.2, harmaccelLfd, lambda = 1e-8)
```

Compute the smoothed canonical correlations :

```
cca.smoothed <- cca.fd(fd.s.1, fd.s.2, ncan=3,
                  ccafdPar1, ccafdPar2)

par(mfrow=c(2,2))
plot.fd(cca.smoothed\$weight1[1], main="(a)",
           ylab="HON weight function")
plot.fd(cca.smoothed\$weight2[1], main="(b)",
           ylab="KAK weight function")
plot.fd(cca.unsmoothed\$weight1[1], main="(c)",
           ylab="HON weight function",
           xlab="Time (proportion of a day)")
plot.fd(cca.unsmoothed\$weight2[1], main="(d)",
           ylab="KAK weight function",
           xlab="Time (proportion of a day)")
```

Canonical correlations are often used to see which samples of functions are most strongly associated, an application of this type is presented in Section 4.4. In such applications, provided clear cut differences exists, any reasonable choice of λ, or several values of λ, will lead to informative comparisons. The same is true for choosing the orders p and q. In physical applications, these orders can be chosen to restrict the analysis to meaningful principal components.

4.4 Functional canonical correlation analysis of a magnetometer data

This section is based on the work of Maslova *et al.* (2009) who proposed a new method computing an index of magnetic storm activity. Magnetic storms belong to the most important phenomena in near Earth space due to the energy involved and their impact on the operation of satellite based telecommunication and navigation systems.

The data are magnetometer observations, an example is shown in Figure 1.1. When a magnetic storm occurs, the H-component drops for a period of 2-3 days at observatories close to the magnetic equator, reflecting a strong magnetic field generated by a magnetospheric ring current that forms during storms. Figure 4.3 shows a magnetometer record at Honolulu during a storm. Similar curves are observed at other equatorial observatories, but each of them looks different, mostly because at a given universal time, different observatories may be at local day- or nighttime, or dawn or dusk. The position of an observatory relative to the sun has a noticeable impact on the shape of the magnetogram. The change in the shape of the magnetometer records due to the daily rotation of the Earth is called the Sq (Solar quiet) variation. An important direction of research in space physics, going back to Sugiura (1964), has been concerned with developing an index curve that would measure the strength of a magnetic storm globally, as different storm signatures are observed at different observatories. Computing a global index involves averaging over several equatorial observatories after removing the Sq variation from each record. The technical details are quite complex, and there is no universal agreement in the space physics community on the best way to construct a good global index.

Maslova *et al.* (2009) proposed a new method of removing the Sq variation from each record. Without discussing the technicalities, the idea is that the component that is removed changes from day to day. For an older method, WISA, this component was constant over the period of 2-4 weeks. The new method was proposed in two variants, referred to below as 1) "with" and 2) "without centering". There are thus three methods to compare. To perform the comparison, the Sq variation is removed by every method from the record at every observatory. What is left, should reflect the effect of a global ring current, not the location of the station relative to the Sun. Thus if the removal is successful, the remainders, called preindices, at the pairs of stations should be highly correlated.

We present only a small part of the validation study reported by Maslova *et al.* (2009). We consider four observatories, known as the "Dst Observatories", which are listed in Table 4.1. These four observatories yield six pairs listed in Table 4.2. For each pair, we compute the sample FCC's as described in Section 4.3. The smoothing parameter λ is chosen so that all correlations fall into a relatively large subinterval of [0,1], so that a visual comparison is facilitated. We are not interested so much in the values of the sample FCC's as in their order for the three methods. The results are shown in Figure 4.4. High sample FCC's indicate that the preindices obtained by one of the methods are good because they measure the same field generated by

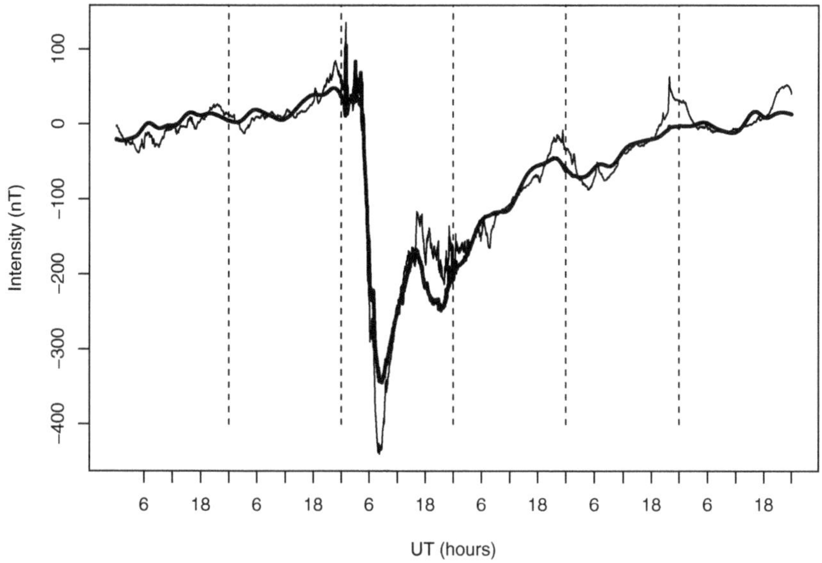

Fig. 4.3 H-component of the magnetogram recorded at Honolulu Mar 29 – Apr 3 (thin line) together with the global index developed by Maslova *et al.* (2009) (thick line). The dashed lines separate UT days. The drop reflects a magnetic storm.

Table 4.1 Dst Geomagnetic observatories and their coordinates.

s	Name	Colatitude	Longitude
1	Hermanus (HER)	124.43	19.23
2	Kakioka (KAK)	53.77	140.18
3	Honolulu (HON)	68.68	202.00
4	San Juan (SJG)	71.89	293.85

a global ring current. Figure 4.4 shows that the preindices constructed with the new method always have higher sample FCC's than those obtained with an older WISA method. The new method with centering is generally better than the new method without centering. A more detailed analysis confirms these assertions.

4.5 Square root of the covariance operator

In this section we review certain properties of the covariance operator C which will be used in Section 4.6

Table 4.2 Pairs of four Dst stations (first set) used to compare methodologies.

Combination #	Stations
1	HON & KAK
2	HON & SJG
3	HON & HER
4	KAK & SJG
5	KAK & HER
6	SJG & HER

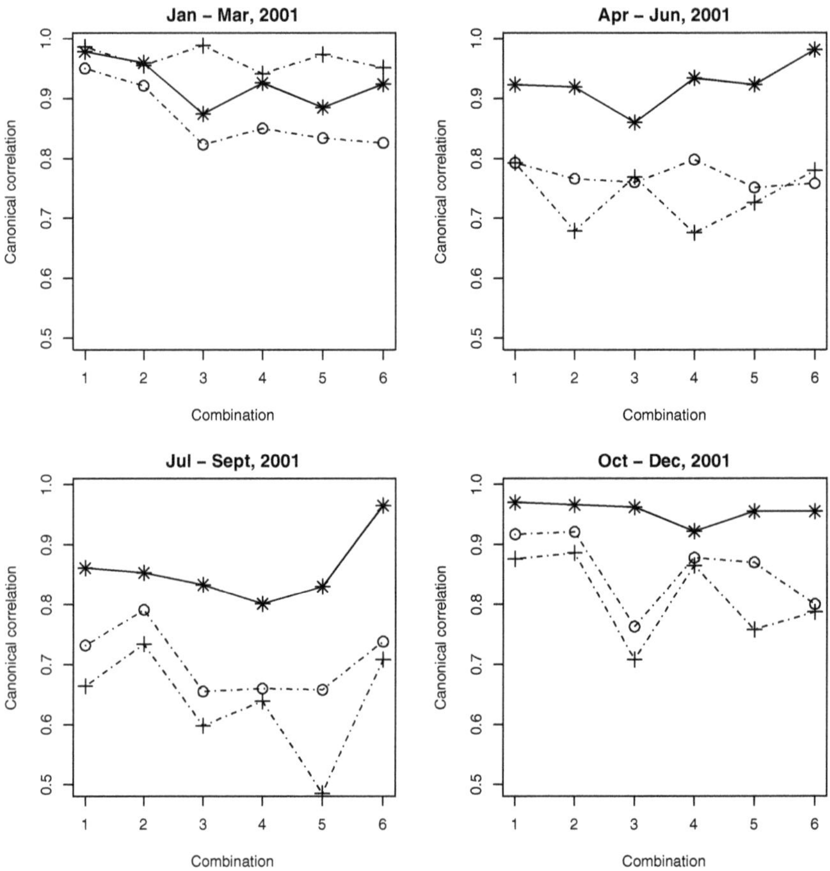

Fig. 4.4 Canonical correlations for the new method (star), new method without centering (cross) and WISA (circle), applied to all combinations of four Dst stations (see Table 4.2).

Recall from Chapter 2 that C admits the decomposition

$$C(x) = \sum_{j=1}^{\infty} \lambda_j \langle x, v_j \rangle v_j, \quad x \in L^2, \tag{4.15}$$

in which the λ_j are nonnegative, the v_j form a basis and satisfy $C(v_j) = \lambda_j v_j$.

An operator R is called a square root of C if $RR = C$. Every covariance (in fact, every positive–definite) operator has a unique positive–definite square root. It is defined by

$$C^{1/2}(x) = \sum_{j=1}^{\infty} \lambda_j^{1/2} \langle x, v_j \rangle v_j, \quad x \in L^2.$$

Direct verification shows that $C^{1/2}$ is symmetric and positive–definite.

Suppose A is an operator defined on a subspace $\mathcal{D}(A)$ with range $\mathcal{R}(A)$. An operator B with domain $\mathcal{R}(A)$ is called the inverse of A if $B(A(x)) = x$ for all $x \in \mathcal{D}(A)$ and $A(B(y)) = y$ for all $y \in \mathcal{R}(A)$. If A has an inverse, it is unique and is denoted A^{-1}. An operator A is invertible if and only if $A(x) = 0$ implies $x = 0$.

We see that C and $C^{1/2}$ (defined on the whole of L^2) are invertible if and only if

$$\lambda_j > 0 \quad \text{for each } j \geq 1. \tag{4.16}$$

Condition (4.16) is assumed in the following.

If condition (4.16) holds, then

$$\mathcal{R}(C^{1/2}) = \left\{ y \in L^2 : \sum_{j=1}^{\infty} \lambda_j^{-1} \langle y, v_j \rangle^2 < \infty \right\}. \tag{4.17}$$

Indeed, if $y \in \mathcal{R}(C^{1/2})$, then for some $x \in L^2$

$$y = C^{1/2}x = \sum_{i=1}^{\infty} \lambda_i^{1/2} \langle x, v_i \rangle v_i,$$

and so

$$\sum_{j=1}^{\infty} \lambda_j^{-1} \langle y, v_j \rangle^2 = \sum_{j=1}^{\infty} \lambda_j^{-1} \left\langle \sum_{i=1}^{\infty} \lambda_i^{1/2} \langle x, v_i \rangle v_i, v_j \right\rangle^2$$

$$= \sum_{j=1}^{\infty} \lambda_j^{-1} \left(\lambda_j^{1/2} \langle x, v_j \rangle \right)^2 = \sum_{j=1}^{\infty} \langle x, v_j \rangle^2 = \|x\|^2 < \infty.$$

Conversely, if $\sum_{j=1}^{\infty} \lambda_j^{-1} \langle y, v_j \rangle^2 < \infty$, then $x = \sum_{j=1}^{\infty} \lambda_j^{-1/2} \langle y, v_j \rangle v_j$ is a well–defined element of L^2, and a direct verification shows that $C^{1/2}(x) = y$. The inverse of $C^{1/2}$ is thus defined by

$$C^{-1/2}(y) = \sum_{j=1}^{\infty} \lambda_j^{-1/2} \langle y, v_j \rangle v_j, \quad y \in \mathcal{R}(C^{1/2}). \tag{4.18}$$

Notice that under assumption (4.16) each v_k is in $\mathcal{R}(C^{1/2})$, and since $\mathcal{R}(C^{1/2})$ is a linear subspace, so are all finite linear combinations of the v_k. However, in contrast to the Euclidean space R^p, these finite linear combinations do not fill the whole of

L^2. Define, for example, $y = \sum_{k=1}^{\infty} \lambda_k^{1/2} v_k$. Since $\sum_{k=1}^{\infty} \lambda_k < \infty$, $y \in L^2$. However, y is not in $\mathcal{R}(C^{1/2})$ because

$$\sum_{j=1}^{\infty} \lambda_j^{-1} \langle y, v_j \rangle^2 = \sum_{j=1}^{\infty} \lambda_j^{-1} \lambda_j = \infty.$$

We conclude this section by recalling some facts about Hilbert–Schmidt operators in $\mathcal{L}(\mathcal{H}_1, \mathcal{H}_2)$ which will be needed to establish the existence of FCC's. Suppose $\{v_i\}$ is a basis in $\mathcal{H}_1$, and $\{u_j\}$ is a basis in $\mathcal{H}_2$. If $A \in \mathcal{L}(\mathcal{H}_1, \mathcal{H}_2)$, then

$$A(v_i) = \sum_{j=1}^{\infty} \langle A(v_i), u_j \rangle u_j = \sum_{j=1}^{\infty} a_{ji} u_j.$$

The coefficients a_{ji} determine the operator A. If $\sum_{i,j=1}^{\infty} a_{ji}^2 < \infty$, A is called Hilbert–Schmidt. The sum does not depend on the choice of bases, and its square root is the Hilbert–Schmidt norm. The space of Hilbert–Schmidt operators in $\mathcal{L}(\mathcal{H}_1, \mathcal{H}_2)$ is denoted $\mathcal{S}(\mathcal{H}_1, \mathcal{H}_2)$. If $A_1 \in \mathcal{S}(\mathcal{H}_1, \mathcal{H}_2)$ and $A_2 \in \mathcal{S}(\mathcal{H}_2, \mathcal{H}_3)$, then $A_2 A_1 \in \mathcal{S}(\mathcal{H}_1, \mathcal{H}_3)$, and $\|A_2 A_1\|_{\mathcal{S}} \le \|A_1\|_{\mathcal{S}} \|A_2\|_{\mathcal{S}}$.

If A is an integral operator of the form

$$A(x)(t) = \int_{T_1} a(t,s) x(s) ds, \quad x \in \mathcal{H}_1,$$

then A is Hilbert–Schmidt if and only if $\int_{T_2} \int_{T_1} a^2(t,s) dt\, ds < \infty$. In that case,

$$\int_{T_2} \int_{T_1} a^2(t,s) dt\, ds = \sum_{i,j=1}^{\infty} a_{ji}^2 < \infty.$$

4.6 Existence of the functional canonical components

Recall the notation introduced in Sections 4.1 and 4.2. The central message of this section is that the whole spaces $\mathcal{H}_1$ and $\mathcal{H}_2$ are too large to define a functional CCA. It it possible only on smaller subspaces. In practice, this is reflected by the required smoothing in the sample FCCA, as discussed in Section 4.3. It is difficult to capture this idea in a theoretical framework. We present the approach of He *et al.* (2003) who propose to restrict the spaces $\mathcal{H}_1$ and $\mathcal{H}_2$ by imposing conditions on the magnitude of the eigenvalues of C_{11} and C_{22} and their interactions, see Assumption 4.1. A more general framework for functional CCA was developed by Eubank and Hsing (2008). Cupidon *et al.* (2007) is also relevant.

We would like to construct operator analogs of the matrices $\mathbf{M}_X$ and $\mathbf{M}_Y$ defined in Section 4.1. Define $\mathcal{R}_1 = \mathcal{R}(C_{11}^{1/2}) \subset \mathcal{H}_1$ and $\mathcal{R}_2 = \mathcal{R}(C_{22}^{1/2}) \subset \mathcal{H}_2$. We need the following condition

$$C_{12}(\mathcal{H}_2) \subset \mathcal{R}_1, \quad C_{21}(\mathcal{H}_1) \subset \mathcal{R}_2. \tag{4.19}$$

To establish a convenient sufficient condition for (4.19), consider the expansions

$$X = \sum_{i=1}^{\infty} \xi_i v_i, \ \xi_i = \langle X, v_i \rangle, \quad Y = \sum_{j=1}^{\infty} \zeta_j u_j, \ \zeta_j = \langle Y, u_j \rangle,$$

where the eigenfunctions v_i and u_j satisfy $C_{11} v_i = \lambda_i v_i$, $C_{22} u_j = \gamma_j u_j$. We assume that

$$\lambda_i > 0, \quad \gamma_j > 0 \qquad \text{for each } i, j > 0. \tag{4.20}$$

Next define the correlation coefficients

$$r_{ji} = \frac{E[\xi_i \zeta_j]}{\sqrt{E\xi_i^2 E\zeta_j^2}} = \frac{E[\xi_i \zeta_j]}{\lambda_i^{1/2} \gamma_j^{1/2}}. \tag{4.21}$$

Proposition 4.1. *If* (4.20) *holds and the coefficients* r_{ji} *in* (4.21) *satisfy*

$$\sum_{i,j=1}^{\infty} r_{ji}^2 < \infty, \tag{4.22}$$

then (4.19) *holds.*

Proof. We focus on the first relation $C_{12}(\mathcal{H}_2) \subset \mathcal{R}_1$. We must show that if $x \in C_{12}(\mathcal{H}_2)$ and (4.22) holds, then

$$\sum_{i=1}^{\infty} \lambda_i^{-1} \langle x, v_i \rangle^2 < \infty. \tag{4.23}$$

If $x = C_{12}(y)$ for some $y \in \mathcal{H}_2$, then

$$x = E[\langle Y, y \rangle X] = \left[\left\langle \sum_j \zeta_j u_j, y \right\rangle \sum_k \xi_k v_k \right] = \sum_{j,k} E[\zeta_j \xi_k] \langle u_j, y \rangle v_k. \tag{4.24}$$

Consequently, since $E\xi_i^2 = \lambda_i$, $E\zeta_j^2 = \gamma_j$,

$$\langle x, v_i \rangle = \sum_j E[\xi_i \zeta_j] \langle u_j, y \rangle = \sum_j r_{ji} \lambda_i^{1/2} \gamma_j^{1/2} \langle u_j, y \rangle.$$

Therefore, by the Cauchy–Schwartz inequality,

$$\langle x, v_i \rangle^2 \le \lambda_i \left(\sum_j r_{ji}^2 \gamma_j \right) \left(\sum_j \langle u_j, y \rangle^2 \right) = \lambda_i \|y\|^2 \sum_j r_{ji}^2 \gamma_j,$$

and so we obtain

$$\sum_{i=1}^{\infty} \lambda_i^{-1} \langle x, v_i \rangle^2 \le \|y\|^2 \sum_{i,j=1}^{\infty} r_{ji}^2 \gamma_j.$$

Thus, a sufficient condition for $C_{12}(\mathcal{H}_2) \subset \mathcal{R}_1$ is $\sum_{i,j=1}^{\infty} r_{ji}^2 \gamma_j < \infty$ and, analogously, a sufficient condition for $C_{21}(\mathcal{H}_1) \subset \mathcal{R}_2$ is $\sum_{i,j=1}^{\infty} r_{ji}^2 \lambda_i < \infty$. Since $\gamma_j \le \gamma_1$, $\lambda_i \le \lambda_1$, both these conditions are implied by (4.22). □

If assumption (4.19) holds, we can define a correlation operator

$$R = C_{11}^{-1/2} C_{12} C_{22}^{-1/2} : \mathcal{R}_2 \to \mathcal{H}_1.$$

Its adjoint operator $R^* : \mathcal{H}_1 \to \mathcal{R}_2$ is uniquely defined by

$$\langle R(y), x \rangle = \langle y, R^*(x) \rangle, \quad y \in \mathcal{R}_2, \ x \in \mathcal{H}_1.$$

Lemma 4.1. *If condition* (4.22) *holds, then* $R \in \mathcal{S}(\mathcal{R}_2, \mathcal{H}_1)$, $R^* \in \mathcal{S}(\mathcal{H}_1, \mathcal{R}_2)$, *and*

$$R(u_j) = \sum_{k=1}^{\infty} r_{jk} v_k, \quad R^*(v_i) = \sum_{k=1}^{\infty} r_{ki} u_k.$$

Proof. Observe that

$$C_{22}^{-1/2}(u_j) = \sum_k \gamma_k^{-1/2} \langle u_j, u_k \rangle u_k = \gamma_j^{-1/2} u_j.$$

By (4.24),

$$C_{12} C_{22}^{-1/2}(u_j) = \gamma_j^{-1/2} C_{12}(u_j) = \gamma_j^{-1/2} \sum_{i,k} E[\zeta_i \xi_k] \langle u_i, u_j \rangle v_k$$
$$= \gamma_j^{-1/2} \sum_k E[\xi_k \zeta_j] v_k.$$

Consequently,

$$R(u_j) = \gamma_j^{-1/2} \sum_k E[\xi_k \zeta_j] C_{11}^{-1/2}(v_k)$$
$$= \gamma_j^{-1/2} \sum_k E[\xi_k \zeta_j] \lambda_k^{-1/2} v_k = \sum_k r_{jk} v_k.$$

Next, notice that

$$\langle R(u_j), v_i \rangle = \left\langle \sum_k r_{jk} v_k, v_i \right\rangle = r_{ji} = \langle u_j, R^*(v_i) \rangle,$$

implying $R^*(v_i) = \sum_k r_{ki} u_k$. □

We now define the operator

$$M_Y = R^* R : \mathcal{R}_2 \to \mathcal{R}_2.$$

Direct verification shows that M_Y is symmetric and positive–definite, and by Lemma 4.1, it is a Hilbert–Schmidt operator, as a composition of two Hilbert–Schmidt operators. The operator M_Y thus admits decomposition (2.4), which we write down as

$$M_Y(y) = \sum_{k=1}^{\infty} \rho_k^2 \langle y, f_k \rangle f_k, \quad y \in \mathcal{R}_2,$$

with orthonormal eigenfunctions $f_k \in \mathcal{R}_2$. All eigenvalues ρ_k^2 are positive as $\langle M_Y(y), y\rangle = 0$ implies $y = 0$. This is because $\langle M_Y(y), y\rangle = \|Ry\|^2$, and R is invertible on $\mathcal{R}_2$.

To ensure the existence of the FCC's, we need to strengthen condition (4.22) to the following assumption:

Assumption 4.1. *Condition* (4.20) *holds and*

$$\sum_{i,j=1}^{\infty} \lambda_i^{-1} r_{ji}^2 < \infty \quad \text{and} \quad \sum_{i,j=1}^{\infty} \gamma_j^{-1} r_{ji}^2 < \infty.$$

To understand why Assumption 4.1 is needed, define analogously to the multivariate setting $e_k = \rho_k^{-1} R(f_k)$. We would like e_k to be an element of $\mathcal{R}_1$ so that we can define the weight function $a_k = C_{11}^{-1/2} e_k$. Observe that by Lemma 4.1,

$$\begin{aligned}
\sum_{i=1}^{\infty} \lambda_i^{-1} \langle R(f_k), v_i\rangle^2 &= \sum_{i=1}^{\infty} \lambda_i^{-1} \left\langle \sum_{j=1}^{\infty} \langle f_k, u_j\rangle R(u_j), v_i \right\rangle^2 \\
&= \sum_{i=1}^{\infty} \lambda_i^{-1} \left(\sum_{j=1}^{\infty} \langle f_k, u_j\rangle r_{ji} \right)^2 \\
&\le \sum_{i=1}^{\infty} \lambda_i^{-1} \sum_{j=1}^{\infty} \langle f_k, u_j\rangle^2 \sum_{j=1}^{\infty} r_{ji}^2 \\
&\le \|f_k\|^2 \sum_{i,j=1}^{\infty} \lambda_i^{-1} r_{ji}^2.
\end{aligned}$$

Consequently $R(f_k) \in \mathcal{R}_1$, if Assumption 4.1 holds. In fact, the same argument is valid for any $y \in \mathcal{R}_2$, so we have $R(\mathcal{R}_2) \subset \mathcal{R}_1$. Thus, under Assumption 4.1 the domain of the conjugate operator R^* is $\mathcal{R}_1$. Changing the roles of the spaces $\mathcal{H}_1$ and $\mathcal{H}_2$, we see that $S = C_{22}^{-1/2} C_{21} C_{11}^{-1/2}$ maps $\mathcal{R}_1$ into $\mathcal{R}_2$, By an analog of Lemma 4.1,

$$S(v_j) = \sum_{k=1}^{\infty} s_{jk} u_k, \quad s_{ji} = \gamma_i^{-1/2} \lambda_j^{-1/2} E[\zeta_i \xi_j].$$

Thus, $\langle u_j, S(v_i)\rangle = \langle u_j, R^*(v_i)\rangle$, and so we conclude that $S = R^*$. Finally define

$$M_X = RR^* : \mathcal{R}_1 \to \mathcal{R}_1$$

and observe that $M_X(\rho_k^{-1} R(f_k)) = \rho_k^2(\rho_k^{-1} R(f_k))$, and that $\|e_k\|^2 = 1$. We summarize these calculations in the following proposition:

Proposition 4.2. *If Assumption 4.1 holds, then*

$$R = C_{11}^{-1/2} C_{12} C_{22}^{-1/2} : \mathcal{R}_2 \to \mathcal{R}_1,$$
$$S = C_{22}^{-1/2} C_{21} C_{11}^{-1/2} : \mathcal{R}_1 \to \mathcal{R}_2,$$
$$S = R^*, \; R = S^*.$$

The operators M_Y and M_X have the same eigenvalues ρ_k^2, all eigenvalues are positive, and the normalized eigenfunctions e_k of M_X are related to the normalized eigenfunctions f_k of M_Y via

$$e_k = \rho_k^{-1} R(f_k), \quad f_k = \rho_k^{-1} R^*(e_k).$$

We are now able to define the weight functions

$$a_k = C_{11}^{-1/2}(e_k), \quad b_k = C_{22}^{-1/2}(f_k).$$

These functions are well defined elements of, respectively, $\mathcal{H}_1$ and $\mathcal{H}_2$ because $e_k \in \mathcal{R}(C_{11}^{1/2})$ and $f_k \in \mathcal{R}(C_{22}^{1/2})$. The following theorem shows that $(\rho_k, a_k, b_k, \langle a_k, X\rangle, \langle b_k, Y\rangle)$ are the functional canonical components defined in Section 4.2. We say that the pair of random variables (A_j, B_j) is uncorrelated with (A_k, B_k) if condition (4.5) holds.

Theorem 4.1. *If Assumption 4.1 holds, then*

(i) $\langle a_k, C_{12}(b_k)\rangle = \rho_k$, $\langle a_k, C_{11}(a_k)\rangle = 1$, $\langle b_k, C_{22}(b_k)\rangle = 1$.
(ii) For any $a \in \mathcal{H}_1, b \in \mathcal{H}_2$ such that $\langle a, C_{11}a\rangle = 1$, $\langle b, C_{22}b\rangle = 1$, and such that $(\langle a, X\rangle, \langle b, Y\rangle)$ is uncorrelated with $(\langle a_j, X\rangle, \langle b_j, Y\rangle)$ for $j < k$, $\langle a, C_{12}b\rangle \le \rho_k$.
(iii) If $j \neq k$, the pairs $(\langle a_j, X\rangle, \langle b_j, Y\rangle)$ and $(\langle a_k, X\rangle, \langle b_k, Y\rangle)$ are uncorrelated.

Proof. The equalities in part (i) are easy to verify. For example

$$\langle a_k, C_{12}(b_k)\rangle = \left\langle C_{11}^{-1/2}(e_k), C_{12} C_{22}^{-1/2}(f_k)\right\rangle = \langle e_k, R(f_k)\rangle = \langle e_k, \rho_k e_k\rangle = \rho_k.$$

To lighten the notation, we verify part (ii) for $k = 2$. For functions a and b satisfying the assumptions of part (ii) define $x = C_{11}^{1/2}(a)$, $y = C_{22}^{1/2}(b)$. Then

$$\langle a, C_{12}(b)\rangle = \left\langle C_{11}^{-1/2}(x), C_{12} C_{22}^{-1/2}(y)\right\rangle = \langle x, R(y)\rangle \le \|x\| \|R(y)\|.$$

Condition $\langle a, C_{11}a\rangle = 1$, is equivalent to $\|x\| = 1$, so it remains to verify that $\|R(y)\| \le \rho_2$. By Theorem 3.2

$$\rho_2^2 = \sup\left\{\langle M_Y(y), y\rangle : y \in \mathcal{R}_2, \|y\| = 1, \langle y, f_1\rangle = 0\right\}.$$

Since $\langle M_Y(y), y\rangle = \|R(y)\|^2$, and y also has unit norm, it is enough to check that $\langle y, f_1\rangle = 0$. This holds because

$$\begin{aligned}
\langle y, f_1\rangle &= \left\langle C_{22}^{1/2}(b), f_1\right\rangle = \left\langle b, C_{22}^{1/2}(f_1)\right\rangle = \left\langle b, C_{22} C_{22}^{-1/2}(f_1)\right\rangle \\
&= \langle b, C_{22}(b_1)\rangle = \langle b, E[\langle Y, b_1\rangle Y]\rangle \\
&= E[\langle b, \langle Y, b_1\rangle Y\rangle] = E[\langle b, Y\rangle \langle Y, b_1\rangle] = 0.
\end{aligned}$$

Part (iii) is easy to verify. For example

$$E[\langle a_j, X\rangle \langle b_k, Y\rangle] = E[\langle b_k, \langle X, a_j\rangle Y\rangle] = \langle b_k, C_{21}(a_j)\rangle$$
$$= \left\langle C_{22}^{-1/2}(f_k), C_{21}C_{11}^{-1/2}(e_j)\right\rangle = \langle f_k, R^*(e_j)\rangle = \langle f_k, \rho_k f_j\rangle = 0. \qquad \square$$

Assumption 4.1 states that in order for the FCC's to exist, the correlations of the scores ξ_i and ζ_j must tend to zero very fast. This is trivially the case if $r_{ji} = 0$ if $i > p$ or $j > q$, for some integers p and q. We thus conclude this section by considering the case when the random functions X and Y admit the finite expansions

$$X = \sum_{i=1}^{p} \xi_i v_i, \quad Y = \sum_{j=1}^{q} \zeta_j u_j,$$

with orthonormal systems $v_1, \dots, v_p \in \mathcal{H}_1$, $u_1, \dots, u_q \in \mathcal{H}_2$. Strictly speaking, this case is not covered by the theory developed in this section because condition (4.16) fails, but it is, in fact, much simpler, and the FCC's are directly related to the multivariate canonical components of the vectors

$$\boldsymbol{\xi} = [\xi_1, \dots, \xi_p]^T, \quad \boldsymbol{\zeta} = [\zeta_1, \dots, \zeta_q]^T.$$

Define the linear spans

$$\mathcal{R}_1 = \mathrm{sp}\{v_1, \dots, v_p\}, \quad \mathcal{R}_2 = \mathrm{sp}\{u_1, \dots, u_q\}.$$

Consider the operators C_{ij}, $i, j = 1, 2$, defined in Section 4.2, but with their domains restricted to the appropriate subspaces $\mathcal{R}_i, i = 1, 2$. For example, if $y = \sum_{j=1}^{p} y_j u_j$, $y_j = \langle y, u_i\rangle$, then

$$C_{12}(y) = E[\langle Y, y\rangle X] = E\left[\left\langle \sum_{j=1}^{q} \zeta_j u_j, \sum_{j'=1}^{q} y_{j'} u_{j'}\right\rangle \sum_{i=1}^{p} \xi_i v_i\right]$$
$$= E\left[\sum_{j=1}^{q} \zeta_j y_j \sum_{i=1}^{p} \xi_i v_i\right] = \sum_{i=1}^{p} \left(E[\xi_i \zeta_j] y_j\right) v_i.$$

Thus, the ith coefficient of $C_{12}(y)$ in the basis $\{v_1, \dots, v_p\}$ coincides with the ith component of $\mathbf{C}_{12}\mathbf{y}$, where $\mathbf{C}_{12} = E[\boldsymbol{\xi}\boldsymbol{\zeta}^T]$ and $\mathbf{y} = [y_1, \dots, y_q]^T$. If the matrices $\mathbf{C}_{11} = E[\boldsymbol{\xi}\boldsymbol{\xi}^T]$ and $\mathbf{C}_{22} = E[\boldsymbol{\zeta}\boldsymbol{\zeta}^T]$ are nonsingular, then the canonical components $(\rho_k, \mathbf{a}_k, \mathbf{b}_k, A_k, B_k)$ of the random vectors $\boldsymbol{\xi}$ and $\boldsymbol{\zeta}$ are defined as in Section 4.1. Direct verification then shows that $(\rho_k, a_k, b_k, A_k, B_k)$ are the FCC's of X and Y, where

$$a_k = \mathbf{a}_k^T \mathbf{v}, \quad \mathbf{v} = [v_1, \dots, v_p]^T \quad \text{and} \quad b_k = \mathbf{b}_k^T \mathbf{u}, \quad \mathbf{u} = [v_u, \dots, u_q]^T. \quad (4.25)$$

Chapter 5
Two sample inference for the mean and covariance functions

Due to possibly different FPC's structures, working with two functional samples may be difficult. An important contribution has been made by Benko *et al.* (2009) who developed bootstrap procedures for testing the equality of mean functions, the FPC's, and the eigenspaces spanned by them. In this chapter, we present asymptotic procedures for testing the equality of the means and the covariance operators in two independent samples. Section 5.1 focuses on testing the equality of mean functions. It shows that instead of statistics which have chi–square limits, those that converge to weighted sums of squares of independent standard normals can also be used. In other chapters we focus on statistics converging to chi–square distributions, but analogous versions converging to weighted sums of normals can be readily constructed.

In Section 5.1 we present the procedures for testing the equality of the mean functions, together with the theorems that justify them asymptotically; the proofs of these theorems are presented in Section 5.3. Finite sample performance is examined in Section 5.2. The theory presented in Section 5.4 is based on the work of Panaretos *et al.* (2010), which contains some further extensions and numerical applications.

5.1 Equality of mean functions

We consider two samples $X_1, \ldots, X_N$ and $X_1^*, \ldots, X_M^*$. We assume that they satisfy the model

$$X_i(t) = \mu(t) + \varepsilon_i(t), \quad 1 \le i \le N \tag{5.1}$$

and

$$X_i^*(t) = \mu^*(t) + \varepsilon_i^*(t), \quad 1 \le i \le M. \tag{5.2}$$

We wish to test the null hypothesis

$$H_0: \ \mu = \mu^* \quad \text{in } L^2$$

against the alternative that H_0 is false. We assume:

$$\text{the two samples are independent,} \tag{5.3}$$

$$\begin{aligned}&\varepsilon_1, \ldots, \varepsilon_N \text{ are independent and identically distributed with}\\ &E\varepsilon_1(t) = 0 \quad \text{and} \quad E\|\varepsilon_1\|^4 < \infty\end{aligned} \tag{5.4}$$

and similarly

$$\begin{aligned}&\varepsilon_1^*, \ldots, \varepsilon_M^* \text{ are independent and identically distributed with}\\ &E\varepsilon_1^*(t) = 0 \quad \text{and} \quad E\|\varepsilon_1^*\|^4 < \infty.\end{aligned} \tag{5.5}$$

Note that the ε_i^* are not assumed to have the same distribution as the ε_i.

Since

$$\bar{X}_N(t) = \frac{1}{N}\sum_{i=1}^{N} X_i(t) \quad \text{and} \quad \bar{X}_M^*(t) = \frac{1}{M}\sum_{i=1}^{M} X_i^*(t)$$

are unbiased estimators for $\mu(t)$ and $\mu^*(t)$, respectively, it is natural to reject the null hypothesis if

$$U_{N,M} = \frac{NM}{N+M}\int_0^1 (\bar{X}_N(t) - \bar{X}_M^*(t))^2 dt$$

is large. Our first method is based directly on $U_{N,M}$.

Method I: We start with establishing the convergence of $U_{N,M}$ under H_0.

Theorem 5.1. *If H_0 and (5.3)-(5.5) hold, and*

$$\frac{N}{N+M} \to \theta \quad \textit{with some } 0 \le \theta \le 1, \tag{5.6}$$

then

$$U_{N,M} \xrightarrow{d} \int_0^1 \Gamma^2(t)dt, \tag{5.7}$$

where $\{\Gamma(t), 0 \le t \le 1\}$ is a Gaussian process satisfying $E\Gamma(t) = 0$ and

$$E[\Gamma(t)\Gamma(s)] = (1-\theta)c(t,s) + \theta c^*(t,s), \tag{5.8}$$

with $c(t,s) = \mathrm{Cov}(X_1(t), X_1(s))$ and $c^*(t,s) = \mathrm{Cov}(X_1^*(t), X_1^*(s))$.

The distribution of the limit in (5.7) depends on the unknown covariance functions c and c^*. According to the Karhunen–Loève expansion (2.8), we can assume that

$$\Gamma(t) = \sum_{k=1}^{\infty} \tau_k^{1/2} N_k \varphi_k(t),$$

where the N_k are independent standard normal random variables, $\tau_1 \ge \tau_2 \ge \cdots$ and $\varphi_1, \varphi_2, \ldots$ are the eigenvalues and eigenfunctions of the operator determined by $(1-\theta)c + \theta c^*$. Clearly,

$$\int_0^1 \Gamma^2(t)dt = \sum_{k=1}^{\infty} \tau_k N_k^2,$$

so to provide a reasonable approximation for $\int_0^1 \Gamma^2(t)dt$, we only need to estimate the τ_k's. This can be done easily using $\hat{\tau_k}$'s, the eigenvalues of the empirical covariance function

$$\hat{z}_{N,M}(t,s) = \frac{M}{M+N}\frac{1}{N}\sum_{i=1}^{N}(X_i(t)-\bar{X}_N(t))(X_i(s)-\bar{X}_N(s))$$
$$+\frac{N}{M+N}\frac{1}{M}\sum_{i=1}^{M}(X_i^*(t)-\bar{X}_M^*(t))(X_i^*(s)-\bar{X}_M^*(s)).$$

The sum $\sum_{k=1}^{d}\hat{\tau}_k N_k^2$ provides an approximation to the limit in (5.7) if d is large enough. The choice of d is discussed in Section 5.2.

The asymptotic consistency of Method I follows form thefollowing result.

Theorem 5.2. *Suppose* (5.3)-(5.5) *and* (5.6) *hold, and*

$$\int_0^1 (\mu(t)-\mu^*(t))^2 dt > 0,$$

then $U_{N,M} \xrightarrow{P} \infty$.

The next method is essentially a projection version of the procedure based on $U_{N,M}$. It is easier to implement in R because it does not require the numerical evaluation of the integral defining $U_{N,M}$.

Method II: Now we use projections onto the space determined by the leading eigenfunctions of the operator $Z = (1-\theta)C + \theta C^*$. We assume that the eigenvalues of Z satisfy

$$\tau_1 > \tau_2 > \cdots > \tau_d > \tau_{d+1}. \tag{5.9}$$

The corresponding eigenfunctions are $\varphi_1, \ldots, \varphi_{d+1}$. We want to project the observations onto the space spanned by $\varphi_1, \ldots, \varphi_d$.. Since these functions are unknown, we are using the corresponding eigenfunctions of $\hat{Z}_{N,M}$, denoted by $\hat{\varphi}_1, \ldots, \hat{\varphi}_d$. Now we project $\bar{X}_N - \bar{X}_M^*$ into the linear space spanned by $\hat{\varphi}_1, \ldots, \hat{\varphi}_d$. Let

$$\hat{a}_i = \langle \bar{X}_N - \bar{X}_M^*, \hat{\varphi}_i \rangle, \quad 1 \le i \le d,$$

and introduce $\hat{\mathbf{a}} = (\hat{a}_1, \ldots, \hat{a}_d)^T$. We show that under the conditions of Theorem 5.1 the vector $(NM/(N+M))^{1/2}\hat{\mathbf{a}}$ is approximately d-variate normal up to some random signs. The asymptotic variance of $(NM/(N+M))^{1/2}\hat{\mathbf{a}}$ is $Q = \{Q(i,j), 1 \le i,j \le d\}$, where

$$Q(i,j) = (1-\theta)E\langle X_1-\mu,\varphi_i\rangle\langle X_1-\mu,\varphi_j\rangle + \theta E\langle X_1^*-\mu^*,\varphi_i\rangle\langle X_1^*-\mu^*,\varphi_j\rangle,$$

$1 \le i, j \le d$. It is easy to see that

$$\begin{aligned} Q(i,j) &= \int_0^1 \int_0^1 (1-\theta)E[(X_1(t)-\mu(t))(X_1(s)-\mu(s))]\varphi_i(t)\varphi_j(s)dt\,ds \\ &\quad + \int_0^1 \int_0^1 (1-\theta)E[(X_1^*(t)-\mu^*(t))(X_1^*(s)-\mu^*(s))]\varphi_i(t)\varphi_j(s)dt\,ds \\ &= \int_0^1 \int_0^1 z(t,s)\varphi_i(t)\varphi_j(s)dt\,ds \\ &= \begin{cases} \tau_i, & \text{if } i=j \\ 0, & \text{if } i \ne j. \end{cases} \end{aligned} \tag{5.10}$$

In light of (5.9) and (5.10), testing procedures can be based on the statistics

$$T_{N,M}^{(1)} = \frac{NM}{N+M}\sum_{k=1}^{d} \hat{a}_k^2/\hat{\tau}_k$$

and

$$T_{N,M}^{(2)} = \frac{NM}{N+M}\sum_{k=1}^{d} \hat{a}_k^2.$$

Theorem 5.3. *If H_0 and* (5.3)–(5.5), (5.6) *and* (5.9) *hold, then*

$$T_{N,M}^{(1)} \xrightarrow{d} \chi^2(d) \tag{5.11}$$

and

$$T_{N,M}^{(2)} \xrightarrow{d} \sum_{k=1}^{d} \tau_k N_k^2, \tag{5.12}$$

where $\chi^2(d)$ is a chi–square random variable with d degrees of freedom, and $N_1, N_2, \ldots, N_d$ are independent standard normal random variables.

It is clear that $T_{N,M}^{(2)}$ is a projection version of $U_{N,M}$, we only use the first d terms in the L^2 expansion of $\bar{X}_N - \bar{X}_M^*$. The statistic $T_{N,M}^{(1)}$ is an asymptotically distribution free modification of $T_{N,M}^{(2)}$, and hence of $U_{N,M}$.

The consistency of testing procedures based on $T_{N,M}^{(1)}$ and $T_{N,M}^{(2)}$ can be easily established along the lines of Theorem 5.2.

Theorem 5.4. *If (5.3)-(5.5),* (5.6) *and* (5.9) *hold, and $\mu - \mu^*$ is not orthogonal to the linear span of $\varphi_1, \ldots, \varphi_d$, then $T_{N,M}^{(1)} \xrightarrow{P} \infty$ and $T_{N,M}^{(2)} \xrightarrow{P} \infty$.*

The difference between tests based on $U_{N.M}$ and $T_{N,M}^{(1)}, T_{N,M}^{(2)}$ is that the last two only see the difference between μ and μ^* in a d dimensional subspace. If $\mu = \mu^*$ in this subspace, then H_0 will not be rejected. However, this has little practical relevance if the first d τ_k explain a large percentage of the variance of the difference. The difference in the span of $\phi_{d+1}, \phi_{d+2}, \ldots,$ cannot then be practically distinguished from the randomness in the errors.

5.2 A simulation study and an application

We present the results of a small simulation study aimed at comparing the testing procedures introduced in Section 5.1. Since Method I essentially reduces to the statistic $T^{(2)}_{N,M}$ of Method II, we compare the statistics $T^{(1)}_{N,M}$ and $T^{(2)}_{N,M}$.

We consider sample sizes $N = 50$ and $N = 100$ and $M = N$ as well as $M = 2N$. To compare the sizes, we set $\mu(t) = \mu^*(t) = 0$. Under the alternative, we set $\mu(t) = 0$ and $\mu^*(t) = at(1-t)$. The power is then a function of the parameter a. We consider two setting for the errors: 1) Both the ε_i and the ε_i^* are Brownian bridges; 2) The ε_i are Brownian bridges, and the ε_i^* are Brownian motions. To compute the test statistics we converted the Gaussian processes simulated as increments into functional objects using 49 Fourier basis function. We then computed the test statistics with $d = 5$.

The tests have size very close to the nominal size, almost always within one percent. We did not detect any systematic differences in size between the two tests. The tests have remarkably good power. To illustrate, Figure 5.1 shows fifty trajectories of the Brownian bridge in the left panel and 50 independent trajectories of the Brownian bridge plus $\mu^*(t) = at(1-t)$ with $a = 0.8$ in the right panel. Except one function in the right panel which goes visibly above the other functions, both sets look very similar, and it would be difficult to tell by eye that they have different mean functions. Yet, based on one thousand replications, $T^{(1)}_{50,50}$ rejects the null hypothesis of equal means with probability 0.91 and $T^{(2)}_{50,50}$ with probability 0.98, at the nominal size $\alpha = 5\%$. For such relatively small sizes, the tests suffer from an elevated probability of type I error. For $\alpha = 5\%$, the empirical size is 6.6% for $T^{(1)}_{50,50}$ and 7.3% for $T^{(2)}_{50,50}$. When both M and N exceed 100, the empirical sizes are within 1% of the nominal sizes. Typical results are shown in Table 5.1. The power is higher if the distributions of the errors are the same in both samples.

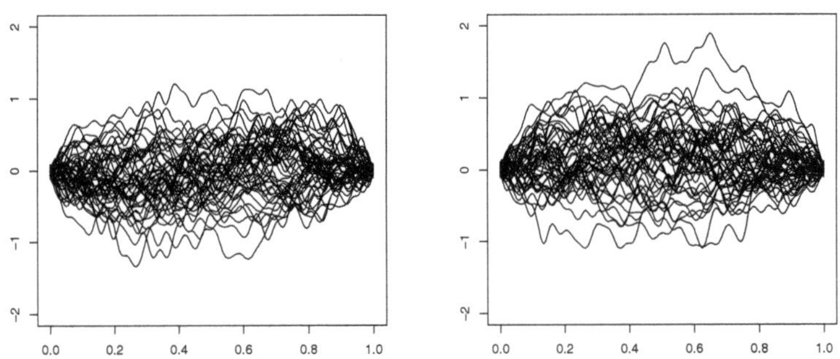

Fig. 5.1 Fifty trajectories of the Brownian bridge (left) and fifty independent trajectories of the Brownian bridge plus $\mu^*(t) = 0.8t(1-t)$ (right). The tests can detect the different means with probability higher than 90%.

Table 5.1 Size ($a = \mathbf{0.0}$) and power ($a > \mathbf{0}$) of the tests based on $T^{(1)}_{100,200}$ and $T^{(2)}_{100,200}$. The sample with $N = \mathbf{100}$ has Brownian bridge errors, the one with $M = \mathbf{200}$ has Brownian motion errors.

	$\alpha = .01$		$\alpha = .05$		$\alpha = .10$	
a	$T^{(1)}_{100,200}$	$T^{(2)}_{100,200}$	$T^{(1)}_{100,200}$	$T^{(2)}_{100,200}$	$T^{(1)}_{100,200}$	$T^{(2)}_{100,200}$
0.0	1.3	1.4	5.6	5.7	10.8	10.8
0.1	1.7	1.7	6.8	6.4	12.2	12.1
0.2	2.6	3.0	9.8	11.0	16.5	18.2
0.3	4.8	6.1	15.8	18.5	24.7	27.3
0.4	9.8	10.7	23.6	27.7	35.6	39.0
0.5	18.5	19.5	37.8	41.5	50.1	54.6
0.6	29.3	29.7	52.3	55.0	64.3	67.2
0.7	42.7	43.1	65.5	67.5	75.6	78.1
0.8	59.2	57.7	79.3	80.3	87.1	88.1
0.9	73.7	71.0	88.7	89.0	93.7	94.7
1.0	85.3	82.0	94.4	94.8	97.2	97.7
1.1	92.8	89.8	97.8	97.3	98.9	98.7
1.2	96.6	94.7	99.3	99.2	99.9	99.7
1.3	98.9	97.7	99.8	99.7	99.9	99.9
1.4	99.5	99.3	100.0	99.9	100.0	100.0
1.5	99.8	99.8	100.0	100.0	100.0	100.0
1.6	100.0	100.0	100.0	100.0	100.0	100.0

Table 5.2 P–values (in percent) of the tests based on statistics $T^{(1)}_{N,M}$ and $T^{(2)}_{N,M}$ applied to medfly data.

d	1	2	3	4	5	6	7	8	9
$T^{(1)}$	1.0	2.2	3.0	5.7	10.3	15.3	3.2	2.7	5.0
$T^{(2)}$	1.0	1.0	1.0	1.1	1.1	1.0	1.0	1.1	1.1

We conclude this section by an application of the two tests to an interesting data set consisting of egg-laying trajectories of Mediterranean fruit flies (medflies). This data set will be revisited in Chapter 10. Müller and Stadtmüller (2005), Section 6, consider 534 egg–laying curves (count of eggs per unit time interval) of medflies who lived at least 30 days. Each function is defined over an interval $[0, 30]$, and its value on day $t \leq 30$ is the count of eggs laid by fly i on that day. The 534 flies are classified into long–lived, i.e. those who lived longer than 44 days, and short–lived, i.e. those who died before the end of the 44th day after birth. In the sample, there are 256 short–lived, and 278 long–lived flies. This classification naturally defines two samples: *Sample 1:* the egg-laying curves $X_i(t),\ 0 < t \leq 30,\ i = 1, 2, \ldots, 256$ of the short–lived flies. *Sample 2:* the egg-laying curves $X_j^*(t),\ 0 < t \leq 30,\ j = 1, 2, \ldots, 278$ of the long–lived flies. The egg-laying curves are very irregular; Figure 10.1 shows ten smoothed curves of short– and long–lived flies. The tests are applied to such smooth trajectories.

Table 5.2 shows the P–values as a function of d. For both samples, $d = 2$ explains slightly over 85% of variance, so this is the value we would recommend

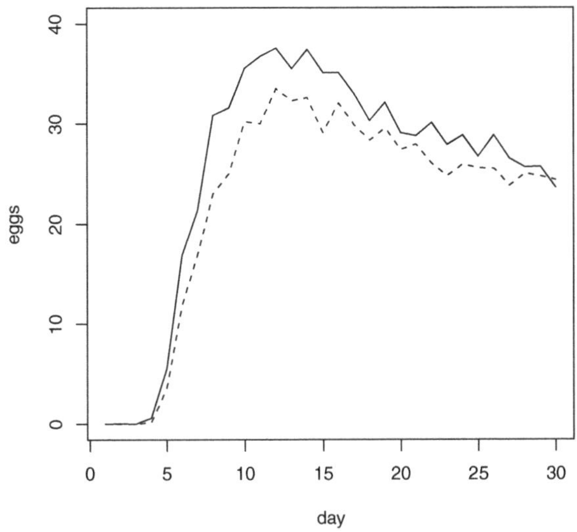

Fig. 5.2 Estimated mean functions for the medfly data: short lived –solid line; long lived –dashed line.

to use. Both tests reject the equality of the mean functions, even though the sample means, shown in Figure 5.2, are not far apart. The P–values for the statistic $T^{(2)}$ are much more stable, equal to about 1%, no matter the value of d. The behavior of the test based on $T^{(1)}$ is more erratic. This indicates that while the test based on $T^{(1)}$ is easier to apply because it uses standard chi–square critical values, the test based on $T^{(2)}$ may be more reliable.

5.3 Proofs of the theorems of Section 5.1

Proof of Theorem 5.1. By Theorem 2.1, there are two independent Gaussian processes Γ_1 and Γ_2 with zero means and covariances C and C^* such that $(N^{-1/2}\sum_{1\le i\le N}(X_i-\mu), M^{-1/2}\sum_{1\le i\le M}(X_i^*-\mu^*))$ converges weakly in L^2 to (Γ_1,Γ_2). This proves (5.7) with $\Gamma=(1-\theta)\Gamma_1+\theta\Gamma_2$. □

Proof of Theorem 5.2. It follows from the proof of Theorem 5.1 that

$$U_{N,M}=\frac{NM}{N+M}\int_0^1(\mu(t)-\mu^*(t))^2dt+O_P(1),$$

which implies the result. □

Proof of Theorem 5.3. The central limit theorem for sums of independent and identically distributed random vectors in R^d yields

$$\left(\frac{NM}{N+M}\right)^{1/2}\left[\langle \bar{X}_N - \bar{X}_M^*, \varphi_1\rangle, \ldots, \langle \bar{X}_N - \bar{X}_M^*, \varphi_d\rangle\right]^T \tag{5.13}$$
$$\stackrel{d}{\to} \mathbf{N}_d(\mathbf{0}, Q),$$

where $\mathbf{N}_d(\mathbf{0}, Q)$ is a d-variate normal random vector with mean $\mathbf{0}$ and covariance matrix Q. Since

$$\int_0^1\int_0^1 (\hat{Z}_{N,M}(t,s) - Z(t,s))^2 dt\, ds = o_P(1),$$

Lemma 2.3 and inequality (2.3) imply

$$\max_{1\le i\le d} |\hat{\tau}_i - \tau| = o_P(1) \tag{5.14}$$

and

$$\max_{1\le i\le d} \|\hat{\varphi}_i - \hat{c}_i\varphi\| = o_P(1), \tag{5.15}$$

where $\hat{c}_1, \ldots, \hat{c}_d$ are random signs. We showed in the proof of Theorem 5.1 that

$$N\int_0^1 (\bar{X}_N(t) - \mu(t))^2 dt = O_P(1)$$

and

$$M\int_0^1 (\bar{X}_M(t) - \mu^*(t))^2 dt = O_P(1),$$

so by (5.15) we have

$$\max_{1\le i\le d}\left(\frac{NM}{N+M}\right)^{1/2} \left|\langle \bar{X}_N - \bar{X}_M^*, \hat{\varphi}_i - \hat{c}_i\varphi\rangle\right| = o_P(1).$$

Now the results in Theorem 5.3 follow from (5.13), (5.14)and from the observation that neither $T_{N,M}^{(1)}$ nor $T_{N,M}^{(2)}$ depend on the random signs $\hat{c}_i$. □

Proof of Theorem 5.4. Following the proof of Theorem 5.3 one can easily verify that

$$\left(\frac{NM}{N+M}\right)^{1/2}\hat{a}_i = \left(\frac{NM}{N+M}\right)^{1/2}\langle \mu - \mu^*, \varphi_i\rangle + O_P(1), \quad 1 \le i \le d.$$

Since (5.15) also holds, both parts of Theorem 5.4 are proven. □

5.4 Equality of covariance operators

We consider two samples: $X_1, X_2, \ldots, X_N$ and $X_1^*, X_2^*, \ldots X_M^*$. The functions in each sample are iid mean zero elements of L^2, and the two samples are independent. Consider the covariance operators

$$C(x) = E[\langle X, x\rangle X], \quad C^*(x) = E[\langle X^*, x\rangle X^*],$$

where X has the same distribution as the X_i, and X^* the same distribution as the X_j^*. We want to test

$$H_0 : C = C^* \quad \text{versus} \quad H_A : C \neq C^*.$$

In Theorem 5.5, we will assume that X and X^* are Gaussian elements of L^2. This means that the equality of the covariances implies the equality in distribution. Thus, under the additional assumption of normality, H_0 states that the X_i have the same distribution as the X_j^*.

Denote by $\hat{C}$ and $\hat{C}^*$ the empirical counterparts of C and C^*, and by $\hat{R}$, the empirical covariance operator of the pooled data, i.e.

$$\begin{aligned}\hat{R}(x) &= \frac{1}{N+M}\left\{\sum_{i=1}^{N}\langle X_i, x\rangle X_i + \sum_{j=1}^{M}\langle X_j^*, x\rangle X_j^*\right\}\\ &= \hat{\theta}\hat{C}(x) + (1-\hat{\theta})\hat{C}^*(x), \quad x \in L^2,\end{aligned}$$

where

$$\hat{\theta} = \frac{N}{N+M}.$$

The operator $\hat{R}$ has $N+M$ eigenfunctions, which are denoted $\hat{\phi}_k$. We also set

$$\hat{\lambda}_k = \frac{1}{N}\sum_{n=1}^{N}\left\langle X_n, \hat{\phi}_k\right\rangle^2, \quad \hat{\lambda}_k^* = \frac{1}{M}\sum_{m=1}^{M}\left\langle X_m^*, \hat{\phi}_k\right\rangle^2.$$

Note that the $\hat{\lambda}_k$ and the $\hat{\lambda}_k^*$ are not the eigenvalues of the operators $\hat{C}$ and $\hat{C}^*$, but rather the sample variances of the coefficients of X and X^* with respect to the orthonormal system $\{\hat{\phi}_k,\ 1 \leq k \leq N+M\}$ formed by the eigenfunctions of the operator $\hat{R}$.

The test statistic is defined by

$$\hat{T} = \frac{N+M}{2}\hat{\theta}(1-\hat{\theta})\sum_{i,j=1}^{p}\frac{\left\langle(\hat{C}-\hat{C}^*)\hat{\phi}_i, \hat{\phi}_j\right\rangle^2}{(\hat{\theta}\hat{\lambda}_i + (1-\hat{\theta})\hat{\lambda}_i^*)(\hat{\theta}\hat{\lambda}_j + (1-\hat{\theta})\hat{\lambda}_j^*)}.$$

Theorem 5.5. *Suppose X and X^* are Gaussian elements of L^2 such that $E\|X\|^4 < \infty$ and $E\|X^*\|^4 < \infty$. Suppose also that $\hat{\theta} \to \theta \in (0,1)$, as $N \to \infty$. Then*

$$\hat{T} \xrightarrow{d} \chi^2_{p(p+1)/2}, \quad N, M \to \infty,$$

where $\chi^2_{p(p+1)/2}$ denotes a chi-square random variable with $p(p+1)/2$ degrees of freedom.

Proof. Introduce the random operators

$$C_i(x) = \langle X_i, x\rangle X_i, \quad C_j^*(x) = \langle X_j^*, x\rangle X_j^*, \quad x \in L^2.$$

The C_i form a sequence of iid elements the Hilbert space $\mathcal{S}$ of the Hilbert–Schmidt operators acting on L^2, and the same is true for the C_j^*. Under H_0, the C_i and the C_j^* have the same mean C. They also have the same covariance operator, which is an operator acting on $\mathcal{S}$ given by

$$\mathfrak{G}(\Psi) = E[\langle C_i - C, \Psi\rangle_{\mathcal{S}} (C_i - C)] = E[\langle C_i, \Psi\rangle_{\mathcal{S}} C_i] - \langle C, \Psi\rangle_{\mathcal{S}} C, \quad \Psi \in \mathcal{S}. \tag{5.16}$$

Under H_0, the second term $\langle C, \Psi\rangle_{\mathcal{S}} C$ is the same for both samples. By (2.2),

$$E[\langle C_i, \Psi\rangle_{\mathcal{S}} C_i] = E\left[\sum_{n=1}^{\infty} \langle C_i(e_n), \Psi(e_n)\rangle C_i\right]$$
$$= \sum_{n=1}^{\infty} E\left[\langle\langle X_i, e_n\rangle X_i, \Psi(e_n)\rangle \langle X_i, e_n\rangle X_i\right] = \sum_{n=1}^{\infty} E\left[\langle X_i, e_n\rangle^2 \langle X_i, \Psi(e_n)\rangle X_i\right].$$

The assumption of Gaussianity and $C = C^*$ imply that the X_i and the X_j^* have the same distribution, so

$$E\left[\langle X_i, e_n\rangle^2 \langle X_i, \Psi(e_n)\rangle X_i\right] = E\left[\langle X_j^*, e_n\rangle^2 \langle X_j^*, \Psi(e_n)\rangle X_j^*\right].$$

We want to apply the CLT in the Hilbert space $\mathcal{S}$ to the operators C_i. By Theorem 2.1, we must verify that $E\|C_i\|^2_{\mathcal{S}} < \infty$. This holds because, by Parseval's equality,

$$E\|C_i\|^2_{\mathcal{S}} = E\sum_{n=1}^{\infty} \|\langle X_i, e_n\rangle X_i\|^2 = E\left[\|X_i\|^2 \sum_{n=1}^{\infty} |\langle X_i, e_n\rangle|^2\right] = E\|X_i\|^4.$$

We therefore obtain,

$$N^{1/2}(\hat{C} - C) \xrightarrow{d} Z_1, \quad M^{1/2}(\hat{C}^* - C^*) \xrightarrow{d} Z_2, \tag{5.17}$$

where Z_1 is a Gaussian element of $\mathcal{S}$ with the same covariance operator as C_1, and Z_2 a Gaussian element with the same covariance operator as C_2. Thus, Z_1 and Z_2 are independent, and, under H_0, both have the covariance operator equal to $\mathfrak{G}$.

For every $1 \le i, j \le p$, introduce the random variables

$$W_{N,M}(i,j) = \left\langle [(N+M)\hat{\theta}(1-\hat{\theta})]^{1/2}(\hat{C} - \hat{C}^*)\hat{c}_i\hat{\phi}_i, \hat{c}_j\hat{\phi}_j \right\rangle,$$

so that

$$\hat{T} = \frac{\sum_{i,j=1}^{p} W^2_{N,M}(i,j)}{2(\hat{\theta}\hat{\lambda}_i + (1-\hat{\theta})\hat{\lambda}_i^*)(\hat{\theta}\hat{\lambda}_j + (1-\hat{\theta})\hat{\lambda}_j^*)} \tag{5.18}$$

Observe that under H_0,

$$W_{N,M}(i,j) = \left\langle \left[(1-\hat{\theta})^{1/2} N^{1/2} (\hat{C} - C) - \hat{\theta}^{1/2} M^{1/2} (\hat{C}^* - C^*) \right] \hat{c}_i \hat{\phi}_i, \hat{c}_j \hat{\phi}_j \right\rangle.$$

By Theorem 2.7, under H_0, $\hat{c}_i \hat{\phi}_i \xrightarrow{P} v_i$, with the v_i being the eigenfunctions of C (and of C^*). Therefore , by (5.17),

$$W_{N,M}(i,j) \xrightarrow{d} \langle Z v_i, v_j \rangle, \quad Z = (1-\theta)^{1/2} Z_1 - \theta^{1/2} Z_2.$$

Since Z_1 and Z_2 are independent random operators in $\mathcal{S}$, we see that the covariance operator of Z is also equal to $\mathfrak{G}$. By (5.18), we therefore obtain

$$\begin{aligned}
\hat{T} &\xrightarrow{d} \frac{\sum_{i,j=1}^{p} \langle Z(v_i), v_j \rangle^2}{2(\theta\lambda_i + (1-\theta)\lambda_i)(\theta\lambda_j + (1-\theta)\lambda_j)} \\
&= \frac{\sum_{i,j=1}^{p} \langle Z(v_i), v_j \rangle^2}{2\lambda_i \lambda_j} \\
&= \sum_{k=1}^{n} \frac{\langle Z(v_k), v_k \rangle^2}{2\lambda_k^2} + \sum_{k<n} \frac{\langle Z(v_k), v_n \rangle^2}{2\lambda_k \lambda_n} + \sum_{k>n} \frac{\langle Z(v_k), v_n \rangle^2}{2\lambda_k \lambda_n} \\
&= \sum_{k=1}^{n} \frac{\langle Z(v_k), v_k \rangle^2}{2\lambda_k^2} + \sum_{k<n} \frac{\langle Z(v_k), v_n \rangle^2 + \langle Z(v_n), v_k \rangle^2}{2\lambda_k \lambda_n}. \qquad (5.19)
\end{aligned}$$

To identify the distribution of the right-hand side of (5.19), it is convenient to represent the operator Z in terms of the operators V_{ij} defined by

$$V_{ij}(x) = \langle v_i, x \rangle \, v_j.$$

By Lemma 5.1,

$$Z \stackrel{d}{=} \sqrt{2} \sum_{i=1}^{\infty} \lambda_i \zeta_{ii} V_{ii} + \sum_{i<j} \sqrt{\lambda_i \lambda_j} \zeta_{ij} (V_{ij} + V_{ji}), \qquad (5.20)$$

where the ζ_{ij} are iid standard normal.

Since the v_i form a basis, (5.20) implies that

$$Z(v_k) = \sqrt{2} \lambda_k \zeta_{kk} v_k + \sum_{k<j} \sqrt{\lambda_k \lambda_j} \zeta_{kj} v_j + \sum_{i<k} \sqrt{\lambda_i \lambda_k} \zeta_{ik} v_i,$$

and so

$$\langle Z(v_k), v_n \rangle = \begin{cases} \sqrt{2} \lambda_k \zeta_{kk} & \text{if } k = n, \\ \sqrt{\lambda_k \lambda_n} \zeta_{kn} & \text{if } k < n, \\ \sqrt{\lambda_k \lambda_n} \zeta_{nk} & \text{if } k > n. \end{cases} \qquad (5.21)$$

Using (5.19) and (5.21), we see that

$$\hat{T} \xrightarrow{d} \sum_{k=1}^{p} \zeta_{kk}^2 + \sum_{k<p} \frac{\zeta_{kn}^2 + \zeta_{nk}^2}{2} \stackrel{d}{=} \sum_{k=1}^{p} \zeta_{kk}^2 + \sum_{k<p} \zeta_{kn}^2 \stackrel{d}{=} \chi^2_{p(p+1)/2}. \qquad \square$$

Lemma 5.1. *Under the assumptions of Theorem 5.5,*

$$\mathfrak{G} = \sum_{i=1}^{\infty} (\sqrt{2}\lambda_i)^2 \langle V_{ii}, \cdot \rangle_{\mathcal{S}} V_{ii} + \sum_{i<j} \lambda_i \lambda_j \langle V_{ij} + V_{ji}, \cdot \rangle_{\mathcal{S}} (V_{ij} + V_{ji}).$$

Proof. We work with the expansion

$$X_n = \sum_{i=1}^{\infty} \sqrt{\lambda_i} \xi_{ni} v_i,$$

where the λ_i are the eigenvalues of C and $\{\xi_{ni},\ i \geq 1\}$ are independent sequences of iid standard normal random variables.

Direct verification then shows that

$$C_n = \sum_{i,j=1}^{\infty} \sqrt{\lambda_i \lambda_j} \xi_{ni} \xi_{nj} V_{ij}, \quad C = \sum_{i=1}^{\infty} \lambda_i V_{ii}.$$

Let Ψ be an arbitrary Hilbert–Schmidt operator. Then, by (5.16),

$$\mathfrak{G}(\Psi) = \sum_{i,j,\ell,k} E[\xi_{ni}\xi_{nj}\xi_{n\ell}\xi_{nk}] \sqrt{\lambda_i \lambda_k \lambda_\ell \lambda_k} \langle V_{ij}, \Psi \rangle_{\mathcal{S}} V_{\ell k} - \sum_{i,j} \lambda_i \lambda_j \langle V_{ii}, \Psi \rangle_{\mathcal{S}} V_{jj}.$$

The expected value $E[\xi_{ni}\xi_{nj}\xi_{n\ell}\xi_{nk}]$ is zero unless there are two pairs of equal indices, or all indices are equal. We therefore have

$$\begin{aligned}
\mathfrak{G}(\Psi) &= \sum_{i \neq j} \lambda_i \lambda_j \left[\langle V_{ii}, \Psi \rangle_{\mathcal{S}} V_{jj} + \langle V_{ij}, \Psi \rangle_{\mathcal{S}} V_{ij} + \langle V_{ij}, \Psi \rangle_{\mathcal{S}} V_{ji} \right] \\
&\quad + 3 \sum_i \lambda_i^2 \langle V_{ii}, \Psi \rangle_{\mathcal{S}} V_{ii} - \sum_i \lambda_i^2 \langle V_{ii}, \Psi \rangle_{\mathcal{S}} V_{ii} - \sum_{i \neq j} \lambda_i \lambda_j \langle V_{ii}, \Psi \rangle_{\mathcal{S}} V_{jj} \\
&= 2 \sum_i \lambda_i^2 \langle V_{ii}, \Psi \rangle_{\mathcal{S}} V_{ii} + \sum_{i \neq j} \lambda_i \lambda_j \left[\langle V_{ij}, \Psi \rangle_{\mathcal{S}} V_{ij} + \langle V_{ij}, \Psi \rangle_{\mathcal{S}} V_{ji} \right].
\end{aligned}$$

The proof is completed by rearranging the terms. □

5.5 Bibliographical notes

In Section 5.1 we assume that the observations in each sample are independent. If the functions are obtained from a time record, for example daily or annual curves, then the assumption of independence need not hold. Horváth *et al.* (2011) extend the methodology and theory of Section 5.1 to dependent errors ε_i and ε_i^*. Instead of the covariance kernel $z_{N,M}(t,s)$, a kernel corresponding to suitably defined long–run covariances must be used. The dependence is quantified by the notion of L^p–m–approximability introduced in Chapter 16. Gromenko and Kokoszka (2011) develop a test for the equality of the mean functions of the curves from two disjoint spatial regions. They emphasize computational issues arising in small sample sizes of spatially dependent curves.

The two sample problem for the covariance operators if the assumption of normality is violated is studied by studied by Fremdt *et al.* (2011) and Kraus and Panaretos (2011). Boente *et al.* (2011) develop a bootstrap test to test the equality of covariance operators.

The two sample problem when the equality of the whole distributions is tested is studied by Hall and Keilegom (2007) who emphasize the role of smoothing in two sample problems for functional data.

Laukaitis and Račkauskas (2005) consider the model $X_{g,i}(t) = \mu_g(t) + \varepsilon_{g,i}(t),\ g = 1, 2, \ldots, G$, with innovations $\varepsilon_{g,i}$ and group means μ_g, and test $H_0 : \mu_1(t) = \cdots = \mu_G(t)$. Other related contributions are Cuevas *et al.* (2004), Delicado (2007) and Ferraty *et al.* (2007).

Chapter 6
Detection of changes in the mean function

In this chapter, we present a methodology for the detection of changes in the mean of functional observations. At its core is a significance test for testing the null hypothesis of a constant functional mean against the alternative of a changing mean. We also show how to locate the change points if the null hypothesis is rejected. Our methodology is readily implemented using the R package `fda`. The null distribution of the test statistic is asymptotically pivotal with a well-known asymptotic distribution going back to the work of Kiefer (1959).

In Section 6.1, we provide some background and motivation. After formulating the assumptions in Section 6.2, we describe the test procedure in Section 6.3. The finite sample performance is investigated in Section 6.4, which also contains an illustrative application to the detection of changes in mean patters of annual temperatures. The proofs of the theorems of Section 6.3 are presented in Section 6.5.

6.1 Introduction

Throughout the book we typically assume that the observations X_i have mean zero. This is clearly not true in applications, so a suitable assumption is that $X_i = \mu + Y_i$, where $EY_i = 0$. These equalities are in the space L^2, which, in particular, means that $EX_i(t) = \mu(t)$ for almost all $t \in [0, 1]$. The various procedures discussed in this book refer then to the mean adjusted variables $X_i - \mu$ which are estimated by $X_i - \bar{X}$. In particular, the FPC's v_k, are those of $X - \mu$, and we have the following L^2 expansion

$$X_i(t) = \mu(t) + \sum_{k=1}^{\infty} \xi_{ki} v_k(t), \quad 1 \le i \le N.$$

The FPC's v_k and their eigenvalues are then estimated using the sample covariance operator

$$\hat{C}(x) = N^{-1} \sum_{i=1}^{N} \langle X_i - \bar{X}_N, x \rangle (X_i - \bar{X}_N), \quad x \in L^2.$$

The above approach is however not valid if the observations X_i do not have the same mean. If the data are collected sequentially, like annual temperature curves, intra-day price curves, or daily magnetometer curves, then it is possible that the mean function changes over time. The simplest type of change is that the mean function changes abruptly from one deterministic curve to another. Such an assumption is clearly a convenient idealization. However, as has been shown for scalar observations, procedures aimed at detecting such simple "jump changes" also have power to detect more complex changes. The model for an abrupt change is $X_i = \mu_1 + Y_i,\ 1 \le i \le k^*,\ X_i = \mu_2 + Y_i,\ k^* < i \le N$, where k^* is an unknown change point. Assuming $k^*/N \to \theta$, a simple verification shows that then $\hat{C}$ is close to $C_Y + \theta(1-\theta)\langle \Delta, \cdot \rangle \Delta$, where $\Delta = \mu_1 - \mu_2$. The eigenfunctions of $\hat{C}$ will then no longer estimate the eigenfunctions of C_Y, the covariance operator of the Y_i. In general, if the mean function changes, inference based on the FPC's will no longer be valid.

It is important to distinguish between a change point problem and the problem of testing for the equality of means discussed in Chapter 5. In the latter setting, it is known which population or group each observation belongs to. In the change point setting, we do not have any partition of the data into several sets with possibly different means. The change can occur at any point, and we want to test if it occurs or not, and if it does, to estimate the point of change.

In this chapter, we assume that the observations are independent. This assumption is often approximately satisfied, and allows to focus on the aspect of the methodology directly related to change point detection. The case of dependent observations is considered in Chapter 16. We note that even if the mean zero Y_i are independent, but the mean μ changes, a test of independence, like the one studied in Chapter 7, will show that the $X_i = \mu_i + Y_i$ are dependent. This phenomenon is well–known for scalar observations, and is referred to as spurious dependence.

Change point methodology is often applied to time series of average annual temperatures at specific locations, or to series derived from such data, like the land surface or marine global temperature series described and used in several examples in Shumway and Stoffer (2006). Figure 6.1 shows the series of average annual temperatures in Central England from 1780 to 2007. Longer records of temperature in England, reaching into 1600's, are available, but we focus on the more recent period because starting from late 1700's daily temperatures have been recorded, and the annual curves can be viewed as functional observations. One such curve is shown in Figure 6.2. Detecting a change point in mean in a series like the one shown in Figure 6.1 should not be taken literary to mean that the mean actually abruptly changes from one value to another in a specific year. It means that the assumption of a constant mean value for the whole series is not acceptable. The estimated change point then shows a rough break point after which the temperatures are higher on average. Different model formulations are obviously possible. One can postulate a straight line regression model for the annual mean and test for changes in slope.

In this chapter, we focus on the change in mean problem in the functional setting. In this case, the mean is a function, and the change can be not only in the average level of this function, but also in its shape. For the daily temperature data, a change in

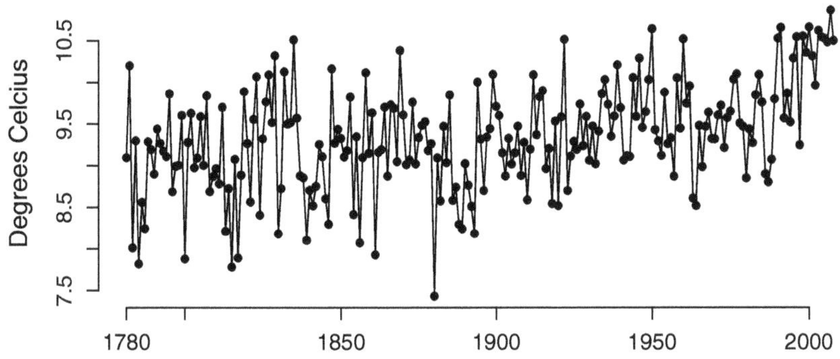

Fig. 6.1 Annual average temperatures in central England 1780–2007.

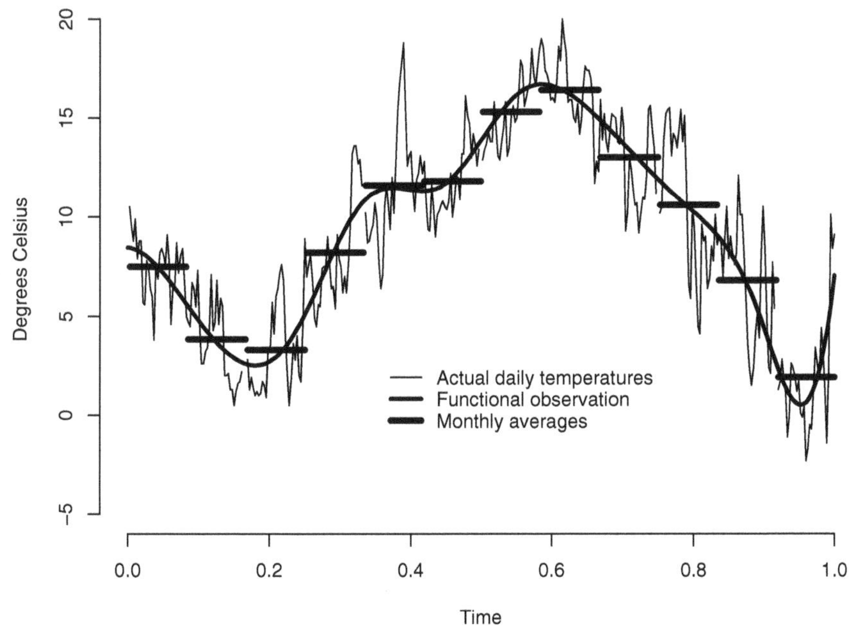

Fig. 6.2 Daily temperatures in **1916** with monthly averages and functional object obtained by smoothing with B-splines.

shape may mean, for example, that while the overall annual average stays the same, summers may become warmer and winters colder. In Section 6.4 we show that in the functional setting more subtle changes can be detected than in the multivariate setting which studies average monthly temperatures. The difference between the multivariate and the functional data is illustrated in Figure 6.2.

6.2 Notation and assumptions

We assume that the observations $X_i \in L^2$ are independent, and we want to test if their mean remains constant in i. Thus we test the null hypothesis

$$H_0: \ EX_1 = EX_2 = \cdots = EX_N.$$

Note that under H_0, we do not specify the value of the common mean.

The test we construct has a particularly good power against the alternative in which the data can be divided into several consecutive segments, and the mean is constant within each segment, but changes from segment to segment. The simplest case of only two segments (one change point) is specified in Assumption 6.4.

Under the null hypothesis, we can represent each functional observation as

$$X_i(t) = \mu(t) + Y_i(t), \qquad EY_i(t) = 0. \tag{6.1}$$

The following assumption specifies conditions on $\mu(\cdot)$ and the errors $Y_i(\cdot)$ needed to establish the asymptotic distribution of the test statistic.

Assumption 6.1. *The mean $\mu(\cdot)$ is in L^2. The errors $Y_i(\cdot)$ are iid mean zero random elements of L^2 which satisfy*

$$E\|Y_i\|^2 = \int EY_i^2(t)dt < \infty. \tag{6.2}$$

Assumption 6.1 implies that the covariance function

$$c(t,s) = E[Y_i(t)Y_i(s)] \quad t,s \in [0,1] \tag{6.3}$$

is square integrable, i.e. is in $L^2([0,1] \times [0,1])$. Consequently, it implies the following expansions:

$$c(t,s) = \sum_{1 \le k < \infty} \lambda_k v_k(t) v_k(s) \tag{6.4}$$

and

$$Y_i(t) = \sum_{1 \le \ell < \infty} \xi_{\ell,i} v_\ell(t). \tag{6.5}$$

The v_k are eigenfunctions of the covariance operator with kernel (6.3). The sequences $\{\xi_{\ell,i},\ \ell = 1, 2, \ldots\}$ are independent, and within each sequence the $\xi_{\ell,i}$

are uncorrelated with mean zero and variance λ_ℓ. The infinite sum in (6.5) converges in L^2 with probability one.

Recall that the estimated eigenelements are defined by

$$\int \hat{c}(t,s)\hat{v}_\ell(s)ds = \hat{\lambda}_\ell \hat{v}_\ell(t), \quad \ell = 1,2,\ldots, \tag{6.6}$$

where

$$\hat{c}(t,s) = \frac{1}{N}\sum_{1\le i\le N}\left(X_i(t) - \bar{X}_N(t)\right)\left(X_i(s) - \bar{X}_N(s)\right)$$

and

$$\bar{X}_N(t) = \frac{1}{N}\sum_{1\le i\le N} X_i(t).$$

To control the distance between the estimated and the population eigenelements, we need the following assumptions:

Assumption 6.2. *The eigenvalues λ_ℓ satisfy, for some $d > 0$*

$$\lambda_1 > \lambda_2 > \cdots > \lambda_d > \lambda_{d+1}.$$

Assumption 6.3. *The Y_i in* (6.1) *satisfy*

$$E\|Y_i\|^4 = \int EY_i^4(t)dt < \infty.$$

By Theorem 2.7, for each $k \le d$:

$$\limsup_{N\to\infty} NE\left[\|\hat{c}_k v_k - \hat{v}_k\|^2\right] < \infty, \quad \limsup_{N\to\infty} NE\left[|\lambda_k - \hat{\lambda}_k|^2\right] < \infty. \tag{6.7}$$

We establish the consistency of the test under the alternative of one change point formalized in Assumption 6.4. A similar argument can be developed if there are several change points, but the technical complications then obscure the main idea explained in Sections 6.3 and 6.5 (in particular the functions (6.9) and (6.16) would need to be modified). The more general case is studied empirically in Section 6.4.

Assumption 6.4. *The observations follow the model*

$$X_i(t) = \begin{cases}\mu_1(t) + Y_i(t), & 1 \le i \le k^*,\\ \mu_2(t) + Y_i(t), & k^* < i \le N,\end{cases} \tag{6.8}$$

in which the Y_i satisfy Assumption 6.1, the mean functions μ_1 and μ_2 are in $L^2(\mathcal{T})$, and

$$k^* = [n\theta] \quad \text{for some} \quad 0 < \theta < 1.$$

We will see in the proof of Theorem 6.2 that under Assumption 6.4 the sample covariances of the functional observations converge to the function

$$\tilde{c}(t,s) = c(t,s) + \theta(1-\theta)(\mu_1(t) - \mu_2(t))(\mu_1(s) - \mu_2(s)). \tag{6.9}$$

This is a symmetric, square integrable function, and it is easy to see that for any $x, y \in L^2$,

$$\iint \tilde{c}(t,s)x(t)x(s)dt\,ds \geq 0,$$

so $\tilde{c}(t,s)$ is a covariance function. Consequently, it has orthonormal eigenfunctions w_k and nonnegative eigenvalues γ_k satisfying

$$\int \tilde{c}(t,s)w_k(s)ds = \gamma_k w_k(t). \tag{6.10}$$

The quantities $\tilde{c}(t,s)$, w_k and γ_k are used in Section 6.3 to describe the distribution of the test statistic under the alternative of a single change point.

6.3 Detection procedure

To explain the idea of the test procedure, denote

$$\hat{\mu}_k(t) = \frac{1}{k}\sum_{1\leq i\leq k} X_i(t), \quad \widetilde{\mu}_k(t) = \frac{1}{N-k}\sum_{k<i\leq N} X_i(t).$$

If the mean is constant, the difference $\Delta_k(t) = \hat{\mu}_k(t) - \widetilde{\mu}_k(t)$ is small for all $1 \leq k < N$ and all $t \in [0,1]$. However, $\Delta_k(t)$ can become large due to chance variability if k is close to 1 or to N. It is therefore usual to work with the sequence

$$P_k(t) = \sum_{1\leq i\leq k} X_i(t) - \frac{k}{N}\sum_{1\leq i\leq N} X_i(t) = \frac{k(N-k)}{N}\,[\hat{\mu}_k(t) - \widetilde{\mu}_k(t)]$$

in which the variability at the end points is attenuated by a parabolic weight function. If the mean changes, the difference $P_k(t)$ is large for some values of k and of t. Since the observations are in an infinite dimensional domain, we work with the projections of the functions $P_k(\cdot)$ on the principal components of the data. These projections can be expressed in terms of scores which can be easily computed using the **R** package `fda`.

Consider thus the scores corresponding to the largest d eigenvalues:

$$\hat{\xi}_{\ell,i} = \int [X_i(t) - \bar{X}_N(t)]\hat{v}_\ell(t)dt, \quad i = 1,2,\ldots,N, \;\; \ell = 1,2,\ldots,d.$$

Observe that the value of $P_k(t)$ does not change if the $X_i(t)$ are replaced by $X_i(t) - \bar{X}_N(t)$. Consequently, setting $k = [Nx],\ x \in (0,1)$, we obtain

$$\int \left\{ \sum_{1 \le i \le Nx} X_i(t) - \frac{[Nx]}{N} \sum_{1 \le i \le N} X_i(t) \right\} \hat{v}_\ell(t)dt = \sum_{1 \le i \le Nx} \hat{\xi}_{\ell,i} - \frac{[Nx]}{N} \sum_{1 \le i \le N} \hat{\xi}_{\ell,i}. \tag{6.11}$$

Identity (6.11) shows that scores can be used for testing the constancy of the mean function.

The following theorem can be used to derive a number of test statistics. To state it, introduce the statistic

$$T_N(x) = \frac{1}{N} \sum_{\ell=1}^{d} \hat{\lambda}_\ell^{-1} \left(\sum_{1 \le i \le Nx} \hat{\xi}_{\ell,i} - x \sum_{1 \le i \le N} \hat{\xi}_{\ell,i} \right)^2 \tag{6.12}$$

and let $B_1(\cdot), \ldots, B_d(\cdot)$ denote independent standard Brownian bridges.

Theorem 6.1. *Suppose Assumptions 6.1, 6.2,6.3 hold. Then, under H_0,*

$$T_N(x) \xrightarrow{d} \sum_{1 \le \ell \le d} B_\ell^2(x) \quad (0 \le x \le 1),$$

in the Skorokhod topology of $D[0,1]$.

Theorem 6.1 is proved in Section 6.5.

By Theorem 6.1, $U(T_N) \xrightarrow{d} U(\sum_{1 \le \ell \le d} B_\ell^2(\cdot))$, for any continuous functional $U : D[0,1] \to R$. Applying integral or max functionals, or their weighted versions, leads to useful statistics. We focus on the integral of the squared function, i.e. the Cramér–von–Mises functional, which is known to produce effective tests. We thus consider the convergence $\int_0^1 T_N(x)dx \xrightarrow{d} \int_0^1 \sum_{1 \le l \le d} B_\ell^2(x)dx$, which can be rewritten as

$$S_{N,d} := \frac{1}{N^2} \sum_{l=1}^{d} \hat{\lambda}_\ell^{-1} \sum_{k=1}^{N} \left(\sum_{1 \le i \le k} \hat{\xi}_{\ell,i} - \frac{k}{N} \sum_{1 \le i \le N} \hat{\xi}_{\ell,i} \right)^2 \xrightarrow{d} \int_0^1 \sum_{1 \le \ell \le d} B_\ell^2(x)dx. \tag{6.13}$$

The distribution of the random variable

$$K_d = \int_0^1 \sum_{1 \le \ell \le d} B_\ell^2(x)dx \tag{6.14}$$

was derived by Kiefer (1959). Denoting by $c_d(\alpha)$ its $(1-\alpha)$th quantile, the test rejects H_0 if $S_{N,d} > c_d(\alpha)$. The critical values $c_d(\alpha)$ are presented in Table 6.1.

A multivariate analog of statistic (6.13) considered in Horváth *et al.* (1999) is

$$M_{N,d} = \frac{1}{N^2} \sum_{k=1}^{N} \left(\frac{k}{N} \frac{N-k}{N} \right)^2 \boldsymbol{\Delta}(k) \hat{\boldsymbol{D}}_d^{-1} \boldsymbol{\Delta}^T(k), \tag{6.15}$$

where $\boldsymbol{\Delta}(k)$ is the difference of the mean vectors (of dimension d) computed from the first k and the last $N-k$ data vectors, and $\hat{\boldsymbol{D}}_d$ is the $d \times d$ matrix of estimated residual vectors. If d is large, the inverse of $\hat{\boldsymbol{D}}_d$ is unstable. In statistic (6.13), this inverse is "replaced" by inverses of the d largest eigenvalues $\hat{\lambda}_\ell$, and the whole statistic is properly "diagonalized" so that only the most important variability of the data is considered, while the high dimensional noise is ignored.

We now turn to the behavior of the test under the alternative. We will show that it is consistent, i.e. $S_{N,d} \xrightarrow{P} \infty$. In fact, we can obtain the rate of divergence: under H_A, $S_{n,d}$ grows linearly with N. We formulate these results under the assumption of one change point.

Under Assumption 6.4, for $1 \le k \le d$, introduce the functions

$$g_k(x) = \begin{cases} x(1-\theta)\int (\mu_1(t) - \mu_2(t))w_k(t)dt, & 0 < x \le \theta \\ \theta(1-x)\int (\mu_1(t) - \mu_2(t))w_k(t)dt, & \theta < x < 1. \end{cases} \tag{6.16}$$

Theorem 6.2. *Under Assumption 6.4,*

$$\sup_{0 \le x \le 1} \left| N^{-1}T_N - \mathbf{g}^T(x)\Sigma^*\mathbf{g}(x) \right| = o_P(1),$$

where

$$\mathbf{g}(x) = [g_1(x), \ldots, g_d(x)]^T \quad \text{and} \quad \Sigma^* = \begin{bmatrix} 1/\gamma_1 & 0 & \cdots & 0 \\ 0 & 1/\gamma_2 & \cdots & 0 \\ \vdots & \vdots & \vdots & \vdots \\ 0 & 0 & \cdots & 1/\gamma_d \end{bmatrix}.$$

Theorem 6.2 is proven in Section 6.5.

It follows that the test statistic (6.13) satisfies the law of large numbers under the alternative, i.e.

$$\frac{1}{N} S_{N,d} \xrightarrow{P} \sum_{1 \le k \le d} \frac{1}{\gamma_k} \int_0^1 g_k^2(x)dx.$$

If $\int_0^1 g_k^2(x)dx > 0$ for some $1 \le k \le d$, then $S_{N,d} \xrightarrow{P} \infty$.

To understand when the test is consistent, introduce the jump function $\Delta(t) = \mu_1(t) - \mu_2(t)$. By (6.16), the condition $\int_0^1 g_k^2(x)dx > 0$ is equivalent to $\int_0^1 \Delta(s)w_k(s)ds \neq 0$. Thus the test will have no power if

$$\int_0^1 \Delta(s)w_k(s)ds = 0, \quad \text{for all } 1 \le k \le d. \tag{6.17}$$

This lead us to the following corollary.

Corollary 6.1. *If Assumption 6.4 holds, and the jump function $\Delta(t) = \mu_1(t) - \mu_2(t)$ is not orthogonal to the subspace spanned by the first d eigenfunctions of the covariance kernel $\tilde{c}(t,s)$ (6.9), then $S_{N,d} \xrightarrow{P} \infty$, as $N \to \infty$.*

To estimate the change point, we plot the function $T_N(x)$ in (6.12) against $0 \le x \le 1$, and estimate θ by the value of x which maximizes $T_N(x)$. The intuition behind this estimator is clear from (6.12) and (6.11). To ensure uniqueness, we formally define this estimator as

$$\hat{\theta}_N = \inf\left\{x : T_N(x) = \sup_{0\le y\le 1} T_N(y)\right\}. \tag{6.18}$$

Its weak consistency is established in the following proposition

Proposition 6.1. *If the assumptions of Corollary 6.1 hold, then* $\hat{\theta}_N \xrightarrow{P} \theta$.

Proof. The argument x maximizing $T_N(x)$ clearly maximizes $A_N(x) = N^{-1}T_N(x)$. Theorem 6.2 states that $\sup_{0\le x\le 1} |A_N(x) - A(x)| \xrightarrow{P} 0$, where

$$A(x) = \mathbf{g}^T(x)\Sigma^*\mathbf{g}(x) = \begin{cases} x^2(1-\theta)^2 A, & 0 \le x \le \theta \\ \theta^2(1-x)^2 A, & \theta < x < 1, \end{cases}$$

with

$$A = \sum_{1\le\ell\le d} \frac{1}{\gamma_\ell}\left(\int \Delta(t)w_\ell(t)dt\right)^2. \tag{6.19}$$

Under the assumptions of Corollary 6.1, $A > 0$, and it is easy to verify that $A(x)$ has then a unique maximum at $x = \theta$. □

The asymptotic distribution of the estimator $\hat{\theta}_N$ is studied in Aue *et al.* (2009). They show that if $A > 0$ in (6.19), and the assumptions of Corollary 6.1 hold, then $N(\hat{\theta}_N - \theta)$ converges in distribution to the location of the supremum of a two–sided random walk with a drift. If the size of the change depends on depends on the sample size, i.e. $\Delta(t) = \Delta_N(t)$, and the size of the change goes to zero as $N \to \infty$, but not faster than $N^{-1/2}$, then there is a sequence $c(N) \to \infty$ such that $c(N)(\hat{\theta}_N - \theta)$ converges in distribution to the location of the supremum of a two–sided random walk with a drift. By Lemmas 2.2 and 2.3 and Theorem 2.6, even under the alternative the eigenvalues and the eigenfunctions (up to a random sign) stabilize.

An important aspect of the procedure is the choice of the number d of the eigenfunctions v_k. This issue is common to all FDA procedures using functional PCA, and several approaches have been proposed. These include an adaptation of the *scree plot* of Cattell (1966), see Section 9.4, the *pseudo AIC* and the *cross–validation*. All these methods are implemented in the MATLAB PACE package developed at the University of California at Davis. We use the *cumulative percentage variance* approach, which is explained in Section 6.4. A general recommendation for the cumulative percentage variance method is to use d which explains 85% of the variance.

6.4 Finite sample performance and application to temperature data

In this section, we report the results of a simulation study that examines the finite sample performance of the test. Recall that the test rejects if $S_{N,d}$ of (6.13) exceeds the $(1 - \alpha)$th quantile of K_d of (6.14). For $d \le 5$, these quantiles were computed by Kiefer (1959) using a series expansion of the CDF of K_d. Horváth *et al.* (1999) used these expansions to find the critical values for $d = 12$ and noticed that the critical values obtained by simulating K_d by discretizing the integral are slightly different, but actually lead to more accurate tests. To cover a fuller range of the d values, Table 6.1 gives simulated critical values for $d = 1, \ldots, 30$, computed by discretizing the integral over 1,000 points and running 100,000 replications.

The simulation study consists of two parts. First we use standard Gaussian processes as the errors Y_i and a number of rather arbitrary mean functions μ. This part assesses the test in some generic cases analogous to assuming a normal distribution of scalar observations. In the second part, we use mean functions and errors derived from monthly temperature data. No assumptions on the marginal distribution of the Y_i's or the shape of the μ's are made. This part assesses the test in a specific, practically relevant setting.

Gaussian processes. To investigate the empirical size, without loss of generality, $\mu(t)$ was chosen to be equal to zero and two different cases of $Y_i(t)$ were considered,

Table 6.1 Simulated critical values of the distribution of K_d.

Nominal size			d			
	1	2	3	4	5	6
10%	0.345165	0.606783	0.842567	1.065349	1.279713	1.485200
5%	0.460496	0.748785	1.001390	1.239675	1.469008	1.684729
1%	0.740138	1.072101	1.352099	1.626695	1.866702	2.125950
	7	8	9	10	11	12
10%	1.690773	1.897365	2.096615	2.288572	2.496635	2.686238
5%	1.895557	2.124153	2.322674	2.526781	2.744438	2.949004
1%	2.342252	2.589244	2.809778	3.033944	3.268031	3.491102
	13	14	15	16	17	18
10%	2.884214	3.066906	3.268958	3.462039	3.650724	3.837678
5%	3.147604	3.336262	3.544633	3.740248	3.949054	4.136169
1%	3.708033	3.903995	4.116829	4.317087	4.554650	4.734714
	19	20	21	22	23	24
10%	4.024313	4.214800	4.404677	4.591972	4.778715	4.965613
5%	4.327286	4.532917	4.718904	4.908332	5.101896	5.303462
1%	4.974172	5.156282	5.369309	5.576596	5.759427	5.973941
	25	26	27	28	29	30
10%	5.159057	5.346543	5.521107	5.714145	5.885108	6.083306
5%	5.495721	5.688849	5.866095	6.068351	6.242770	6.444772
1%	6.203718	6.393582	6.572949	6.771058	6.977607	7.186491

namely the trajectories of the standard Brownian motion (BM), and the Brownian bridge (BB). These processes were generated by transforming cumulative sums of independent normal variables computed on a grid of 10^3 equispaced points in $[0, 1]$. Following Ramsay and Silverman (2005) (Chapter 3) discrete trajectories were converted to functional observations (functional objects in R) using B-spline and Fourier bases and various numbers of basis functions. No systematic dependence either on the type of the basis or on the number of basis functions was found. The results reported in this section were obtained using B-spline basis with 800 basis functions. We used a wide spectrum of N and d, but to conserve space, we present the results for $N = 50, 150, 200, 300, 500$ and $d = 1, 2, 3, 4$. All empirical rejection rates are based on $1,000$ replications.

Table 6.2 shows the empirical sizes based on critical values reported in Table 6.1. The empirical sizes are fairly stable. Except for a very few cases of small sample sizes, all deviations from the nominal significance levels do not exceed two standard errors computed using the normal approximation $\sqrt{p(1-p)/R}$, where p is a nominal level and R the number of repetitions. Table 6.2 shows that for these Gaussian processes, the empirical size does not depend appreciably either on n or on d.

In the power study, several cases that violate the null were considered. We report the power for $k^* = [N/2]$. Several other values of k^* were also considered, and only a small loss of power was observed for $N/4 < k^* \leq 3N/4$. A few different mean functions μ before and after change were used, namely $\mu_i(t) = 0, t, t^2, \sqrt{t}, e^t, \sin(t), \cos(t), i = 1, 2$, for instance $\mu_1(t) = t$ and $\mu_2(t) = \cos(t)$, etc.

Table 6.2 Empirical size (in percent) of the test using the B-spline basis.

Process	d=1			d=2			d=3			d=4		
	10%	5%	1%	10%	5%	1%	10%	5%	1%	10%	5%	1%
					$N = 50$							
BM	10.3	4.6	0.1	9.9	4.8	0.7	8.4	3.3	0.6	9.7	4.8	0.8
BB	11.2	5.5	0.8	10.6	4.9	1.1	8.4	4.0	0.9	8.5	4.3	1.2
					$N = 100$							
BM	12.2	5.6	1.3	9.8	5.6	0.9	9.3	4.6	0.9	9.0	5.4	0.9
BB	12.4	5.7	0.7	10.2	4.2	0.6	9.9	4.6	1.0	8.3	4.1	0.8
					$N = 150$							
BM	10.8	5.7	1.3	9.7	4.6	1.2	11.8	6.2	0.8	10.8	5.3	1.1
BB	10.5	5.0	1.2	9.8	4.4	1.1	10.4	6.2	0.7	10.5	5.1	1.2
					$N = 200$							
BM	9.7	5.4	0.8	9.2	4.3	0.7	9.3	5.8	1.3	10.8	5.5	0.9
BB	9.2	5.1	0.8	10.8	5.6	1.2	10.0	5.2	1.0	9.6	5.2	1.0
					$N = 300$							
BM	10.3	5.2	1.5	11.1	6.1	0.6	10.1	4.5	0.6	9.9	5.5	0.7
BB	10.4	5.6	1.1	9.4	4.8	0.9	9.9	4.1	0.8	10.5	5.3	1.3
					$N = 500$							
BM	11.6	6.3	1.3	10.6	6.9	1.5	10.9	5.7	1.4	9.0	4.4	0.6
BB	11.7	5.1	1.3	9.7	5.8	1.4	10.3	5.3	1.1	10.0	5.4	1.1

Table 6.3 Empirical power (in percent) of the test using B-spline basis. Change point at $k^* = [n/2]$.

Process	d=1			d=2			d=3		
	10%	5%	1%	10%	5%	1%	10%	5%	1%
				$N = 50$					
BM; BM + $\sin(t)$	81.5	70.8	43.7	72.6	60.0	33.2	67.7	54.9	27.3
BM; BM + t	88.4	78.0	54.1	84.7	74.0	45.4	77.5	64.3	36.0
BB; BB + $\sin(t)$	99.8	99.4	97.4	100	100	99.9	100	100	100
BB; BB + t	99.9	99.8	98.9	100	100	99.9	100	100	100
				$N = 100$					
BM; BM + $\sin(t)$	97.4	95.3	86.3	96.4	91.0	76.5	93.5	88.0	68.7
BM; BM + t	99.0	97.5	91.2	98.7	97.1	87.6	97.5	94.9	83.8
BB; BB + $\sin(t)$	100	100	100	100	100	100	100	100	100
BB; BB + t	100	100	100	100	100	100	100	100	100
				$N = 150$					
BM; BM + $\sin(t)$	99.9	99.5	96.6	99.6	98.6	95.1	98.9	97.4	90.3
BM; BM + t	100	99.8	98.7	99.8	99.7	98.8	99.9	99.7	97.8
BB; BB + $\sin(t)$	100	100	100	100	100	100	100	100	100
BB; BB + t	100	100	100	100	100	100	100	100	100
				$N = 200$					
BM; BM + $\sin(t)$	100	99.9	99.1	100	99.8	99.0	99.9	99.7	98.2
BM; BM + t	100	100	100	100	100	99.9	100	100	99.3
BB; BB + $\sin(t)$	100	100	100	100	100	100	100	100	100
BB; BB + t	100	100	100	100	100	100	100	100	100

Table 6.3 presents selected results of the power study. It shows that the test has overall good power. For small samples, $N \leq 100$, in cases where the BB was used the power is slightly higher than for those with the BM. Nonetheless, for $N \geq 150$ the power approaches 100% for both processes and all choices of other parameters. The power decreases as the number of principal components d increases. This can be explained as follows: the critical values of $S_{N,d}$ increase with d, but the change point is mainly captured by a few initial leading principal components explaining the major part of the variance.

Analysis of central England temperatures. The goal of this section is twofold: to investigate the performance of the test in a real world setting, and to demonstrate the advantages of the functional approach for high–dimensional data. The data consists of 228 years (1780 to 2007) of average daily temperatures in central England. They were published by the British Atmospheric Data Centre, and compiled by Nick Humphreys at the University of Utah. The original data can thus be viewed as 228 curves with 365 measurements on each curve. These data were converted to functional objects in R using 12 B-spline basis functions. Multivariate observations were obtained as in Horváth *et al.* (1999) by computing monthly averages resulting in 228 vectors of dimension $d = 12$. (We could not even compute statistics (6.15) for vectors of dimension 365 because R reported that $\hat{\boldsymbol{D}}$ was singular.) These two procedures are illustrated in Figure 6.2. Even though we used 12 B-splines and 12 averages, the resulting data look quite different, especially in spring and fall, when the temperatures change most rapidly. Gregorian months form a somewhat arbitrary fixed partition of the data, while the splines adapt to their shapes which differ from year to year.

To compute statistic (6.13), we used $d = 8$ eigenfunctions which explain 84% of variability. If the test indicates a change, we estimate it by the estimator $\hat{\theta}_N$ (6.18). This divides the data set into two subsets. The procedure is then repeated for each subset until periods of constant mean functions are obtained. We proceed in exactly the same manner using statistic (6.15). We refer to these procedures, respectively, as FDA and MDA approaches. The resulting segmentations are shown in Tables 6.4 and 6.5.

The functional approach identified two more change point, 1850 and 1992, which roughly correspond to the industrial revolution and the advent of rapid global warming. The multivariate approach "almost" identified these change points with the P–values in iterations 4 and 5 being just above the significance level of 5%. This may indicate that the functional method has better power, perhaps due to its greater flexibility in capturing the shape of the data. This conjecture is investigated below. Figure 6.3 shows average temperatures in the last four segments, and clearly illustrates the warming trend.

Table 6.4 Segmentation procedure of the data into periods with constant mean function.

Iteration	Segment	Decision	$S_{N,d}$ $M_{N,d}$	P-value	Estimated change point
		England temperatures ($d = 8$) (FDA approach)			
1	1780 - 2007	Reject	8.020593	0.00000	1926
2	1780 - 1925	Reject	3.252796	0.00088	1808
3	1780 - 1807	Accept	0.888690	0.87404	-
4	1808 - 1925	Reject	2.351132	0.02322	1850
5	1808 - 1849	Accept	0.890845	0.87242	-
6	1850 - 1925	Accept	1.364934	0.41087	-
7	1926 - 2007	Reject	2.311151	0.02643	1993
8	1926 - 1992	Accept	0.927639	0.84289	-
9	1993 - 2007	Accept	1.626515	0.21655	-
		England temperatures ($d = 12$) (MDA approach)			
1	1780 - 2007	Reject	7.971031	0.00000	1926
2	1780 - 1925	Reject	3.576543	0.00764	1815
3	1780 - 1814	Accept	1.534223	0.81790	-
4	1815 - 1925	Accept	2.813596	0.07171	-
5	1926 - 2007	Accept	2.744801	0.08662	-

Table 6.5 Summary and comparison of segmentation. Beginning and end of data period in bold.

Approach	Change points					
FDA	**1780**	1808		1850	1926	1993 **2007**
MDA	**1780**		1815		1926	**2007**

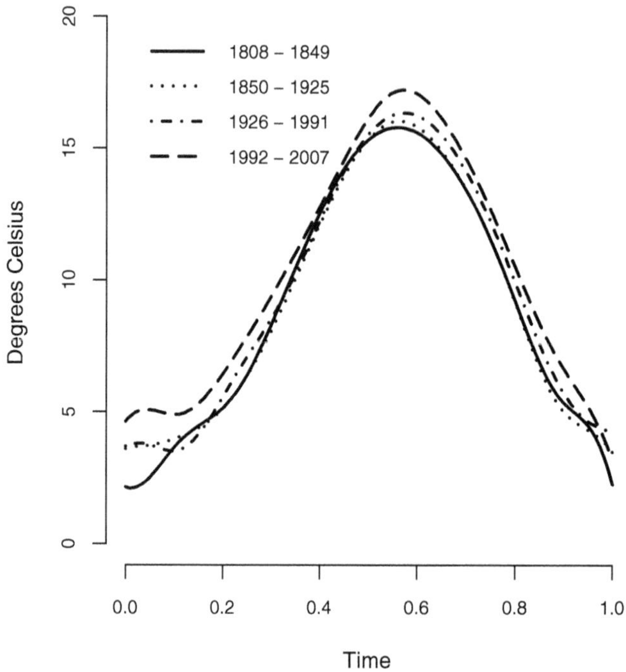

Fig. 6.3 Average temperature functions in the estimated partition segments.

The analysis presented above assumes a simple functional change point model for the daily temperatures. Obviously, one cannot realistically believe that the mean curves change abruptly in one year, this is merely a modeling assumption useful in identifying patterns of change in mean temperature curves. Well-established alternative modeling approaches have been used to study the variability of temperatures. For example, Hosking (1984) fitted a fractionally differenced ARMA(1,1) model to the series of *annual* average temperatures in central England in 1659–1976. It is generally very difficult to determine on purely statistical grounds if a change–point or a long–range dependent model is more suitable for any particular finite length record, see Berkes *et al.* (2006) and Jach and Kokoszka (2008) for recent methodology, discussion and references. It is often more useful to choose a modeling methodology which depends on specific goals, and this is the approach we use. One way of checking an approximate adequacy of our model is to check if the residuals obtained after subtracting the mean in each segment are approximately independent and identically distributed. This can be done by applying the test developed by Gabrys and Kokoszka (2007) which is a functional analog of the well–known test of Hosking

(1980) and Li and McLeod (1981) (see also Hosking (1981, 1989). The P-value of 8% indicates the acceptance of the hypothesis that the residuals are iid.

Keeping these caveats in mind, we use the partitions obtained above to generate realistic synthetic data with and without change–points. We use them to evaluate and compare the size and power properties of the FDA and MDA tests, and to validate our findings. We compute the residuals of every observation in a constant mean segment by subtracting the average of the segment, i.e. $\hat{\mathbf{Y}}_{is} = \mathbf{X}_{is} - \hat{\mu}_s$, where $s = 1, \ldots, S$ denotes the segment, and $i = 1, \ldots, I_s$ indexes observations in the sth segment. The $\hat{\mathbf{Y}}_{is}$ are functional residuals, and their average in each segment is clearly the zero function.

To assess the empirical size, we simulate "temperature-like" data by considering two cases. *Case I*: for every constant mean segment s, we produce synthetic observations by adding to its mean function $\hat{\mu}_s$ errors drawn from the empirical distribution of the residuals of that segment, i.e. synthetic (bootstrap) observations in the sth segment are generated via $\mathbf{X}^*_{is} = \hat{\mu}_s + \hat{\mathbf{Y}}_{i^*s}$, where i^* indicates that $\hat{\mathbf{Y}}_{i^*s}$ is obtained by drawing with replacement from $\left\{\hat{\mathbf{Y}}_{is},\ i = 1, \ldots, I_s\right\}$. *Case II*: We compute residuals in each segment and pool them together. We use this larger set of residuals to create new observations by adding to the average of a segment the errors drawn with replacement from that pool of residuals. For each segment, we generate 1000 of these bootstrap sequences. Table 6.6 shows the the resulting empirical sizes. As the sample size increases, the FDA rejection rates approach nominal sizes, while the MDA test is much more conservative. For the 1993–2007 segment, the size is not reported because the matrix $\boldsymbol{D}$ was (numerically) singular for most bootstrap replications.

We next investigate the power. Three cases are considered. *Case I*: For each segment, we produce synthetic observations using the bootstrap procedure and sampling residuals from a corresponding period. This means that the errors in each segment come from possibly different distributions. *Case II*: We pool together two,

Table 6.6 Empirical size of the test for models derived from the temperature data.

Segment	Number of functions	10%	5%	1%	10%	5%	1%
			Case I			*Case II*	
		FDA approach ($d = 8$)					
1780 - 1807 (Δ_1)	28	8.0	3.0	0.1	7.6	2.5	0.2
1808 - 1849 (Δ_2)	42	9.5	3.9	0.4	9.7	4.1	0.4
1850 - 1925 (Δ_3)	76	10.0	4.7	0.7	10.2	4.3	0.7
1926 - 1992 (Δ_4)	66	8.8	3.7	0.8	9.2	4.1	1.0
1993 - 2007 (Δ_5)	16	3.8	0.3	0.0	3.3	0.1	0.0
		MDA approach ($d = 12$)					
1780 - 1807 (Δ_1)	28	3.0	0.5	0.0	2.8	0.4	0.0
1808 - 1849 (Δ_2)	42	5.3	2.3	0.1	5.4	1.3	0.0
1850 - 1925 (Δ_3)	76	6.9	1.9	0.0	9.1	4.2	0.6
1926 - 1992 (Δ_4)	66	7.9	3.3	0.5	7.4	2.7	0.2
1993 - 2007 (Δ_5)	16	-	-	-	0.0	0.0	0.0

three, four, or five sets of residuals (depending on how many constant mean segments we consider) and sample from that pool to produce new observations. This means that the errors in each segment come from the same distribution. *Case III*: We slightly modify *Case II* by combining all residuals from all segments into one population and use it to produce new observations. In both *Case II* and *Case III*, the theoretical assumptions of Section 6.2 are satisfied, cf. Assumption 6.4, i.e. the means change, but the errors come from the same population. Table 6.7 shows the power of the test for FDA approach and Table 6.8 presents results of discrete MDA method. As seen in Table 6.7, the differences between the three cases are of the order of the chance error. Table 6.7 shows that the test has excellent power, even in small samples, both for single and multiple change points. As for the Gaussian processes, power is slightly higher if there is a change point around the middle of the sample. Comparing Tables 6.7 and 6.8, it is seen that in FDA approach dominates the MDA approach. There are a handful of cases, indicated with *, when MDA performed better, but their frequency and the difference size suggests that this may be attributable to the chance error.

6.5 An approximation theorem for functional observations and proofs of Theorems 6.1 and 6.2

A key element of the proofs is bound (6.31), which follows from a functional central limit theorem in a Hilbert space, see e.g. Kuelbs (1973). A result of this type is needed because the observations $X_i(\cdot)$ are elements of a Hilbert space, and to detect a change point, we must monitor the growth of the partial sums $\sum_{1\le i\le Nx} X_i(t)$ which are a function of $0 < x < 1$. Lemma 6.2 is particularly noteworthy because it shows that the eigenvalues and the eigenfunctions also converge under the alternative.

For the sake of completeness we provide a new and simple proof of the result of Kuelbs (1973) in a form most suitable for the application to the proofs of Theorems 6.1 and 6.2. We start with an L^2 version of the Kolmogorov inequality. Let $\{Z_i(t), 0 \le t \le 1, 1 \le i \le N\}$ be independent identically distributed random functions with values in $L^2[0,1]$ satisfying

$$EZ_1(t) = 0 \quad \text{and} \quad \int EZ_1^2(t)dt < \infty. \tag{6.20}$$

Lemma 6.1. *If (6.20) holds, then for all $\epsilon > 0$ we have*

$$\epsilon P\left\{\max_{1\le k\le N} \int \left(\sum_{i=1}^{k} Z_i(t)\right)^2 dt \ge \epsilon\right\} dt \le E\int \left(\sum_{i=1}^{N} Z_i(t)\right)^2 dt. \tag{6.21}$$

Table 6.7 Empirical power of the test for change-point models derived from temperature data (FDA approach).

Segment	Sample size	Change point(s) θ	Nominal level								
			Case *I*			Case *II*			Case *III*		
			10%	5%	1%	10%	5%	1%	10%	5%	1%
			England (d = 8) (FDA approach)								
Δ_1, Δ_2	70	.41	85.6	76.8	49.7	86.4	76.9	46.3	87.0	75.7	45.3
Δ_1, Δ_3	104	.28	86.2	75.8	47.4	88.6	78.8	50.6	93.1	83.3	58.1
Δ_1, Δ_4	94	.31	100	100	98.7	100	100	99.3	99.8	99.7	96.3
Δ_1, Δ_5	44	.66	100	99.9	93.4	100	99.8	92.7	99.8	99.6	92.2
Δ_2, Δ_3	118	.36	87.9*	78.5	52.8	88.0	78.9	52.1	88.6	79.6	54.0
Δ_2, Δ_4	108	.40	99.7	99.0	95.6	100	99.6	96.7	100	99.3	95.7
Δ_2, Δ_5	58	.74	99.2	97.8	86.3	99.4	98.6	85.8	99.6	98.7	86.6
Δ_3, Δ_4	142	.54	99.9*	99.5*	99.1	100	100	98.9*	99.6	99.1	96.6
Δ_3, Δ_5	92	.84	99.1	96.7	82.9	99.4	97.4	84.4	98.9	95.4	79.6
Δ_4, Δ_5	82	.82	93.0	85.0	58.8	94.0	86.3	57.0	77.9	64.9	32.6
$\Delta_1, \Delta_2, \Delta_3$	146	.20 .49	99.1	97.9	89.6	99.2	97.0	89.9	99.3	98.5	94.2
$\Delta_1, \Delta_2, \Delta_4$	136	.21 .52	100	100	100	100	100	100	100	100	100
$\Delta_1, \Delta_2, \Delta_5$	86	.34 .83	100	100	99.7	99.9	99.9	99.2	100	100	99.7
$\Delta_2, \Delta_3, \Delta_4$	184	.23 .65	100	100	99.9	100	100	99.9	100	99.9*	99.9
$\Delta_2, \Delta_3, \Delta_5$	134	.32 .89	100	99.3*	96.4	99.9	99.8	97.4	100	99.7	97.7
$\Delta_3, \Delta_4, \Delta_5$	158	.49 .91	100	100	100	100	100	100	100	100	100
$\Delta_1, \Delta_2, \Delta_3, \Delta_4$	212	.14 .33 .69	100	100	100	100	100	100	100	100	100
$\Delta_1, \Delta_2, \Delta_3, \Delta_5$	162	.18 .44 .91	100	100	99.9	100	100	99.9	100	100	100
$\Delta_2, \Delta_3, \Delta_4, \Delta_5$	200	.22 .60 .93	100	100	100	100	100	100	100	100	100
$\Delta_1, \Delta_2, \Delta_3, \Delta_4, \Delta_5$	228	.13 .31 .64 .93	100	100	100	100	100	100	100	100	100

Table 6.8 Empirical power of the test for change-point models derived from temperature data (MDA approach).

Segment	Sample size	Change point(s) θ	Nominal level								
			Case *I*			Case *II*			Case *III*		
			10%	5%	1%	10%	5%	1%	10%	5%	1%
			England (d = 8) (FDA approach)								
Δ_1, Δ_2	70	.41	82.9	70.2	38.2	85.2	73.4	39.3	76.2	59.6	26.8
Δ_1, Δ_3	104	.28	79.7	63.9	32.6	79.4	64.8	30.5	81.1	67.4	35.1
Δ_1, Δ_4	94	.31	100	99.4	95.8	99.9	99.0	96.0	99.3	96.9	82.0
Δ_1, Δ_5	44	.66	98.4	93.8	54.5	99.0	93.0	55.8	98.5	91.8	49.0
Δ_2, Δ_3	118	.36	88.3	75.9	46.8	86.7	75.6	43.5	82.3	70.7	41.7
Δ_2, Δ_4	108	.40	97.3	93.3	77.5	97.8	95.6	78.1	98.3	95.7	80.7
Δ_2, Δ_5	58	.74	93.9	85.5	50.4	94.7	85.2	48.3	96.3	90.9	57.9
Δ_3, Δ_4	142	.54	100	100	98.5	100	99.8	99.0	99.5	98.9	94.6
Δ_3, Δ_5	92	.84	98.2	93.9	71.2	99.1	94.2	71.3	96.7	90.2	58.2
Δ_4, Δ_5	82	.82	78.4	63.1	28.0	79.4	63.4	26.4	60.9	44.1	15.7
$\Delta_1, \Delta_2, \Delta_3$	146	.20 .49	97.5	93.2	76.9	97.7	93.1	77.9	97.4	94.9	80.2
$\Delta_1, \Delta_2, \Delta_4$	136	.21 .52	100	100	100	100	100	99.9	100	100	99.9
$\Delta_1, \Delta_2, \Delta_5$	86	.34 .83	100	99.8	96.2	99.9	99.7	95.7	100	99.8	97.4
$\Delta_2, \Delta_3, \Delta_4$	184	.23 .65	100	100	99.1	100	99.9	98.7	100	100	99.5
$\Delta_2, \Delta_3, \Delta_5$	134	.32 .89	99.8	99.4	93.7	99.6	99.3	93.8	99.7	98.6	92.1
$\Delta_3, \Delta_4, \Delta_5$	158	.49 .91	100	100	100	100	100	100	100	100	100
$\Delta_1, \Delta_2, \Delta_3, \Delta_4$	212	.14 .33 .69	100	100	99.9	100	100	100	100	100	100
$\Delta_1, \Delta_2, \Delta_3, \Delta_5$	162	.18 .44 .91	100	100	99.1	100	99.9	99.1	100	100	98.9
$\Delta_2, \Delta_3, \Delta_4, \Delta_5$	200	.22 .60 .93	100	100	100	100	100	100	100	100	100
$\Delta_1, \Delta_2, \Delta_3, \Delta_4, \Delta_5$	228	.13 .31 .64 .93	100	100	100	100	100	100	100	100	100

Proof. Let $\mathcal{F}_k$ be the σ algebra generated by $\{Z_\ell(t), 0 \le t \le 1, 1 \le \ell \le k\}$. By the independence of the Z_i's and assumption (6.20), we have

$$E\left\{\int\left(\sum_{i=1}^{k+1} Z_i(t)\right)^2 dt \,\middle|\, \mathcal{F}_k\right\} \ge \int\left(\sum_{i=1}^{k} Z_i(t)\right)^2 dt,$$

and therefore $\{\int\left(\sum_{i=1}^{k} Z_i(t)\right)^2 dt, \mathcal{F}_k\}$ is a non-negative submartingale. Hence by Doob's maximal inequality (cf. Hall and Heyde (1980), p. 14), we have that

$$\begin{aligned}
&\epsilon P\left\{\max_{1\le k\le N}\int\left(\sum_{i=1}^{k} Z_i(t)\right)^2 dt \ge \epsilon\right\} dt \\
&\le E\left\{\int\left(\sum_{i=1}^{N} Z_i(t)\right)^2 dt I\left[\max_{1\le k\le N}\int\left(\sum_{1\le i\le k} Z_i(t)\right)^2 dt > \epsilon\right]\right\} \\
&\le E\left\{\int\left(\sum_{i=1}^{N} Z_i(t)\right)^2 dt\right\},
\end{aligned}$$

completing the proof. □

The following result was obtained by Kuelbs (1973), and now we present a proof based on projections.

Theorem 6.3. . *If $Y_1, Y_2, \ldots, Y_N$ satisfy Assumption 6.1, then we can define a sequence of Gaussian processes $\Gamma_N(x,t)$ such that $E\Gamma_N(x,t) = 0$, $E\Gamma_N(x,t)\Gamma_N(x',t') = \min(x,x')c(t,t')$ and*

$$\sup_{0\le x\le 1}\int\left(N^{-1/2}\sum_{1\le i\le Nx} Y_i(t) - \Gamma_N(x,t)\right)^2 dt = o_P(1) \quad (N\to\infty). \tag{6.22}$$

We also note that for all N

$$\left\{\Gamma_N(x,t), 0\le x,t\le 1\right\} \stackrel{d}{\to} \left\{\sum_{\ell=1}^{\infty}\lambda_\ell^{1/2} W_\ell(x) v_\ell(t), 0\le x,t\le 1\right\},$$

where W_ℓ are independent, identically distributed Wiener processes.

Proof. Let $M \ge 1$ and define

$$Y_{i.M}(t) = \sum_{\ell=1}^{M}\langle Y_i, v_\ell\rangle\, v_\ell(t)$$

and

$$\bar{Y}_{i.M}(t) = Y_i(t) - Y_{i.M}(t) = \sum_{\ell=M+1}^{\infty} \langle Y_i, v_\ell \rangle \, v_\ell(t).$$

First we show that for all $\epsilon > 0$ we have

$$\lim_{M\to\infty} \limsup_{N\to\infty} P\left\{ \max_{1\le k\le N} \int \left(N^{-1/2} \sum_{i=1}^{k} \bar{Y}_{i.M}(t) \right)^2 dt \ge \epsilon \right\} = 0. \tag{6.23}$$

Using Lemma 6.1 we get that

$$\begin{aligned} P&\left\{ \max_{1\le k\le N} \int \left(N^{-1/2} \sum_{i=1}^{k} \bar{Y}_{i.M}(t) \right)^2 dt \ge \epsilon \right\} \\ &\le \frac{1}{\epsilon} E \int \left(N^{-1/2} \sum_{i=1}^{k} \bar{Y}_{i.M}(t) \right)^2 dt \\ &= \frac{1}{\epsilon} \int (E\bar{Y}_{1.M}(t))^2 dt \\ &= \frac{1}{\epsilon} \sum_{\ell=M+1}^{\infty} \lambda_\ell, \end{aligned}$$

proving (6.23).

It is easy to see that

$$N^{-1/2} \sum_{i=1}^{k} Y_{i,M}(t) = \sum_{\ell=1}^{M} \left(N^{-1/2} \sum_{i=1}^{k} \langle Y_i, v_\ell \rangle \right) v_\ell(t).$$

Next we note that the vectors $\mathbf{Y}_i = (\langle Y_i, v_1 \rangle, \ldots, \langle Y_i, v_M \rangle)^T, 1 \le i \le N$ are independent and identically distributed random vectors with $E\mathbf{Y}_i = \mathbf{0}$ and $E\mathbf{Y}_i\mathbf{Y}_i^T = \mathrm{diag}(\lambda_1, \ldots, \lambda_M)$. By Donsker's theorem for any N we can define M independent Wiener processes $W_{1,N}, \ldots W_{M,N}$ such that

$$\max_{1\le \ell\le M} \max_{1\le k\le N} \left| N^{-1/2} \sum_{i=1}^{k} \langle Y_i, v_\ell \rangle - \lambda_\ell^{1/2} W_{\ell,N}(k/N) \right| = o_P(1). \tag{6.24}$$

Using the continuity of the Wiener process we conclude that for all $M \ge 1$

$$\max_{1\le \ell\le M} \sup_{0\le x\le 1} \left| W_{\ell,N}([Nx]/N) - W_{\ell,N}(x) \right| = o_P(1). \tag{6.25}$$

Thus by (6.24) and (6.25)we have for all $M \geq 1$ that

$$\sup_{0\leq x\leq 1}\int\left(\sum_{\ell=1}^{M}\left(N^{-1/2}\sum_{i=1}^{[Nx]}\langle Y_i, v_\ell\rangle-\lambda_\ell^{1/2}W_{\ell,N}(x)\right)v_\ell(t)\right)^2 dt \tag{6.26}$$

$$= \sup_{0\leq x\leq 1}\sum_{\ell=1}^{M}\left(N^{-1/2}\sum_{i=1}^{[Nx]}\langle Y_i, v_\ell\rangle-\lambda_\ell^{1/2}W_{\ell,N}(x)\right)^2$$

$$= o_P(1).$$

The process $\Gamma_N(x,t)$ is defined by

$$\Gamma_N(x,t)=\sum_{\ell=1}^{\infty}\lambda_\ell^{1/2}W_{\ell,N}(x)v_\ell(t),$$

where $\{W_{\ell,N}\}$ are suitably chosen independent Wiener processes.

To finish the proof it is enough to prove that

$$E\sup_{0\leq x\leq 1}\int\left(\sum_{\ell=M+1}^{\infty}\lambda_\ell^{1/2}W_{\ell,N}(x)v_\ell(t)\right)^2 dt\to 0\quad (M\to\infty).$$

(Note that the expected value above does not depend on N.) By the orthonormality of the v_ℓ's we have that

$$E\sup_{0\leq x\leq 1}\int\left(\sum_{\ell=M+1}^{\infty}\lambda_\ell^{1/2}W_{\ell,N}(x)v_\ell(t)\right)^2 dt = E\sup_{0\leq x\leq 1}\sum_{\ell=M+1}^{\infty}\lambda_\ell W_{\ell,N}^2(x)$$

$$\leq E\left(\sup_{0\leq x\leq 1}W_{1,N}^2(x)\right)\sum_{\ell=M+1}^{\infty}\lambda_\ell$$

$$\to 0\quad (M\to\infty),$$

which completes the proof of the theorem. □

Proof of Theorem 6.1. We will work with the unobservable projections

$$\tilde{\beta}_{k,i}=\int Y_i(t)\hat{v}_k(t)dt,\quad \beta_{k,i}=\int Y_i(t)v_k(t)dt,\quad \beta_{k,i}^*=\hat{c}_k\beta_{k,i}$$

and the vectors

$$\boldsymbol{\beta}_i=[\beta_{1,i},\ldots,\beta_{d,i}]^T,\quad \boldsymbol{\beta}_i^*=[\beta_{1,i}^*,\ldots,\beta_{d,i}^*]^T,\quad 1\leq i\leq N.$$

Since the Y_i are iid functions with mean zero, the $\boldsymbol{\beta}_i$ are iid mean zero vectors in R^d. A simple calculation using the orthonormality of the v_k shows that each $\boldsymbol{\beta}_i$ has a diagonal covariance matrix

$$\Sigma_d = \begin{bmatrix} \lambda_1 & 0 & \cdots & 0 \\ 0 & \lambda_2 & \cdots & 0 \\ \vdots & \vdots & \vdots & \vdots \\ 0 & 0 & \cdots & \lambda_d \end{bmatrix}$$

The functional central limit theorem, thus implies that

$$N^{-1/2} \sum_{1 \le i \le Nx} \boldsymbol{\beta}_i \stackrel{d}{\to} \boldsymbol{\Delta}_d(x) \quad (0 \le x \le 1), \tag{6.27}$$

where the convergence is in the Skorokhod space $D^d[0,1]$. The process $\{\boldsymbol{\Delta}_d(x),\ 0 \le x \le 1\}$ takes values in R^d, has zero mean and covariance matrix Σ_d. Convergence (6.27) implies in turn that

$$\frac{1}{N}\left[\sum_{1\le i\le Nx} \boldsymbol{\beta}_i - x\sum_{1\le i\le N} \boldsymbol{\beta}_i\right]^T \Sigma_d^{-1} \left[\sum_{1\le i\le Nx} \boldsymbol{\beta}_i - x\sum_{1\le i\le N} \boldsymbol{\beta}_i\right] \stackrel{d}{\to} \sum_{1\le i\le d} B_i^2(x) \tag{6.28}$$

in the Skorokhod space $D[0,1]$.

The matrix Σ_d is estimated by $\widehat{\Sigma}_d$. By (6.7) and Assumption 6.2, $\widehat{\Sigma}_d^{-1} \stackrel{P}{\to} \Sigma_d{}^{-1}$, so (6.28) yields

$$\frac{1}{N}\left[\sum_{1\le i\le Nx} \boldsymbol{\beta}_i - x\sum_{1\le i\le N} \boldsymbol{\beta}_i\right]^T \widehat{\Sigma}_d^{-1} \left[\sum_{1\le i\le Nx} \boldsymbol{\beta}_i - x\sum_{1\le i\le N} \boldsymbol{\beta}_i\right] \stackrel{d}{\to} \sum_{1\le i\le d} B_i^2(x). \tag{6.29}$$

Note that

$$\sum_{1\le i\le Nx} \beta^*_{k,i} - x\sum_{1\le i\le N} \beta^*_{k,i} = \hat{c}_k\left(\sum_{1\le i\le Nx} \beta_{k,i} - x\sum_{1\le i\le N} \beta_{k,i}\right).$$

Since $\hat{c}_k^2 = 1$, we can replace the $\boldsymbol{\beta}_i$ in (6.29) by the $\boldsymbol{\beta}_i^*$, and obtain

$$\frac{1}{N}\left[\sum_{1\le i\le Nx} \boldsymbol{\beta}^*_i - x\sum_{1\le i\le N} \boldsymbol{\beta}^*_i\right]^T \widehat{\Sigma}_d^{-1} \left[\sum_{1\le i\le Nx} \boldsymbol{\beta}^*_i - x\sum_{1\le i\le N} \boldsymbol{\beta}^*_i\right] \stackrel{d}{\to} \sum_{1\le i\le d} B_i^2(x). \tag{6.30}$$

We now turn to the effect of replacing the $\beta^*_{i,k}$ by $\tilde{\beta}_{i,k}$. Observe that

$$\sup_{0<x<1}\left|N^{-1/2}\sum_{1\le i\le Nx}\beta^*_{i,k}-N^{-1/2}\sum_{1\le i\le Nx}\tilde{\beta}_{i,k}\right|$$
$$=\sup_{0<x<1}\left|\int\left(N^{-1/2}\sum_{1\le i\le Nx}Y_i(t)\right)(\hat{c}_k v_k(t)-\hat{v}_k(t))\,dt\right|$$
$$\le\sup_{0<x<1}\left[\int\left(N^{-1/2}\sum_{1\le i\le Nx}Y_i(t)\right)^2 dt\right]^{1/2}\left[\int(\hat{c}_k v_k(t)-\hat{v}_k(t))^2\,dt\right]^{1/2}.$$

The first factor is bounded in probability, i.e.

$$\sup_{0<x<1}\int\left(N^{-1/2}\sum_{1\le i\le Nx}Y_i(t)\right)^2 dt=O_P(1). \tag{6.31}$$

Relation (6.31) follows from the weak convergence in $D([0,1],L^2(\mathcal{T}))$ of the partial sum process $\sum_{1\le i\le Nx}Y_i,\ x\in[0,1]$, see Theorem 6.3.

Combining (6.31) and (6.7), we obtain

$$\sup_{0<x<1}\left|N^{-1/2}\sum_{1\le i\le Nx}\beta^*_{i,k}-N^{-1/2}\sum_{1\le i\le Nx}\tilde{\beta}_{i,k}\right|\stackrel{P}{\to}0,$$

which in turn implies that

$$\left\|\left[\sum_{1\le i\le Nx}\boldsymbol{\beta}^*_i-x\sum_{1\le i\le N}\boldsymbol{\beta}^*_i\right]-\left[\sum_{1\le i\le Nx}\hat{\boldsymbol{\eta}}_i-x\sum_{1\le i\le N}\hat{\boldsymbol{\eta}}_i\right]\right\|=o_P(N^{-1/2}), \tag{6.32}$$

where the norm is the Euclidean norm in R^d. Relations (6.30) and (6.32) yield the claim in Theorem 6.1. □

Proof of Theorem 6.2. Theorem 6.2 follows from relation (6.36) and Lemma 6.3. To establish them, we need the following Lemma.

Lemma 6.2. *Under Assumption 6.4, for every $1\le k\le d$, as $N\to\infty$,*

$$\hat{\lambda}_k\stackrel{P}{\to}\gamma_k, \tag{6.33}$$

$$\int[\hat{v}_k(t)-\hat{c}_k w_k(t)]^2dt\stackrel{P}{\to}0, \tag{6.34}$$

where $\hat{v}_k,\hat{\lambda}_k$ are defined by (6.6), w_k,γ_k by (6.10) and $\hat{c}_k=\operatorname{sign}\int_{\mathcal{T}}v_k(t)\hat{v}_k(t)dt$.

Proof. It is easy to see that

$$\bar{X}_N(t) = \bar{Y}_N(t) + \frac{k^*}{N}\mu_1(t) + \frac{N-k^*}{N}\mu_2(t)$$

and, denoting $\Delta(t) = \mu_1(t) - \mu_2(t)$,

$$\begin{aligned}
\hat{c}_N(t,s) &= \frac{1}{N}\left(\sum_{1\le i\le k^*} + \sum_{k^*<i\le N}\right)(X_i(t) - \bar{X}_N(t))(X_i(s) - \bar{X}_N(s)) \\
&= \frac{1}{N}\sum_{1\le i\le k^*}\left(Y_i(t) - \bar{Y}_N(t) + \mu_1(t) - \frac{k^*}{N}\mu_1(t) - \frac{N-k^*}{N}\mu_2(t)\right) \\
&\quad\times\left(Y_i(s) - \bar{Y}_N(s) + \mu_1(s) - \frac{k^*}{N}\mu_1(s) - \frac{N-k^*}{N}\mu_2(s)\right) \\
&\quad+\frac{1}{N}\sum_{k^*<i\le N}\left(Y_i(t) - \bar{Y}_N(t) + \mu_2(t) - \frac{k^*}{N}\mu_1(t) - \frac{N-k^*}{N}\mu_2(t)\right) \\
&\quad\times\left(Y_i(s) - \bar{Y}_N(s) + \mu_2(s) - \frac{k^*}{N}\mu_1(s) - \frac{N-k^*}{N}\mu_2(s)\right) \\
&= \frac{1}{N}\sum_{1\le i\le k^*}\left(Y_i(t) - \bar{Y}_N(t) + \frac{N-k^*}{N}\Delta(t)\right)\left(Y_i(s) - \bar{Y}_N(s) + \frac{N-k^*}{N}\Delta(s)\right) \\
&\quad+\frac{1}{N}\sum_{k^*<i\le N}\left(Y_i(t) - \bar{Y}_N(t) - \frac{k^*}{N}\Delta(t)\right)\left(Y_i(s) - \bar{Y}_N(s) - \frac{k^*}{N}\Delta(s)\right).
\end{aligned}$$

Rearranging terms, we obtain

$$\begin{aligned}
\hat{c}_N(t,s) &= \frac{1}{N}\sum_{i=1}^{N}\left(Y_i(t) - \bar{Y}_N(t)\right)\left(Y_i(s) - \bar{Y}_N(s)\right) \\
&\quad+\frac{k^*}{N}\left(1 - \frac{k^*}{N}\right)\Delta(t)\Delta(s) + r_N(t,s),
\end{aligned}$$

where

$$\begin{aligned}
r_N(t,s) &= \left(1 - \frac{k^*}{N}\right)\frac{1}{N}\sum_{1\le i\le k^*}\left[\left(Y_i(t) - \bar{Y}_N(t)\right)\Delta(s) + \left(Y_i(s) - \bar{Y}_N(s)\right)\Delta(t)\right] \\
&\quad+\frac{k^*}{N}\frac{1}{N}\sum_{k^*<i\le N}\left[\left(Y_i(t) - \bar{Y}_N(t)\right)\Delta(s) + \left(Y_i(s) - \bar{Y}_N(s)\right)\Delta(t)\right].
\end{aligned}$$

Using the law of large numbers (Theorem 2.2), we obtain $\iint r_N^2(t,s)dt\,ds \xrightarrow{P} 0$ and, by Theorem 2.6,

$$\iint \left[\hat{c}_N(t,s) - \tilde{c}_N(t,s)\right]^2 \xrightarrow{P} 0. \tag{6.35}$$

Hence Lemmas 2.2 and 2.3 imply, respectively, (6.33) and (6.34). □

As an immediate corollary to (6.33), we obtain

$$\widehat{\Sigma}_d^{-1} \xrightarrow{P} \Sigma^*. \tag{6.36}$$

Lemma 6.3. *Under Assumption 6.4,*

$$\sup_{0\le x\le 1}\left|\frac{1}{N}\left[\sum_{1\le i\le Nx}\hat{\xi}_{k,i} - x\sum_{1\le i\le N}\hat{\xi}_{k,i}\right] - \hat{c}_k g_k(x)\right| = o_P(1),$$

with the functions g_k defined by (6.16).

Proof. Denote

$$\hat{g}_k(x) = \frac{1}{N}\left[\sum_{1\le i\le Nx}\hat{\xi}_{k,i} - x\sum_{1\le i\le N}\hat{\xi}_{k,i}\right], \quad x\in[0,1],$$

and observe that

$$\hat{\xi}_{k,i} = \int Y_i(t)\hat{v}_k(t)dt + \int \mu_1(t)\hat{v}_k(t)dt - \int \bar{X}_N(t)\hat{v}_k(t)dt, \quad \text{if } 1\le i\le k^*$$

and

$$\hat{\xi}_{k,i} = \int Y_i(t)\hat{v}_k(t)dt + \int \mu_2(t)\hat{v}_k(t)dt - \int \bar{X}_N(t)\hat{v}_k(t)dt, \quad \text{if } k^* < i\le N.$$

We will use the relation

$$\sup_{0<x<1}\left|\sum_{1\le i\le Nx}\int Y_i(t)\hat{v}_k(t)dt\right| = O_P(N^{1/2}), \tag{6.37}$$

which follows from (6.31).

Suppose first that $0 < x \le \theta$. Then, by (6.37) and (6.34), uniformly in $x\in[0,1]$,

$$\begin{aligned}\hat{g}_k(x) &= x(1-\theta)\left[\int \mu_1(t)\hat{v}_k(t)dt - \int \mu_2(t)\hat{v}_k(t)dt\right] + o_P(N^{-1/2})\\ &= x(1-\theta)\hat{c}_k\left[\int \mu_1(t)w_k(t)dt - \int \mu_2(t)w_k(t)dt\right] + o_P(1).\end{aligned}$$

If $x > \theta$, then, uniformly in $x\in[0,1]$,

$$\hat{g}_k(x) = \theta(1-x)\hat{c}_k\left[\int \mu_1(t)w_k(t)dt - \int \mu_2(t)w_k(t)dt\right] + o_P(1). \qquad \square$$

6.6 Bibliographical notes

The problem of change point detection in a sequence of scalar observations has been extensively studied, and it is not possible to review the literature. Mathematical foundations most closely related to the approach developed in this chapter are presented in Csörgő and Horváth (1997). Brodsky and Darkhovsky (1993) offer a different perspective. Some references to the change point problem in the multivariate. setting are Srivastava and Worsley (1986), Shumway and Stoffer (1991) (Kalman filter), Horváth *et al.* (1999), Lavielle and Teyssiére (2006), Zamba and Hawkins (2006) and Qu and Perron (2007).

The problem of detecting a change in the mean of a sequence of Banach–space valued random elements is theoretically studied by Rackauskas and Suquet (2006). Motivated by detecting an epidemic change (the mean changes and then returns to its original value), they propose a statistic based on increasingly fine dyadic partitions of the index interval, and derived its limit, which is nonstandard.

Chapter 7
Portmanteau test of independence

Most inferential tools of functional data analysis rely on the assumption of iid functional observations. In designed experiments this assumption can be ensured, but for observational data, especially derived from time series, it requires a verification. In this chapter, based on the paper of Gabrys and Kokoszka (2007), we describe a simple portmanteau test of independence for functional observations whose idea is as follows. The functional observations $X_n(t),\ n = 1, 2, \ldots, N$, are approximated by the first p terms of the principal component expansion

$$X_n(t) \approx \sum_{k=1}^{p} X_{kn} v_k(t), \quad n = 1, 2, \ldots, N. \tag{7.1}$$

where the X_{kn} are the scores. For the sake of an intuitive argument, assume first that the populations FPC's $v_k(t)$ are known. Testing the iid assumption for the curves $X_n(\cdot)$ reduces then to testing this assumption for the random vectors $[X_{1n}, \ldots, X_{pn}]^T$. For this purpose, the method proposed by Chitturi (1976) can be used: we find multivariate analogs of correlations and an analog of the "sum of squares" which has a χ^2 asymptotic distribution. In reality, the $v_k(t)$ must be replaced by the EFPC's. This transition is delicate in the problem of testing for independence because the EFPC's depend on all observations.

The test studied in this chapter is analogous to the Ljung–Box test which is extensively used in time series analysis. It essentially tests if the curves are uncorrelated. As is common in time series analysis, evidence against independence can be found if the test is applied to some transformations of the functional observations, for example to the curves $X_n^2(t)$.

We note that Székely *et al.* (2007) and Székely and Rizzo (2009) proposed a test of independence of two variables X and Y, which can be of arbitrary dimension, and so can also be elements of a Hilbert space. Their test is based on a measure of dependence, known as the "correlation of distances" which is derived from differences of characteristic functions. To apply such a test, a sample of iid pairs $(X_i, Y_i), i = 1, 2, \ldots, N$, is required, and so a different inferential problem is solved than that studied in this chapter.

The chapter is organized as follows. In Section 7.1, we formulate the test procedure together with mathematical assumptions and theorems establishing its asymptotic validity. The proofs of the theorems of Section 7.1 are presented in Section 7.4. Sections 7.2 and 7.3 are devoted, correspondingly, to the study of the finite sample performance of the test and its application to two types of functional data: credit card sales and geomagnetic records. Section 7.5 contains some lemmas on Hilbert space valued random elements which are used in Section 7.4, and may be useful in other similar contexts. In Section 7.6, we develop the required theory for random matrices.

7.1 Test procedure

We observe mean zero random functions $\{X_n(t),\ t \in [0,1],\ n = 1,2,\dots N\}$ and want to test

$$H_0 : \text{the } X_n(\cdot) \text{ are independent and identically distributed}$$

versus

$$H_A : H_0 \text{ does not hold.}$$

We approximate the $X_n(t)$ by

$$\hat{X}_n(t) = \sum_{k=1}^{p} \hat{X}_{kn}\hat{v}_k(t),$$

where

$$\hat{X}_{kn} = \int X_n(t)\hat{v}_k(t)dt = \int \hat{X}_n(t)\hat{v}_k(t)dt. \tag{7.2}$$

In practice, the number p must be selected so that the first p EFPC's explain a large fraction of the sample variance, see Section 3.3.

To establish the null distribution of the test statistic, we require the following assumption:

Assumption 7.1. *The observations $X_1, X_2, \dots X_N$ are iid in L^2, have mean zero, and satisfy*

$$E\|X_n\|^4 = E\left[\int X_n^2(t)dt\right]^2 < \infty. \tag{7.3}$$

The eigenvalues of the (population) covariance operator satisfy (2.12).

We will work with the random vectors

$$\hat{\mathbf{X}}_n = [\hat{X}_{1n}, \hat{X}_{2n}, \dots, \hat{X}_{pn}]^T \tag{7.4}$$

and analogously defined (unobservable) vectors

$$\mathbf{X}_n = [X_{1n}, X_{2n}, \ldots, X_{pn}]^T, \tag{7.5}$$

where

$$X_{kn} = \int X_n(t) v_k(t) dt. \tag{7.6}$$

Under H_0, the $\mathbf{X}_n$ are iid mean zero random vectors in R^p for which we denote

$$v(i,j) = E[X_{in} X_{jn}], \quad \mathbf{V} = [v(i,j)]_{i,j=1,\ldots,p}.$$

The matrix $\mathbf{V}$ is thus the $p \times p$ covariance matrix of the $\mathbf{X}_n$. By $\mathbf{C}_h$ we denote the sample autocovariance matrix whose entries are

$$c_h(k,l) = \frac{1}{N} \sum_{n=1}^{N-h} X_{kn} X_{l,n+h}, \quad 0 \le h < N.$$

Notice that we do not use the "hat" ^ in the definition of the above sample covariances because they cannot be computed from the data. When we work with vectors (7.4), rather than (7.5), we use the "hat".

Denote by $r_{f,h}(i,j)$ and $r_{b,h}(i,j)$ the (i,j) entries of $\mathbf{C}_0^{-1}\mathbf{C}_h$ and $\mathbf{C}_h\mathbf{C}_0^{-1}$, respectively, and introduce the random variable

$$Q_N = N \sum_{h=1}^{H} \sum_{i,j=1}^{p} r_{f,h}(i,j) r_{b,h}(i,j). \tag{7.7}$$

Analogously to the way Q_N (7.7) is constructed from the vectors $\mathbf{X}_n$, $n = 1, \ldots, N$, we construct the statistic $\hat{Q}_N$ from the vectors $\hat{\mathbf{X}}_n$, $n = 1, \ldots, N$.

The following theorem establishes the limit null distribution of the test statistic $\hat{Q}_N$.

Theorem 7.1. *If Assumption 7.1 holds, then $\hat{Q}_N \xrightarrow{d} \chi^2_{p^2 H}$ (Chi–square distribution with $p^2 H$ degrees of freedom).*

Theorem 7.1 is proven in Section 7.4.

Lemma 7.1 identifies an alternative expression for the statistic $\hat{Q}_N$, which is convenient in calculations. It is proven in Section 7.4, but it essentially follows from the fact that unlike the general multivariate case, in our case, the matrix $\hat{\mathbf{C}}_0$ is diagonal.

Lemma 7.1. *The statistic $\hat{Q}_N$ has the form*

$$\hat{Q}_N = N \sum_{h=1}^{H} \sum_{i,j=1}^{p} \hat{c}_h^2(i,j) \hat{\lambda}_i^{-1} \hat{\lambda}_j^{-1}. \tag{7.8}$$

Lemma 7.1 also shows that $\hat{Q}_N$ does not depend on the sign of the EFPC's $\hat{v}_k$. Supressing the "hat", set $v'_k = c_k v_k$, where $c_k^2 = 1$. Using the "prime" to denote quantities computed with v'_k rather than v_k, observe that

$$c'_h(k,l) = \frac{1}{N} \sum_{n=1}^{N-h} \langle X_n, c_k v_k \rangle \langle X_{n+h}, c_l v_l \rangle = c_k c_l c_h(k,l).$$

Out of many possible directional alternatives, we focus on the functional AR(1) model which is treated in detail in Chapter 13. Suppose then that

$$X_{n+1} = \Psi(X_n) + \varepsilon_{n+1} \tag{7.9}$$

with iid mean zero innovations $\varepsilon_n \in L^2$. We assume that $\{X_n\}$ is a stationary solution to equation (7.9), which exists under mild assumptions on Ψ, see Chapter 13.

Introduce the $p \times p$ matrix $\boldsymbol{\Psi}$ with entries

$$\psi_{lk} = \langle v_l, \Psi(v_k) \rangle, \ l,k = 1,2,\dots p,$$

where the v_k are the eigenfunctions of the covariance operator of X_1. Clearly, if Ψ is not zero, then ψ_{lk} is not zero for some l and k, and so the matrix $\boldsymbol{\Psi}$ is not zero for sufficiently large p.

The following theorem establishes the consistency against the AR(1) model (7.9).

Theorem 7.2. *Suppose the functional observations X_n follow a stationary solution to equations* (7.9), *conditions* (2.12) *and* (7.3) *hold, and p is so large that the $p \times p$ matrix $\boldsymbol{\Psi}$ is not zero. Then $\hat{Q}_N \xrightarrow{P} \infty$.*

Theorem 7.2 is proven in Section 7.4. The idea is to show that if $\boldsymbol{\Psi}$ is not zero, then $N^{-1} Q_N$ tends in probability to a positive constant. More generally, formula (7.8) shows that the test is consistent whenever the first p estimated eigenvalues are uniformly bounded (in N), and for some $1 \le h \le H$, at least one covariance $\hat{c}_h(i,j)$ is uniformly (in N) bounded away from zero. These conditions hold if the $\hat{\lambda}_j$ converge to a finite limit and one of the $\hat{c}_h(i,j)$ converges to a nonzero limit, as $N \to \infty$.

7.2 Finite sample performance

In this section we investigate the finite sample properties of the test using some generic models and sample sizes typical of applications discussed in Section 7.3.

To investigate the empirical size, we generated independent trajectories of the standard Brownian motion (BM) on [0, 1] and the standard Brownian bridge (BB). This was done by transforming cumulative sums of independent normal variables computed on a grid of m equispaced points in [0, 1]. We used values of m ranging from 10 to 1440, and found that the empirical size basically does not depend on m (the tables of this section use $m = 100$).

To compute the principal components $\hat{v}_k$ and the corresponding eigenvalues using the R package fda, the functional data must be represented (smoothed) using a specified number of functions from a basis. We worked with Fourier and B splines functional bases. we used 8, 16, and 80 basis functions. All results are based on one thousand replications.

Table 7.1 shows empirical sizes for the Brownian bridge and and the Fourier basis for several values of the lag $H = 1,\ 3,\ 5$, the number of principal components $p = 3, 4, 5$ and sample sizes $N = 50, 100, 300$. The standard errors in this table are between 0.5 and 1 percent. In most cases,the empirical sizes are within two standard errors of the nominal size, and the size improves somewhat with increasing N. The same is true for the BM and B splines; no systematic dependence on the type of data or basis is seen, which accords with the nonparametric nature of the test.

In a power study, we focus on the AR(1) model (7.9), which can be more explicitly written as:

$$X_n(t) = \int \psi(t,s) X_{n-1}(s) ds + \varepsilon_n(t), \quad t \in [0,1], \ n = 1, 2, \ldots, N. \qquad (7.10)$$

A sufficient condition for the assumptions of Theorem 7.2 to hold is

$$\|\Psi\|_{\mathcal{S}}^2 = \iint \psi^2(t,s) dt\, ds < 1. \qquad (7.11)$$

In our study, the innovations ε_n in (7.10) are either standard BM's or BB's. We used two kernel functions: the Gaussian kernel

$$\psi(t,s) = C \exp\left\{\frac{t^2 + s^2}{2}\right\}, \quad (t,s) \in [0,1]^2,$$

and Wiener kernel

$$\psi(t,s) = C \min(s,t), \quad (t,s) \in [0,1]^2.$$

Table 7.1 Empirical size (in percent) of the test using Fourier basis. The simulated observations are Brownian bridges.

Lag	p=3			p=4			p=5		
	10%	5%	1%	10%	5%	1%	10%	5%	1%
				N=50					
1	7.7	2.5	0.6	7.4	2.8	0.3	7.9	3.5	0.4
3	6.8	2.5	0.3	6.7	3.3	0.6	4.9	2.0	0.3
5	4.9	2.0	0.0	3.6	1.4	0.2	4.0	1.7	0.2
				N=100					
1	9.0	5.1	0.4	8.9	3.9	0.6	10.0	3.9	0.8
3	8.1	3.5	0.6	8.3	4.0	0.9	7.5	3.2	0.4
5	8.8	3.6	0.6	6.6	2.7	0.3	6.7	2.4	0.3
				N=300					
1	9.8	4.6	1.2	9.4	4.0	0.9	10.1	4.7	0.6
3	9.3	4.8	1.0	9.1	4.7	0.9	10.0	5.4	0.8
5	7.2	3.7	1.0	8.2	3.8	0.7	10.6	5.5	1.2

Table 7.2 Empirical power of the test using Fourier basis. The observations follow the AR(1) model (7.10) with Gaussian kernel with $\|\Psi\|_{\mathcal{S}} = 0.5$ and iid standard Brownian motion innovations.

Lag	p=3			p=4			p=5		
	10%	5%	1%	10%	5%	1%	10%	5%	1%
				N=50					
1	44.7	33.8	17.7	41.9	29.4	12.6	38.5	26.1	9.2
3	35.2	27.0	13.3	34.0	24.7	10.8	33.2	21.6	8.7
5	26.7	20.0	11.0	24.4	15.8	8.1	21.5	14.3	6.0
				N=100					
1	71.2	64.2	51.4	74.4	66.5	48.1	77.7	68.0	46.1
3	67.9	61.0	44.9	67.5	58.6	42.8	68.4	56.9	38.1
5	62.3	54.6	38.6	59.0	49.9	32.3	55.1	45.5	27.9
				N=300					
1	98.7	98.2	96.7	99.2	98.9	97.2	99.8	99.5	98.5
3	97.6	97.1	95.5	99.0	98.4	96.8	99.2	98.3	96.6
5	96.8	95.9	92.8	98.1	97.0	93.8	98.4	97.3	94.4

The constants C were chosen so that $\|\Psi\|_{\mathcal{S}} = 0.3, 0.5, 0.7$. We used both Fourier and B spline basis.

The power against this alternative is expected to increase rapidly with N, as the test statistic is proportional to N. This is clearly seen in Table 7.2. The power also increases with $\|\Psi\|_{\mathcal{S}}$; for $\|\Psi\|_{\mathcal{S}} = 0.7$ and the Gaussian kernel, it is practically 100% for $N = 100$ and all choices of other parameters.

There are two less trivial observations: The power is highest for lag $H = 1$. This is because for the AR(1) process the "correlation" between X_n and X_{n-1} is largest at this lag. By increasing the maximum lag H, the value of $\hat{Q}_N$ generally increases, but the critical value increases too (degrees of freedom increase by p^2 for a unit increase in H). The power also depends on how the action of the operator Ψ is "aligned" with the eigenvectors v_k. If the inner products $\langle v_i, \Psi v_k \rangle$ are large in absolute value, the power is high. Thus, with all other parameters being the same, the power in Table 7.3 is greater than in Table 7.2 because of the different covariance structure of the Brownian bridge and the Brownian motion. In all cases, the power for the Wiener kernel is slightly lower than for the Gaussian kernel.

7.3 Application to credit card transactions and diurnal geomagnetic variation

In this section, we apply our test to two data sets which we have encountered in earlier chapters. The first data set consists of the number of transactions with credit cards issued by Vilnius Bank, Lithuania. The second, is a daily geomagnetic variation.

Table 7.3 Empirical power of the test using Fourier basis. The observations follow the AR(1) model (7.10) with Gaussian kernel with $\|\Psi\|_S = 0.5$ and iid standard Brownian bridge innovations.

Lag	p=3			p=4			p=5		
	10%	5%	1%	10%	5%	1%	10%	5%	1%
				N=50					
1	98.3	97.0	92.1	98.4	96.4	87.6	99.1	97.3	87.6
3	95.2	90.3	77.4	92.1	86.2	69.6	89.9	85.1	63.2
5	86.9	80.2	61.7	78.5	71.7	51.4	75.2	63.9	40.4
				N=100					
1	100	100	100	100	100	100	100	100	100
3	100	100	99.9	100	100	99.7	100	99.9	99.8
5	99.9	99.3	98.7	99.9	99.8	98.6	99.7	99.5	97.8

Table 7.4 P-values for the functional AR(1) residuals of the credit card data X_n.

Lag, H	p=1	p=2	p=3	p=4	p=5	p=6	p=7
			BF=40				
1	69.54	22.03	13.60	46.29	80.35	96.70	99.20
2	35.57	38.28	7.75	47.16	64.92	95.00	99.04
3	54.44	53.63	25.28	52.61	71.33	86.84	94.93
			BF=80				
1	57.42	18.35	53.30	89.90	88.33	95.40	99.19
2	35.97	23.25	23.83	45.07	55.79	46.39	70.65
3	36.16	36.02	26.79	30.21	56.81	34.51	47.00

Suppose $D_n(t_i)$ is the number of credit card transactions in day n, $n = 1, \dots, 200$, (03/11/2000 – 10/02/2001) between times t_{i-1} and t_i, where $t_i - t_{i-1} = 8$ min, $i = 1, \dots, 128$. For our analysis, the transactions were normalized to have time stamps in interval [0, 1], which thus corresponds to one day. To remove the weekly periodicity, we work with the differences $X_n(t_i) = D_n(t_i) - D_{n-7}(t_i)$, $n = 1, 2, \dots, 193$. Figure 1.7 displays the first three weeks of these data. A characteristic pattern of an AR(1) process with clusters of positive and negative observations is clearly seen. Two consecutive days are shown in the left–most panel of Figure 1.6 together with functional objects obtained by smoothing with 40 and 80 Fourier basis functions. As expected, the test rejects the null hypothesis at 1% level for both smooths, and all lag values $1 \le H \le 5$ and the number of principal components equal to 4, 5, 10 and 20.

Next we applied the test to the residuals $\hat{\varepsilon}_n = X_n - \hat{\Psi}(X_{n-1})$. We estimated the autoregressive function ψ using the function `linmod` from the `R` package `fda`. Table 7.4 displays the P-values for the sequence of the residuals. They support the model functional AR(1) model proposed by Laukaitis and Račkauskas (2002). We note that the above validation of the AR(1) model is not fully asymptotically justified. Even for real–valued time series, see Ljung and Box (1978) and Section 4.4 of Lütkepohl (2005), sums of squared residual autocorrelations do not converge to a chi–square distribution, but it is a good approximation. The number of degrees of freedom in the chi–square approximation is slightly less than $p^2 H$. A different method of testing the fit of the functional AR(1) model is developed in Chapter 15.

We now turn to the ground-based magnetometer records. We focus on the horizontal (H) component measured at Honolulu in 2001. It is the component of the magnetic field tangent to the Earth's surface and pointing toward the magnetic North; its variation best reflects the changes in the large currents flowing in the magnetic equatorial plane, as already discussed in Section 4.4. The top panel of Figure 7.1 shows two weeks of these data. Following Xu and Kamide (2004), we subtracted the linear change over a day to obtain the curves like those showed in the bottom panel of Figure 7.1. More precisely, we connected the first and last observation in

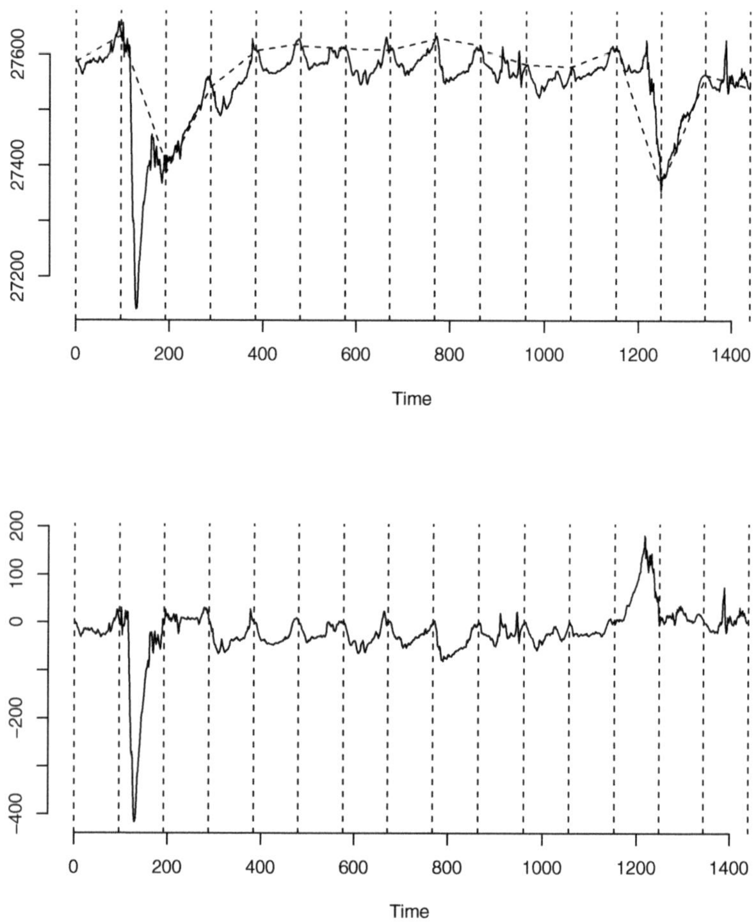

Fig. 7.1 Top: horizontal intensity (nT) measuret at Honolulu 30/3/2001 - 13/4/2001 with the straight lines connecting first and last measurements in each day. Bottom: the same after substracting the lines.

Table 7.5 P-values (in percent) for the magnetometer data split by season.

Lag	Feb, Mar, Apr, May		Jun, Jul, Aug, Sep	
H	p=4	p=5	p=4	p=5
1	13.44	6.51	1.03	1.23
3	3.37	2.99	31.72	42.59

a given day by a line, and subtracted this line from the data. After centering over the period under study, we obtained the mean zero functional observations we work with. The analysis was conducted using Fourier base functions.

Testing one year magnetometer data with lags $H = 1, 2, 3$ and different numbers of principal components $p = 3, 4, 5$, yields P–values very close to zero. This indicates that while principal component analysis, advocated by Xu and Kamide (2004), may be a useful exploratory tool to study daily variation over the whole year, one must be careful when using any inferential tools based on it, as they typically require independent and identically distributed observations (a simple random sample), see e.g. Section 5.2 of Seber (1984). We also applied the test to smaller subsets of data roughly corresponding to boreal Spring and Summer. The P–values, reported in Table 7.5, show that the transformed data can to a reasonable approximation be viewed as a functional simple random sample, at least with respect to the second order properties. The discrepancy in the outcome of the test when applied to the whole year and to a season is probably due to the annual change of the position of the Honolulu observatory relative to the Sun whose energy drives the convective currents mainly responsible for the daily variation.

The two examples discussed in this section show that our test can detect departures from the assumption of independence (credit card data) or from the assumption of identical distribution (magnetometer data), and confirm both assumptions when they are expected to hold. In our examples, the results of the test do not depend much on the choice of the smoothing basis.

7.4 Proofs of the results of Section 7.1

Proof of Lemma 7.1:. Direct verification shows that

$$\sum_{i,j=1}^{p} \hat{r}_{f,h}(i,j)\hat{r}_{b,h}(i,j) = \operatorname{tr}\left\{\hat{\mathbf{C}}_h^T \hat{\mathbf{C}}_0^{-1} \hat{\mathbf{C}}_h \hat{\mathbf{C}}_0^{-1}\right\} \tag{7.12}$$

and that

$$\sum_{i,j=1}^{p} \hat{c}_h^2(i,j)\hat{\lambda}_i^{-1}\hat{\lambda}_j^{-1} = \operatorname{tr}\left\{\hat{\mathbf{C}}_h^T \hat{\boldsymbol{\Lambda}}^{-1} \hat{\mathbf{C}}_h \hat{\boldsymbol{\Lambda}}^{-1}\right\},$$

so it suffices to verify that $\hat{\mathbf{C}}_0 = \hat{\boldsymbol{\Lambda}}$.

Assuming that the sample mean function has been subtracted from the data, we have $\hat{X}_{in} = \int X_n(t)\hat{v}_i(t)dt$. Therefore the (i, j) entry of $\hat{\mathbf{C}}_0$ is

$$
\begin{aligned}
\hat{c}_0(i,j) &= N^{-1} \sum_{n=1}^{N} \hat{X}_{in}\hat{X}_{jn} \\
&= N^{-1} \sum_{n=1}^{N} \int X_n(t)\hat{v}_i(t)dt \int X_n(s)\hat{v}_j(s)ds \\
&= \int \hat{v}_j(s) \left(N^{-1} \sum_{n=1}^{N} \int X_n(t)\hat{v}_i(t)dt X_n(s) \right) ds \\
&= \int \hat{v}_j(s)\hat{C}(\hat{v}_i)(s)ds \\
&= \int \hat{v}_j(s)(\hat{\lambda}_i\hat{v}_i)ds = \hat{\lambda}_i\delta_{ij}.
\end{aligned}
$$

□

Proof of Theorem 7.1:. By Theorem 7.6, it is enough to show that $\hat{Q}_N - Q_N \xrightarrow{P} 0$. Recall from Section 7.1 that the value of $\hat{Q}_N$ does not change if we replace $\hat{v}_k$ by $v_{kN} = \hat{c}_k\hat{v}_k$, where $\hat{c}_k$ is defined in Section 2.5. In the following, we replace $\hat{v}_k$ by v_{kN}.

By (7.7), relation $\hat{Q}_N - Q_N \xrightarrow{P} 0$ will follow if we show that

$$\hat{\mathbf{C}}_0 - \mathbf{C}_0 \xrightarrow{P} 0 \tag{7.13}$$

and

$$N^{1/2}(\hat{\mathbf{C}}_h - \mathbf{C}_h) \xrightarrow{P} 0, \quad h \geq 1. \tag{7.14}$$

Recall that

$$c_h(k,l) = \frac{1}{N} \sum_{n=1}^{N-h} X_{kn}X_{l,n+h}; \quad \hat{c}_h(k,l) = \frac{1}{N} \sum_{n=1}^{N-h} \hat{X}_{kn}\hat{X}_{l,n+h}.$$

Relation (7.13) follows from Theorems 2.4 and 2.7 because

$$
\begin{aligned}
\hat{c}_0(k,l) - c_0(k,l) &= \left\langle \hat{C}(v_{kN}), v_{lN} \right\rangle - \left\langle \hat{C}(v_k), v_l \right\rangle \\
&= \left\langle \hat{C}(v_{kN} - v_k), v_{lN} \right\rangle + \left\langle \hat{C}(v_k), v_{lN} - v_l \right\rangle,
\end{aligned}
$$

which implies $|\hat{c}_0(k,l) - c_0(k,l)| \leq \|\hat{C}\|_S O_P(N^{-1/2})$.

To prove (7.14), we work with the decomposition $\hat{c}_h(k,l) - c_h(k,l) = M_1 + M_2$, where

$$M_1 = \frac{1}{N} \sum_{n=1}^{N-h} (X_{kn} - \hat{X}_{kn})X_{l,n+h}; \quad M_2 = \frac{1}{N} \sum_{n=1}^{N-h} \hat{X}_{kn}(X_{l,n+h} - \hat{X}_{l,n+h}).$$

We will first show that $N^{1/2}M_1 \xrightarrow{P} 0$. Observe that

$$
\begin{aligned}
N^{1/2}M_1 &= N^{-1/2} \sum_{n=1}^{N-h} \langle X_n, v_k - v_{kN} \rangle \langle X_{n+h}, v_l \rangle \\
&= \left\langle N^{-1/2} \sum_{n=1}^{N-h} \langle X_{n+h}, v_l \rangle X_n, v_k - v_{kN} \right\rangle \\
&= \langle S_N, Y_N \rangle ,
\end{aligned}
$$

where

$$
S_N := N^{-1/2} \sum_{n=1}^{N-h} \langle X_{n+h}, v_l \rangle X_n; \quad Y_N = v_k - v_{kN}.
$$

Note that by (2.13),

$$
\begin{aligned}
&E| \langle S_N, Y_N \rangle | \le E[\|S_N\| \, \|Y_N\|] \le (E\|S_N\|^2)^{1/2}(E\|Y_N\|^2)^{1/2} \\
&= O(N^{-1/2})(E\|S_N\|^2)^{1/2}.
\end{aligned}
$$

To show that $N^{1/2}M_1 \xrightarrow{P} 0$, it thus remains to verify that $E\|S_N\|^2$ is bounded. Notice that

$$
\begin{aligned}
E\|S_N\|^2 &= N^{-1}E\| \sum_{n=1}^{N-h} \langle X_{n+h}, v_l \rangle X_n \|^2 \\
&= N^{-1}E \sum_{m,n=1}^{N-h} \langle X_{m+h}, v_l \rangle \langle X_{n+h}, v_l \rangle \langle X_m, X_n \rangle \\
&= N^{-1} \sum_{n=1}^{N-h} E[\langle X_{n+h}, v_l \rangle]^2 E\|X_n\|^2 \\
&\le \left[E\|X_n\|^2\right]^2.
\end{aligned}
$$

To show that $N^{1/2}M_2 \xrightarrow{P} 0$, decompose M_2 as $M_2 = M_{21} + M_{22}$, where

$$
\begin{aligned}
M_{21} &= \frac{1}{N} \sum_{n=1}^{N-h} \langle X_n, v_k \rangle \langle X_{n+h}, v_l - v_{lN} \rangle ; \\
M_{22} &= \frac{1}{N} \sum_{n=1}^{N-h} \langle X_n, v_{kN} - v_k \rangle \langle X_{n+h}, v_l - v_{lN} \rangle .
\end{aligned}
$$

By the argument developed for M_1, $N^{1/2}M_{21} \xrightarrow{P} 0$, so we must show $N^{1/2}M_{22} \xrightarrow{P} 0$. This follows from Lemma 7.4. □

Proof of Theorem 7.2:. We will verify below that

$$c_1(k,l) \xrightarrow{P} \psi_{lk}\lambda_k \tag{7.15}$$

and

$$\hat{c}_1(k,l) - c_1(k,l) \xrightarrow{P} 0. \tag{7.16}$$

Choose $1 \le l,k \le p$ such that $\psi_{lk} \neq 0$. Then by (7.8), (7.16), (7.15) and Theorem 13.2, $N^{-1}\hat{Q}_N \xrightarrow{P} q > 0$, completing the proof.

We now verify (7.15), and then the relation

$$\hat{\mathbf{C}}_h - \mathbf{C}_h \xrightarrow{P} 0, \tag{7.17}$$

from which (7.16) follows.

Observe that

$$\begin{aligned}
c_1(k,l) &= N^{-1}\sum_{n=1}^{N-1} \langle v_k, X_n\rangle \langle v_l, X_{n+1}\rangle \\
&= N^{-1}\sum_{n=1}^{N-1} \langle v_k, X_n\rangle \langle v_l, \Psi(X_n)\rangle + N^{-1}\sum_{n=1}^{N-1} \langle v_k, X_n\rangle \langle v_l, \varepsilon_{n+1}\rangle \\
&\xrightarrow{P} E\left[\langle v_k, X_n\rangle \langle v_l, \Psi(X_n)\rangle\right] = \sum_{j=1}^{\infty} \psi_{lj} E[\langle v_k, X_n\rangle \langle v_j, X_n\rangle] \\
&= \sum_{j=1}^{\infty} \psi_{lj} \langle C(v_k), v_j\rangle = \sum_{j=1}^{\infty} \psi_{lj}\lambda_k \langle v_k, v_j\rangle \\
&= \psi_{lk}\lambda_k.
\end{aligned}$$

To prove (7.17), we use the notation introduced in the proof of Theorem 7.1. We must show that $M_1 \xrightarrow{P} 0$ and $M_2 \xrightarrow{P} 0$. We will display the argument only for M_1. Observe that

$$M_1 = \left\langle N^{-1}\sum_{n=1}^{N-h} \langle X_{n+h}, v_l\rangle X_n, v_k - v_{kN} \right\rangle.$$

By Theorem 13.2, $\|v_k - v_{kN}\| \xrightarrow{P} 0$. Since,

$$E\|N^{-1}\sum_{n=1}^{N-h} \langle X_{n+h}, v_l\rangle X_n\| \le E\| \langle X_{n+h}, v_l\rangle X_n\| \le E\|X_n\|^2,$$

it follows that $M_1 \xrightarrow{P} 0$. □

7.5 Auxiliary lemmas for H-valued random elements

Consider the empirical lag-h autocovariance operator

$$C_{N,h}(x) = \frac{1}{N} \sum_{n=1}^{N-h} \langle X_n, x \rangle X_{n+h}. \tag{7.18}$$

Recall that the Hilbert-Schmidt norm of a Hilbert-Schmidt operator S is defined by

$$\|S\|_{\mathcal{S}}^2 = \sum_{j=1}^{\infty} \|S(e_j)\|^2,$$

where $\{e_1, e_2, \ldots\}$ is any orthonormal basis.

Lemma 7.2. *Suppose the X_i are iid random elements in a separable Hilbert space with $E\|X_0\|^2 < \infty$, then for $h \geq 1$,*

$$E\|C_{N,h}\|_{\mathcal{S}}^2 = \frac{N-h}{N^2} \left(E\|X_0\|^2\right)^2.$$

Proof. Observe that

$$\begin{aligned}
\|C_{N,h}\|_{\mathcal{S}}^2 &= \sum_{j=1}^{\infty} \|C_{N,h}(e_j)\|^2 \\
&= \sum_{j=1}^{\infty} \left\langle \frac{1}{N} \sum_{n=1}^{N-h} \langle X_n, e_j \rangle X_{n+h}, \frac{1}{N} \sum_{m=1}^{N-h} \langle X_m, e_j \rangle X_{m+h} \right\rangle \\
&= \sum_{j=1}^{\infty} \frac{1}{N^2} \sum_{m,n=1}^{N-h} \langle X_m, e_j \rangle \langle X_n, e_j \rangle \langle X_{m+h}, X_{n+h} \rangle .
\end{aligned}$$

It follows from the independence of the X_n that

$$\begin{aligned}
E\|C_{N,h}\|_{\mathcal{S}}^2 &= \frac{1}{N^2} \sum_{n=1}^{N-h} \sum_{j=1}^{\infty} E[\langle X_n, e_j \rangle]^2 E[\langle X_{n+h}, X_{n+h} \rangle]^2 \\
&= E\|X_0\|^2 \frac{1}{N^2} \sum_{n=1}^{N-h} E\left[\sum_{j=1}^{\infty} [\langle X_n, e_j \rangle]^2 \right] \\
&= \left[E\|X_0\|^2\right]^2 \frac{N-h}{N^2}.
\end{aligned}$$

□

Lemma 7.3. *Suppose $\{U_N\}$ and $\{V_N\}$ are random sequences in a Hilbert space such that $\|U_N\| \xrightarrow{P} 0$ and $\|V_N\| = O_P(1)$ i.e. $\lim_{C\to\infty} \limsup_{N\to\infty} P(\|V_N\| > C) = 0$. Then*

$$\langle U_N, V_N \rangle \xrightarrow{P} 0.$$

Proof. The Lemma follows from the corresponding property of real random sequences and the inequality $|\langle U_N, V_N\rangle| \le \|U_N\|\|V_N\|$. □

Lemma 7.4. *Suppose X_n, Z_N, Y_N are random elements in a separable Hilbert space. We assume*

$$E\|Y_N\|^2 = O(N^{-1}), \quad E\|Z_N\|^2 = O(N^{-1}); \tag{7.19}$$

$$X_n \sim iid, \quad E\|X_n\|^2 < \infty. \tag{7.20}$$

Then, for $h \ge 1$,

$$N^{-1/2}\sum_{n=1}^{N-h}\langle X_n, Y_N\rangle\langle X_{n+h}, Z_N\rangle \xrightarrow{P} 0.$$

Proof. Observe that

$$N^{-1/2}\sum_{n=1}^{N-h}\langle X_n, Y_N\rangle\langle X_{n+h}, Z_N\rangle = \left\langle C_{N,h}(Y_N), N^{1/2}Z_N\right\rangle,$$

with the operator $C_{N,h}$ defined in (7.18). Since $P(N^{1/2}\|Z_N\| > C) \le C^{-2}NE\|Z_N|^2$, $N^{1/2}\|Z_N\| = O_P(1)$. Thus, by Lemma 7.3, it remains to verify that $C_{N,h}(Y_N) \xrightarrow{P} 0$. Since the Hilbert–Schmidt norm is not less than the uniform operator norm $\|\cdot\|_{\mathcal{L}}$, see Section 2.1, we obtain from Lemma 7.2:

$$\begin{aligned} E\|C_{N,h}(Y_N)\| &\le E[\|C_{N,h}\|_{\mathcal{L}}\|Y_N\|] \le E[\|C_{N,h}\|_{\mathcal{S}}\|Y_N\|] \\ &\le \left(E\|C_{N,h}\|_{\mathcal{S}}^2\right)^{1/2}\left(E\|Y_N\|^2\right)^{1/2} = O(N^{-1/2})O(N^{-1/2}) = O(N^{-1}). \end{aligned}$$ □

7.6 Limit theory for sample autocovariance matrices

In this section, we present some results on limits of autocovariance matrices. These results were used in previous sections, and are generally known, but are presented here with detailed proofs for completeness and ease of reference. If the matrix $\mathbf{V}$ is the covariance matrix of the vectors $\mathbf{X}_n$ defined in Section 7.1, then, an argument analogous to that used in the proof of Lemma 7.1 shows that $\mathbf{V} = \mathrm{diag}(\lambda_1, \lambda_2, \ldots, \lambda_p)$, and many arguments presented below could be simplified. However, if the $\mathbf{X}_n$ were obtained by projecting on another system, rather than on the FPC's of the observations, then $\mathbf{V}$ would no longer be diagonal. We therefore present these useful results in the general case.

Consider random vectors $\mathbf{X}_1, \ldots, \mathbf{X}_N$, where $\mathbf{X}_t = [X_{1t}, X_{2t}, \ldots, X_{pt}]^T$. We assume that the $\mathbf{X}_t$, $t = 1, 2, \ldots$ are iid mean zero with finite variance and denote

$$v(i,j) = E[X_{it}X_{jt}], \quad \mathbf{V} = [v(i,j)]_{i,j=1,\ldots,p}. \tag{7.21}$$

By $\mathbf{C}_h$ we denote the sample autoccovariance matrix with entries

$$c_h(k,l) = \frac{1}{N}\sum_{t=1}^{N-h} X_{kt}X_{l,t+h}, \quad h \geq 0.$$

In order to find the asymptotic distribution of $\mathbf{C}_h$ we use Theorem 6.4.2 in Brockwell and Davis (1991) which we state here for ease of reference.

Theorem 7.3. *Suppose $\{Y_t\}$ is a strictly stationary m–dependent sequence with mean zero and finite variance. Denote*

$$v = \gamma(0) + 2\sum_{j=1}^{m}\gamma(j), \quad \gamma(j) = E[Y_tY_{t+j}].$$

If $v \neq 0$, then

$$N^{-1/2}\sum_{t=1}^{N} Y_t \xrightarrow{d} N(0,v)$$

and

$$v = \lim_{N\to\infty} N\,\mathrm{Var}\left[\frac{1}{N}\sum_{t=1}^{N} Y_t\right].$$

We first find the asymptotic distribution of $\mathbf{C}_0$. We will show that $N^{1/2}(\mathbf{C}_0 - V)$ tends to a matrix $\mathbf{Z}_0$ whose entries $Z_0(k,l)$ are jointly Gaussian mean zero.

Observe that

$$\sum_{k,l=1}^{p} a_{kl}[c_0(k,l) - v(k,l)] = \frac{1}{N}\sum_{t=1}^{N} Y_t,$$

where

$$Y_t = \sum_{k,l=1}^{p} a_{kl}(X_{kt}X_{lt} - v(k,l)).$$

The Y_t are iid with mean zero and variance

$$\begin{aligned} EY_t^2 &= E\left[\sum_{k,l=1}^{p} a_{kl}(X_{kt}X_{lt} - v(k,l))\right]^2 \\ &= \sum_{k,l,i,j=1}^{p} a_{kl}a_{ij}E\left[(X_{kt}X_{lt} - v(k,l))(X_{it}X_{jt} - v(i,j))\right] \\ &= \sum_{k,l,i,j=1}^{p} a_{kl}a_{ij}[\eta(k,l,i,j) - v(i,j)v(k,l)], \end{aligned}$$

where

$$\eta(k,l,i,j) = E[X_{kt}X_{lt}X_{it}X_{jt}]. \tag{7.22}$$

Thus, by the CLT,

$$\begin{aligned} N^{1/2} & \sum_{k,l=1}^{p} a_{kl}[c_0(k,l) - v(k,l)] \\ & \xrightarrow{d} N\left(0, \sum_{k,l,i,j=1}^{p} a_{kl}a_{ij}[\eta(k,l,i,j) - v(i,j)v(k,l)]\right). \end{aligned} \tag{7.23}$$

Convergence (7.23) is equivalent to

$$N^{1/2}(\mathbf{C}_0 - \mathbf{V}) \xrightarrow{d} \mathbf{Z}_0, \tag{7.24}$$

where $\mathbf{Z}_0$ is a random matrix with jointly Gaussian entries $Z_0(k,l),\ k,l = 1,\ldots,p$, with mean zero and covariances

$$E[Z_0(k,l)Z_0(i,j)] = \eta(k,l,i,j) - v(i,j)v(k,l). \tag{7.25}$$

We now find the asymptotic distribution of $\mathbf{C}_h$ for $h \geq 1$. Note that

$$\sum_{k,l=1}^{p} a_{kl}c_h(k,l) = \frac{1}{N}\sum_{t=1}^{N-h} Y_t,$$

where

$$Y_t = \sum_{k,l=1}^{p} a_{kl}X_{kt}X_{l,t+h}.$$

The Y_t are identically distributed with mean zero and are h–dependent. Observe that

$$EY_t^2 = \sum_{k,l,i,j=1}^{p} a_{kl}a_{ij}v(k,i)v(l,j)$$

and $E[Y_tY_{t+s}] = 0$ for $s \geq 1$. Thus, by Theorem 7.3,

$$N^{1/2}\sum_{k,l=1}^{p} a_{kl}c_h(k,l) \xrightarrow{d} N(0, \sum_{k,l,i,j=1}^{p} a_{kl}a_{ij}v(k,i)v(l,j))$$

what is equivalent to

$$N^{1/2}\mathbf{C}_h \xrightarrow{d} \mathbf{Z}_h, \tag{7.26}$$

where $\mathbf{Z}_h$ is a random matrix with jointly Gaussian mean zero entries $Z_h(k,l),\ k,l = 1,\ldots,p$, with

$$E[Z_h(k,l)Z_h(i,j)] = v(k,i)v(l,j) \quad (h \geq 1). \tag{7.27}$$

The above calculations suggest the following result:

Theorem 7.4. *If the* $\mathbf{X}_t$ *are iid with finite fourth moment, then*

$$N^{1/2}[\mathbf{C}_0 - V, \mathbf{C}_1, \ldots, \mathbf{C}_H] \xrightarrow{d} [\mathbf{Z}_0, \mathbf{Z}_1, \ldots, \mathbf{Z}_H],$$

where the $\mathbf{Z}_h$, $h = 0, 1, \ldots, H$, *are independent mean zero Gaussian matrices with covariances* (7.25) *and* (7.27).

Proof. To lighten the notation, we present the proof for $H = 1$. We must thus show that for any numbers a_{0kl}, a_{1kl}, $k, l = 1, \ldots, p$,

$$\begin{aligned} N^{1/2} \sum_{k,l=1}^{p} & [a_{0kl}(c_0(k,l) - v(k,l)) + a_{1kl}c_1(k,l)] \\ & \xrightarrow{d} \sum_{k,l=1}^{p} [a_{0kl}Z_0(k,l) + a_{1kl}Z_1(k,l)]. \end{aligned} \tag{7.28}$$

Since $\mathbf{Z}_0$ and $\mathbf{Z}_1$ are independent,

$$\begin{aligned} E & \left\{ \sum_{k,l=1}^{p} [a_{0kl}Z_0(k,l) + a_{1kl}Z_1(k,l)] \right\}^2 \\ & = \sum_{k,l,i,j=1}^{p} [a_{0kl}a_{0ij}(\eta(k,l,i,j) - v(i,j)v(k,l)) + a_{1kl}a_{1ij}v(k,i)v(l,j)]. \end{aligned} \tag{7.29}$$

We must thus show that the left–hand side of (7.28) converges to a normal random variable with mean zero and variance (7.29). Observe that the left–hand side of (7.28) is equal to $N^{-1/2}\sum_{t=1}^{N} Y_t$, where

$$Y_t = \sum_{k,l=1}^{p} [a_{0kl}(X_{kt}X_{lt} - v(k,l)) + a_{1kl}X_{kt}X_{l,t+1}].$$

The Y_t are identically distributed with mean zero and are 1–dependent. Direct verification shows that the variance of Y_t is equal to (7.29) and autocovariances of the Y_t at positive lags vanish. Convergence (7.28) follows therefore from Theorem 7.3. □

We now want to find the asymptotic distribution of $\mathbf{C}_0^{-1}$. We first state a proposition which is a matrix version of the delta method and essentially follows from Proposition 6.4.3 in Brockwell and Davis (1991) by writing the matrices as column vectors, e.g. we write a 2×2 matrix with entries a_{ij} as $[a_{11}, a_{12}, a_{21}, a_{22}]^T$.

Proposition 7.1. *Suppose* $\mathbf{A}_N$ *is a sequence of* $p \times q$ *matrices such that for some deterministic matrix* $\boldsymbol{\mu}$ *of the same dimension*

$$c_N^{-1}(\mathbf{A}_N - \boldsymbol{\mu}) \xrightarrow{d} \mathbf{Z} \quad (c_N \to 0), \tag{7.30}$$

where $\mathbf{Z}$ *is a mean zero Gaussian matrix. Suppose* $g : \mathbf{A} \mapsto g(\mathbf{A})$ *is a function that maps* $R^p \times R^q$ *into* $R^r \times R^s$*, i.e.* $g(\mathbf{A}) = [g_{ij}(\mathbf{A})]_{i=1,\ldots,r,\ j=1,\ldots,s}$*. If* g *has continuous derivatives in a neighborhood of* $\boldsymbol{\mu}$*, then*

$$c_N^{-1}(g(\mathbf{A}_N) - g(\boldsymbol{\mu})) \stackrel{d}{\to} \nabla g(\boldsymbol{\mu})(\mathbf{Z}), \tag{7.31}$$

where $\nabla g(\boldsymbol{\mu})(\mathbf{Z})$ *is a* $r \times s$ *Gaussian matrix with* (i, j)*–entry*

$$[\nabla g(\boldsymbol{\mu})(\mathbf{Z})](i, j) = \sum_{k=1}^{p} \sum_{l=1}^{q} \left[\frac{\partial g_{ij}(\boldsymbol{\mu})}{\partial z(k,l)} \right] Z(k,l). \tag{7.32}$$

Consider the function $g(\mathbf{A}) = \mathbf{A}^{-1}$ from $R^p \times R^p$ into itself. The derivative of this function at an invertible matrix $\mathbf{V}$, $\nabla g(\mathbf{V})$, is the linear operator

$$\mathbf{H} \mapsto -\mathbf{V}^{-1}\mathbf{H}\mathbf{V}^{-1}, \tag{7.33}$$

see e.g. Noble (1969), p. 24, Exercise 1.50. We want to identify the partial derivatives $\partial g_{ij}(\mathbf{V})/\partial z(k,l)$ appearing in (7.32). Let $u(k,l)$ be the (k,l)–entry of $\mathbf{V}^{-1}$. Direct verification shows that the (i, j)–entry of $\mathbf{V}^{-1}\mathbf{Z}\mathbf{V}^{-1}$ is $\sum_{k,l=1}^{p} u(i,k)u(l, j)z(k,l)$, so

$$\frac{\partial g_{ij}(\mathbf{V})}{\partial z(k,l)} = -u(i,k)u(l, j). \tag{7.34}$$

From (7.24) and Proposition 7.1, we thus obtain

$$N^{1/2}(\mathbf{C}_0^{-1} - \mathbf{V}^{-1}) \stackrel{d}{\to} \mathbf{Y}_0, \tag{7.35}$$

where $\mathbf{Y}_0$ is a mean zero Gaussian matrix with (i, j)–entry

$$Y_0(i, j) = -\sum_{k,l=1}^{p} u(i,k)u(l, j)Z_0(k,l). \tag{7.36}$$

We now want to find the limit of $N^{1/2}\mathbf{C}_0^{-1}\mathbf{C}_h$, $h \geq 1$. We view $[\mathbf{C}_0 - \mathbf{V}, \mathbf{C}_h]^T$ as a $(2p) \times p$ matrix and apply Proposition 7.1. By Theorem 7.4, $N^{1/2}[\mathbf{C}_0 - \mathbf{V}, \mathbf{C}_h]^T \stackrel{d}{\to} [\mathbf{Z}_0, \mathbf{Z}_h]^T$. Consider the function $g(\mathbf{A}_1, \mathbf{A}_2) = \mathbf{A}_1^{-1}\mathbf{A}_2$. By Proposition 7.1, with $\boldsymbol{\mu} = [\mathbf{V}, \mathbf{0}]^T$,

$$N^{-1/2}\mathbf{C}_0^{-1}\mathbf{C}_h \stackrel{d}{\to} \nabla g(\boldsymbol{\mu})([\mathbf{Z}_0, \mathbf{Z}_h]^T).$$

We must find the explicit form of $\nabla g(\boldsymbol{\mu})$. The map $[\mathbf{A}_1, \mathbf{A}_2]^T \mapsto \mathbf{A}_1\mathbf{A}_2$ has derivative

$$[\mathbf{H}_1, \mathbf{H}_2]^T \mapsto \mathbf{H}_1\mathbf{A}_2 + \mathbf{A}_1\mathbf{H}_2,$$

see e.g. Noble (1969), p. 24, Exercise 1.50. Combining this with (7.33), we obtain

$$\nabla g([\mathbf{A}_1, \mathbf{A}_2]^T)([\mathbf{H}_1, \mathbf{H}_2]^T) = -\mathbf{A}_1^{-1}\mathbf{H}_1\mathbf{A}_1^{-1}\mathbf{A}_2 + \mathbf{A}_1^{-1}\mathbf{H}_2.$$

It thus follows that $\nabla g(\boldsymbol{\mu})([\mathbf{H}_1, \mathbf{H}_2]^T) = \mathbf{V}^{-1}\mathbf{H}_2$, and so we obtain

$$N^{1/2}\mathbf{C}_0^{-1}\mathbf{C}_h \stackrel{d}{\to} \mathbf{V}^{-1}\mathbf{Z}_h, \quad h \geq 1. \tag{7.37}$$

Using the same technique as in the proof of Theorem 7.4, relation (7.37) can be extended to the following theorem:

Theorem 7.5. *If the $\mathbf{X}_t$ are iid with finite fourth moment, then*

$$N^{1/2}\mathbf{C}_0^{-1}[\mathbf{C}_1, \ldots, \mathbf{C}_H] \stackrel{d}{\to} \mathbf{V}^{-1}[\mathbf{Z}_1, \ldots, \mathbf{Z}_H], \tag{7.38}$$

where the $\mathbf{Z}_h$, $h = 0, 1, \ldots, H$, are independent mean zero Gaussian matrices with covariances (7.27).

Denote by $r_{f,h}(i,j)$ and $r_{b,h}(i,j)$ the (i,j) entries of $\mathbf{C}_0^{-1}\mathbf{C}_h$ and $\mathbf{C}_h\mathbf{C}_0^{-1}$, respectively. Introduce the statistic

$$Q_N = N\sum_{h=1}^{H}\sum_{i,j=1}^{p} r_{f,h}(i,j)r_{b,h}(i,j). \tag{7.39}$$

Theorem 7.6. *If the $\mathbf{X}_t$ are iid with finite fourth moment, then $Q_N \stackrel{d}{\to} \chi^2_{p^2H}$.*

Proof. Similarly to (7.38), it can be verified that

$$N^{1/2}[\mathbf{C}_1, \ldots, \mathbf{C}_H]\mathbf{C}_0^{-1} \stackrel{d}{\to} [\mathbf{Z}_1, \ldots, \mathbf{Z}_H]\mathbf{V}^{-1}, \tag{7.40}$$

and that convergence (7.38) and (7.40) are joint. Since the matrices $[\mathbf{C}_0^{-1}\mathbf{C}_h, \mathbf{C}_h\mathbf{C}_0^{-1}]$ are asymptotically independent, it suffices to verify that

$$N\sum_{i,j=1}^{p} r_{f,h}(i,j)r_{b,h}(i,j) \stackrel{d}{\to} \chi^2_{p^2}. \tag{7.41}$$

To lighten the notation, in the remainder of the proof we suppress the index h (the limit distributions do not depend on h).

Denote by $\rho_f(i,j)$ and $\rho_b(i,j)$, respectively, the entries of matrices $\mathbf{V}^{-1}\mathbf{Z}$ and $\mathbf{Z}\mathbf{V}^{-1}$. By (7.38) and (7.40), it suffices to show that

$$\sum_{i,j=1}^{p} \rho_f(i,j)\rho_b(i,j) \stackrel{d}{=} \chi^2_{p^2}. \tag{7.42}$$

Denote by $\tilde{\mathbf{Z}}$ the column vector of length p^2 obtained by expanding the matrix $\mathbf{Z}$ row by row. Then the covariance matrix of $\tilde{\mathbf{Z}}$ is the $p^2 \times p^2$ matrix $\mathbf{V}\otimes\mathbf{V}$. By formula (23) on p. 600 of Anderson (1984), its inverse is $(\mathbf{V}\otimes\mathbf{V})^{-1} = \mathbf{V}^{-1}\otimes\mathbf{V}^{-1} = \mathbf{U}\otimes\mathbf{U}$. It thus follows from theorem 3.3.3 of Anderson (1984) that

$$\tilde{\mathbf{Z}}'(\mathbf{U}\otimes\mathbf{U})\tilde{\mathbf{Z}} \stackrel{d}{=} \chi^2_{p^2}. \tag{7.43}$$

It remains to show that the LHS of (7.42) is equal to the LHS of (7.43). The entry $Z(i,k)$ of the vector $\tilde{\mathbf{Z}}^T$ multiplies the row $u(i,\cdot)u(k,\cdot)$ of $\mathbf{U}\otimes\mathbf{U}$; the entry $Z(j,l)$ of $\tilde{\mathbf{Z}}$ multiplies the column $u(\cdot,j)u(\cdot,l)$. Consequently,

$$\begin{aligned}\tilde{\mathbf{Z}}'(\mathbf{U}\otimes\mathbf{U})\tilde{\mathbf{Z}} &= \sum_{i,j,k,l=1}^{p} u(i,j)u(k,l)Z(i,k)Z(j,l)\\ &= \sum_{i,l=1}^{p}\sum_{j=1}^{p} u(i,j)Z(j,l)\sum_{k=1}^{p} Z(i,k)u(k,l)\\ &= \sum_{i,l=1}^{p} \rho_f(i,l)\rho_b(i,l),\end{aligned}$$

completing the proof. □

Part II
The functional linear model

Chapter 8
Functional linear models

In this chapter we review some important ideas related to the functional linear model. Like its multivariate counterpart, this model has been developed in various directions, and has been found to be extremely useful in a broad range of applications. The relevant research is very rich and multifaceted, and we do not aim at a full review of the very extensive literature on this subject. Our objective in this chapter is to explain briefly the general ideas and point to some recent advances. Some additional references are given in Section 8.7. Our choice of topics is partially motivated by the the methodology presented in Chapters 9, 11 and 10. Practically all inferential tool for the functional linear model have been developed under the assumption that the regressor/response pairs, (X_i, Y_i), are independent. They must therefore be applied with care to functional data obtained over time or space.

8.1 Introduction

The linear regression is perhaps the most useful and widely used statistical model. The simplest linear model is the familiar straight line regression

$$Y_i = \beta_0 + \beta_1 x_i + \varepsilon_i, \quad i = 1, 2, \ldots, N,$$

in which all random variables are scalars, and the regressors x_i are typically assumed to be known scalars. In a functional linear model, some of these quantities are curves, and analogs of the coefficients β_0 and β_1 must be then appropriately defined.

To provide a motivating example, we start with a problem studied in Chiou *et al.* (2004) and based on an experiment reported in Carey *et al.* (2002) in which 1200 female medflies were fed one of 12 dietary doses ranging from full diet to 30% of full diet. For each medfly, the count of eggs laid every day was recorded, and so the egg-laying trajectories were obtained. Some of those are shown in Figure 8.1. As expected, the total count of eggs increases with the dietary dose, but a biological question of interest is whether this increase is due to a systematic increase at all

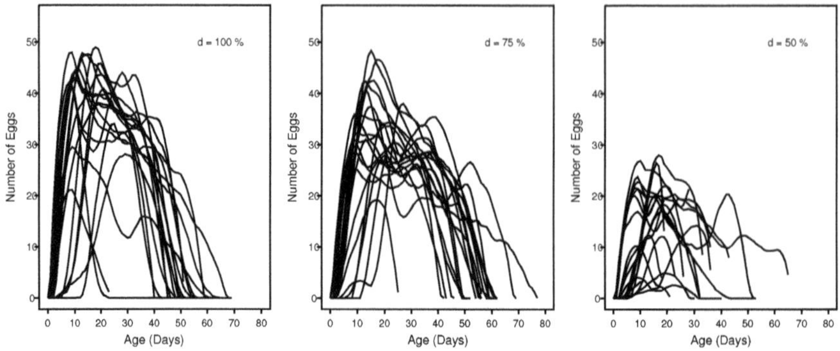

Fig. 8.1 Smoothed egg-laying trajectories of twenty randomly selected med flies at dose levels 100%, 75% and 50%. Source: Chiou *et al.* (2004).

ages, or whether the different diet levels lead to different patterns of egg-laying. For example, on a reacher diet, flies could start laying eggs earlier, and continue to lay them well into a mature age, or produce a lot more eggs at the prime reproductive age. To study this question, it is convenient to consider a linear model in which the dose levels are scalar regressors and the egg–laying curves are functional responses.

We can distinguish three cases, in which either the responses or the regressors, or both are curves. We assume for simplicity that the responses and the regressors have mean zero. In all formulations, we assume that the errors ε_i are independent of the explanatory variables (regressors) X_i.

The fully functional model:

$$Y_i(t) = \int \psi(t,s)X_i(s)ds + \varepsilon_i(t). \tag{8.1}$$

In this model, the responses Y_i are curves, and so are the regressors X_i. It is further studied in Section 8.3 and Chapters 9, 11, 10.

The scalar response model:

$$Y_i = \int \psi(s)X_i(s)ds + \varepsilon_i, \tag{8.2}$$

in which the regressors are curves, but the responses are scalars. The properties and extensions of this model are reviewed in Section 8.4.

The functional response model:

$$Y_i(t) = \psi(t)x_i + \varepsilon_i(t), \tag{8.3}$$

in which the responses are curves, but the regressors are known scalars. Extensions of this model are described in Section 8.5.

Models (8.1), (8.2) and (8.3) are just prototypes intended to illustrate the general idea. The main issue is that the functions ψ are infinite dimensional objects which must be estimated from a finite sample. Without any restrictions on ψ, a perfect fit is possible (all residuals are zero), and the resulting estimates $\hat{\psi}$ are erratic, noise type functions, which do not provide useable insights. We encountered a similar problem in Section 4.3. The parameter ψ is therefore often estimated by restricting the action of the corresponding operators to subspaces spanned by the EFPC's of the data. As we have seen in Chapter 3, the EFPC's summarize the main features of the data. This estimation approach thus removes a noise–like variability. Another approach is to impose a roughness penalty on the estimates, which has a similar effect of removing noise and producing interpretable estimates. Models (8.1), (8.2) and (8.3) have been modified in various directions, depending on applications at hand, and suitable estimation methods have been developed. Before we discuss some of these extensions, we first review in Section 8.2 the fundamental idea of the standard linear model.

8.2 Standard linear model and normal equations

The standard linear model, see e.g. Chapter 3 of Seber and Lee (2003), takes the form

$$\mathbf{Y} = \mathbf{X}\boldsymbol{\beta} + \boldsymbol{\varepsilon},$$

where

$\mathbf{Y}$ is the $N \times 1$ vector of responses;
$\mathbf{X}$ is the $N \times p$ regression matrix, typically assumed to be of rank p;
$\boldsymbol{\beta}$ is the $p \times 1$ parameter vector;
$\boldsymbol{\varepsilon}$ is the $N \times 1$ vector of mean zero errors.

The least squares estimator of $\boldsymbol{\beta}$ minimizes the Euclidean norm of the difference $Y - \mathbf{X}\boldsymbol{\beta}$. Set $\boldsymbol{\theta} = \mathbf{X}\boldsymbol{\beta}$, and denote by $\hat{\boldsymbol{\theta}}$ the projection of $\mathbf{Y}$ on the subspace $L_X \subset R^n$ spanned by the columns of $\mathbf{X}$. Thus $\hat{\boldsymbol{\theta}}$ is the unique vector minimizing the length of $\mathbf{Y} - \boldsymbol{\theta}$ over $\boldsymbol{\theta} \in L_X$. The vector $\mathbf{Y} - \hat{\boldsymbol{\theta}}$ is orthogonal to L_X, so $\mathbf{X}^T(\mathbf{Y} - \hat{\boldsymbol{\theta}}) = 0$, i.e. $\mathbf{X}^T\hat{\boldsymbol{\theta}} = \mathbf{X}^T\mathbf{Y}$. If $\mathbf{X}$ is of rank p, there is a unique $\hat{\boldsymbol{\beta}}$ such that $\hat{\boldsymbol{\theta}} = \mathbf{X}\hat{\boldsymbol{\beta}}$, in which case $\hat{\boldsymbol{\beta}}$ satisfies the *normal equations*

$$\mathbf{X}^T\mathbf{X}\boldsymbol{\beta} = \mathbf{X}^T\mathbf{Y}.$$

It can be shown that if $\mathbf{X}$ is of rank p, then $\mathbf{X}^T\mathbf{X}$ is nonsingular, and so the least squares estimator of $\boldsymbol{\beta}$ is given by

$$\hat{\boldsymbol{\beta}} = (\mathbf{X}^T\mathbf{X})^{-1}\mathbf{X}^T\mathbf{Y}.$$

We often write the standard model as

$$y_i = x_{i1}\beta_1 + x_{i2}\beta_2 + \dots x_{ip}\beta_p + \varepsilon_i, \quad i = 1, 2, \dots, N, \tag{8.4}$$

and we think of

$$y_i, \quad \mathbf{x}_i = [x_{i1}, x_{i2}, \dots, x_{ip}]^T, \quad \varepsilon_i$$

as realizations of the corresponding random variables

$$y,\ \ \mathbf{x} = [x_1, x_2, \ldots, x_p]^T,\ \ \varepsilon.$$

Then, the population model becomes

$$y = \mathbf{x}^T \boldsymbol{\beta} + \varepsilon. \tag{8.5}$$

8.3 The fully functional model

In Equation (8.1), the value $\psi(t,s)$ reflects the effect of the explanatory function X_i at time s on the response function Y_i at time t. To develop an estimation procedure analogous to the least squares estimation described in Section 8.2, and implemented in the R package `fda`, equation (8.1) is rewritten in the following form:

$$\mathbf{Y}(t) = \int \mathbf{X}(s)\beta(s,t)ds + \boldsymbol{\varepsilon}(t), \tag{8.6}$$

where

$$\beta(s,t) = \psi(t,s),$$

and where

$$\begin{aligned}\mathbf{Y}(t) &= [Y_1(t), Y_2(t), \ldots, Y_N(t)]^T;\\ \mathbf{X}(s) &= [X_1(s), X_2(s), \ldots, X_N(s)]^T;\\ \boldsymbol{\varepsilon}(t) &= [\varepsilon_1(t), \varepsilon_2(t), \ldots, \varepsilon_N(t)]^T.\end{aligned}$$

Suppose $\{\eta_k,\ k \geq 1\}$ and $\{\theta_\ell,\ \ell \geq 1\}$ are some bases, for example Fourier and spline, which need not be orthonormal. The functions η_k are suitable for expanding the functions X_i and the θ_i for expanding the Y_i. The idea of the estimation of the kernel $\beta(\cdot,\cdot)$ is to consider estimates of the form

$$\beta^*(s,t) = \sum_{k=1}^{K}\sum_{\ell=1}^{L} b_{k\ell}\eta_k(s)\theta_\ell(t),$$

in which K and L are relatively small numbers which are used as smoothing parameters; the smaller K and L, the smoother the estimate of $\beta(\cdot,\cdot)$. A least squares estimator is then obtained by finding $b_{k\ell}$ which minimize the residual sum of squares:

$$\sum_{i=1}^{N} \|Y_i - \int X_i(s)\beta^*(s,\cdot)\|^2$$

Consistency properties of this approach are not fully understood, but it gives useful estimates, which can be computed analogously to the standard vector case.

To provide a heuristic derivation of the normal equations, introduce the column vectors

$$\boldsymbol{\eta}(s) = [\eta_1(s), \eta_2(s), \ldots, \eta_K(s)]^T, \quad \boldsymbol{\theta}(t) = [\theta_1(t), \theta_2(t), \ldots, \theta_L(t)]^T$$

and the $K \times L$ matrix

$$\mathbf{B} = [b_{k,\ell},\ 1 \le k \le K,\ 1 \le \ell \le L].$$

In this notation, $\beta^*(s,t) = \boldsymbol{\eta}^T(s)\mathbf{B}\boldsymbol{\theta}(t)$, and inserting to (8.6), we obtain

$$\mathbf{Y}(t) = \int \mathbf{X}(s)\boldsymbol{\eta}^T(s)\mathbf{B}\boldsymbol{\theta}(t)ds + \boldsymbol{\varepsilon}(t).$$

Introducing the $N \times K$ matrix $\mathbf{Z}^*$ defined by

$$\mathbf{Z}^* = \int \mathbf{X}(s)\boldsymbol{\eta}^T(s)ds,$$

we obtain an approximate identity

$$\mathbf{Y}(t) = \mathbf{Z}^*\mathbf{B}\boldsymbol{\theta}(t) + \boldsymbol{\varepsilon}(t). \tag{8.7}$$

Next, introducing the $L \times L$ matrix

$$\mathbf{J} = \int \boldsymbol{\theta}(t)\boldsymbol{\theta}^T(t)dt,$$

we see that (8.7) implies that

$$\int \mathbf{Y}(t)\boldsymbol{\theta}^T(t)dt = \mathbf{Z}^*\mathbf{B}\mathbf{J} + \int \boldsymbol{\varepsilon}(t)\boldsymbol{\theta}^T(t)dt.$$

To obtain an analog of the normal equations of Section 8.2, multiply by $\mathbf{Z}^{*T}$ and ignore the error terms. This gives an approximate identity

$$\mathbf{Z}^{*T}\mathbf{Z}^*\mathbf{B}\mathbf{J} = \mathbf{Z}^{*T}\int \mathbf{Y}(t)\boldsymbol{\theta}^T(t)dt, \tag{8.8}$$

which we must solve for $\mathbf{B}$.

If $\mathbf{A}$ is a $p \times q$ matrix, we denote by vec($\mathbf{A}$) a column vector of length pq obtained by stacking the columns of $\mathbf{A}$ under each other starting from the left. One can then show that for any matrices $\mathbf{A}, \mathbf{X}, \mathbf{B}$ for which the product $\mathbf{AXB}$ is defined

$$\mathrm{vec}(\mathbf{AXB}) = (\mathbf{B}^T \otimes \mathbf{A})\mathrm{vec}(\mathbf{X}).$$

We can therefore rewrite (8.8) as

$$(\mathbf{J}^T \otimes [\mathbf{Z}^{*T}\mathbf{Z}^*])\mathrm{vec}(\mathbf{B}) = \mathrm{vec}\left(\mathbf{Z}^{*T}\int \mathbf{Y}(t)\boldsymbol{\theta}^T(t)dt\right)$$

If the matrices $\mathbf{J}$ and $\mathbf{Z}^{*T}\mathbf{Z}^*$ are nonsingular, a unique solution exists:

$$\mathrm{vec}(\mathbf{B}) = (\mathbf{J}^T \otimes [\mathbf{Z}^{*T}\mathbf{Z}^*])^{-1}\mathrm{vec}\left(\mathbf{Z}^{*T}\int \mathbf{Y}(t)\boldsymbol{\theta}^T(t)dt\right),$$

see Lemma 4.3.1. of Horn and Johnson (1991).

An alternative approach to the estimation of $\beta(\cdot,\cdot)$, discussed in Section 16.4.2 of Ramsay and Silverman (2005), is to allow large K and L, but to introduce a roughness penalty on the estimates. Asymptotic properties of this approach are known in the scalar response case discussed in Section 8.4.

Using EFPC's rather than fixed bases offers another approach to the estimation of $\psi(\cdot,\cdot)$. Methods of this type are based on Lemma 8.1. To formulate it, consider two L^2–valued mean zero random functions X and Y, and their expansions

$$X(s) = \sum_{i=1}^{\infty} \xi_i v_i(s), \quad Y(t) = \sum_{j=1}^{\infty} \zeta_j u_j(t), \tag{8.9}$$

where the v_j are the FPC's of X and the u_j the FPC's of Y, see Section 3.3, and

$$\xi_i = \langle X, v_i \rangle, \quad \zeta_j = \langle Y, u_j \rangle.$$

Lemma 8.1. *Suppose $X, Y, \varepsilon \in L^2$ are mean zero, ε is independent of X, and the following linear equation holds*

$$Y(t) = \int \psi(t,s)X(s)ds + \varepsilon(t), \tag{8.10}$$

with the kernel $\psi(\cdot,\cdot)$ satisfying

$$\iint \psi^2(t,s)dt\,ds < \infty. \tag{8.11}$$

Then

$$\psi(t,s) = \sum_{k=1}^{\infty}\sum_{\ell=1}^{\infty} \frac{E[\xi_\ell \zeta_k]}{E[\xi_\ell^2]} u_k(t) v_\ell(s),$$

where the convergence is in $L^2([0,1]\times[0,1])$.

Proof. Since $\{v_i\}$ and $\{u_j\}$ are bases in L^2, the functions $\{v_i(s)u_j(t),\ 0 \le s,t \le 1\}$ form a basis in $L^2([0,1]\times[0,1])$, so $\psi(t,s)$ admits a unique representation

$$\psi(t,s) = \sum_{k=1}^{\infty}\sum_{\ell=1}^{\infty} \psi_{k\ell} u_k(t) v_\ell(s), \tag{8.12}$$

and by (8.11) the coefficients $\psi_{k\ell}$ satisfy

$$\sum_{k=1}^{\infty}\sum_{\ell=1}^{\infty} \psi_{k\ell}^2 = \iint \psi^2(t,s)dt\,ds < \infty. \tag{8.13}$$

Inserting (8.9) and (8.12) into (8.10), and using the orthonormality of the v_i, we obtain

$$\sum_{j=1}^{\infty} \zeta_j u_j(t) = \sum_{k=1}^{\infty} \sum_{i=1}^{\infty} \psi_{ki} \xi_i u_k(t) + \varepsilon(t).$$

Multiplying by $u_\ell(t)$ and integrating, we further obtain

$$\zeta_\ell = \sum_{i=1}^{\infty} \psi_{\ell i} \xi_i + \langle u_\ell, \varepsilon \rangle . \tag{8.14}$$

Finally, multiplying by ξ_k, taking the expectation, and using the independence of X and ε, we arrive at

$$E[\zeta_\ell \xi_k] = \psi_{\ell k} E[\xi_k^2],$$

from which the claim follows. □

Remark 8.1. Observe that $E[\xi_\ell^2] = \lambda_\ell$, the eigenvalue corresponding to v_ℓ. The eigenfunctions v_ℓ belonging to zero eigenvalues can be omitted from representation (8.9) without changing it (it is an L^2 representation), so we may assume that $E[\xi_\ell^2] > 0$ for each $\ell \geq 1$.

Lemma 8.1 implies that if X and Y satisfy (8.10) with ψ satisfying (8.11), then

$$\sum_{k=1}^{\infty} \sum_{\ell=1}^{\infty} \frac{(E[\xi_\ell \zeta_k])^2}{\lambda_\ell^2} < \infty. \tag{8.15}$$

It can be conversely assumed that (8.10) and (8.15) hold, and then the implication that ψ satisfies (8.11) will follow. This is the approach adopted by Yao *et al.* (2005b).

Lemma 8.1 and Remark 8.1 suggest the following estimator:

$$\hat{\psi}_{KL}(t,s) = \sum_{k=1}^{K} \sum_{\ell=1}^{L} \hat{\lambda}_\ell^{-1} \hat{\sigma}_{\ell k} \hat{u}_k(t) \hat{v}_\ell(s),$$

where $\hat{\sigma}_{\ell k}$ is an estimator of $E[\xi_\ell \zeta_k]$. The simplest estimator is

$$\hat{\sigma}_{\ell k} = \frac{1}{N} \sum_{i=1}^{N} \langle X_i, \hat{v}_\ell \rangle \langle Y_i, \hat{u}_k \rangle . \tag{8.16}$$

It is clear that for this estimator, $\hat{\psi}_{KL}(t,s)$ does not depend on the signs of the $\hat{v}_\ell$ and $\hat{u}_k$.

If the curves $X_n, Y_n,\ n = 1, 2, \ldots, N$ are observed at sparse, irregular times, and are subject to measurement error, Yao *et al.* (2005b) propose the following procedure to calculate $\hat{\sigma}_{\ell k}$. In the notation of Section 4.2, observe that

$$c_{21}(t,s) = E[X(s)Y(t)] = \sum_{i=1}^{\infty} \sum_{j=1}^{\infty} E[\xi_i \zeta_j] v_i(s) u_j(t),$$

and so

$$E[\xi_\ell \zeta_k] = \iint c_{21}(t,s) v_\ell(s) u_k(t) ds dt.$$

The surface $c_{21}(\cdot,\cdot)$ is estimated by two–dimensional scatter plot smoothing, and the resulting estimate is denoted by $\hat{c}_{21}(\cdot,\cdot)$. We then set

$$\hat{\sigma}_{\ell k} = \iint \hat{c}_{21}(t,s) \hat{v}_\ell(s) \hat{u}_k(t) ds dt. \tag{8.17}$$

To establish the consistency of $\hat{\psi}_{KL}(t,s)$, we must assume that K and L are functions of the sample size N. Then, under regularity conditions,

$$\iint \left[\hat{\psi}_{KL}(t,s) - \psi(t,s)\right]^2 dt\, ds \xrightarrow{P} 0, \quad (K, L \to \infty),$$

see Theorem 1 of Yao *et al.* (2005b) for the details.

We conclude this section with a brief description of an extension of the functional linear model proposed by Müller *et al.* (2008). To explain the idea, note that by (8.14), $E[\zeta_\ell | X] = \sum_{i=1}^{\infty} \psi_{\ell i} \xi_i$. Therefore,

$$E[Y|X] = \sum_{\ell=1}^{\infty} E[\zeta_\ell | X] u_\ell = \sum_{\ell=1}^{\infty} \sum_{i=1}^{\infty} \psi_{\ell i} \xi_i u_\ell.$$

This means that the prediction of Y is a linear combination of the scores ξ_i. To obtain a greater modeling flexibility, Müller *et al.* (2008) propose merely an additive structure, i.e. postulate that

$$E[Y|X] = \sum_{\ell=1}^{\infty} \sum_{i=1}^{\infty} f_{\ell i}(\xi_i) u_\ell,$$

where the functions $f_{\ell i}$ are assumed to be smooth. Assuming that the scores ξ_i are independent, they develop a model fitting approach which is easy to implement.

8.4 The scalar response model

Model (8.2) can be estimated by a simplified version of the procedure described in Section 8.3, regularization with a roughness penalty is also often useful, see Chapter 15 of Ramsay and Silverman (2005) for examples of applications and further discussion.

An asymptotic theory for the estimation with roughness penalty was developed by Li and Hsing (2007). To explain the idea of their results, denote by $\{\phi_k,\ k = 1, 2, \ldots\}$ the (normalized) Fourier basis and by $g^{(m)}$ the mth derivative of a function

g on $[0,1]$. An estimator that involves both restricting the number of basis functions and a smoothness penalty is obtained by minimizing

$$\sum_{i=1}^{N}\left[Y_i - \int g^*(s)X_i(s)ds\right]^2 + \lambda \int \left[g^{*(m)}(s)\right]^2 ds.$$

An estimate of $\psi(s)$ is $g^*(s) = \sum_{k=1}^{K} b_k\phi_k(s)$. Using the orthogonality of the functions in the Fourier basis and their derivatives, we see that this reduces to finding $b_1, b_2, \ldots, b_K$ which minimize

$$\sum_{i=1}^{N}\left[Y_i - \sum_{k=1}^{K}\langle X_i, \phi_k\rangle b_k\right]^2 + \lambda \sum_{k=1}^{K} b_k^2 \int \left[\phi_k^{(m)}(s)\right]^2 ds.$$

We denote the resulting estimator by

$$\hat{\psi}_{K,\lambda}(s) = \sum_{k=1}^{K} \hat{b}_k\phi_k(s).$$

An asymptotic setting for the estimation with large K and the roughness penalty only, can be obtained informally by setting $K = \infty$, and formally by assuming that the potential estimates g are in a subspace of L^2 of sufficiently smooth and periodic functions. We therefore introduce the following definition:

Definition 8.1. The space $W_{2,\mathrm{per}}^m \subset L^2$ consists of m times differentiable functions, such that $g^{(m)} \in L^2$, and for $0 \le \nu \le m-1$, $g^{(\nu)}$ is absolutely continuous, and $g^{(\nu)}(0) = g^{(\nu)}(1)$.

The space $W_{2,\mathrm{per}}^m$ is an example of a Sobolev space, i.e. a space in which integrability conditions are imposed not only on functions but also on their derivatives. In order to develop a rigorous theory involving smoothing of functional data by a roughness penalty, it is necessary to work with such spaces. In this setting, the estimator $\hat{\psi}_{\infty,\lambda}$ is defined as the function $g \in W_{2,\mathrm{per}}^m$ which minimizes

$$\sum_{i=1}^{N}\left[Y_i - \int g(s)X_i(s)ds\right]^2 + \lambda \int \left[g^{(m)}(s)\right]^2 ds.$$

Li and Hsing (2007) show that if $m \ge 2$, then

$$E\|\hat{\psi}_{\infty,\lambda} - \psi\|^2 = O_P\left(N^{-1/2} + \lambda + N^{-1}\lambda^{-1/(2m)}\right),$$

provided the smoothing parameter λ tends to zero with N, but not too fast, see Theorem 5 of Li and Hsing (2007) for the details.

For a general basis $\{\phi_k\}$, e.g. a nonorthogonal spline basis, an estimator of the coefficient function $\psi(s)$ of the form $\sum_{k=1}^{K} b_k\phi_k(s)$ is obtained by minimizing

$$\sum_{i=1}^{N}\left[Y_i - \sum_{k=1}^{K}\langle X_i, \phi_k\rangle b_k\right]^2 + \lambda\left[\sum_{k=1}^{K} b_k \int \phi_k^{(m)}(s)ds\right]^2, \tag{8.18}$$

with $m = 2$ being the typical choice. In this approach, the emphasis is on choosing an appropriate smoothing parameter λ, while the number K of basis is assumed to be large.

An alternative approach regularizes the estimates of ψ by projecting the regressors onto the p leading EFPC's (those corresponding to the largest eigenvalues), i.e. by using the approximation $X_n \approx \sum_{i=1}^{p} \langle X_n, \hat{v}_i \rangle \hat{v}_i$, in which p is a small number. The coefficient function $\psi(s)$ is then estimated by $\sum_{i=1}^{p} \hat{\psi}_i \hat{v}_i(s)$, with the $\hat{\psi}_i$ being the values of the ψ_i which minimize

$$\sum_{n=1}^{N} \left[Y_n - \sum_{i=1}^{p} \langle X_n, \hat{v}_i \rangle \psi_i \right]^2. \tag{8.19}$$

Reiss and Ogden (2007) proposed several hybrid methods which combine the above two approaches, i.e. projecting onto the EFPC's of the regressors and using the roughness penalty. We describe only one of them, called FPCR_R by the authors, which appears to be most effective. The acronym FPCR stands for *Functional Principal Component Regression*, the subscript R indicated that a roughness penalty is applied to the *regression* rather than the components, the latter method being denoted FPCR_C. We focus on the general idea, the precise formulas and the computational details are given in Reiss and Ogden (2007).

As in (8.18), we seek coefficients b_k such that we can obtain a good and informative approximation

$$Y_i \approx \sum_{k=1}^{K} \tilde{X}_{ik} b_k, \quad \text{where } \tilde{X}_{ik} = \langle X_i, \phi_k \rangle .$$

This brings us to the framework of the standard linear model of Section 8.2. Denote by

$$\tilde{\mathbf{v}}_j = [\tilde{v}_{j1}, \tilde{v}_{j2}, \ldots, \tilde{v}_{jK}]^T, \quad 1 \le j \le K,$$

the multivariate principal components of the vectors

$$\tilde{\mathbf{X}}_i = [\tilde{X}_{i1}, \tilde{X}_{i2}, \ldots, \tilde{X}_{iN}]^T, \quad 1 \le i \le N.$$

The $\tilde{\mathbf{v}}_j$ are the normalized eigenvectors of the sample covariance matrix of the $\tilde{\mathbf{X}}_i$, they coincide with the vectors $\mathbf{u}_j$ of Theorem 3.1, see also Chapter 8 of Johnson and Wichern (2002) for further details. The coefficient vector $\mathbf{b} = [b_1, b_2, \ldots, b_K]^T$ is projected on the first p $\tilde{\mathbf{v}}_j$ (those corresponding to the largest eigenvalues), what yields

$$b_k \approx \sum_{i=1}^{p} \beta_j \tilde{v}_{jk}, \quad 1 \le k \le K.$$

The β_j are estimated by minimizing

$$\sum_{n=1}^{N} \left| Y_n - \sum_{j=1}^{p} \beta_j \sum_{k=1}^{k} \tilde{X}_{jk} \tilde{v}_{jk} \right|^2 + \lambda \left[\sum_{j=1}^{p} \beta_j \sum_{k=1}^{K} \tilde{v}_{jk} \int \phi_k^{(m)}(s) ds \right]^2.$$

Selection of p and λ is discussed in Reiss and Ogden (2007) and Reiss and Ogden (2009a). Denoting the resulting estimates by $\hat{\beta}_i$, the estimate of $\psi(s)$ is then

$$\hat{\psi}(s) = \sum_{k=1}^{K} b_k \phi_k(s) = \sum_{j=1}^{p} \hat{\beta}_j \sum_{k=1}^{K} \tilde{v}_{jk} \phi_k(s).$$

The FPCR$_\text{R}$ method seeks to attain a greater flexibility by including a much higher number of components; the number K of the $\tilde{\mathbf{v}}_j$ we start with, is much larger than the number p of the $\hat{v}_i$ in (8.19). This can be done without overfitting by the inclusion of a roughness penalty. The usual methods use either only deterministic spline functions, or only the EFPC's $\hat{v}_k$. Greater flexibility is often needed when the data are densely observed curves like magnetometer or financial data discussed in this book.

Motivated by applications to classification problems, Müller and Stadtmüller (2005) proposed the model

$$Y_i = g\left(\psi_0 + \int \psi(t) X_i(t) dt\right) + \varepsilon_i,$$

in which g is a link function. Returning to the med fly egg-laying curves introduced in Section 8.1, suppose the $X_i(t),\ 0 \le t \le 30$, are the egg–laying curves of 534 flies that lived for at least 30 days, and define

$$Y_i = \begin{cases} 1 \text{ if fly } i \text{ lived full 44 days} \\ 0 \text{ if fly } i \text{ lived less than 44 days,} \end{cases}$$

If $Y_i = 1$, we say that fly i is long–lived. The link function g may be estimated from the data, but Müller and Stadtmüller (2005) obtained almost equally good results with the usual logit link:

$$g(\eta) = \frac{e^\eta}{1 + e^\eta}.$$

Müller and Stadtmüller (2005) explain how to compute the estimates $\hat{\psi}_0$ and $\hat{\psi}$. These allow us to calculate

$$\hat{\eta}_i = \hat{\psi}_0 + \int \hat{\psi}(t) X_i(t) dt.$$

If $\hat{\eta}_i > 0$ $(g(\hat{\eta}_i) > 1/2)$, we classify fly i as long lived.

Li *et al.* (2010) extended the model of Müller and Stadtmüller (2005) to allow interactions between the functional regressors X_i and some additional covariates. To focus on the central idea, suppose that the link function g is an identity function, $g(x) = x$, and that there is only one additional scalar covariate z_i. In this case, the model of Li *et al.* (2010) becomes

$$Y_i = r(z_i) \int \psi(s) X_i(s) ds + \beta z_i + \varepsilon_i, \tag{8.20}$$

where $r(\cdot)$ is an unknown smooth function, and β is an unknown parameter. In this model, the impact of the regressors X_i on the responses Y_i is modified by the value of z_i in both the multiplicative, via $r(z_i)$, and the additive, via βz_i, manner.

To make this model identifiable, some additional conditions must be imposed. Notice that replacing r by ar and ψ by $a^{-1}\psi$, we obtain the same model for any $a \neq 0$. If we assume that $r \geq 0$, this lack of identifiability can be addressed by requiring that $\int \psi^2(s)ds = 1$.

Parameter estimation in model (8.20) is interesting, and we describe the general idea. Suppose that ψ admits the expansion $\psi = \sum_{j=1}^{p} \alpha_j e_j$, where the e_j are initial elements of a basis system. The condition $\int \psi^2(s)ds = 1$ is then equivalent to $\sum_{j=1}^{p} \alpha_j^2 = 1$. The parameters to be estimated are

$$\boldsymbol{\theta} = [\alpha_1, \alpha_2, \ldots, \alpha_p, \beta]^T \quad \text{and} \quad r(\cdot).$$

They are estimated by an iterative procedure. For a fixed $\boldsymbol{\theta}$, $r(\cdot)$ is estimated by local linear smoothing. For a fixed $r(\cdot)$, $\boldsymbol{\theta}$ is estimated by weighted least squares. These steps are repeated one after another until the differences in the estimates become negligible, i.e. until convergence is achieved. Local linear smoothing assumes that if z is close to z_k, then $r(z) \approx a_{0k} + a_{1k}(z - z_k)$. Thus, in a neighborhood of z_k, (8.20) becomes

$$Y_i = \{a_{0k} + a_{1k}(z_i - z_k)\} \sum_{j=1}^{p} \alpha_j \langle e_j, X_i \rangle + \beta z_i + \varepsilon_i.$$

To estimate a_{0k} and a_{1k}, we fix $\boldsymbol{\theta}$ (starting with a reasonable initial value) and minimize

$$R_k = \sum_{i=1}^{N} w_i(k) \left[Y_i - \{a_{0k} + a_{1k}(z_i - z_k)\} \sum_{j=1}^{p} \alpha_j \langle e_j, X_i \rangle + \beta z_i \right]^2, \qquad (8.21)$$

where the weights $w_i(k)$ decrease as $|z_i - z_k|$ increases. These weights are obtained as

$$w_i(k) = \left[\sum_{\ell=1}^{N} K\left(\frac{|z_\ell - z_k|}{h} \right) \right]^{-1} K\left(\frac{|z_i - z_k|}{h} \right),$$

where K is a kernel function and h is a smoothing bandwidth. Once the estimates $(\hat{a}_{0k}, \hat{a}_{1k})$, $k = 1, 2, \ldots, N$, have been obtained, we estimate $\boldsymbol{\theta}$ by minimizing $\sum_{k=1}^{N} \hat{R}_k$, where $\hat{T}_K$ is equal to R_k with a_{0k}, a_{1k} replaced by $\hat{a}_{0k}, \hat{a}_{1k}$. The resulting estimate $\hat{\boldsymbol{\theta}}$ is used to reestimate a_{0k}, a_{1k}, etc.

8.5 The functional response model

Model (8.3) is too simple for most applications. A useful extension is to consider more than one parameter functions, what leads to the specification

$$Y_i(t) = \sum_{j=1}^{L} x_{ij} \psi_j(t) + \varepsilon_i(t), \quad i = 1, 2, \dots N,$$

which, analogously to (8.6), can be written as $\mathbf{Y}(t) = \mathbf{X}\boldsymbol{\psi}(t) + \boldsymbol{\varepsilon}(t)$. The parameter $\boldsymbol{\psi}(t) = [\psi_1(t), \dots, \psi_L(t)]^T$ can be estimated by using a version of the procedure described in Section 8.3. Chapter 13 of Ramsay and Silverman (2005) contains two interesting applications of this model.

Chiou *et al.* (2004) propose a model in which the intercept function depends on the regressor. Their formulation is suitable for experiments in which multiple responses are available for every level of the explanatory variable, like the med fly data described in Section 8.1, where there are almost 100 responses for every diet level. To introduce that model set $\mu(t) = EY(t)$ and denote by $\theta(x)$ the value of θ which minimizes

$$\int \{E[Y(t)|X = x] - \mu(t)\theta\}^2 \, dt.$$

Direct verification shows that

$$\theta(x) = \left(\int \mu^2(t) dt \right)^{-1} \int \mu(t) E[Y(t)|X = x] dt.$$

If we assume that $Y_i(t) = \mu(t)\theta(x_i) + \varepsilon_i(t)$, we obtain the multiplicative model of Chiou *et al.* (2003) which can be easily estimated by using the sample analogs of the expectations occurring above and some smoothing. To improve the predictions of the functions Y_i, Chiou *et al.* (2004) propose the model

$$Y_i(t) = \mu(t)\theta(x_i) + \sum_{k=1}^{K} \alpha_k(x_i)\psi_k(x_i, t) + \varepsilon_i(t), \tag{8.22}$$

in which the $\psi_k(x, \cdot)$ are the FPC's of the functions $R(x, t) = Y_i(t) - \mu(t)\theta(x)$. To estimate this model, the $\psi_k(x_i, t)$ are estimated by the EFPC's of the residuals $\hat{R}(x, t) = Y_i(t) - \hat{\mu}(t)\hat{\theta}(x)$, and the link functions α_k by the general methods developed in Chiou and Müller (1998).

8.6 Evaluating the goodness–of–fit

In this section we discuss several diagnostic methods for functional regression models. We first review the relevant ideas in the standard setting of Section 8.2. Our objective is to verify if model (8.5) is appropriate.

An elementary approach is to plot the responses y_i against the regressors x_{ij} for $j = 1, 2, \ldots, p$. If model (8.5) is correct, all these scatter plots should approximately follow a line with some spread around it, and have roughly the shape of an ellipse.

There are many possible departures from the model (8.5), an in–depth study is presented in Chapter 10 of Seber and Lee (2003). Here we focus only on two important cases. By (8.5), the conditional expectation $E[y|\mathbf{x}] = \mathbf{x}^T \boldsymbol{\beta}$ is a linear function of $\mathbf{x}$. If, in fact $E[y|\mathbf{x}] = \mu(\mathbf{x})$ is not a linear function, then model (8.5) is not appropriate. Also if (8.5) holds, then $\text{Var}[y|\mathbf{x}] = \text{Var}[\varepsilon]$ is constant. If $\text{Var}[y|\mathbf{x}] = w(\mathbf{x})$, where $w(\cdot)$ is not a constant function, then model (8.5) is not appropriate either.

Focusing first on the conditional expectation $E[y|\mathbf{x}]$, suppose the data follow the model

$$y = \beta_1 g(x_1) + \beta_2 x_2 + \ldots \beta_p x_p + \varepsilon,$$

where $g(\cdot)$ is a nonlinear function. This relation can be rewritten as

$$y = \beta_1 x_1 + \beta_2 x_2 + \ldots \beta_p x_p + (\beta_1(g(x_1) - x_1) + \varepsilon),$$

i.e. as a linear regression in which the error terms have a mean which depends on the value of x_1 in a nonlinear manner. Estimating this regression by the least squares method, we obtain the fit

$$y_i = x_{i1}\hat{\beta}_1 + x_{i2}\hat{\beta}_2 + \ldots x_{ip}\hat{\beta}_p + \hat{e}_i.$$

For example, if the model is $y_i = \beta g(x_i) + \varepsilon_i$, but we think it is $y_i = \beta x_i + \varepsilon_i$, the least squares estimate is $\hat{\beta} = (\sum x_i^2)^{-1} \sum y_i x_i$. The residual then is

$$\hat{e}_i = y_i - \hat{\beta} g(x_i) = \varepsilon_i + (\beta - \hat{\beta}) g(x_i).$$

Thus, if $g(\cdot)$ is nonlinear, the plot of the $\hat{e}_i$ versus the x_i will reveal this nonlinearity.

It can be hoped that if y depends in a nonlinear manner on some coordinates x_j, $j = 1, 2, \ldots, p$, then this nonlinearity will be revealed by one of the plots of the $\hat{e}_i$ against the x_{ij}. If the ε_i in (8.4) do not have a constant variance, then the residulas $\hat{\varepsilon}_i = y_i - (x_{i1}\hat{\beta}_1 + x_{i2}\hat{\beta}_2 + \ldots x_{ip}\hat{\beta}_p)$ should exibit uneven spread with respect to some variable. If $\text{Var}[\varepsilon_i]$ depends in some manner on the $\mathbf{x}_i$, then the plots of the $\hat{\varepsilon}_i$ against the x_{ij} should reveal it.

If model (8.5) is correct, it is useful to check how well the data are described by it. A commonly used measure is the *coefficient of determination* defined as

$$\hat{R}^2 = \frac{\sum_{i=1}^N (\hat{y}_i - \bar{y}_N)^2}{\sum_{i=1}^N (y_i - \bar{y}_N)^2},$$

where $\bar{y}_N = N^{-1} \sum_{i=1}^N y_i$ and $\hat{y}_i = \mathbf{x}_i^T \hat{\boldsymbol{\beta}}$. It measures the proportion of the total sample variance of the responses explained by the model. The population coefficient of determination is defined as

$$R^2 = \frac{\text{Var}[E[y|\mathbf{x}]]}{\text{Var}[y]}. \tag{8.23}$$

We now discuss how these approaches can be adapted to functional linear models.

Scatter plot analysis. The informal graphical methods can be extended to the functional setting as follows. Consider, for example, the fully functional model (8.1). Then, by (8.14),

$$\zeta_\ell = \sum_{j=1}^{p} \psi_{\ell j} \xi_j + \eta_\ell(p), \tag{8.24}$$

where

$$\eta_\ell(p) = \sum_{j=p+1}^{\infty} \psi_{\ell j} \xi_j + \langle u_\ell, \varepsilon \rangle .$$

Equation (8.24) resembles (8.5) with the response ζ_ℓ and the regressors ξ_j, but the errors $\eta_\ell(p)$ are no longer independent of the regressors. Nevertheless, in light of (8.13), the sum $\sum_{j=p+1}^{\infty} \psi_{\ell j} \xi_j$ can be expected to be small, so we may hope that if model (8.1) is appropriate, then the scatter plots of the $\hat{\zeta}_{i\ell}$ against the $\hat{\xi}_{ij}$, $i = 1, 2, \ldots, N$, will approximatelly follow a line. Recall that

$$\hat{\zeta}_{i\ell} = \int Y_i(t) \hat{u}_\ell(t) dt, \quad \hat{\xi}_{ij} = \int X_i(s) \hat{v}_j(s) ds$$

are, respectively, the scores of the Y_i and the X_i in (8.1). When the dependence is not linear, these plots exhibit different patterns. For example, if

$$Y_i(t) = H_2(X_i(t)) + \varepsilon_i(t),$$

where $H_2(x) = x^2 - 1$, the scatterplot of the first FPC clearly shows a quadratic trend, see Figure 9.4. In applications, we consider only the first few values of ℓ and j, see Chiou and Müller (2007) for examples.

As in the standard regression model, one can also work with the residuals

$$\hat{\varepsilon}_i(t) = Y_i(t) - \int \hat{\psi}(t, s) X_i(s) ds, \quad i = 1, 2, \ldots, N,$$

where $\hat{\psi}(t, s)$ is an estimator of $\psi(t, s)$. In principle, any estimator described in Section 8.3 can be used. If model (8.1) is correct, the $\hat{\varepsilon}_i$ should be close to the ε_i, and so the scores of $\hat{\varepsilon}_i$ should be independent of the $\hat{\xi}_{ij}$. The scatter plots of the scores of the residuals $\hat{\varepsilon}_i$ against the $\hat{\xi}_{ij}$ should therefore exhibit no obvious patters.

Finally, one can also check the goodness-of-fit by plotting the scores of the residuals $\hat{\varepsilon}_i$ against the scores of fitted values $\hat{Y}_i(t) = \int \hat{\psi}(t, s) X_i(s) ds$. The points should lie in a horizontal band.

These methods can be easily modified for the models of Sections 8.4 and 8.5. Cook (1994) provides interesting insights into the interpretation of the scatter plots mentioned above in the standard regression setting.

Functional coefficient of determination. Using Lemma 8.1, it is not difficult to compute the pointwise functional population coefficient of determination, cf. (8.23), defined for model (8.1) by

$$R^2(t) = \frac{\mathrm{Var}[E[Y(t)|X]]}{\mathrm{Var}[Y(t)]}.$$

Since $\mathrm{Var}[E[Y(t)|X]] \leq \mathrm{Var}[Y(t)]$, $0 \leq R^2(t) \leq 1$.

Lemma 8.2. *If assumptions of Lemma 8.1 hold, then*

$$R^2(t) = \frac{\sum_{m=1}^{\infty}\sum_{k=1}^{\infty}\sum_{\ell=1}^{\infty} E[\xi_m\zeta_k]E[\xi_m\zeta_\ell]\lambda_m^{-1}u_k(t)u_\ell(t)}{\sum_{j=1}^{\infty}\gamma_j u_j^2(t)}. \tag{8.25}$$

Proof. By (8.9),

$$\begin{aligned}\mathrm{Var}[Y(t)] &= E\left[\left(\sum_{j=1}^{\infty}\zeta_j u_j(t)\right)^2\right] \\ &= \sum_{j=1}^{\infty}\sum_{j'=1}^{\infty} E[\zeta_j\zeta_{j'}]u_j(t)u_{j'}(t) \\ &= \sum_{j=1}^{\infty}\gamma_j u_j^2(t).\end{aligned}$$

Since $E[Y(t)|X] = \int \psi(t,s)X(s)ds$, we obtain

$$\begin{aligned}\mathrm{Var}[E[Y(t)|X]] &= E\left[\left(\int \psi(t,s)X(s)ds\right)^2\right] \\ &= E\iint \psi(t,s)\psi(t,s')X(s)X(s')dsds'.\end{aligned}$$

Thus, by Lemma 8.1,

$$\begin{aligned}&\mathrm{Var}[E[Y(t)|X]] \\ &= E\left\{\iint \sum_{k=1}^{\infty}\sum_{\ell=1}^{\infty}\frac{E[\xi_\ell\zeta_k]}{E[\xi_\ell^2]}u_k(t)v_\ell(s)\right. \\ &\qquad \left.\times \sum_{k'=1}^{\infty}\sum_{\ell'=1}^{\infty}\frac{E[\xi_{\ell'}\zeta_{k'}]}{E[\xi_{\ell'}^2]}u_{k'}(t)v_{\ell'}(s')X(s)X(s')dsds'\right\} \\ &= \sum_{k=1}^{\infty}\sum_{\ell=1}^{\infty}\frac{E[\xi_\ell\zeta_k]}{\lambda_\ell}u_k(t) \\ &\qquad \times \sum_{k'=1}^{\infty}\sum_{\ell'=1}^{\infty}\frac{E[\xi_{\ell'}\zeta_{k'}]}{\lambda_{\ell'}}u_{k'}(t)E\left\{\int v_\ell(s)X(s)ds\int v_{\ell'}(s')X(s')ds'\right\}.\end{aligned}$$

Observe that

$$\begin{aligned}E\left\{\int v_\ell(s)X(s)ds\int v_{\ell'}(s')X(s')ds'\right\} &= E\left[\langle v_\ell, X\rangle\langle v_{\ell'}, X\rangle\right] \\ &= \langle C(v_\ell), v_{\ell'}\rangle = \lambda_\ell\delta_{\ell\ell'}.\end{aligned}$$

Therefore

$$\mathrm{Var}[E[Y(t)|X]] = \sum_{\ell=1}^{\infty}\sum_{k=1}^{\infty}\sum_{k'=1}^{\infty} \frac{E[\xi_\ell \zeta_k]}{\lambda_\ell}\frac{E[\xi_\ell \zeta_{k'}]}{\lambda_{\ell'}}\lambda_\ell \, u_k(t)u_{k'}(t),$$

and so we obtain

$$\frac{\mathrm{Var}[E[Y(t)|X]]}{\mathrm{Var}[Y(t)]} = \frac{\sum_{\ell=1}^{\infty}\sum_{k=1}^{\infty}\sum_{k'=1}^{\infty} E[\xi_\ell \zeta_k]E[\xi_\ell \zeta_{k'}]\lambda_\ell^{-1} \, u_k(t)u_{k'}(t)}{\sum_{j=1}^{\infty}\gamma_j u_j^2(t)}.$$

Setting $m = \ell, \ell = k'$, we obtain (8.25). □

The coefficient $R^2(t)$ (8.25) quantifies the degree to which the functional linear model explains the variability of the response curves at a fixed point t. To define a global measure of the degree of linear association, we can either integrate $R^2(t)$ or integrate the numerator and the denominator separately, to obtain

$$\tilde{R}^2 = \int R^2(t)dt$$

and

$$R^2 = \frac{\int \mathrm{Var}[E[Y(t)|X]]dt}{\int \mathrm{Var}[Y(t)]dt} = \frac{\sum_{m=1}^{\infty}\sum_{k=1}^{\infty}(E[\xi_m\zeta_k])^2\lambda_m^{-1}}{\sum_{j=1}^{\infty}\gamma_j}.$$

A closed form formula for $\tilde{R}^2$ is not available. Both $\tilde{R}^2$ and R^2 are between 0 and 1. (If the function Y is defined on an interval $[a,b]$ rather than $[0,1]$, then we define $\tilde{R}^2 = (b-a)^{-1}\int_a^b R^2(t)dt$.)

Sample analogs of $R^2(t)$, $\tilde{R}^2$ and R^2 are defined by replacing the population eigenfunctions and eigenvalues by their sample counterparts, and truncating the infinite sums, for example,

$$\hat{R}^2 = \frac{\sum_{m=1}^{M}\sum_{k=1}^{K}\hat{\sigma}_{mk}^2\hat{\lambda}_m^{-1}}{\sum_{j=1}^{J}\hat{\gamma}_j}$$

where $\hat{\sigma}_{mk}$ is an estimator of $E[\xi_m\zeta_k]$, e.g. estimator (8.16).

An application of the coefficients $\tilde{R}^2$ and R^2 to clinical data is discussed in Section 5 of Yao *et al.* (2005b).

8.7 Bibliographical notes

The functional linear model is introduced in its various forms in Chapters 12–17 of Ramsay and Silverman (2005). Additional case studies are described in Chapters 8, 9 and 12 of Ramsay and Silverman (2002). Examples of R code are discussed in Chapters 9 and 10 of Ramsay *et al.* (2009), which also give a quick application oriented introduction.

An important application of the functional linear model is the prediction of the response Y given a new observation of the explanatory variable X. This can be done without postulating a linear relationship. Such nonparametric approaches are discussed in Chapters 5, 6 and 7 of Ferraty and Vieu (2006). The general idea is to nonparametrically estimate the a function r in a relation $Y_i \approx r(X_i)$. Such methods have been developed for scalar responses. Model (8.20) can be viewed as a hybrid nonparametric/linear model. The book of Shi and Choi (2011) studies Bayesian methods for Gaussian functional regression.

Another important application of FDA is in classification problems; an example is given in Section 1.4. In addition to a gene's temporal expression profile, other factors or covariates may be important. Motivated by such settings, Ma and Zhong (2008) consider what can be called a functional nonparametric mixed effect model of the form $Y_i(t) = \mu(X_i(t)) + \mathbf{Z}_i(t)\mathbf{b}_i + \varepsilon_i(t)$, where μ is a smooth function, $\mathbf{b}_i$ is a mean zero column random vector of dimension m with a covariance matrix $\mathbf{B}$, and $\mathbf{Z}_i(t) = [Z_{i1}(t), Z_{i2}(t), \ldots, Z_{im}(t)]$ is a design matrix. The estimation of μ is formulated using the notion of the reproducing kernel Hilbert space (RKHS) which is necessary to accommodate smoothness properties of the estimates, a point not addressed in this book. To explain briefly, note that smoothness connects the values of a function evaluated at neighboring points. In the space L^2, the value $x(t)$ at any given $t \in [0,1]$ is not relevant, and the functional $L^2 \ni x \mapsto x(t)$ is not continuous. It can be defined as a continuous functional on a smaller space of functions in L^2 with square integrable second derivatives, similar to the Sobolev space defined in Section 8.4. It is an example of a RKHS with a suitably defined inner product $\langle \cdot, \cdot \rangle_{\text{RKHS}}$. On that space, by Riesz' representation theorem, $x(t) = \langle x, R_t \rangle_{\text{RKHS}}$, for some element R_t of the RKHS. The value $R_t(s)$ of the function R_t at point $s \in [0,1]$ is typically denoted $R(t,s)$, and the function $R(\cdot, \cdot)$ is called the reproducing kernel. An interested reader is referred to Gu (2002).

A central issue for functional data is dimension reduction appropriate for a given problem. Li and Hsing (2010) assume a general model $Y_i = f(\langle \beta_1, X_i \rangle, \ldots, \langle \beta_K, X_i \rangle, \varepsilon_i)$, in which the responses Y_i are scalars, and the predictors X_i are functions; f is an arbitrary function and $\beta_1, \ldots, \beta_K$ are linearly independent functions. The functions f and β_k are unknown, and K is also unknown. The problem of interest is to test for and estimate the value of K, which is called the dimension of the effective dimension reduction space.

We now list several other references related to the issues discussed in this chapter. Cuevas *et al.* (2002) discuss the functional model in which the explanatory variables are fixed rather than random functions; we focus in this book on the latter case. Malfait and Ramsay (2003) emphasize that in many situations the general model (8.1) is inappropriate because the response $Y_i(t)$ can depend only on the values of $X_i(s)$ for $s \leq t$. McKeague and Sen (2010) study a scalar response impact point model $Y = \alpha + \beta X(\theta) + \varepsilon$ in which Y depends on the function X only through an unknown point $\theta \in (0, 1)$. In an application, θ corresponds to a gene location that impacts the response Y. Febrero-Bande *et al.* (2010) study the detection of influential data points in a functional model with scalar responses. Chiou *et al.* (2004) discuss functional response models and give interesting data examples. Cardot *et al.* (2003b) discuss estimation with splines, while Cardot *et al.* (2003c) present an interesting application to predicting land use from remote sensing data. Cai and Hall (2006) study theoretical foundations of prediction in the scalar linear model. Reiss and Ogden (2010) introduce a linear model with images as explanatory variables.

Chapter 9
Test for lack of effect in the functional linear model

In this chapter, we study the fully functional linear model (8.1) and test the nullity of the operator Ψ, i.e.

$$H_0 : \Psi = 0 \quad \text{versus} \quad H_A : \Psi \neq 0.$$

We thus test the null hypothesis that the curves X_n have no effect on the curves Y_n. This is analogous to testing $H_0 : \beta_1 = 0$ in the straight line regression, $y_i = \beta_0 + \beta_1 x_i + \varepsilon_i$. In the functional setting, the slope corresponds to a linear operator which transforms functions into functions. Just as in the case of straight line regression, the nullity of Ψ does not mean that there is no dependence between the curves X_n and Y_n, but that if there is a dependence, it cannot be described by a functional linear model.

The usual t–test for the slope of the regression line is equivalent to an F–test. The F–test is a standard tool for testing the significance of the coefficients in the scalar linear model $y_i = \beta_0 + \beta_1 x_{i,1} + \ldots \beta_{p-1} x_{i,p-1} + \varepsilon_i$. The F–test, valid for normal ε_i, is asymptotically equivalent to a χ^2–test, see e.g. Chapter 4 of Seber and Lee (2003). The test we propose is a χ^2–test in which projections on the EFPC's play the role of the regressors. We impose only moment conditions on the distribution of the regressor and error curves.

This chapter is organized as follows. In Section 9.1 we provide some background and motivation for the test procedure described in Section 9.2. Its finite sample performance is assessed in Section 9.3, followed by a detailed application to magnetometer data in Section 9.4. The asymptotic results and their extensions are stated in Section 9.5, with the proofs presented in Section 9.6.

9.1 Introduction and motivation

The test procedure described in this chapter was motivated by a question of space physics. The most important magnetospheric phenomenon observed at high latitudes, i.e. in the polar regions, are the substorms, which manifest themselves in

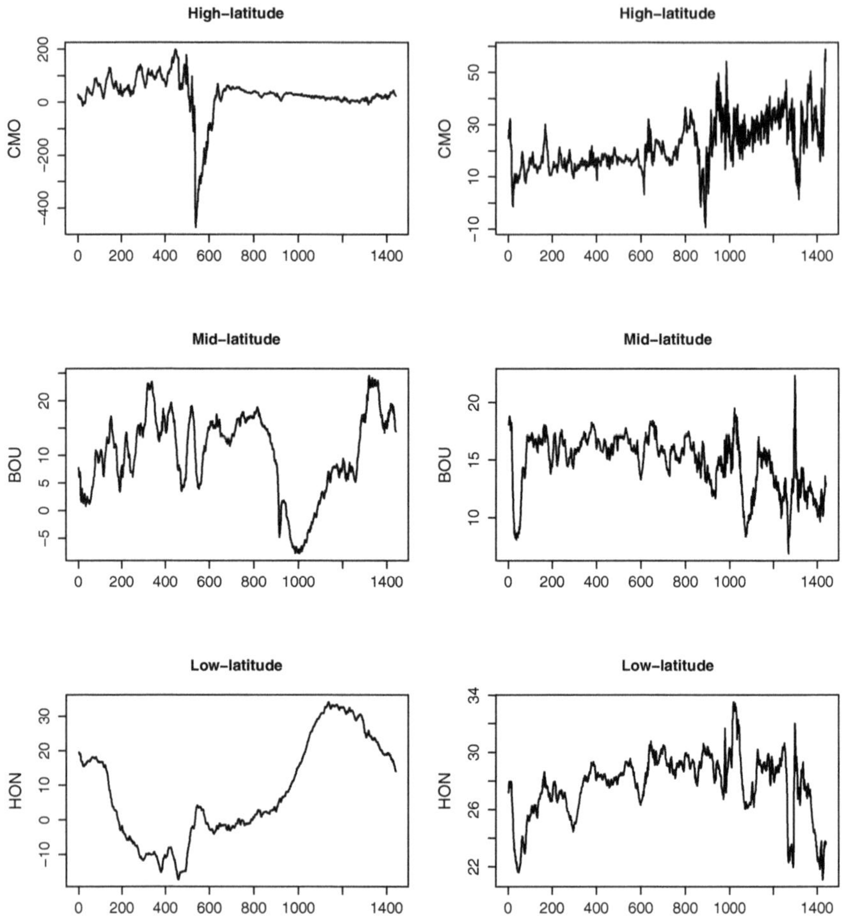

Fig. 9.1 Horizontal intensities of the magnetic field measured at a high-, mid- and low-latitude stations (College, Alaska; Boulder, Colorado; Honolulu, Hawaii) during a substorm (left column) and a quiet day (right column). The top left panel shows a typical signature of a substorm. Note the different vertical scales for high-latitude records. Each graph is a record over one day, which we view as a single functional observation.

a spectacular manner as the Northern Lights, the *aurora borealis*. There has been some debate if the currents flowing in mid and low magnetic latitudes are "correlated" with the substorms. All magnetospheric currents are observed indirectly through continuous records measured by terrestrial magnetometers. Examples of such records, cut into one day pieces, are shown in Figure 9.1. The left top panel shows a day with a substorm, the right top panel a day without a substorm. Comparing the bottom left and right panels, little difference can be found. Some difference, at least in the range, can be seen in the middle panels. There is thus a need for a quantitative statistical tool for testing if the substorms have any (linear) effect on lower latitude records. This problem is described in detail in Section 9.4.

Testing the null hypothesis of no effect exhibits new features in the functional setting due to the fact that the data are infinitely dimensional, and every dimension reduction technique restricts the domain of Ψ, and so leads to a loss of information about Ψ. These issues are addressed in different contexts in Cuevas *et al.* (2002) and Cardot *et al.* (2003). The testing procedure we propose is similar to that developed in Cardot *et al.* (2003) who consider scalar responses Y_n. It turns out that the more symmetric fully functional formulation actually leads to a somewhat simpler test statistic which can be readily computed using the principal components decompositions of the the Y_n and the X_n. Our test statistic has χ^2 limiting distribution which is a good approximation for sample sizes around 50. The research presented in this chapter is based on the papers Kokoszka *et al.* (2008), and Maslova *et al.* (2010b).

9.2 The test procedure

We assume that the response variables Y_n, the explanatory variables X_n and the errors ε_n are zero mean random elements of the Hilbert space L^2. Denoting by X (Y) a random function with the same distribution as each X_n (Y_n), we introduce the operators:

$$C(x) = E[\langle X, x\rangle X], \quad \Gamma(x) = E[\langle Y, x\rangle Y], \quad \Delta(x) = E[\langle X, x\rangle Y]$$

and denote their empirical counterparts by $\widehat{C}, \widehat{\Gamma}, \widehat{\Delta}$, e.g.

$$\widehat{C}(x) = \frac{1}{N}\sum_{n=1}^{N} \langle X_n, x\rangle X_n.$$

We define the eigenelements of C and Γ by

$$C(v_k) = \lambda_k v_k, \quad \Gamma(u_j) = \gamma_j u_j.$$

Empirical eigenelements are defined correspondingly and denoted by $(\hat{\lambda}_k, \hat{v}_k), (\hat{\gamma}_j, \hat{u}_j)$.

The testing procedure involves restrictions of the operators defined above to certain finite dimensional subspaces. This is a dimension reduction procedure which necessarily involves some loss of information about the action of Ψ. The subspace $\mathcal{V}_p = \mathrm{sp}\{v_1, \ldots, v_p\}$ contains the best approximations to the X_n which are linear combinations of the first p FPC's, see Section 3.2. Similarly, $\mathcal{U}_q = \mathrm{sp}\{u_1, \ldots, u_q\}$

is a good approximation to $\mathrm{sp}\{Y_1, \ldots, Y_n\}$. Since, by (8.1), $\Delta = \Psi C$, we have, for $k \leq p$,

$$\Psi(v_k) = \lambda_k^{-1} \Delta(v_k). \tag{9.1}$$

Thus, Ψ vanishes on $\mathrm{sp}\{v_1, \ldots, v_p\}$ if and only if $\Delta(v_k) = 0$ for each $k = 1, \ldots, p$. (We postulate in Assumption 9.2 that $\lambda_k > 0$.) Observe that

$$\Delta(v_k) \approx \widehat{\Delta}(v_k) = \frac{1}{N} \sum_{n=1}^{N} \langle X_n, v_k \rangle Y_n.$$

Since $\mathrm{sp}\{Y_1, \ldots, Y_N\}$ is well approximated by $\mathcal{U}_q$, a test can be developed by checking if

$$\left\langle \widehat{\Delta}(v_k), u_j \right\rangle = 0, \quad k = 1, \ldots, p, \; j = 1, \ldots, q. \tag{9.2}$$

If such a test accepts H_0, it means that for every $x \in \mathcal{V}_p$, $\Psi(x)$ is not in $\mathcal{U}_q$. Intuitively, it means that up to a small error arising from the approximations by the principal components and a random error, no function Y_n, $n = 1, 2, \ldots, N$, can be expressed as a linear combination of functions X_n, $n = 1, 2, \ldots, N$.

A test statistic should thus involve squares of the inner products in (9.2). Theorem 9.1 states that the statistic

$$\hat{T}_N(p, q) = N \sum_{k=1}^{p} \sum_{j=1}^{q} \hat{\lambda}_k^{-1} \hat{\gamma}_j^{-1} \left\langle \widehat{\Delta}(\hat{v}_k), \hat{u}_j \right\rangle^2 \tag{9.3}$$

converges in distribution to a chi–squared distribution with pq degrees of freedom. Since $\lambda_k = E \langle X, v_k \rangle^2$ and $\gamma_j = E \left\langle Y, u_j \right\rangle^2$, the statistics $\hat{T}_N(p, q)$ is essentially a normalized sum of squared correlations.

If H_0 fails, then $\Psi(v_k) \neq 0$ for some $k \geq 1$. If we impose conditions only on the first p largest eigenvalues, the test will be consitent only if Ψ does not vanish on one of the v_k, $k = 1, 2, \ldots, p$. The test has no power if Ψ does not vanish on the orthogonal complement of $\mathrm{sp}\{v_1, \ldots, v_p\}$. Further, to ensure consistency, one of the v_k, $k = 1, 2, \ldots, p$ must be mapped into $\mathrm{sp}\{u_1, \ldots, u_q\}$. These restrictions are intuitively appealing because we want to test if the main sources of the variability of the responses Y can be explained by the main sources of the variability of the explanatory variables X. These ideas are formalized in Theorem 9.2 which establishes the consistency of the test.

In linear regression setting, it is often of interest to test if specific covariates have no effect on the responses. In our setting, we could ask if specific FPC's v_k have no effect. It is easy to see from the proof of Theorem 9.1, see Lemma 9.1 in particular, that if we want to test if FPC's $v_{i(1)}, \ldots, v_{i(p')}$ have no effect, we must modify the statistic (9.3) by including only these components. The limit χ^2 distribution will then have $p'q$ degrees of freedom. A further obvious modification can be made if we want to check if there is an effect in the subspace spanned by some FPC's of the responses Y_k. Modifications of this type are useful if some principal components have obvious interpretations. This is sometimes the case in space physics applications, see Xu and Kamide (2004), but in the case of when the X_n are high–latitude records, see Section 9.4, the v_k cannot, at this point, be readily interpreted.

We now present an algorithmic summary of the testing procedure, some aspects of which are elaborated on in Section 9.4.

Summary of the testing procedure. 1. Check the linearity assumption using FPC score predictor-response plots, see Section 9.4.

2. Select the number of important PC's, p and q using both the scree test and CPV, see Section 9.4

3. Compute the test statistics $\hat{T}_N(p,q)$ (9.3). Note that

$$\langle \widehat{\Delta}(\hat{v}_k), \hat{u}_j \rangle = \left\langle \frac{1}{N} \sum_{n=1}^{N} \langle X_n, \hat{v}_k \rangle Y_n, \hat{u}_j \right\rangle = \frac{1}{N} \sum_{n=1}^{N} \langle X_n, \hat{v}_k \rangle \langle Y_n, \hat{u}_j \rangle,$$

where $\langle X_n, \hat{v}_k \rangle$ is the kth score of the X_n, and $\langle Y_n, \hat{u}_j \rangle$ is jth score of the Y_n. These scores and the eigenvalues $\hat{\lambda}_k$ and $\hat{\gamma}_j$ are output of functions available in the R package `fda`.

4. If $\hat{T}_N(p,q) > \chi^2_{pq}(\alpha)$, reject the null hypothesis of no linear effect. The critical value $\chi^2_{pq}(\alpha)$ is the $(1-\alpha)$th quantile of the chi-squared distribution with pq degrees of freedom.

9.3 A small simulation study

In this section, we present the results of a small simulation study intended to evaluate the empirical size and power of the test in standard Gaussian settings.

We used $R = 1000$ replications of samples of processes ε_n, X_n and Y_n, $n = 1, 2, \ldots, N$. In order to evaluate the empirical size, we generated samples of pairs (ε_n, Y_n) with independent components. To find the empirical power, we generated samples of pairs (ε_n, X_n) with independent components, and calculated Y_n according to (8.1). As ε_n, X_n and Y_n, we used Brownian bridge and motion processes in various combinations. The computations were performed using the R package `fda`. We used both Fourier and spline bases.

Since the Brownian bridge and motion have very regular Karhunen-Loève decompositions, see e.g. Bosq (2000), p. 26, it is not surprising that the size and power of the test do not depend appreciably on p and q. Figures 9.2 and 9.3 illustrate this point. The horizontal axes represent various combinations of p and q; 1 stands for $p = 1$ and $q = 1$; 2 for $p = 1$, $q = 2$; 3 for $p = 1$, $q = 3$, etc. All combinations of $p \leq 4$, $q \leq 4$ were considered in the size study and $p \leq 6, q \leq 6$ in the power study. The results for Brownian bridges and motions and Fourier and spline bases are practically the same. For this reason, we present the results only in cases when all processes are Brownian bridges, and the analysis was performed with the Fourier basis.

Naturally, the bigger the sample size the closer the empirical size of the test is to the nominal size. Nevertheless, there is little or no improvement in the size of the test starting from $N = 40 - 80$; these values can therefore be considered sufficient to obtain reasonable size; with $N = 40$ the test being slightly conservative.

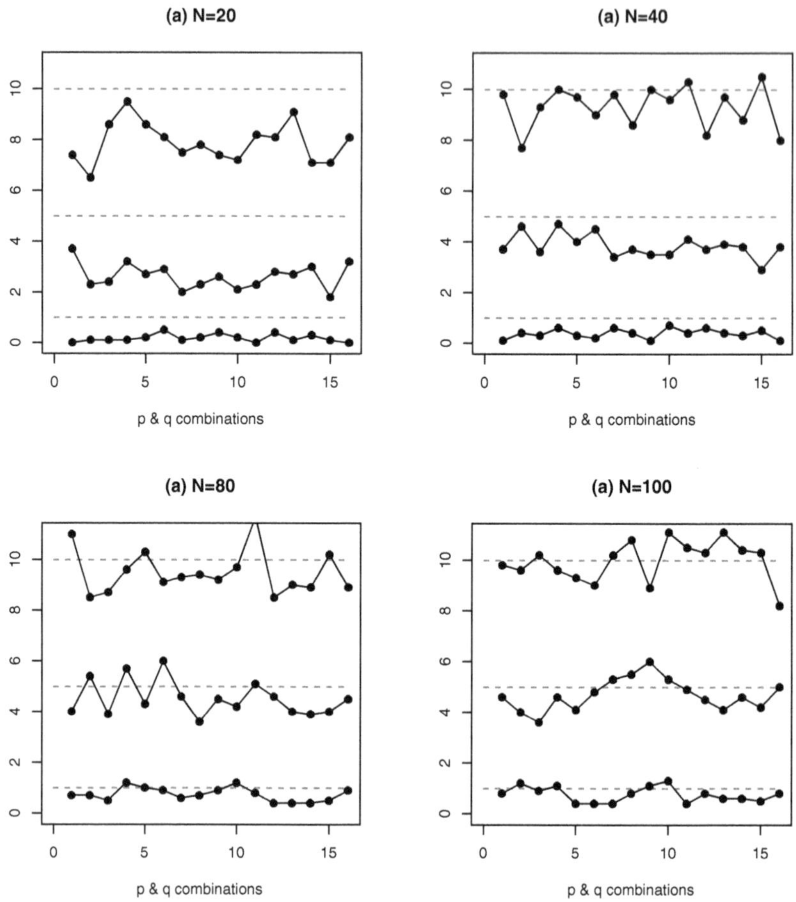

Fig. 9.2 Empirical size of the test for $\alpha = 1\%, 5\%, 10\%$ (indicated by dotted lines) for different combinations of p and q. Here ε_n and Y_n, $n = 1, 2, \ldots, N$ are two independent Brownian Bridges.

To evaluate the empirical power, we used the Gaussian kernel

$$\psi(s,t) = C \exp\left\{(t^2 + s^2)/2\right\}, \quad t \in [0,1],\ s \in [0,1] \tag{9.4}$$

with constants C such that $\|\Psi\| < 1$, i.e. $|C| < 1$ (the norm in this section is the Hilbert–Schmidt norm). Panels (a) and (b) of Figure 9.3 present power when the dependence between X_n and Y_n is quite strong, $\|\Psi\| = 0.75$. For $N = 80$, the power is practically 100% if $\|\Psi\| = 0.75$. The right column of Figure 9.3 shows the power of the test when $\|\Psi\| = 0.5$. In this case power increases slower with N.

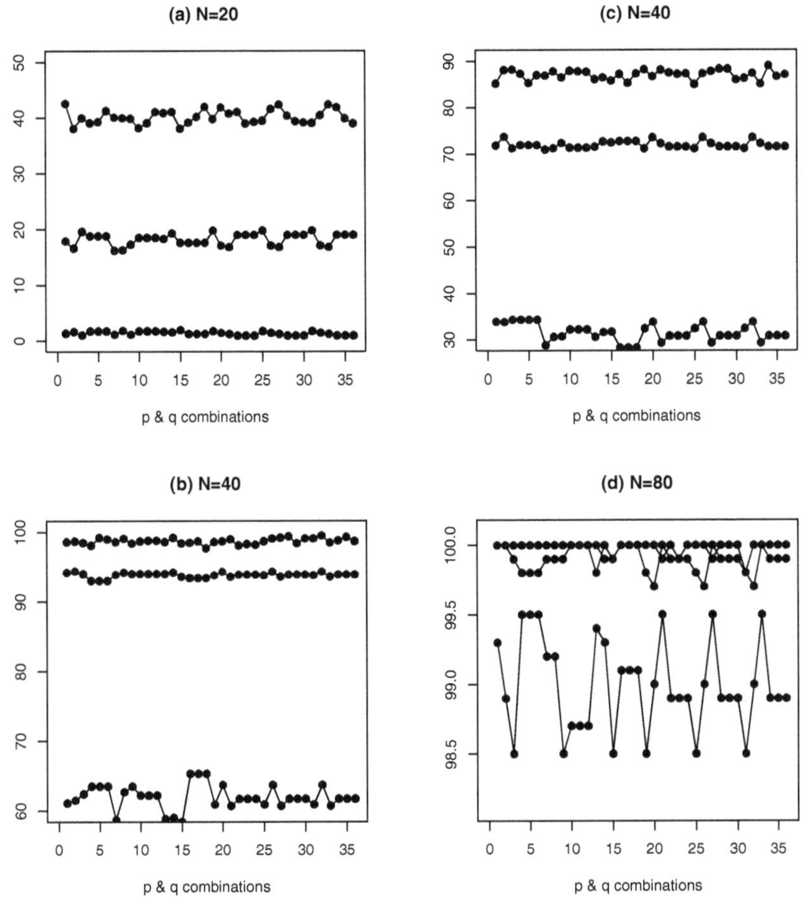

Fig. 9.3 Empirical power of the test for different combinations of principal components and different sample sizes N. Here X_n and ε_n are Brownian Bridges. In panels (a), (b) $\|\Psi\| = 0.75$; in panels (c), (d) $\|\Psi\| = 0.5$.

9.4 Application to magnetometer data

About a hundred terrestrial geomagnetic observatories form a network, INTERMAGNET, designed to monitor the magnetic fields generated by electrical currents flowing in the magnetosphere and ionosphere (M-I). Modern digital magnetometers record three components of the magnetic field in five second resolution, but the data made available by INTERMAGNET (http://www.intermagnet.org) consist of one minute averages (1440 data points per day per component per observatory). Figure 9.1 shows examples of magnetometer records. We work with the Horizontal (H) component of the magnetic field. This is the component lying in the Earth's tangent plane and pointing toward the magnetic North. It most directly reflects the variation of the M–I currents we wish to study. The M–I currents form a complex interactive system which at present is only partially understood, see Kamide *et al.* (1998) and Daglis *et al.* (2003). The magnetometer records contain intertwined signatures of many currents, and an effort has been under way to deconvolute the signatures of various currents. So far this has been done by preprocessing records from every individual station, and then combining the filtered signals from stations at the same magnetic latitude (e.g. equatorial stations, or auroral stations), see Jach *et al.* (2006) for a recent example of such an approach. Better understanding of the M–I system can however be obtained only by modeling interactions between the various currents.

It is believed, see e.g. Rostoker (2000), that the auroral currents may have an indirect impact on the equatorial and mid-latitude currents. The question of interest is whether the auroral geomagnetic activity reflected in the high–latitude curves has an effect on the processes in the equatorial belt reflected by the mid– and low–latitude curves. This question is of particular interest for days during which a high–latitude activity known as a substorm occurs. Its most spectacular manifestation are the Norther Lights caused by high–energy electrojets flowing for a few hours in the auroral belt. The top left panel of Figure 9.1 shows a signature of a substorm. It is believed that there is energy transfer between the auroral electrojets and lower latitude currents, but the direct physical mechanisms which might be responsible for this interaction are a matter of debate. The question can be cast into the setting of the functional linear model (8.1) in which the X_n are centered high–latitude records and Y_n are centered mid– or low–latitude records. This postulates an approximate statistical model for the data and allows us to the the null hypothesis $\Psi = 0$. If the null is true, we conclude that the high–latitude curves X_n have no linear effect on the lower latitude curves. If the null is rejected, this indicates the existence of an effect, which can be approximately linear (in the functional sense).

Detailed description of the data. We analyze one-minute averages of the horizontal intensity of the magnetic field from four sets of stations given in Table 9.1. Only one high–latitude station is used because substorms last for a few hours at night local time, and we want to study their effect as the longitudal distance increases. The mid– and low–latitude observatories are roughly aligned along the same longitude. The

Table 9.1 Geomagnetic observatories used in this study.

Latitude	I	II	III	IV
High	College (CMO)	–	–	–
Mid	Boulder (BOU)	Fredericksburg (FRD)	Tihany (THY)	Memambetsu (MMB)
Low	Honolulu (HON)	San Juan (SJG)	Hermanus (HER)	Kakioka (KAK)

functional data consists of daily curves in UT (Universal Time), with 1440 observations per curve. Figure 9.1 provides examples of such curves.

Several types of data sets are analyzed. The first set consists of all days with substorms from January until August, 2001 (set A). Then, the same analysis is performed on the so called *medium strength* substorms (defined by the dynamic range of 400-700 nT) during the same period (set B). Substorms often occur during much larger disturbances known as geomagnetic storms. In order to eliminate possible confounding effects of storms, we removed all days n such that a storm was taking place on days $n-1$, n, or $n+1$ (set A*). We also removed such days from the list of medium strength substorms (set B*). To eliminate the possibility of confounding by next day's storm, we also considered only isolated substorm days, i.e. substorm days followed by at least two days without any substorms (set I). Finally, to provide an additional validation of our findings, we select the substorms that took place during three month periods: January – March (A1), March – May (A2), and June – August (A3). The main reason of performing a separate analysis on medium strength substorms is that very strong substorms can be viewed as outliers and may distort the overall pattern of dependence. They are also typically generated by a different physical mechanism than medium strength substorms: the strong substorms are connected to the instability associated with the release of energy stored in the magnetosphere, and the medium substorms are associated with the direct pushing of the enhanced solar wind. Comparing the substorms over three month periods guards against the violation of the assumption of iid observations. Due to the annual rotation of the Earth, the locations of the stations relative to the Sun change over time. Hence, the substorms that happened long time apart might follow different statistical distributions. There were 101 substorm days from January until August during 2001, 81 substorm of which did not have any storms around; 41 substorms were medium strength, 35 medium strength substorms after removing the ones close to the storms; 43 isolated substorms occurred during 2001. We observed 40 substorm days from January until March, 42 – from March until May, and 42 – from June until August.

Details of test application and interpretation. In order to perform the test, the minute-by-minute data were converted into functional objects in R using B–spline basis with 149 basis functions. The number of basis functions is not crucial, the only requirement being that the smoothed curves should look almost identical to the original, while some noise can be eliminated.

In order to ensure that the test gives reliable results, the approximate validity of the functional linear model must be checked. For this purpose, a technique introduced by Chiou and Müller (2007), which relies on a visual examination of scatter plots of scores, can be used. If the model is valid, score plots are roughly football–shaped. When the dependence is not linear, these plots exhibit different patterns. The number of plots is pq, where p and q are as in Section 9.2. They show the interaction of the kth PC of the X_n ($k = 1, \ldots, p$) and jth PC of the Y_n ($j = 1, \ldots, q$). To illustrate this technique, consider a non-linear model: $Y_n(t) = H_2(X_n(t)) + \varepsilon_n(t)$, where $H_2(x) = x^2 - 1$ is the Hermite polynomial of rank 2. For this model, the

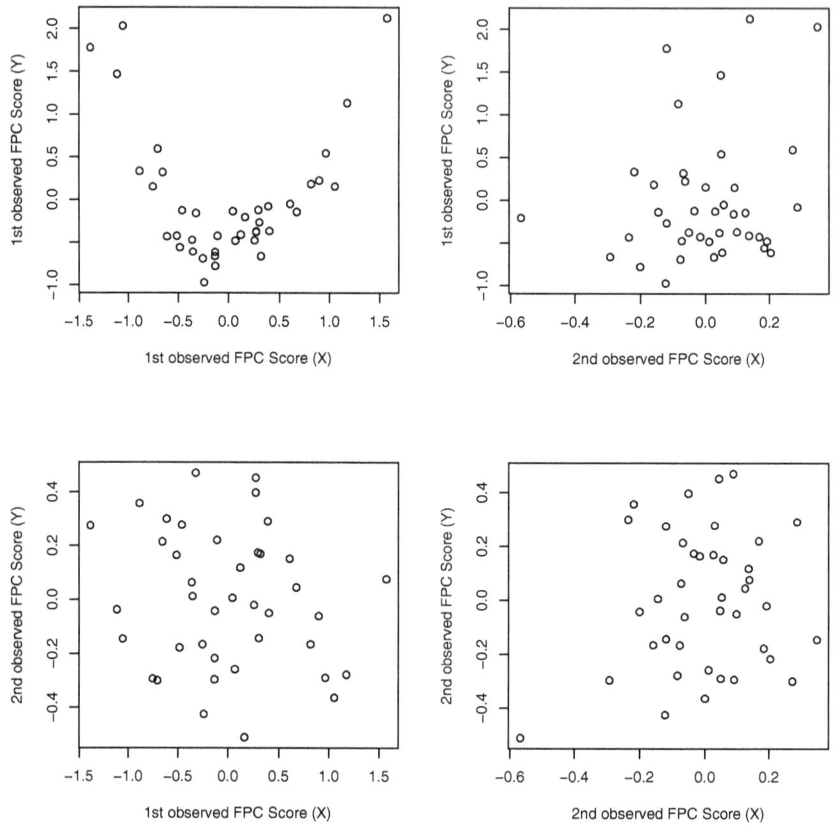

Fig. 9.4 Functional predictor-response plots of FPC scores of response functions versus FPC scores of predictor functions for $Y_n(t) = H_2(X_n(t)) + \varepsilon_n(t)$, where $H_2(x) = x^2 - 1$, $n = 1, \ldots, 40$.

plot in the top left corner of Figure 9.4 exhibits a quadratic trend. For the functional linear model to be valid all these plots should be "pattern–free". Figure 9.5 shows examples of these plots for magnetometer data. We used CMO medium strength substorm records as X, and THY with no lag – as Y. These scatter plots indicate linear relationship with some outliers. Since we do not require Gaussianity, only finite fourth moment, these outliers need not invalidate our conclusions. In case of other pairs of functional data, the score plots look similar. We conclude that a linear model is approximately appropriate for our application.

We now describe how to choose the most important FPC's that will be used in the test. One of the ways to pick them is to use the scree test, which is a graphical method first proposed by Cattell (1966). To apply the scree method one plots the

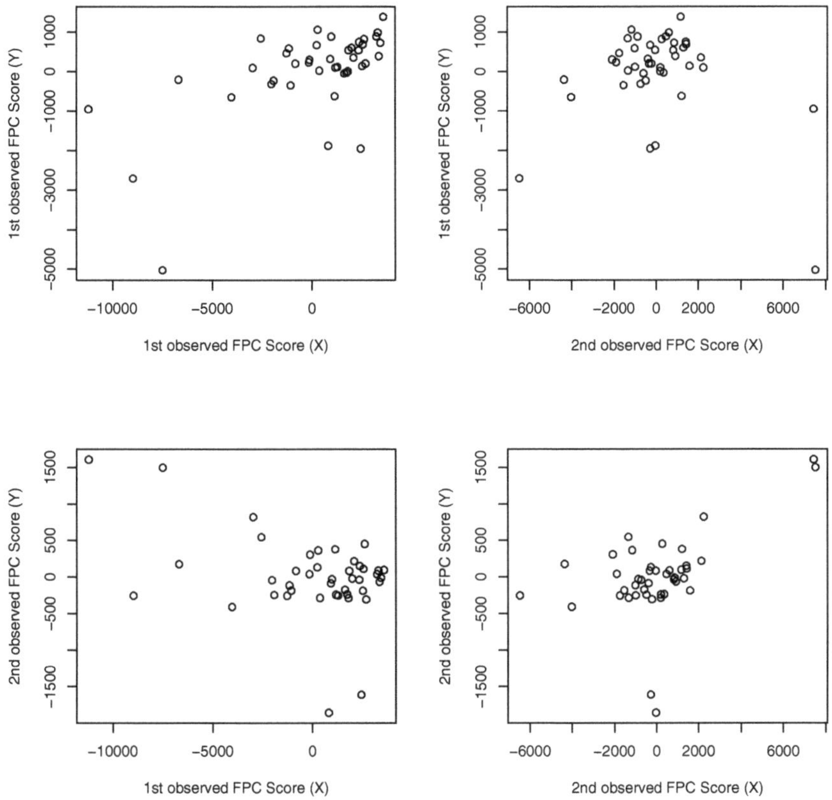

Fig. 9.5 Functional predictor-response plots of FPC scores of response functions versus FPC of explanatory functions for magnetometer data (CMO vs THY0)

successive eigenvalues, see Figure 9.6, and find the place where the smooth decrease of eigenvalues appears to level off. To the right of this point one finds only "factorial scree" ("scree" is a geological term referring to the debris which collects on the lower part of a rocky slope). Table 9.2 provides the number of most important principal components and corresponding percentage of total variability explained by them for all substorms that occurred from January until August, 2001. For other data sets under consideration the general pattern is similar. One can also see from Figure 9.7 that each subsequent component picks up variation that declines in smoothness. For example, the 10th principal components resemble random noise and explain a small percentage of variability, that is why they are not included in the analysis.

When applying the test to magnetometer data, in most cases there is a clear rejection or acceptance for all combinations of the most important principal components. In those cases, we can either reject "1" or fail to reject "0" the null hypothesis

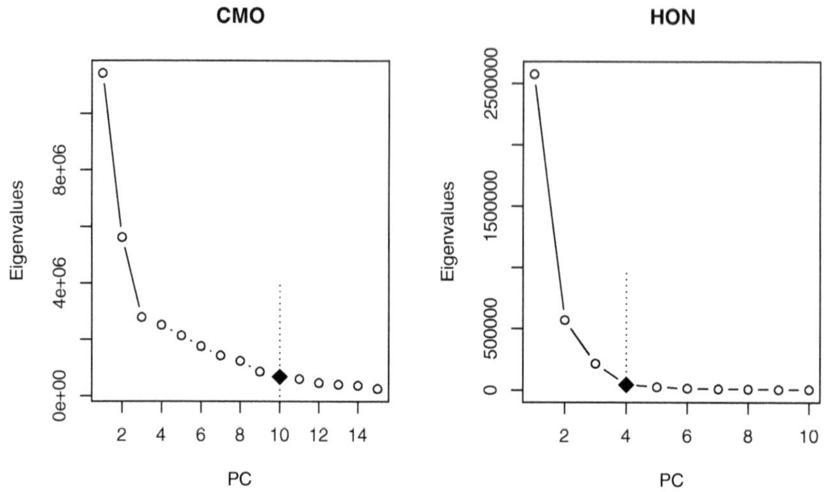

Fig. 9.6 Eigenvalues for different principal components of the substorm days that occurred from March until May, 2001, from College(CMO), Honolulu (HON) stations. The black diamond denotes the number of most important principal components selected by the scree test.

Table 9.2 Number of principal components retained by the scree test, and percentage of total variability explained, during substorm days that occurred from January until August, 2001.

Stations	PC	%	Stations	PC	%	Stations	PC	%	Stations	PC	%
CMO	10	82.52									
BOU0	5	91.36	FRD0	4	90.83	THY0	5	92.17	MMB0	4	92.30
BOU1	4	86.40	FRD1	4	89.55	THY1	5	89.49	MMB1	4	91.01
BOU2	4	91.17	FRD2	4	92.32	THY2	4	91.57	MMB2	4	94.59
BOU3	4	91.74	FRD3	4	92.68	THY3	4	91.51	MMB3	4	95.60
HON0	4	96.56	SJG0	5	97.08	HER0	4	95.07	KAK0	4	94.33
HON1	3	94.91	SJG1	4	94.57	HER1	4	94.31	KAK1	4	93.80
HON2	4	97.44	SJG2	3	92.73	HER2	4	95.89	KAK2	4	96.39
HON3	4	97.79	SJG3	4	96.42	HER3	4	95.53	KAK3	3	94.66

with a reasonable confidence. We use the nominal 95% confidence level in this Section. However, there are some cases when it is not clear what conclusion to draw. We denote such cases "1?" – inclined toward rejecting the null hypothesis, "0?"– inclined toward failing to reject the null, "1?0?"– inconclusive. Figure 9.8 gives examples of such cases. We plot rejection regions up to the number of important principal components. Grey areas mean that we reject H_0, white – fail to reject H_0. The conclusion is clear when all, or almost all, rectangles are of the same color. We can then conclude that X has an effect on Y (all grey) or there is no effect (all white). Left panel of Figure 9.8 gives an example when it is not clear what to conclude. However, based on our previous experience we are most likely to reject the

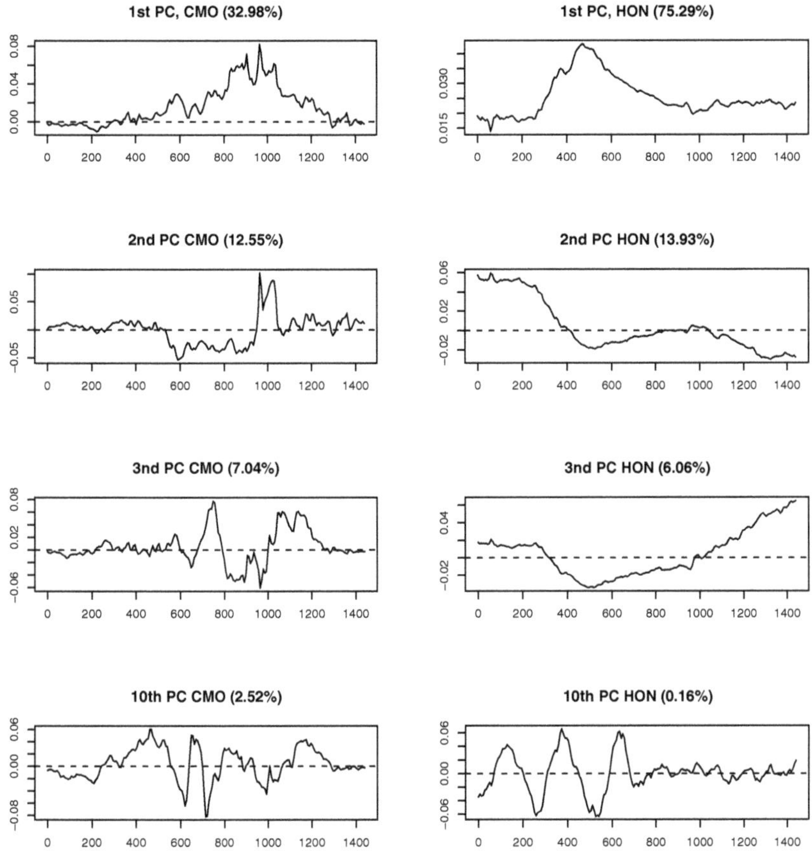

Fig. 9.7 EFPC's of the substorm days that occurred from January until August, 2001, from College(CMO) and Honolulu (HON) stations.

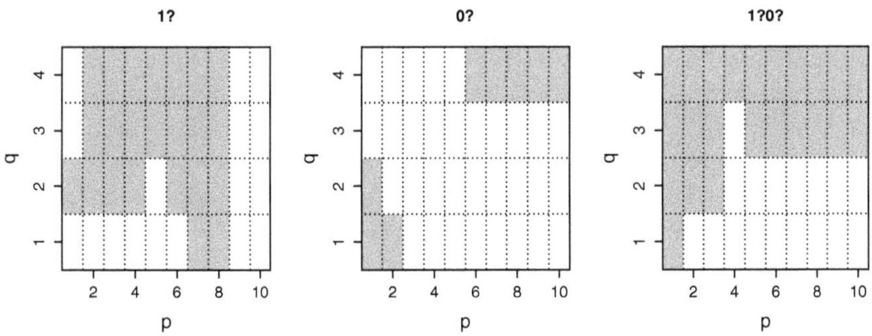

Fig. 9.8 Examples of rejection/acceptance plots at 5% level which are difficult to interpret. Grey area – reject H_0, white – fail to reject H_0.

null hypothesis. In the case shown in the middle panel, the conclusion is also not clear, but we lean toward accepting the independence of X and Y. Finally, the right panel presents an example where it is rather unclear what to conclude.

Results and conclusions. We now discuss the results of the application of the test. We consider high-latitude records from College, Alaska, (CMO) as X, and let Y be the observations from all eight mid- and low-latitude stations during the same UT time as the CMO data. We also analyze responses one, two and three days after substorms were recorded at the CMO station. Such a setting should allow us to see if there is a longitudinal effect of substorms; how long this effect, if any, lasts; and what the global influence of a substorm is.

Column A in Table 9.3 presents the test results for all the substorms that occurred from January to August. We see that the effect of substorms observed at CMO is statistically significant at all mid- and low- latitude stations at the same UT (e.g. BOU0, HON0). This is true for one-day lag data as well (e.g. BOU1, HON1), but for the lag of two days the results are inconclusive. We conclude that the effect of substorms observed at CMO persists for about 48 hours, at all longitudes and latitudes. In the column labeled A* we provide the test results for the set of the substorms where none of the events occurred close to storms. As one can see, the results are similar to the ones in column A. This means that the observed effect is not attributable to an impact of storms on high-latitude currents. We also analyzed the effects of isolated substorms, i.e. there were at least 2 quiet days after such substorms (see column I in Table 9.3). As one can see, there is significant linear dependence between records observed at high latitude and mid-, low-latitude during substorm days, as well as the next day. This means that the next day effect cannot be attributed to the confounding effect of substorms on consecutive days. Next, we analyze the effect of medium strength substorms. Table 9.3, column B, presents the test results. We can see that the medium strength substorm effect is weaker than in case of all substorms. The effect of medium strength substorms appears significant on the same day, but on the following days is absent. It fades out faster for further longitudes. We draw the same conclusion from column B* which includes test results on the medium strength substorms that were not effected by the storm activity. Table 9.4 gives the results for the three sets of substorms in three month periods. In column A1 the results for the substorm days from January to March, 2001 are presented. The conclusions are similar to the ones we got for all substorms from January until August (see Table 9.3, column A). The dependence seems to last for two days. We come to the same conclusion dealing with the other two sets of the substorms, the ones that occurred in Spring and Summer 2001 (see columns A1 and A2 of Table 9.4), the second day dependence being weaker in summer. This agrees with the earlier analysis, as there are fewer strong substorms in summer months.

We conclude that there is a pattern that suggests that there is a dependence between high- and mid-, and high- and low-latitude records with no and one day lag. There is no significant dependence for data with two- and three-day lags.

We conclude this section with a discussion of the physical meaning of our findings. The ground magnetic effects of a localized auroral current system in the

Table 9.3 Results of the test for all substorm days (A), substorm days excluding days around the day with a storm (A*); medium strength substorms (B), medium strength substorms excluding storm days (B*) that occurred from January to August, 2001; (I) isolated substorms that occurred from January to December, 2001.

									Mid-latitude										
A	A*	I	B	B*	A	A*	I	B	B*	A	A*	I	B	B*	A	A*	I	B	B*
		BOU0					BOU1					BOU2					BOU3		
1	1	1	1?	1?	1	1	1	0	1?	1?0?	1?0?	0	0	0	0	1?0?	0	0?	1?0?
		FRD0					FRD1					FRD2					FRD3		
1	1	1	1?	1?0?	1	1	1	0	0?	0?	0?	0	0	0	0	0?	0	0?	1?0?
		THY0					THY1					THY2					THY3		
1	1	1	1?	1?	1	1	1	0?	1?0?	1?0?	1?0?	0	0	0	0	0?	0	0?	1?0?
		MMB0					MMB1					MMB2					MMB3		
1	1	1	0?	1?	1	1	1	0	1?0?	1?	1?	0?	0	0	0?	0	0	0?	1?0?
									Low-latitude										
A	A*	I	B	B*	A	A*	I	B	B*	A	A*	I	B	B*	A	A*	I	B	B*
		HON0					HON1					HON2					HON3		
1	1	1	1?	1?	1	1	1	1?0?	1?	1?0?	0?	0	0	0	0?	0?	0	0?	1?0?
		SJG0					SJG1					SJG2					SJG3		
1	1	1	1?	1?	1	1	1	0	0?	0?	0	0	0	0	0	0	0	0?	1?0?
		HER0					HER1					HER2					HER3		
1	1	1	0?	1?	1	1	1	0?	0?	1?0?	1?0?	0	0	0?	0	0?	0	0?	1?0?
		KAK0					KAK1					KAK2					KAK3		
1	1	1	1?	1?	1	1	1	1?	1?	1?	1?	1?0?	0	0	0	0?	0	0?	1?0?

Table 9.4 Results of the test for substorm days that occurred in 2001 from January to March (A1), March to May (A2), June to August (A3).

Mid-latitude											
A1	A2	A3	A1	A2	A3	A1	A2	A3	A1	A2	A3
BOU0			BOU1			BOU2			BOU3		
1	1	1	1	1	1?0?	0	0	0	0	0	0
FRD0			FRD1			FRD2			FRD3		
1	1	1	1	1	1?0?	0	0	0	0	0	0
THY0			THY1			THY2			THY3		
1	1	1	1	1	1	0	0	0	0	0	0
MMB0			MMB1			MMB2			MMB3		
1	1	1	1	1	1?0?	1?0?	0	0?	0?	0	0

Low-latitude											
A1	A2	A3	A1	A2	A3	A1	A2	A3	A1	A2	A3
HON0			HON1			HON2			HON3		
1	1	1	1	1	1?0?	0?	0	0	0	0	0
SJG0			SJG1			SJG2			SJG3		
1	1	1	1	1	1?	0	0	0	0	0	0
HER0			HER1			HER2			HER3		
1	1	1	1	1	1	0	0	0	0	0	0
KAK0			KAK1			KAK2			KAK3		
1	1	1	1	1	1	1?0?	0	0	0?	0	0

ionosphere normally become insignificant for a location at the Earth's surface 400-500 km away from the center of the current system. Therefore the substorm auroral currents in the ionosphere would not be expected to have significant *direct* effects on the H–component measurements at mid–latitudes and most certainly not at low (equatorial) latitudes. The influence is likely not directly from the auroral electrojects, but the full current curcuit in the M-I system that drives the auroral electorojects during substorms. Conceptually, this would not be entirely unexpected. However, what is unexpected is that on subsequent day, after a 24 hour lag, the mid– and low–latitude field is still affected by prior day's substorm activity defined by high–latitude magnetic fields. The result is dependent on the strength of the substorms, i.e. only the effect of strong substorms extends to low latitudes on the second day. The interpretation of this result is not readily apparent. These statistical findings may imply some physical connections between the substorm electrodynamics and the physical processes in other regions of the M-I system that we are not aware of at the present time.

9.5 Asymptotic theory

Our first assumption specifies independence and moment conditions.

Assumption 9.1. *The triples $(Y_n, X_n, \varepsilon_n)$ form a sequence of independent identically distributed random elements such that ε_n is independent of X_n and*

$$EX_n = 0 \quad \text{and} \quad E\varepsilon_n = 0; \tag{9.5}$$

$$E\|X_n\|^4 < \infty \quad \text{and} \quad E\|\varepsilon_n\|^4 < \infty. \tag{9.6}$$

The next assumption extends condition (2.12) to both response and explanatory variables.

Assumption 9.2. *The eigenvalues of the operators C and Γ satisfy, for some $p > 0$ and $q > 0$,*

$$\lambda_1 > \lambda_2 > \ldots \lambda_p > \lambda_{p+1}, \quad \gamma_1 > \gamma_2 > \ldots \gamma_q > \gamma_{q+1}. \tag{9.7}$$

Under these assumptions, we can quantify the behavior of the test under H_0 and H_A.

Theorem 9.1. *Under H_0 and Assumptions 9.1 and 9.2, $\hat{T}_N(p,q) \xrightarrow{d} \chi^2_{pq}$, as $N \to \infty$.*

Theorem 9.2. *If Assumptions 9.1 and 9.2 hold, and $\langle \Psi(v_k), u_j \rangle \neq 0$ for some $k \leq p$ and $j \leq q$, then $\hat{T}_N(p,q) \xrightarrow{P} \infty$, as $N \to \infty$.*

Jiofack and Nkiet (2010) showed that statistic (9.3) can be used outside the context of the functional linear model to test the independence of two functional samples. The null hupothesis is then formulated in the following assumption.

Assumption 9.3. *The pairs (Y_n, X_n) form a sequence of mean zero independent identically distributed random elements such that Y_n is independent of X_n and $E\|X_n\|^4 < \infty$, $E\|Y_n\|^4 < \infty$.*

They proved the following result together with an analog of Theorem 9.2.

Theorem 9.3. *If Assumptions 9.2 and 9.3 hold, then $\hat{T}_N(p,q) \xrightarrow{d} \chi^2_{pq}$, as $N \to \infty$.*

Jiofack and Nkiet (2010) also showed that statistic (9.3) can be used to test a broader null hypothesis which corresponds to a lack of correlation rather than independence. In the functional context, zero correlation can be defined by the condition $\Delta = 0$, which means that for any $x, y \in L^2$, $E[\langle X, x\rangle \langle Y, y\rangle] = 0$. To formulate the null hypothesis in this context, we introduce the following assumption.

Assumption 9.4. *The pairs (Y_n, X_n) form a sequence of mean zero independent identically distributed random elements such that $E\|X_n\|^4 < \infty$, $E\|Y_n\|^4 < \infty$. The operator Δ is equal to zero.*

If only $\Delta = 0$ is assumed, rather than the independence of the two samples, then the statistic $\hat{T}_N(p,q)$ no longer converges to the chi–squared distribution. This is essentially because without assuming independence, the fourth order moments

$$\kappa_{ijk\ell} = E[\langle X, v_i\rangle \langle X, v_k\rangle \langle Y, u_j\rangle \langle y, u_\ell\rangle]$$

need not vanish if $i \neq k$ or $j \neq \ell$. To describe the limit distribution, we introduce the $pq \times pq$ matix

$$\mathbf{K} = \begin{bmatrix} \kappa_{1111} & \kappa_{1112} & \cdots & \kappa_{11pq} \\ \kappa_{1211} & \kappa_{1212} & \cdots & \kappa_{12pq} \\ \vdots & \vdots & \cdots & \vdots \\ \kappa_{pq11} & \kappa_{pq12} & \cdots & \kappa_{pqpq} \end{bmatrix}. \tag{9.8}$$

We also introduce the $pq \times pq$ diagonal matix

$$\mathbf{H} = \mathrm{diag}(\lambda_1, \lambda_2, \ldots, \lambda_p) \otimes \mathrm{diag}(\gamma_1, \gamma_2, \ldots, \gamma_q), \tag{9.9}$$

where $\otimes$ denotes the Kronecker product. For the properties of the Kronecker product we refer to Chapter 4 of Horn and Johnson (1991). For example, if $p = q = 2$, then

$$\mathbf{H} = \begin{bmatrix} \lambda_1\gamma_1 & 0 & 0 & 0 \\ 0 & \lambda_1\gamma_2 & 0 & 0 \\ 0 & 0 & \lambda_2\gamma_1 & 0 \\ 0 & 0 & 0 & \lambda_2\gamma_2 \end{bmatrix}.$$

With this notation in place, we state the following result.

Theorem 9.4. *If Assumptions 9.2 and 9.4 hold, then* $\hat{T}_N(p,q) \xrightarrow{d} \mathbf{G}^T\mathbf{H}^{-1}\mathbf{G}$, *where* $\mathbf{G}$ *is a mean zero Gaussian vector in* R^{pq} *with covariance matrix* $\mathbf{K}$, *and* $\mathbf{H}$ *is defined by* (9.9).

Note that if X is independent of Y, then $\mathbf{K} = \mathbf{H}$, and so $\mathbf{G}^T\mathbf{H}^{-1}\mathbf{G} \stackrel{d}{=} \chi^2_{pq}$, see e.g. Theorem 2.9 of Seber and Lee (2003). Jiofack and Nkiet (2010) explain how to implement the test for a general matrix $\mathbf{K}$.

9.6 Proofs of Theorems 9.1 and 9.2

The test statistics (9.3) does not change if we replace $\hat{v}_k$ by $\hat{c}_k\hat{v}_k$ and $\hat{u}_j$ by $\hat{c}_j\hat{u}_j$, see Section 2.5. To lighten the notation in this section, we therefore write $\hat{v}_k$ in place of $\hat{c}_k\hat{v}_k$ and $\hat{u}_j$ in place of $\hat{c}_j\hat{u}_j$.

Proof of Theorem 9.1. Theorem 9.1 follows from Corollary 9.1, which is arrived at through a series of simple lemmas. Lemma 9.1 shows that the χ^2 limit holds for the population eigenelements. The remaining lemmas show that the differences between the empirical and population eigenelements have asymptotically negligible effect.

Lemma 9.1. *Under the assumptions of Theorem 9.1,*

$$\sqrt{N}\left\{\left\langle \widehat{\Delta}(v_k), u_j \right\rangle,\ 1 \le j \le q, 1 \le k \le p\right\} \xrightarrow{d} \left\{\eta_{kj}\sqrt{\lambda_k \gamma_j},\ 1 \le j \le q, 1 \le k \le p\right\}, \tag{9.10}$$

with $\eta_{kj} \sim N(0,1)$. *Moreover,* η_{kj} *and* $\eta_{k'j'}$ *are independent if* $(k,j) \ne (k',j')$.

Proof. Under H_0, $\sqrt{N}\left\langle \widehat{\Delta}(v_k), u_j \right\rangle = N^{-1/2}\sum_{n=1}^{N} \langle X_n, v_k \rangle \langle \varepsilon_n, u_j \rangle$. The summands have mean zero and variance $\lambda_k \gamma_j$, so (9.10) follows.

To verify that η_{kj} and $\eta_{k'j'}$ are independent if $(k,j) \ne (k',j')$, it suffices to show that $\sqrt{N}\left\langle \widehat{\Delta}(v_k), u_j \right\rangle$ and $\sqrt{N}\left\langle \widehat{\Delta}(v_{k'}), u_{j'} \right\rangle$ are uncorrelated. Observe that

$$\begin{aligned}
&E\left[\sqrt{N}\left\langle \widehat{\Delta}(v_k), u_j \right\rangle, \sqrt{N}\left\langle \widehat{\Delta}(v_{k'}), u_{j'} \right\rangle\right] \\
&= \frac{1}{N} E\left[\sum_{n=1}^{N} \langle X_n, v_k \rangle \langle \varepsilon_n, u_j \rangle \sum_{n'=1}^{N} \langle X_{n'}, v_{k'} \rangle \langle \varepsilon_{n'}, u_{j'} \rangle\right] \\
&= \frac{1}{N} \sum_{n,n'=1}^{N} E\left[\langle X_n, v_k \rangle \langle X_{n'}, v_{k'} \rangle\right] E\left[\langle \varepsilon_n, u_j \rangle \langle \varepsilon_{n'}, u_{j'} \rangle\right] \\
&= \frac{1}{N} \sum_{n=1}^{N} E\left[\langle X_n, v_k \rangle \langle X_n, v_{k'} \rangle\right] E\left[\langle \varepsilon_n, u_j \rangle \langle \varepsilon_n, u_{j'} \rangle\right] \\
&= \langle C(v_k), v_{k'} \rangle \langle \Gamma u_j, u_{j'} \rangle = \gamma_k \delta_{kk'} \gamma_j \delta_{jj'}.
\end{aligned}$$

□

Recall that the Hilbert–Schmidt norm of a Hilbert–Schmidt operator S is defined by $\|S\|_{\mathcal{S}}^2 = \sum_{j=1}^{\infty} \|S(e_j)\|^2$, where $\{e_1, e_2, \dots\}$ is any orthonormal basis, and that it dominates the operator norm: $\|S\|_{\mathcal{L}} \le \|S\|_{\mathcal{S}}$.

Lemma 9.2. *Under the assumptions of Theorem 9.1,*

$$E\|\widehat{\Delta}\|_{\mathcal{S}}^2 = N^{-1} E\|X\|^2 E\|\varepsilon_1\|^2.$$

Proof. Observe that

$$\|\widehat{\Delta}(e_j)\|^2 = N^{-2} \sum_{n,n'=1}^{N} \langle X_n, e_j \rangle \langle X_{n'}, e_j \rangle \langle Y_n, Y_{n'} \rangle.$$

Therefore, under H_0,

$$E\|\widehat{\Delta}\|_{\mathcal{S}}^2 = N^{-2}\sum_{j=1}^{\infty}\sum_{n,n'=1}^{N} E\left[\langle X_n, e_j\rangle\langle X_{n'}, e_j\rangle\langle \varepsilon_n, \varepsilon_{n'}\rangle\right]$$
$$= N^{-2}\sum_{j=1}^{\infty}\sum_{n=1}^{N} E\langle X_n, e_j\rangle^2 E\|\varepsilon_n\|^2$$
$$= N^{-1}E\|\varepsilon_1\|^2\sum_{j=1}^{\infty}\langle X, e_j\rangle^2 = N^{-1}E\|\varepsilon_1\|^2 E\|X\|^2. \quad \square$$

Lemma 9.3. *Under the assumptions of Theorem 9.1,*

$$\sqrt{N}\left\{\left\langle\widehat{\Delta}(\hat{v}_k), \hat{u}_j\right\rangle,\ 1 \le j \le q, 1 \le k \le p\right\}$$
$$\xrightarrow{d} \left\{\eta_{kj}\sqrt{\lambda_k\gamma_j},\ 1 \le j \le q, 1 \le k \le p\right\} \tag{9.11}$$

with η_{kj} equal to those in Lemma 9.1.

Proof. By Lemma 9.1, it suffices to verify that

$$\sqrt{N}\left\langle\widehat{\Delta}(\hat{v}_k), \hat{u}_j\right\rangle - \sqrt{N}\left\langle\widehat{\Delta}(v_k), u_j\right\rangle \xrightarrow{P} 0. \tag{9.12}$$

Relation (9.12), will follow from

$$\sqrt{N}\left\langle\widehat{\Delta}(v_k), \hat{u}_j - u_j\right\rangle \xrightarrow{P} 0 \tag{9.13}$$

and

$$\sqrt{N}\left\langle\widehat{\Delta}(\hat{v}_k - v_k), \hat{u}_j\right\rangle \xrightarrow{P} 0. \tag{9.14}$$

To verify (9.13), note that by (2.13), $\sqrt{N}(\hat{u}_j - u_j) = O_P(1)$, and by Lemma 9.2, $E\|\widehat{\Delta}v_k\| \le E\|\widehat{\Delta}\|_{\mathcal{S}} = O(N^{-1/2})$. Thus (9.13) follows from Lemma 7.3.

To use the same argument for (9.14) (with (2.13)), we note that

$$\sqrt{N}\left\langle\widehat{\Delta}(\hat{v}_k - v_k), \hat{u}_j\right\rangle = \sqrt{N}\left\langle\hat{v}_k - v_k, \tilde{\Delta}(\hat{u}_j)\right\rangle,$$

where $\tilde{\Delta}(x) = N^{-1}\sum_{n=1}^{N}\langle Y_n, x\rangle X_n$. Lemma 9.2 shows that under H_0, $E\|\tilde{\Delta}\|_{\mathcal{S}} = E\|\widehat{\Delta}\|_{\mathcal{S}}$. $\square$

By (2.13), $\hat{\lambda}_k \xrightarrow{P} \lambda_k$ and $\hat{\gamma}_j \xrightarrow{P} \gamma_j$, so we obtain

Corollary 9.1. *Under the assumptions of Theorem 9.1,*

$$\sqrt{N}\left\{\hat{\lambda}_k^{-1/2}\hat{\gamma}_j^{-1/2}\left\langle\widehat{\Delta}(\hat{v}_k), \hat{u}_j\right\rangle,\ 1 \le j \le q,\ 1 \le k \le p\right\}$$
$$\xrightarrow{d} \left\{\eta_{kj},\ 1 \le j \le q,\ 1 \le k \le p\right\}, \tag{9.15}$$

with η_{kj} equal to those in Lemma 9.1.

Proof of theorem 9.2. Denote

$$\hat{S}_N(p,q) = \sum_{k=1}^{p} \sum_{j=1}^{q} \hat{\lambda}_k^{-1} \hat{\gamma}_j^{-1} \left\langle \widehat{\Delta}(\hat{v}_k), \hat{u}_j \right\rangle^2 .$$

By Lemma 9.6 and (2.13), $\hat{S}_N(p,q) \stackrel{P}{\rightarrow} S(p,q) > 0$. Hence $\hat{T}_N(p,q) = N\hat{S}_N(p,q) \stackrel{P}{\rightarrow} \infty$.

To establish Lemma 9.6, it is convenient to split the argument into two simple lemmas: Lemma 9.4 and Lemma 9.5.

Lemma 9.4. *If Y_n, $n \geq 1$, are identically distributed, then $E\|\widehat{\Delta}\| \leq E\|Y\|^2$.*

Proof. For arbitrary $u \in L^2$ with $\|u\| \leq 1$,

$$\|\widehat{\Delta} u\| \leq N^{-1} \sum_{n=1}^{N} |\langle Y_n, u\rangle| \|Y_n\| \leq N^{-1} \sum_{n=1}^{N} \|Y_n\|^2 .$$

Since the Y_n are identically distributed, the claim follows. □

Lemma 9.5. *If Assumption 9.1 holds, then for any functions $v, u \in L^2$,*

$$\left\langle \widehat{\Delta}(v), u \right\rangle \stackrel{P}{\rightarrow} \langle \Delta(v), u\rangle .$$

Proof. The result follows from the Law of Large Numbers after noting that

$$\left\langle \widehat{\Delta}(v), u \right\rangle = \frac{1}{N} \sum_{n=1}^{N} \langle X_n, v\rangle \langle Y_n, u\rangle$$

and

$$E\left[\langle X_n, v\rangle \langle Y_n, u\rangle\right] = E\left[\langle \langle X_n, v\rangle Y_n, u\rangle\right] = \langle \Delta(v), u\rangle . \qquad \square$$

Lemma 9.6. *If Assumptions 9.1 and 9.2 hold, then $\left\langle \widehat{\Delta}(\hat{v}_k), \hat{u}_j \right\rangle \stackrel{P}{\rightarrow} \left\langle \Delta(v_k), u_j \right\rangle$, $j \leq q, k \leq p$.*

Proof. By Lemma 9.5, it suffices to show

$$\left\langle \widehat{\Delta}(v_k),\ \hat{u}_j - u_j \right\rangle \stackrel{P}{\rightarrow} 0$$

and

$$\left\langle \widehat{\Delta}(\hat{v}_k) - \widehat{\Delta}(v_k),\ \hat{u}_j \right\rangle \stackrel{P}{\rightarrow} 0.$$

These relations follow from Lemma 7.3, relations (2.13) and Lemma 9.4. □

Chapter 10
Two sample inference for regression kernels

In Chapter 5, we studied two sample procedures for the mean function and the covariance operator. This chapter is devoted to testing the equality of the regression operators in two functional linear models. We are concerned with the following problem: We observe two samples: sample 1: (X_i, Y_i), $1 \le i \le N$, and sample 2: (X_j^*, Y_j^*), $1 \le j \le M$. The explanatory variables X_i and X_j^* are functions, whereas the responses Y_i and Y_j^* can be either functions or scalars (the Y_i and Y_j^* are either both functions, or both scalars). We model the dependence of the Y_i (Y_j^*) on the X_i (X_j^*) by the functional regression models

$$Y_i = \Psi X_i + \varepsilon_i, \quad Y_j^* = \Psi^* X_j^* + \varepsilon_j^*,$$

where Ψ and Ψ^* are linear operators whose domain is a function space, and which take values either in the same function space or in the real line. We wish to test if the operators Ψ and Ψ^* are equal.

In Section 10.1, we provide motivation and background for the methodology developed in this chapter. The testing procedures are derived in Sections 10.2 and 10.3, respectively, for scalar and functional responses. As with the usual two sample tests for the equality of means, we make a distinction between the simpler case of "equal variances" and the more complex case of "unequal variances". We thus have four testing procedures, which are summarized in Section 10.4. A reader interested only the description of the test can start with Section 10.4, and refer to Sections 10.2 and 10.3 for further details, as needed. Section 10.5 presents the results of a small simulation study. Applications to medfly and magnetometer data are presented in Section 10.6. Asymptotic results and their proofs are collected in Section 10.7. This chapter is based on the paper of Horváth *et al.* (2009).

10.1 Motivation and introduction

We begin this section with a motivating example, which is continued in Section 10.6.

Egg–laying curves of Mediterranean fruit flies. Müller and Stadtmüller (2005), Section 6, consider 534 egg-laying curves (count of eggs per unit time interval) of medflies who lived at least 30 days. Each function is defined over an interval $[0, 30]$, and its value on day $t \leq 30$ is the count of eggs laid by fly i on that day. The 534 flies are classified into long–lived, i.e. those who lived longer than 44 days, and short–lived, i.e. those who died before the end of the 44th day after birth. In the sample, there are 256 short–lived, and 278 long–lived flies. This classification naturally defines two samples: *Sample 1:* the egg-laying curves $X_i(t),\ 0 < t \leq 30,\ i = 1, 2, \ldots, 256$ of the short–lived flies, and the corresponding total number of eggs Y_i. *Sample 2:* the egg-laying curves $X_j^*(t),\ 0 < t \leq 30,\ j = 1, 2, \ldots, 278$ of the long–lived flies, and the corresponding total number of eggs Y_j^*. The egg-laying curves are very irregular; Figure 10.1 shows ten smoothed curves of short– and long–lived flies.

The smoothed egg-laying curves are considered as regressors, X_i for the short–lived, and X_j^* for the long–lived flies. The responses are the lifetime count of eggs, Y_i and Y_j^*, respectively. The average of the Y_j^* is obviously larger than that of the Y_i, but a question of interest is whether after adjusting for the means, the structure of the dependence of the Y_j^* on the curves $X_j^*(t)$ is different from the dependence of the Y_i on the curves $X_i(t)$. We thus consider two linear models:

$$Y_i - \bar{Y} = \int \psi(t)(X_i(t) - \bar{X}(t))dt + \varepsilon_i, \quad i = 1, 2, \ldots, 256,$$

$$Y_j^* - \bar{Y}^* = \int \psi^*(t)X_j^*(t) - \bar{X}^*(t))dt + \varepsilon_j^*, \quad j = 1, 2, \ldots, 278.$$

and wish to test $H_0 : \psi = \psi^*$.

The above linear models describe the dependence structure of the data remarkably well. We applied the graphical test of Chiou and Müller (2007), described in

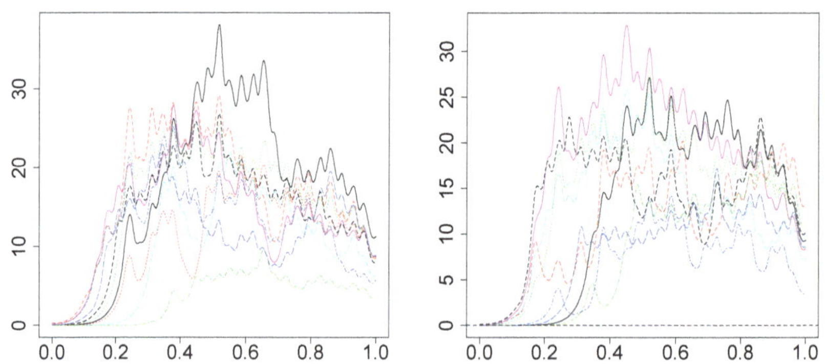

Fig. 10.1 Ten randomly selected smoothed egg–laying curves of short-lived medflies (left panel), and ten such curves for long–lived medfies (right panel).

Section 9.4, in which the responses are graphed against the scores of the initial functional principal components. All graphs show nice elliptical shapes. In Section 10.6, we apply the test derived in Section 10.2 to check if it is reasonable to assume that $\psi = \psi^*$.

Like classical two sample procedures in various forms, the tests of this chapter are likely to be applicable to a wide range of problems, where estimating two significantly different functional linear regressions on subsamples of a larger sample may reveal additional features. In our setting, the role of regression parameter vectors (or matrices) is played by integral operators acting on a function space. The complexity of the test statistics increases as we move from scalar to functional responses and relax assumptions on the covariance structure of the regressors. Even in the multivariate setting, except for comparing mean responses, the problem of comparing the regression coefficients for two models based on two different samples is not trivial, and we could not find a ready reference for it.

In the remainder of this chapter, we do not deal with the errors caused by replacing the FPC's by the EFPC's: the test statistics do not depend on the signs of the EFPC's, and the $O_P(N^{-1/2})$ distances can be handled by the application of Theorem 2.7. The formulas appearing in this chapter are rather complex, and developing arguments analogous to those in Section 5.4 and other chapters would take up too much space and obscure the main ideas. To obtain computable test statistics, we also neglect terms arising from EFPC's with small eigenvalues. These terms are not asymptotically negligible, but they are practically negligible, as established by the simulations presented at the end of Section 10.3.

In the remainder of this chapter, we assume that the mean functions and the means of the responses have been subtracted, and so we consider the scalar response model

$$Y_i = \int_0^1 \psi(s) X_i(s) ds + \varepsilon_i \tag{10.1}$$

and the functional response model

$$Y_i(t) = \int_0^1 \psi(t,s) X_i(s) ds + \varepsilon_i(t). \tag{10.2}$$

Precise model assumptions are formulated in Section 10.7.

Our objective is to test

$$H_0 : \|\psi - \psi^*\| = 0$$

against

$$H_A : \|\psi - \psi^*\| \neq 0,$$

where the norm is in $L^2([0,1])$ for model (10.1) and in $L^2([0,1] \times [0,1])$ for model (10.2).

10.2 Derivation of the test for scalar responses

In this Section, and Section 10.3, we refer to theorems stated and proven in Section 10.7. To understand the procedures, it is however not necessary to study these results, which provide asymptotic justification for the claims we make. To develop a meaningful asymptotic theory, and to ensure that the tests perform well in finite samples, we assume that the two samples sizes are of comparable size. Asymptotically, we postulate that there exists a constant $0 < \zeta < \infty$ such that

$$N/M \to \zeta, \quad N \to \infty. \tag{10.3}$$

Suppose $v_i, i \geq 1$, form a basis in $L^2([0,1])$. Since we now deal with two samples, we may choose two different bases or one common basis. This choice will also depend on what we assume about the variances of the regressors and the errors in the two samples. To focus attention, it is initially convenient to think that the v_i are the FPC's of the regressors X_i.

Since $\psi \in L^2([0,1])$, we can expand it as $\psi(s) = \sum_{i=1}^{\infty} \mu_i v_i(s)$, where $\mu_i = \langle \psi, v_i \rangle$. Consequently, the response variables can be expressed as $Y_i = \sum_{k=1}^{\infty} \mu_k \langle X_i, v_k \rangle + \varepsilon_i$. We truncate the above expansion at $1 \leq p < \infty$, and combine the error made by the truncation with the ε_i. The response is thus given by

$$Y_i = \sum_{k=1}^{p} \mu_k \langle X_i, v_k \rangle + \varepsilon_i', \quad \varepsilon_i' = \varepsilon_i + \sum_{k=p+1}^{\infty} \mu_k \langle X_i, v_k \rangle. \tag{10.4}$$

In terms of matrix and vector notation we have

$$\mathbf{Y} = \mathbf{X}\boldsymbol{\mu} + \boldsymbol{\varepsilon}', \tag{10.5}$$

where, for $1 \leq i \leq N$ and $1 \leq j \leq p$,

$$\mathbf{Y}(i) = Y_i, \quad \mathbf{X}(i,j) = \langle X_i, v_j \rangle, \quad \boldsymbol{\mu}(j) = \mu_j, \quad \boldsymbol{\varepsilon}'(i) = \varepsilon_i'.$$

The least squares estimator for $\boldsymbol{\mu}$ is therefore

$$\hat{\boldsymbol{\mu}} = \left(\mathbf{X}^T\mathbf{X}\right)^{-1}\mathbf{X}^T\mathbf{Y}. \tag{10.6}$$

By Theorem 10.1, $\hat{\boldsymbol{\mu}}$ is a consistent estimator of $\boldsymbol{\mu}$, and for the second sample, the analogously defined $\hat{\boldsymbol{\mu}}^*$ a consistent estimator of $\boldsymbol{\mu}^*$. Thus we can base a test statistic on the difference $\hat{\boldsymbol{\mu}} - \hat{\boldsymbol{\mu}}^*$. To motivate our construction, assume first that the covariance operators of the X_i and X_j^* are equal and the errors ε_i and ε_j^* have equal variances, i.e.

$$\mathrm{E}(X_1(s)X_1(t)) = \mathrm{E}(X_1^*(s)X_1^*(t)) = c(s,t) \tag{10.7}$$

and

$$\mathrm{var}(\varepsilon_1) = \mathrm{var}(\varepsilon_1^*). \tag{10.8}$$

The common covariance operator of the X_i and the X_j^* is denoted by C and its eigenelements by v_i, λ_i, as in Section 2.5. Under these assumptions, we introduce the random variable

$$\Lambda_p = N(1+\zeta)^{-1}(\hat{\boldsymbol{\mu}} - \hat{\boldsymbol{\mu}}^*)^T \boldsymbol{\Sigma}_p^{-1}(\hat{\boldsymbol{\mu}} - \hat{\boldsymbol{\mu}}^*), \tag{10.9}$$

where $\boldsymbol{\Sigma}_p$ is the common asymptotic covariance matrix of $\hat{\boldsymbol{\mu}}$ and $\hat{\boldsymbol{\mu}}^*$ defined by

$$\boldsymbol{\Sigma}_p(i,i) = \lambda_i^{-1}\sigma^2 + \lambda_i^{-2}\mathrm{var}\left(\langle X_1, v_i \rangle \sum_{k=p+1}^{\infty} \mu_k \langle X_1, v_k \rangle \right), \quad i = 1, \ldots, p; \tag{10.10}$$

$$\boldsymbol{\Sigma}_p(i,j) = \lambda_i^{-1}\lambda_j^{-1}\mathrm{E}\left(\langle X_1, v_i \rangle \langle X_1, v_j \rangle \left(\sum_{k=p+1}^{\infty} \mu_k \langle X_1, v_k \rangle \right)^2 \right), \quad i \neq j. \tag{10.11}$$

By Theorem 10.2, $\Lambda_p \xrightarrow{d} \chi^2(p)$, as $N \to \infty$, i.e. Λ_p defined by (10.9), converges to a chi-square random variable with p degrees of freedom. We therefore propose the following test statistic when the covariances are equal

$$\hat{\Lambda}_p = N(1+N/M)^{-1}(\hat{\boldsymbol{\mu}} - \hat{\boldsymbol{\mu}}^*)^T (\hat{\boldsymbol{\Sigma}}_p)^{-1}(\hat{\boldsymbol{\mu}} - \hat{\boldsymbol{\mu}}^*), \tag{10.12}$$

where $\hat{\boldsymbol{\Sigma}}_p$ is the empirical diagonal approximation to the matrix $\boldsymbol{\Sigma}_p$ given by

$$\hat{\boldsymbol{\Sigma}}_p = \hat{\sigma}^2 \begin{bmatrix} \hat{\lambda}_1^{-1} & 0 & \cdots & 0 \\ 0 & \hat{\lambda}_2^{-1} & \cdots & 0 \\ \vdots & \vdots & \vdots & \vdots \\ 0 & 0 & \cdots & \hat{\lambda}_p^{-1} \end{bmatrix},$$

and where $\hat{\sigma}$ is the residual standard deviation from the estimated regression model defined analogously to (10.5), but with both samples pooled together. Thus, the estimates $\hat{\sigma}, \hat{\lambda}_1, \ldots, \hat{\lambda}_p$ are all computed using the pooled sample.

In many applications, the covariance kernels $c(s,t)$ and $c^*(s,t)$ are not necessarily equal. Since the two kernels have different eigenfunctions, we now consider an arbitrary basis $\{w_i\}$ of $L^2([0,1])$. Good choices for the w_i are discussed in Section 10.4. The kernels ψ and ψ^* are expanded as

$$\psi(s) = \sum_{i=1}^{\infty} \mu_i w_i(s), \quad \psi^*(s) = \sum_{j=1}^{\infty} \mu_j^* w_j(s), \tag{10.13}$$

and so

$$Y_i = \sum_{k=1}^{\infty} \mu_k \langle X_i, w_k \rangle + \varepsilon_i, \quad Y_j^* = \sum_{k=1}^{\infty} \mu_k^* \langle X_j^*, w_k \rangle + \varepsilon_j^*.$$

Truncating both sums at p, the response variables can again be expressed as

$$\mathbf{Y} = \mathbf{X}\boldsymbol{\mu} + \boldsymbol{\varepsilon}', \quad \mathbf{Y}^* = \mathbf{X}^*\boldsymbol{\mu}^* + \boldsymbol{\varepsilon}'^*,$$

with all terms analogously defined with respect to our new basis. While this appears similar to our prior calculations, we are expanding with respect to an arbitrary basis which means that $\mathbf{X}$ and $\boldsymbol{\varepsilon}'$ are now potentially correlated. The least squares estimators take, however, the same form

$$\hat{\boldsymbol{\mu}} = (\mathbf{X}^T\mathbf{X})^{-1}\mathbf{X}^T\mathbf{Y}, \quad \hat{\boldsymbol{\mu}}^* = (\mathbf{X}^{*T}\mathbf{X}^*)^{-1}\mathbf{X}^{*T}\mathbf{Y}^*.$$

Thus we can once again compare $\hat{\boldsymbol{\mu}}$ and $\hat{\boldsymbol{\mu}}^*$ to test the null hypothesis. To analyze the asymptotic behavior of these estimates we consider the relation

$$\hat{\boldsymbol{\mu}} = \boldsymbol{\mu} + (\mathbf{X}^T\mathbf{X})^{-1}\mathbf{X}^T\boldsymbol{\varepsilon}'.$$

The vector $\mathbf{X}^T\boldsymbol{\varepsilon}'$ can be expressed as

$$\mathbf{X}^T\boldsymbol{\varepsilon}' = \mathbf{A} + \mathbf{B} + N\mathbf{m},$$

where

$$\mathbf{A} = \begin{bmatrix} \sum_{i=1}^{N} \varepsilon_i \langle X_i, w_1 \rangle \\ \vdots \\ \sum_{i=p}^{N} \varepsilon_i \langle X_i, w_p \rangle \end{bmatrix},$$

$$\mathbf{B} = \begin{bmatrix} \sum_{i=1}^{N} \sum_{k=p+1}^{\infty} \mu_k \left(\langle X_i, w_1 \rangle \langle X_i, w_k \rangle - E[\langle X_1, w_1 \rangle \langle X_1, w_k \rangle]\right) \\ \vdots \\ \sum_{i=1}^{N} \sum_{k=p+1}^{\infty} \mu_k \left(\langle X_i, w_p \rangle \langle X_i, w_k \rangle - E[\langle X_1, w_p \rangle \langle X_1, w_k \rangle]\right) \end{bmatrix},$$

have mean zero and are uncorrelated since the error terms are independent of the explanatory functions. The term $\mathbf{m}$ represents the bias introduced by using an arbitrary basis which is given by

$$\mathbf{m} = \begin{bmatrix} \sum_{k=p+1}^{\infty} \mu_k E[\langle X_1, w_1 \rangle \langle X_1, w_k \rangle] \\ \vdots \\ \sum_{k=p+1}^{\infty} \mu_k E[\langle X_1, w_p \rangle \langle X_1, w_k \rangle] \end{bmatrix}.$$

This yields the form

$$\hat{\mu} = \mu + (\mathbf{X}^T\mathbf{X})^{-1}\mathbf{A} + (\mathbf{X}^T\mathbf{X})^{-1}\mathbf{B} + N(\mathbf{X}^T\mathbf{X})^{-1}\mathbf{m}.$$

Clearly $\mathbf{A}$ and $\mathbf{B}$ are sums of iid random vectors with means zero and finite covariance matrices due to Assumptions 10.1 and 10.2. Thus by the multivariate central limit theorem $N^{-1/2}(\mathbf{A}\ \ \mathbf{B})^T$ is asymptotically normal. We have by the strong law of large numbers that

$$N^{-1}\sum_{i=1}^{N}\langle X_i, w_j\rangle\langle X_i, w_k\rangle \stackrel{a.s.}{\to} E\langle X_1, w_j\rangle\langle X_1, w_k\rangle,$$

for $j = 1, \ldots, p$ and $k = 1, \ldots, p$, or in matrix notation

$$N^{-1}\mathbf{X}^T\mathbf{X} \stackrel{a.s.}{\to} \boldsymbol{\Sigma}_1,$$

where the (j, k) entry of $\boldsymbol{\Sigma}_1$ is $E\langle X_1, w_j\rangle\langle X_1, w_k\rangle$. Thus by Slutsky's Lemma $N^{-1/2}(\hat{\mu} - \mu - N(\mathbf{X}^T\mathbf{X})^{-1}\mathbf{m})$ is asymptotically normal. Since $\mathbf{A}$ has zero mean, we have that the (i, j) entry of its covariance matrix is given by

$$E\sum_{k=1}^{N}\varepsilon_k\langle w_i, X_k\rangle\varepsilon_k\langle w_j, X_k\rangle = N\sigma^2 E\langle w_i, X_1\rangle\langle w_j, X_1\rangle,$$

and therefore

$$\mathrm{cov}(\mathbf{A}) = N\sigma^2\boldsymbol{\Sigma}_1.$$

Turning to $\mathbf{B}$, the (i, j) entry of its covariance matrix is given by

$$N\sum_{k=p+1}^{\infty}\sum_{r=p+1}^{\infty}\mu_k\mu_r E\{(\langle X_1, w_i\rangle\langle X_1, w_k\rangle - E[\langle X_1, w_i\rangle\langle X_1, w_k\rangle]) \\ \times(\langle X_1, w_j\rangle\langle X_1, w_r\rangle - E[\langle X_1, w_j\rangle\langle X_1, w_r\rangle])\}.$$

We will denote the covariance matrix of $\mathbf{B}$ as $N\boldsymbol{\Sigma}_2$. Combining everything, we have by Slutsky's Lemma

$$N^{1/2}(\hat{\mu} - \mu - N(\mathbf{X}^T\mathbf{X})^{-1}\mathbf{m}) \stackrel{d}{\to} N(0, \mathbf{C}),$$

where $\mathbf{C} = \sigma^2\boldsymbol{\Sigma}_1^{-1} + \boldsymbol{\Sigma}_1^{-1}\boldsymbol{\Sigma}_2\boldsymbol{\Sigma}_1^{-1}$.

An identical argument gives, for the second sample,

$$M^{1/2}(\hat{\mu}^* - \mu^* - M(\mathbf{X}^{*T}\mathbf{X}^*)^{-1}\mathbf{m}^*) \stackrel{d}{\to} N(0, \mathbf{C}^*),$$

with all terms analogously defined. Using (10.3), we therefore conclude that

$$N^{1/2}(\hat{\mu} - \hat{\mu}^* - (N(\mathbf{X}^T\mathbf{X})^{-1}\mathbf{m} - M(\mathbf{X}^{*T}\mathbf{X}^*)^{-1}\mathbf{m}^*)) \stackrel{d}{\to} N(0, \mathbf{C} + \zeta\mathbf{C}^*).$$

Neglecting the biases $\mathbf{m}$ and $\mathbf{m}^*$, we thus arrive at the test statistic

$$\hat{\Lambda}_p = N(\hat{\mu} - \hat{\mu}^*)^T (\hat{\sigma}^2 \hat{\boldsymbol{\Sigma}}_1 + (N/M)\hat{\sigma}^{*2}\hat{\boldsymbol{\Sigma}}_1^*)^{-1}(\hat{\mu} - \hat{\mu}^*), \tag{10.14}$$

where $\hat{\sigma}^2$ and $\hat{\sigma}^{*2}$ are residual standard deviations from the regression models for the first and second sample respectively. The matrix $\boldsymbol{\Sigma}_1$ is now estimated with

$$\hat{\boldsymbol{\Sigma}}_1 = N^{-1}\mathbf{X}^T\mathbf{X},$$

with $\hat{\boldsymbol{\Sigma}}_1^*$ defined analogously.

The distributions of statistics (10.12) and (10.14) are approximated by the chi-square distribution with p degrees of freedom. If p is large (in terms of the percentage of variance explained), then all neglected terms are close to 0.

10.3 Derivation for functional responses

Turning to model (10.2), we note that now it is also necessary to choose bases to project the Y_i and the Y_j^* onto. We can then use the results developed in the scalar case.

We first focus on the case of equal variances defined by assumptions (10.7) and, in place of (10.8), by

$$\mathrm{E}(\varepsilon_1(s)\varepsilon_1(t)) = \mathrm{E}(\varepsilon_1^*(s)\varepsilon_1^*(t)). \tag{10.15}$$

Consider an arbitrary orthonormal basis $\{u_i\}_{i=1}^{\infty}$ for $L^2([0,1])$ (on which the Y_i are to be projected), and an analogous basis $\{u_j^*\}_{j=1}^{\infty}$. Though all our results hold for an arbitrary choice for $\{u_i\}_{i=1}^{\infty}$, we will use in our applications the eigenfunctions of the covariance operator for the $\{Y_i\}$, with the $\{u_i^*\}$ defined analogously. Because $\{u_i\}$ and $\{v_i\}$ are both bases for $L^2([0,1])$, it follows that we can construct a basis for $L^2([0,1] \times [0,1])$ using the bivariate functions $u_i(t)v_j(s)$ for $(t,s) \in [0,1] \times [0,1]$, $i, j \geq 1$. We therefore have the expansion $\psi(t,s) = \sum_{k=1}^{\infty}\sum_{l=1}^{\infty} \mu_{kl} u_l(t) v_k(s)$, but we will work with the approximation

$$\hat{\psi}(t,s) = \sum_{k=1}^{p}\sum_{l=1}^{r} \hat{\mu}_{k,l}\hat{u}_l(t)\hat{v}_k(s),$$

where $1 \leq r < \infty$ and $1 \leq p < \infty$ are fixed.

Extending the notation introduced in the case of scalar responses, define the matrices

$$\begin{aligned}
\mathbf{Y}(i,j) &= \langle Y_i, u_j \rangle, \quad i = 1, \ldots, N, \ \ j = 1, \ldots, r, \\
\mathbf{X}(i,j) &= \langle X_i, v_j \rangle, \quad i = 1, \ldots, N, \ \ j = 1, \ldots, p, \\
\boldsymbol{\mu}(i,j) &= \iint \psi(t,s) v_i(s) u_j(t) ds\, dt, \quad i = 1, \ldots p, \ \ j = 1, \ldots r.
\end{aligned}$$

As in the scalar case, we combine any errors made by our approximations with the error of the model, so we also introduce the matrix

$$\boldsymbol{\varepsilon}'(i,j) = \langle \varepsilon_i, u_j \rangle + \sum_{k=p+1}^{\infty} \mathbf{X}(i,k)\boldsymbol{\mu}(k,j), \quad i = 1,\ldots,N, \ \ j = 1,\ldots,r.$$

Projecting the relation $Y_i = \Psi X_i + \varepsilon_i$ onto the u_j, we obtain

$$\langle Y_i, u_j \rangle = \langle \Psi X_i, u_j \rangle + \langle \varepsilon_i, u_j \rangle = \sum_{k=1}^{\infty} \langle X_i, v_k \rangle \langle \Psi v_k, u_j \rangle + \langle \varepsilon_i, u_j \rangle$$

which implies

$$\mathbf{Y} = \mathbf{X}\boldsymbol{\mu} + \boldsymbol{\varepsilon}'. \tag{10.16}$$

The corresponding least squares estimator $\hat{\boldsymbol{\mu}} = (\mathbf{X}^T\mathbf{X})^{-1}\mathbf{X}^T\mathbf{Y}$ consistently estimates the matrix $\boldsymbol{\mu}$. This follows immediately by applying Theorem 10.1 to each column of $\hat{\boldsymbol{\mu}}$. Asymptotic normality follows from Theorem 10.4.

Since $\boldsymbol{\mu}$ is now a matrix, the task of constructing a quadratic form leading to a test statistic is somewhat painful notationally. We start by writing $\boldsymbol{\mu}$ as a column vector of length pr:

$$\begin{aligned}\boldsymbol{\mu}_v^T &= \mathrm{vec}(\boldsymbol{\mu})^T \\ &= (\boldsymbol{\mu}(1,1), \boldsymbol{\mu}(2,1), \ldots, \boldsymbol{\mu}(p,1), \boldsymbol{\mu}(1,2), \ldots, \boldsymbol{\mu}(p-1,r), \boldsymbol{\mu}(p,r)).\end{aligned}$$

In words, $\boldsymbol{\mu}_v$ is constructed by placing the columns of $\boldsymbol{\mu}$ on top of one another. The covariance matrix for the error terms is given by

$$\boldsymbol{\Sigma}_\varepsilon(i,j) = \mathrm{cov}[\langle \varepsilon_1, u_i \rangle, \langle \varepsilon_1, u_j \rangle] = \mathrm{E}[\langle \varepsilon_1, u_i \rangle \langle \varepsilon_1, u_j \rangle], \quad 1 \le i, \ \ j \le r,$$

and the diagonal matrix containing the largest p eigenvalues of C is

$$\boldsymbol{\Gamma}(i,j) = \lambda_i \delta_{ij}, \quad \text{for } 1 \le i,j \le p,$$

where δ_{ij} is Kronecker's delta.

With this notation in place, we consider the random variable

$$\begin{aligned}\Lambda_{pr} &= N(1+\zeta)^{-1}(\hat{\boldsymbol{\mu}}_v - \hat{\boldsymbol{\mu}}_v^*)^T \\ &\quad \times \left(\boldsymbol{\Sigma}_\varepsilon \otimes \boldsymbol{\Gamma}^{-1} + \mathrm{E}\left[\boldsymbol{\Delta}_1 \otimes (\boldsymbol{\Gamma}^{-1}\boldsymbol{\Delta}_2\boldsymbol{\Gamma}^{-1})\right]\right)^{-1} (\hat{\boldsymbol{\mu}}_v - \hat{\boldsymbol{\mu}}_v^*),\end{aligned}$$

where $\boldsymbol{\Delta}_1, \boldsymbol{\Delta}_2$ are defined in (10.22) and (10.23), respectively.

Assuming equal covariances, Theorem 10.4 implies that under H_0, $\Lambda_{pr} \xrightarrow{d} \chi^2(pr)$. An extension of the argument used in the proof of Theorem 10.3 yields that $\Lambda_{pr} \xrightarrow{P} \infty$, under H_A, as long as p and r are so large that $\boldsymbol{\mu} \neq \boldsymbol{\mu}^*$. That such a pair (p,r) exists, follows immediately from the fact that the products $v_i u_j$ form a basis in $L_2([0,1] \times [0,1])$.

A computable approximation to Λ_{pr} is

$$\hat{\Lambda}_{pr} = N(1+N/M)^{-1}(\hat{\mu}_v - \hat{\mu}_v^*)^T \left(\hat{\Sigma}_\varepsilon \otimes \hat{\Gamma}^{-1}\right)^{-1} (\hat{\mu}_v - \hat{\mu}_v^*), \tag{10.17}$$

where $\hat{\Sigma}_\varepsilon$ is the pooled sample covariance matrix of the residuals and $\hat{\Gamma} = \text{diag}(\hat{\lambda}_1, \hat{\lambda}_2, \ldots, \hat{\lambda}_p)$, with the $\hat{\lambda}_i$ being the eigenvalues of the empirical covariance operator of the pooled X_i and X_j^*.

We finally turn to the most complex case of different covariances for the explanatory functions. We now expand both the explanatory and response functions with respect to two arbitrary, potentially different, bases in $L^2[0,1]$, $\{u_i\}$ and $\{w_j\}$, respectively:

$$\psi(t,s) = \sum_{i=1}^{\infty}\sum_{j=1}^{\infty} \mu_{ji} u_i(t) w_j(s), \quad \psi^*(t,s) = \sum_{i=1}^{\infty}\sum_{j=1}^{\infty} \mu_{ji}^* u_i(t) w_j(s).$$

This leads to the relations $\mathbf{Y} = \mathbf{X}\boldsymbol{\mu} + \boldsymbol{\varepsilon}'$, $\mathbf{Y}^* = \mathbf{X}^*\boldsymbol{\mu}^* + \boldsymbol{\varepsilon}^{*\prime}$, with all terms analogously defined as in the equal variance case, but using the bases $\{u_i\}$ and $\{w_j\}$. Thus the least squares estimates are again $\hat{\boldsymbol{\mu}} = (\mathbf{X}^T\mathbf{X})^{-1}\mathbf{X}^T\mathbf{Y}$ and $\hat{\boldsymbol{\mu}}^* = (\mathbf{X}^{*T}\mathbf{X}^*)^{-1}\mathbf{X}^{*T}\mathbf{Y}^*$. ($\boldsymbol{\mu}, \boldsymbol{\mu}^*, \hat{\boldsymbol{\mu}}, \hat{\boldsymbol{\mu}}^*$ are now $p \times r$ matrices) Extending the argument developed in Section 10.2, we arrive at the test statistic

$$\begin{aligned}\hat{\Lambda}_{pr} = {} & N\text{vec}(\hat{\boldsymbol{\mu}} - \hat{\boldsymbol{\mu}}^*)^T(\hat{\Sigma}_\varepsilon \otimes \hat{\Sigma}_1^{-1} \\ & + (N/M)\hat{\Sigma}_\varepsilon^* \otimes \hat{\Sigma}_1^{-1*})^{-1}\text{vec}(\hat{\boldsymbol{\mu}} - \hat{\boldsymbol{\mu}}^*),\end{aligned} \tag{10.18}$$

where the residual covariance matrices Σ_ε and Σ_ε^* are computed for each sample separately. The estimate $\hat{\Sigma}_1^{-1}$ is given by $N^{-1}\mathbf{X}^T\mathbf{X}$, and $\hat{\Sigma}_1^{*-1}$ is defined analogously.

The distribution of statistics (10.17) and (10.18) is approximated by the chi-square distribution with pr degrees of freedom. Selection of p and r is discussed in Section 10.4. If p and/or r are large, the normalized χ^2 distribution can be approximated by a normal distribution, as in Cardot *et al.* (2003), who studied a single scalar response model and tested $\psi = 0$. In our case, due to the complexity of the problem, the rigorous derivation of the normal convergence with $p = p_n$ depending on a sample size would be far more tedious, so it is not pursued. To perform a test, a finite p (and r) must be chosen no matter what approximation is used, and as illustrated in Section 10.6 large p (and r) do not necessarily lead to meaningful results.

10.4 Summary of the testing procedures

In order to apply the tests, we must first verify if a linear functional model approximates the dependence structure of the data reasonably well. This can be done using the techniques of Chiou and Müller (2007) described in Section 9.4. The assumptions of independence and identical distribution of the regressor curves can be verified using the test of Chapter 7. Checking the independence of the errors is more complicated because they are not observable; it is studied in Chapter 11. Before applying the tests, the regressors and the responses must be centered, so that their sample means are zero.

Next, the values of p and r must be chosen. In applications in which the FPC's have a clear interpretation, these values can be chosen so that the action of the operators on specific subspaces spanned by the FPC's of interest is compared. In the absence of such an interpretation, several data driven approaches are available. When the covariances are approximately equal, typically p is chosen so large that $\sum_{k=1}^{p} \hat{\lambda}_k$ exceeds a required percentage of the variance of the X_i (defined as $(N+M)^{-1}(\sum_{i=1}^{N} \int X_i^2(t)dt + \sum_{j=1}^{M} \int X_i^{*2}(t)dt)$ for the centered functions). We choose r analogously for the response functions. When the covariances cannot be assumed equal then we propose, as one possibility, a pooling technique to choose p and r. Pooling the explanatory functions we have

$$(N+M)^{-1}\left(\sum_{i=1}^{N} X_i(s)X_i(t) + \sum_{j=1}^{M} X_j^*(s)X_j^*(t)\right)$$
$$\stackrel{a.s.}{\to} (1+1/\zeta)^{-1}c(s,t) + (1+\zeta)c^*(s,t).$$

We propose taking the w_i to be the eigenfunctions of $(1+1/\zeta)^{-1}c(s,t) + (1+\zeta)c^*(s,t)$ which is itself a covariance kernel. The u_i can be defined in an analogous manner using the response functions. Such a choice will allow smaller values of p (and r) to be taken so that any bias from neglected terms is minimal, but we can still expect reasonable power. The values p and r can be chosen as before, but now with respect to the pooled variance. All these steps can be implemented in the `R` package `fda`, and ready–made functions for the percentage of variance explained by FPC's are available. Other methods of choosing p (or r) are implemented in the MATLAB PACE package developed at the University of California at Davis.

It is often useful to compute the test for a wide range of values of p (and r) and check if a uniform pattern emerges. This approach is illustrated in Section 10.6.

Finally, we compute the test statistic $\hat{\Lambda}$, and reject H_0 if it exceeds the χ^2 density with DF degrees of freedom according to the following table:

Response	**Covariances**	$\hat{\Lambda}$	**DF**
Scalar	Equal	(10.12)	p
Scalar	Different	(10.14)	p
Functional	Equal	(10.17)	pr
Functional	Different	(10.18)	pr

The term "equal covariances" refers to assumptions (10.7), (10.8) in the scalar case, and (10.7), (10.15) in the functional case.

10.5 A small simulation study

Before turning to data examples, we present the results of a small simulation study. We evaluate the performance of the test based on the most general statistic (10.18). The test performs even better in the equal variances case (provided the simulated data have equal variances). We consider the fully functional linear model with integral kernels of the form

$$\psi(s,t) = c \min\{s,t\} \quad \psi^*(s,t) = c^* \min\{s,t\},$$

where c and c^* are constants. We set $N = M = 100$, and use 5 EFPC's for the regressors variables, and 3 EFPC's for the responses. The results are based on 100 replications.

We use standard Brownian motions as error terms, and consider the regressors of the following four types:

(A) Standard Brownian motions in both samples (Gaussian processes, equal covariances).

(B) For the first sample the explanatory functions are standard Brownian motions and for the second sample they are Brownian bridges (Gaussian processes, different covariances).

(C) For both sets of explanatory functions we use

$$X(t) = n^{-1/2} \sum_{k=1}^{\lfloor nt \rfloor} \frac{T_i}{\sqrt{\operatorname{var}(T_i)}},$$

Table 10.1 Empirical rejection rates for the test based on the most general statistic (10.18). From top to bottom, scenarios A, B, C, D described in the text.

	$\alpha/(c,c^*)$	(0,0)	(1,1)	(1,0)	(1.5,0)	(2,0)
(A)	0.10	0.14	0.08	0.50	0.90	0.98
	0.05	0.09	0.03	0.40	0.81	0.98
	0.01	0.03	0.00	0.18	0.63	0.92

	$\alpha/(c,c^*)$	(0,0)	(1,1)	(1,0)	(1.5,0)	(2,0)
(B)	0.10	0.14	0.09	0.45	0.90	0.98
	0.05	0.08	0.06	0.28	0.80	0.95
	0.01	0.00	0.01	0.14	0.60	0.93

	$\alpha/(c,c^*)$	(0,0)	(1,1)	(1,0)	(1.5,0)	(2,0)
(C)	0.10	0.11	0.10	0.47	0.85	0.99
	0.05	0.04	0.05	0.32	0.78	0.95
	0.01	0.02	0.02	0.18	0.60	0.87

	$\alpha/(c,c^*)$	(0,0)	(1,1)	(1,0)	(1.5,0)	(2,0)	(2.5,0)	(3,0)	(3.5,0)
(D)	0.10	0.12	0.09	0.27	0.33	0.49	0.77	0.87	0.93
	0.05	0.04	0.05	0.17	0.22	0.37	0.63	0.78	0.89
	0.01	0.01	0.02	0.05	0.07	0.23	0.48	0.53	0.70

where $\{T_i\}$ are iid t-distributed random variables with 6 degrees of freedom and $n = 200$ (heavy–tailed distribution, equal covariances).

(D) The first set of explanatory functions are defined as in (C). For the second set we consider

$$X^*(t) = X(t)[X(1) - X(t)],$$

where $X(t)$ is defined in (C) (heavy–tailed distribution, different covariances).

As we can see from the tables, the method works fairly well. The empirical sizes are close to the nominal sizes (first two columns of each table), and the power increases with the size of the difference. The power is smaller if the explanatory functions do not have a common distribution, and/or are heavy–tailed.

10.6 Application to medfly and magnetometer data

We now illustrate the application of the test on two examples. The first example is motivated by the work presented in Carey *et al.* (2002), Chiou *et al.* (2004), Müller and Stadtmüller (2005), Chiou and Müller (2007), among others, and studies egg-laying curves of Mediterranean fruit flies (medflies). The second example is an application to the measurements of the magnetic field generated by near Earth space currents.

Egg-laying curves of Mediterranean fruit flies (continued). We applied the test of Section 10.2 (without assuming equal variances) to the medfly data introduced in Section 10.1. Table 10.2 shows the P-values for the five initial FPC's ($p \le 5$). The P-values for larger p do not exceed half a percent. We cannot reject $H_0 : \psi = \psi^*$ if we use the test with $p = 1$, but if $p > 1$, we reject H_0. To understand this result, we must turn to formula (10.13). The test compares estimates of μ_i to those of μ_i^* for $i \le p$. Acceptance of H_0 for $p = 1$ means that the curves $\mu_1 w_1(s)$ and $\mu_1^* w_1(s)$ are not significantly different. Their estimates are shown in the left panel of Figure 10.2. The functions w_i were computed by pooling all explanatory curves, as explained in Section 10.4. The estimated coefficients are $\hat{\mu}_1 = 49.64$, $\hat{\mu}_1^* = 46.60$. By contrast, the estimates $\hat{\mu}_2 = 15.45$, $\hat{\mu}_2^* = 29.88$ are very different, and consequently the curves $\hat{\mu}_2 w_2(s)$ and $\hat{\mu}_2^* w_2(s)$ shown in the right panel of Figure 10.2 look different.

Table 10.2 The values of statistic (10.14) and the corresponding P-values for several values of $p = p^*$.

p	Λ_{pp^*}	P-Value
1	1.3323	0.2484
2	11.3411	0.0034
3	10.6097	0.0140
4	23.8950	0.0001
5	33.1144	0.0000

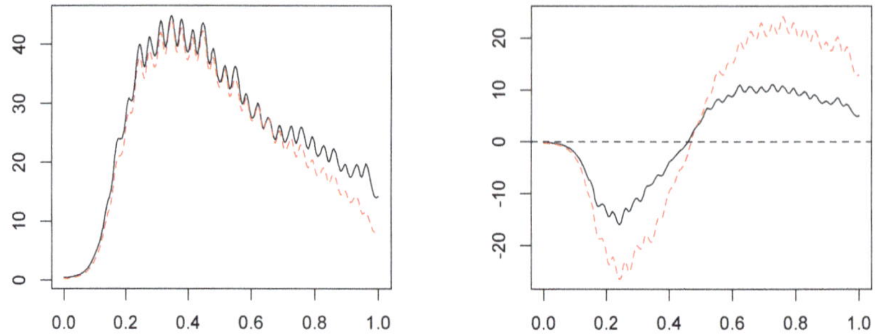

Fig. 10.2 The left panel shows curves $\hat{\mu}_1 w_1(s)$ (solid) and $\hat{\mu}_1^* w_1(s)$ (dotted). The right panel shows correspondingly $\hat{\mu}_2 w_2(s)$ and $\hat{\mu}_2^* w_2(s)$

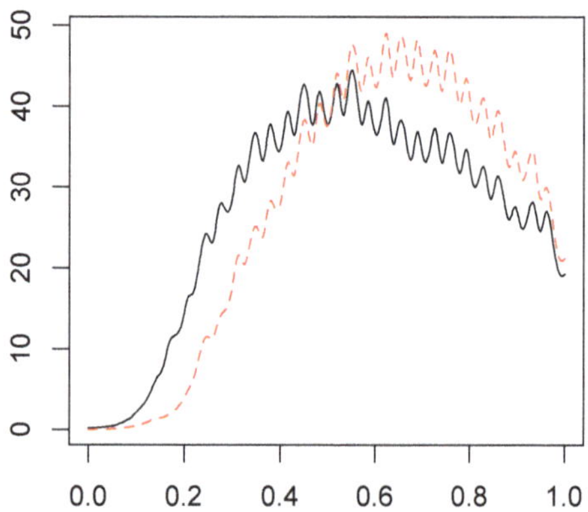

Fig. 10.3 Approximations of the kernel functions ψ (solid) and ψ^* (dotted) with $p = 2$. The curves are the sums of the corresponding curves in the left and right panels of Figure 10.2.

The approximations to ψ and ψ^* which use $p = 2$ FPC's are thus sufficient to detect the difference. They are shown in Figure 10.3.

Comparing the estimates $\hat{\mu}_2$ and $\hat{\mu}_2^*$ or the curves in Figures 10.2 and 10.3 gives a strong hint that the kernels ψ and ψ^* cannot be assumed equal. Our tests allow us to attach statistical significance statements to these conclusions.

Data from terrestrial magnetic observatories. We now apply our methodology to magnetometer data. A comprehensive case study is not our goal, we would rather like to illustrate the steps outlined in Section 10.4 in a practically relevant setting. Broader space physics issues related to this example are explained in Kamide *et al.*

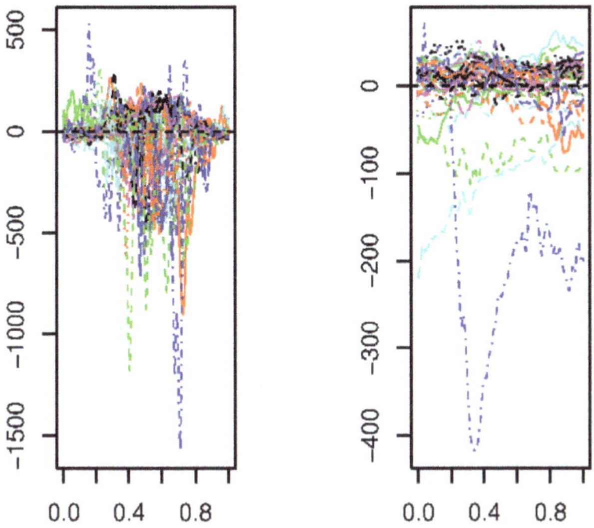

Fig. 10.4 Observations for sample A: left panel CMO (X), right panel HON (Y).

(1998), while Chapters 9, 10, 13 of Kivelson and Russell (1997) provide a detailed background.

A sample of 40 functional regressors and corresponding responses is shown in Figure 10.4. Each curve in Figure 10.4 shows one minute averages in a UT (Universal Time) day of the component of the magnetic field lying in the Earth's tangent plane pointing toward the magnetic North. We thus have 1440 data points per curve. Splitting the magnetometer data into days and treating the daily curves as functional observations is natural because of the daily rotation of the Earth. The curves X_i reflect ionospheric magnetic activity in the polar region known as substorms, which are spectacularly manifested as the northern lights (*aurora borealis*). The curves Y_i reflect magnetospheric activity in the magnetic equatorial region in the same UT day. We consider three samples: A, B, C. Each of them consists of about 40 pairs of curves. All measurements were recorded in 2001, the X_i at College (CMO), Alaska; the Y_i at Honolulu (HON), Hawaii. Sample A contains substorms which took place in January through March, B in April–June, C in July–September. Using the graphical goodness–of–fit test of Chiou and Müller (2007), see Section 8.6, and the test of Chapter 7, Kokoszka *et al.* (2008) verified that the fully functional linear model is a reasonable approximation and that the functional observations can be assumed to be uncorrelated. Moreover, on physical grounds, the data can be assumed to be approximately independent because the M–I system resets itself after each rotation of the Earth, and the effect of larger disturbances of solar origin decay within about two days.

Intuitively, we would expect rejections of the null for all three pairs: A–B, B–C, and A–C, as the position of the axis of the Earth relative to the Sun shifts with each season, and substorms are influenced by the solar wind. This is indeed the

Table 10.3 P–values for testing the equality of regression operators in samples A and B.

p/r	1	2	3	4	5	6	7	8	9	10
1	0.344	0.608	0.231	0.280	0.349	0.380	0.372	0.391	0.351	0.257
2	0.147	0.259	0.274	0.416	0.565	0.422	0.373	0.345	0.339	0.310
3	0.204	0.378	0.399	0.621	0.762	0.592	0.582	0.621	0.654	0.478
4	0.120	0.305	0.299	0.567	0.716	0.619	0.654	0.307	0.315	0.158
5	0.440	0.668	0.555	0.741	0.861	0.730	0.792	0.515	0.453	0.223
6	0.582	0.891	0.798	0.793	0.883	0.554	0.567	0.605	0.218	0.106
7	0.689	0.962	0.950	0.911	0.954	0.749	0.792	0.783	0.566	0.427
8	0.965	0.968	0.972	0.952	0.958	0.815	0.755	0.582	0.432	0.257
9	0.981	0.804	0.962	0.980	0.972	0.821	0.837	0.753	0.722	0.456
10	0.727	0.585	0.903	0.973	0.986	0.972	0.973	0.941	0.935	0.626
11	0.911	0.880	0.991	0.999	0.999	0.998	0.998	0.994	0.995	0.990
12	0.856	0.860	0.989	0.997	0.959	0.962	0.940	0.930	0.845	0.889
13	0.667	0.856	0.982	0.988	0.939	0.950	0.889	0.845	0.784	0.844
14	0.395	0.457	0.798	0.418	0.314	0.445	0.240	0.240	0.201	0.282
15	0.398	0.481	0.847	0.414	0.321	0.456	0.276	0.255	0.170	0.113

case for tests in cases B–C and A–C, for which the P–values are very small: for B–C the largest P-value is 0.034, and for A–C 0.007 (for $p \le 15, r \le 10$). The results for testing samples A and B presented in Table 10.3 indicate the acceptance of H_0. In retrospect, this conclusion is supported by the observation, well–known in the space–physics community, that M–I disturbances tend to be weaker in summer months. Our test thus shows that it is reasonable to assume that the effect of substorms on low–latitude currents is approximately the same in first and second quarter of 2001, but changes in the third quarter (possibly due to weaker substorms).

10.7 Asymptotic theory

We now list he assumptions under which the tests presented in this chapter are valid and present selected asymptotic results. They focus on the simplest case of scalar responses and equal variances, only Theorem 10.4 pertains to functional responses, and is stated for illustration. The asymptotic techniques used in the scalar equal variances case can be extended to the other cases, but the notation becomes more complex, as explained in Section 10.3. The results presented here do not follow from the existing multivariate theory because the regression errors are not independent and include projections on the "left over" FPC's $v_{p+1}, v_{p+2}, \ldots, u_{r+1}, u_{r+2}, \ldots$, etc. Theorems 10.2 and 10.4 are of particular interest, as they state the exact asymptotic distribution of the LSE's in a multivariate regression obtained by projecting a functional regression.

We state the assumptions on the sample $(X_i, Y_i), 1 \le i \le N$. The assumptions on $(X_i^*, Y_i^*), 1 \le i \le M$ are the same. The two samples are assumed independent.

Assumption 10.1. *The observations $\{X_n\}$ are iid mean zero random functions in $L^2([0,1])$ satisfying*

$$\mathrm{E}\|X_n\|^4 = E\left[\int X_n^2(t)dt\right]^2 < \infty.$$

For the linear model with scalar responses, we formulate the following assumption.

Assumption 10.2. *The scalar responses Y_i satisfy*

$$Y_i = \int \psi(s)X_i(s)ds + \varepsilon_i,$$

with iid mean zero errors ε_i satisfying $\mathrm{E}\varepsilon_i^4 < \infty$, and $\psi \in L^2([0,1])$. The errors ε_i and the regressors X_i are independent.

In the case of functional responses, we define an analogous assumption.

Assumption 10.3. *The functional responses $Y_i \in L^2([0,1])$ satisfy*

$$Y_i(t) = \int \psi(t,s)X_i(s)ds + \varepsilon_i(t),$$

with iid mean zero errors ε_i satisfying

$$\mathrm{E}\|\varepsilon_n\|^4 = E\left[\int \varepsilon_n^2(t)dt\right]^2 < \infty,$$

and $\psi \in L^2([0,1] \times [0,1])$. The errors ε_i and the regressors X_i are independent.

Since the following simple lemma is used repeatedly in the proofs, it is stated first for ease of reference.

Lemma 10.1. *Suppose X is a mean zero random element of L^2 satisfying $E\|X\|^2 < \infty$. Then*

$$E[\langle v_i, X\rangle \langle v_j, X\rangle] = \lambda_i \delta_{ij},$$

where δ_{ij} is Kronecker's delta.

Theorem 10.1. *Suppose Assumptions 10.1, 10.2 and condition (2.12) hold. Then,*

$$\hat{\boldsymbol{\mu}} = \left(\mathbf{X}^T\mathbf{X}\right)^{-1}\mathbf{X}^T\mathbf{Y} \stackrel{a.s.}{\to} \boldsymbol{\mu}, \quad \text{as } N \to \infty,$$

where $\boldsymbol{\mu}^T = (\mu_1, \ldots, \mu_p)$ and $\stackrel{a.s.}{\to}$ refers to almost sure convergence.

Proof. To analyze the behavior of $\hat{\boldsymbol{\mu}}$, let us start by considering

$$(\mathbf{X}^T\mathbf{X})(i,j) = \sum_{k=1}^{N} \langle v_i, X_k\rangle \langle v_j, X_k\rangle.$$

Since the X_i are iid, by the strong law of large numbers

$$\frac{1}{N}(\mathbf{X}^T\mathbf{X})(i,j) \stackrel{a.s.}{\to} \mathrm{E}(\langle v_i, X_1\rangle \langle v_j, X_1\rangle), \quad \text{as } N \to \infty.$$

From Lemma 10.1 we have that $\mathrm{E}(\langle v_i, X_1\rangle \langle v_j, X_1\rangle) = \lambda_i \delta_{ij}$. Therefore $N^{-1}(\mathbf{X}^T\mathbf{X})$ converges almost surely to a $p \times p$ diagonal matrix whose diagonal entries are the eigenvalues of C.

Turning to $\mathbf{X}^T\mathbf{Y}$, using (10.4), observe

$$\mathbf{X}^T\mathbf{Y}(i) = \sum_{j=1}^{N} \langle v_i, X_j\rangle Y_j = \sum_{j=1}^{N} \langle v_i, X_j\rangle \sum_{k=1}^{p} \mu_k \langle v_k, X_j\rangle + \sum_{j=1}^{N} \varepsilon'_j \langle v_i, X_j\rangle.$$

Applying again the strong law of large numbers and Lemma 10.1again, we obtain, as $N \to \infty$,

$$N^{-1} \sum_{j=1}^{N} \langle v_i, X_j\rangle \sum_{k=1}^{p} \mu_k \langle v_k, X_j\rangle \stackrel{a.s.}{\to} \mathrm{E} \sum_{k=1}^{p} \mu_k \langle v_i, X_1\rangle \langle v_k, X_1\rangle = \mu_i \lambda_i \delta_{ij}.$$

Lastly, we will show that, as $N \to \infty$, $N^{-1} \sum_{j=1}^{N} \varepsilon'_j \langle v_i, X_j\rangle \stackrel{a.s.}{\to} 0$. Recalling the definition of ε'_i, (10.4), we have

$$N^{-1} \sum_{j=1}^{N} \varepsilon'_j \langle v_i, X_j\rangle = N^{-1} \sum_{j=1}^{N} \varepsilon_j \langle v_i, X_j\rangle + N^{-1} \sum_{j=1}^{N} \sum_{k=p+1}^{\infty} \mu_k \langle v_k, X_j\rangle \langle v_i, X_j\rangle.$$

Since $\{\varepsilon_i\}$ and $\{X_i\}$ are independent, by the strong law of large numbers and Assumption 10.2

$$N^{-1} \sum_{j=1}^{N} \varepsilon_j \langle v_i, X_j\rangle \stackrel{a.s.}{\to} 0.$$

Similarly, using Lemma 10.1 and noting that $i \le p$, we get

$$N^{-1} \sum_{j=1}^{N} \sum_{k=p+1}^{\infty} \mu_k \langle v_k, X_j\rangle \langle v_i, X_j\rangle \stackrel{a.s.}{\to} \mathrm{E} \sum_{k=p+1}^{\infty} \mu_k \langle v_k, X_1\rangle \langle v_i, X_1\rangle = 0. \qquad \square$$

Theorem 10.2. *Suppose Assumptions 10.1, 10.2 and condition* (2.12) *hold. Then, as* $N \to \infty$,

$$\sqrt{N}(\hat{\boldsymbol{\mu}} - \boldsymbol{\mu}) \stackrel{d}{\to} N(0, \boldsymbol{\Sigma}_p),$$

where $N(0, \boldsymbol{\Sigma}_p)$ *is a multivariate normal random vector with mean 0 and covariance matrix* $\boldsymbol{\Sigma}_p$ *defined by* (10.10) *and* (10.11).

Proof. By the definition of $\hat{\boldsymbol{\mu}}$ (10.6),

$$\sqrt{N}(\hat{\boldsymbol{\mu}} - \boldsymbol{\mu}) = \sqrt{N}\left(\left(\mathbf{X}^T\mathbf{X}\right)^{-1} \mathbf{X}^T\mathbf{Y} - \boldsymbol{\mu} \right).$$

Defining $\boldsymbol{\varepsilon}'^T = (\varepsilon'_1, \ldots, \varepsilon'_N)$, the above reduces to

$$\sqrt{N}(\hat{\boldsymbol{\mu}} - \boldsymbol{\mu}) = \left(N^{-1}\mathbf{X}^T\mathbf{X}\right)^{-1} N^{-1/2}\mathbf{X}^T\boldsymbol{\varepsilon}'.$$

By Lemma 10.1, $N^{-1}\mathbf{X}^T\mathbf{X}$ converges almost surely to a diagonal matrix whose diagonal elements are the first p eigenvalues of the covariance operator of the $\{X_i\}$. Therefore we need only focus on the behavior of $N^{-1/2}\mathbf{X}^T\boldsymbol{\varepsilon}'$ and use Slutsky's Theorem to obtain the claimed limiting distribution. Considering the ith coordinate of $N^{-1/2}\mathbf{X}^T\boldsymbol{\varepsilon}'$ we have

$$(N^{-1/2}\mathbf{X}^T\boldsymbol{\varepsilon}')(i) = N^{-1/2}\sum_{j=1}^{N}\langle v_i, X_j\rangle\varepsilon'_j, \quad 1 \le i \le p. \tag{10.19}$$

By Assumption 10.2 the above is a summation of iid random variables. Since each coordinate of $N^{-1/2}\mathbf{X}^T\boldsymbol{\varepsilon}'$ is given by such a sum, Assumption 10.2 implies that $\mathbf{X}^T\boldsymbol{\varepsilon}'$ can be expressed as a sum of iid random vectors. We can apply the multivariate central limit theorem to obtain the claimed multivariate normal limiting distribution if we can show that each entry of the covariance matrix is finite. Therefore we spend the rest of the proof deriving the form for $\boldsymbol{\Sigma}_p$ and showing that its entries are finite. Using the definition of ε'_i, we obtain

$$\begin{aligned} &N^{-1/2}\sum_{j=1}^{N}\langle v_i, X_j\rangle\varepsilon'_j \\ &\quad = N^{-1/2}\left(\sum_{j=1}^{N}\langle v_i, X_j\rangle\varepsilon_j + \sum_{j=1}^{N}\langle v_i, X_j\rangle\sum_{k=p+1}^{\infty}\mu_k\langle v_k, X_j\rangle\right). \end{aligned}$$

Because the $\{X_j\}$ are independent, both sums (with respect to j) are sums of independent and identically distributed random variables. Furthermore, since $\{\varepsilon_j\}$ are independent of all other terms, we also have that the two sums above are uncorrelated. Therefore it follows that

$$\begin{aligned} &\operatorname{var}\left(N^{-1/2}\left(\sum_{j=1}^{N}\langle v_i, X_j\rangle\varepsilon_j + \sum_{j=1}^{N}\langle v_i, X_j\rangle\sum_{k=p+1}^{\infty}\mu_k\langle v_k, X_j\rangle\right)\right) \\ &\quad = \operatorname{var}\left(\langle v_i, X_1\rangle\varepsilon_1\right) + \operatorname{var}\left(\langle v_i, X_1\rangle\sum_{k=p+1}^{\infty}\mu_k\langle v_k, X_1\rangle\right). \end{aligned} \tag{10.20}$$

Considering the first term of (10.20), we have by the independence of X_1 and ε_1 and Lemma 10.1 that

$$\operatorname{var}(\langle v_i, X_1\rangle\varepsilon_1) = \lambda_i\sigma^2 < \infty.$$

Turning to the second term of (10.20), we have by Lemma 10.1

$$\operatorname{var}\left[\langle v_i, X_1\rangle\sum_{k=p+1}^{\infty}\mu_k\langle v_k, X_1\rangle\right] = \mathrm{E}\left[\langle v_i, X_1\rangle\sum_{k=p+1}^{\infty}\mu_k\langle v_k, X_1\rangle\right]^2.$$

Applying the Cauchy-Schwarz inequality it follows that

$$\mathrm{E}\left[\langle v_i, X_1\rangle \sum_{k=p+1}^{\infty} \mu_k \langle v_k, X_1\rangle\right]^2$$
$$\leq \left(\mathrm{E}\left[\langle v_i, X_1\rangle\right]^4 \mathrm{E}\left[\sum_{k=p+1}^{\infty} \mu_k \langle v_k, X_1\rangle\right]^4\right)^{1/2}.$$

As a consequence of Assumption 10.1, we obtain that

$$\mathrm{E}\left[\langle v_i, X_1\rangle\right]^4 < \infty.$$

Using the Cauchy-Schwarz inequality again we have

$$\mathrm{E}\left[\sum_{k=p+1}^{\infty} \mu_k \langle v_k, X_1\rangle\right]^4 \leq \mathrm{E}\left[\sum_{k=p+1}^{\infty} \mu_k^2 \sum_{s=p+1}^{\infty} \langle v_s, X_1\rangle^2\right]^2.$$

Therefore we can infer that

$$\mathrm{E}\left[\langle v_i, X_1\rangle \sum_{k=p+1}^{\infty} \mu_k \langle v_k, X_1\rangle\right]^2$$
$$\leq \left(\sum_{k=p+1}^{\infty} \mu_k^2\right)\left(\mathrm{E}\left[\langle v_i, X_1\rangle\right]^4 \mathrm{E}\left[\sum_{s=p+1}^{\infty} \langle v_s, X_1\rangle^2\right]^2\right)^{1/2}.$$

Using Assumption 10.2 and Bessel's Inequality we obtain that

$$\sum_{k=p+1}^{\infty} \mu_k^2 \leq \|\psi\|^2 < \infty.$$

Similarly, using Assumption 10.1 and Bessel's Inequality we have that

$$\mathrm{E}\left(\sum_{s=p+1}^{\infty} \langle v_s, X_1\rangle^2\right)^2 \leq \mathrm{E}\|X_1\|^4 < \infty.$$

Combining the above with Assumption 10.1 we conclude

$$\mathrm{E}\left[\langle v_i, X_1\rangle \sum_{k=p+1}^{\infty} \mu_k \langle v_k, X_1\rangle\right]^2 < \infty. \tag{10.21}$$

and it follows that the diagonal elements of $\boldsymbol{\Sigma}_p$ are given

$$\boldsymbol{\Sigma}_p(i,i) = \lambda_i^{-1}\sigma^2 + \lambda_i^{-2}\mathrm{var}\left(\langle v_i, X_1 \rangle \sum_{k=p+1}^{\infty} \mu_k \langle v_k, X_1 \rangle \right), \quad i = 1, \ldots, p.$$

Next we examine the joint behavior of the coordinates. Combining (10.21) with the Cauchy-Schwarz inequality we have

$$\mathrm{cov}\left[(\mathbf{X}^T\boldsymbol{\varepsilon}')(i), (\mathbf{X}^T\boldsymbol{\varepsilon}')(j)\right] < \infty \quad i = 1, \ldots, p \quad \text{and} \quad j = 1, \ldots, p.$$

Therefore to finish the proof we need only derive the form for the off diagonal terms of $\boldsymbol{\Sigma}_p$. Using (10.19), Assumption 10.2, and Lemma 10.1, it is easy to verify that for $i \neq j$

$$\begin{aligned}
\mathrm{cov}&\left[(\mathbf{X}^T\boldsymbol{\varepsilon}')(i), (\mathbf{X}^T\boldsymbol{\varepsilon}')(j)\right] = \mathrm{E}((\mathbf{X}^T\boldsymbol{\varepsilon}')(i)(\mathbf{X}^T\boldsymbol{\varepsilon}')(j)) \\
&= \mathrm{E}\left(\sum_{q=1}^{N} \langle v_i, X_q \rangle \sum_{k=p+1}^{\infty} \mu_k \langle v_k, X_q \rangle \sum_{s=1}^{N} \langle v_j, X_s \rangle \sum_{k=p+1}^{\infty} \mu_k \langle v_k, X_s \rangle \right) \\
&= \sum_{q=1}^{N} \mathrm{E}\left(\langle v_i, X_q \rangle \langle v_j, X_q \rangle \left(\sum_{k=p+1}^{\infty} \mu_k \langle v_k, X_q \rangle \right)^2 \right) \\
&= N\mathrm{E}\left(\langle v_i, X_1 \rangle \langle v_j, X_1 \rangle \left(\sum_{k=p+1}^{\infty} \mu_k \langle v_k, X_1 \rangle \right)^2 \right).
\end{aligned}$$

Therefore it follows that the off diagonal terms of $\boldsymbol{\Sigma}_p$ are given by

$$\boldsymbol{\Sigma}_p(i,j) = \lambda_i^{-1}\lambda_j^{-1}\mathrm{E}\left(\langle v_i, X_1 \rangle \langle v_j, X_1 \rangle \left(\sum_{k=p+1}^{\infty} \mu_k \langle v_k, X_1 \rangle \right)^2 \right), \quad i \neq j,$$

which concludes the proof. □

Theorem 10.3. *Suppose Assumptions 10.1, 10.2 and conditions* (2.12), (10.3) (10.7), (10.8) *hold. Suppose further that p is so large that* $\boldsymbol{\mu} \neq \boldsymbol{\mu}^*$. *Then* $\Lambda_p \xrightarrow{P} \infty$, *as* $N \to \infty$.

Proof. We start by expanding Λ_p as

$$\begin{aligned}
\Lambda_p &= N(1+\zeta)^{-1}(\hat{\boldsymbol{\mu}} - \hat{\boldsymbol{\mu}}^*)^T \boldsymbol{\Sigma}_p^{-1}(\hat{\boldsymbol{\mu}} - \hat{\boldsymbol{\mu}}^*) \\
&= N(1+\zeta)^{-1}(\hat{\boldsymbol{\mu}} - \boldsymbol{\mu} - \hat{\boldsymbol{\mu}}^* + \boldsymbol{\mu}^*)^T \boldsymbol{\Sigma}_p^{-1}(\hat{\boldsymbol{\mu}} - \boldsymbol{\mu} - \hat{\boldsymbol{\mu}}^* + \boldsymbol{\mu}^*) \\
&\quad + N(1+\zeta)^{-1}(\boldsymbol{\mu} - \boldsymbol{\mu}^*)^T \boldsymbol{\Sigma}_p^{-1}(\boldsymbol{\mu} - \boldsymbol{\mu}^*) \\
&\quad + 2N(1+\zeta)^{-1}(\boldsymbol{\mu} - \boldsymbol{\mu}^*)^T \boldsymbol{\Sigma}_p^{-1}(\hat{\boldsymbol{\mu}} - \boldsymbol{\mu} - \hat{\boldsymbol{\mu}}^* + \boldsymbol{\mu}^*).
\end{aligned}$$

Therefore we need only consider each term above. From Theorem 10.2 it follows that

$$N(1+\zeta)^{-1}(\hat{\mu}-\mu-\hat{\mu}^*+\mu^*)^T\boldsymbol{\Sigma}_p^{-1}(\hat{\mu}-\mu-\hat{\mu}^*+\mu^*)=O_P(1).$$

and

$$2N(1+\zeta)^{-1}(\mu-\mu^*)^T\boldsymbol{\Sigma}_p^{-1}(\hat{\mu}-\mu-\hat{\mu}^*+\mu^*)=O_P(\sqrt{N}).$$

The last term we need to consider is

$$N(1+\zeta)^{-1}(\mu-\mu^*)^T\boldsymbol{\Sigma}_p^{-1}(\mu-\mu^*).$$

Since $\boldsymbol{\Sigma}_p^{-1}$ is positive definite it follows that

$$(\mu-\mu^*)^T\boldsymbol{\Sigma}_p^{-1}(\mu-\mu^*)>0,$$

and we have

$$N(1+\zeta)^{-1}(\mu-\mu^*)^T\boldsymbol{\Sigma}_p^{-1}(\mu-\mu^*)\to\infty.$$

Furthermore when we divide the above by $\sqrt{N}$ we get

$$N^{1/2}(1+\zeta)^{-1}(\mu-\mu^*)^T\boldsymbol{\Sigma}_p^{-1}(\mu-\mu^*)\to\infty.$$

Therefore $N(1+\zeta)^{-1}(\mu-\mu^*)^T\boldsymbol{\Sigma}_p^{-1}(\mu-\mu^*)$ dominates all the other terms in the limit and the theorem follows. □

Theorem 10.4. *Suppose that Assumptions 10.1, 10.3 and conditions* (2.12), (10.7), (10.3) *and* (10.15) *hold. Then for each fixed $p \geq 1$ and $r \geq 1$, we have*

$$N^{1/2}(\hat{\boldsymbol{\mu}}_v-\boldsymbol{\mu}_v)\xrightarrow{d}N\left(0,\boldsymbol{\Sigma}_\varepsilon\otimes\boldsymbol{\Gamma}^{-1}+E\left[\boldsymbol{\Delta}_1\otimes(\boldsymbol{\Gamma}^{-1}\boldsymbol{\Delta}_2\boldsymbol{\Gamma}^{-1})\right]\right)$$

where $\mathbf{I}_r$ is the $r\times r$ identity matrix, and

$$\boldsymbol{\Delta}_1(j,t)=\left(\sum_{s=p+1}^{\infty}\mu_{sj}\langle v_s,X_1\rangle\right)\left(\sum_{x=p+1}^{\infty}\mu_{xt}\langle v_x,X_1\rangle\right),\tag{10.22}$$

and

$$\boldsymbol{\Delta}_2(i,q)=\langle v_i,X_1\rangle\langle v_q,X_1\rangle.\tag{10.23}$$

Proof. The asymptotic normality follows from an application of the multivariate CLT. The derivation of the exact form of the asymptotic variance involves lengthy technical manipulations, and is omitted to conserve space. □

Chapter 11
Tests for error correlation in the functional linear model

In this chapter, we consider two tests for error correlation in the fully functional linear model, which we call Methods I and II They complement the tools described in Section 8.6 and the graphical goodness of fit checks used in Chapter 9. To construct the test statistics, finite dimensional residuals are computed in two different ways, and then their autocorrelations are suitably defined. From these autocorrelation matrices, two quadratic forms are constructed whose limiting distribution are chi–squared with known numbers of degrees of freedom (different for the two forms). The test statistics can be relatively easily computed using the `R` package `fda`.

The remainder of the chapter is organized as follows. Section 11.2 develops the setting for the least squares estimation needed define the residuals used in Method I. After these preliminaries, both tests are described in Section 11.3. Their finite sample performance is evaluated in Section 11.4 through a simulation study, and further examined in Section 11.5 by applying both methods to magnetometer and financial data. The asymptotic justifications is presented in Section 11.6. This chapter is based on the work of Gabrys *et al.* (2010).

11.1 Motivation and background

For any statistical model, it is important to evaluate its suitability for particular data. For the functional linear model, the methodology of Chiou and Müller (2007), which we use in data examples in Chapters 9 and 10, is very useful. It is equally important to verify model assumptions. An important assumption on the model errors in all functional linear models of Chapter 8 is that these errors are independent and identically distributed. In this chapter, we study two tests aimed at detecting serial correlation in the error functions $\varepsilon_n(t)$ in the fully functional model (8.1). The methodology of Chiou and Müller (2007) was not designed to detect error correlation, and can leave it undetected. Figure 11.1 shows diagnostic plots of Chiou

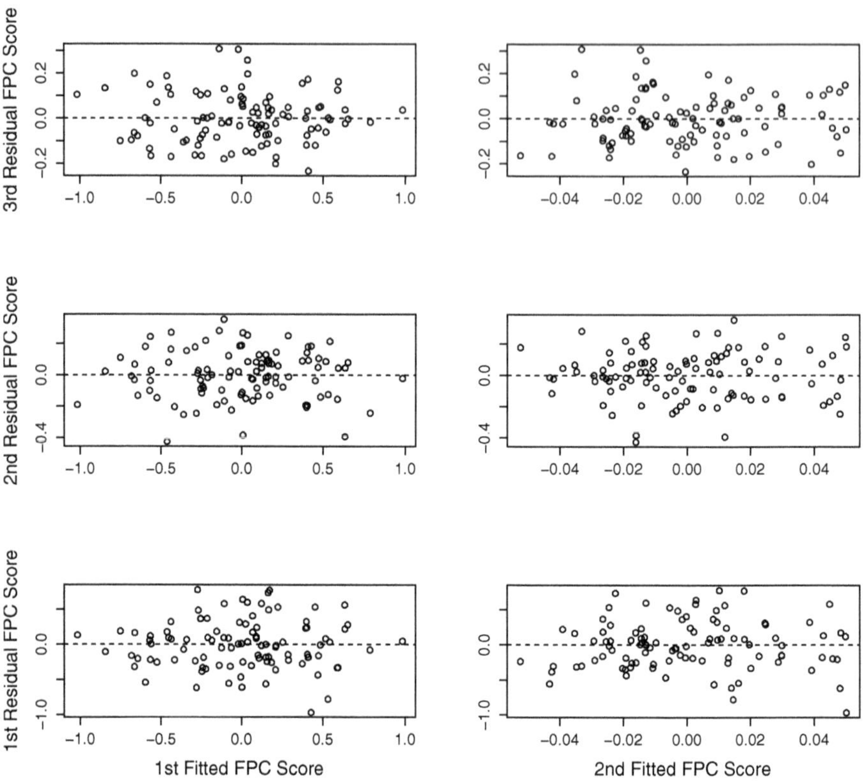

Fig. 11.1 Diagnostic plots of Chiou and Müller (2007) for a synthetic data set simulated according to model (8.1) in which the errors ε_n follow the functional autoregressive model of Chapter 13.

and Müller (2007) obtained for synthetic data that follow a functional linear model with highly correlated errors. These plots exhibit almost ideal football shapes. It is equally easy to construct examples in which our methodology fails to detect departures from model (8.1), but the graphs of Chiou and Müller (2007) immediately show it. The simplest such example is given by $Y_n(t) = X_n^2(t) + \varepsilon_n(t)$ with iid ε_n, see Figure 9.4. Thus, the methods we study in this chapter are complimentary tools designed to test the validity of specification (8.1) with iid errors against the alternative of correlation in the errors.

As in the multivariate regression, error correlation affects various variance estimates, and, consequently, confidence regions and distributions of test statistics. In particular, prediction based on Least Squares estimation is no longer optimal. To illustrate these issues, it is enough to consider the scalar regression model $y_i = \beta_0 + \beta_1 x_i + \varepsilon_i$, $i = 1, 2, \ldots N$, with fixed values x_i. We focus on inference for the slope coefficient β_1, whose least squares estimator is

$$\hat{\beta}_1 = \frac{\sum_{i=1}^{N}(x_i - \bar{x}_N)(y_i - \bar{y}_N)}{\sum_{i=1}^{N}(x_i - \bar{x}_N)^2}.$$

By default, software packages estimate the standard error of $\hat{\beta}_1$ by the square root of the estimated variance

$$\widehat{\mathrm{Var}}(\hat{\beta}_1) = \frac{\hat{\sigma}^2}{\sum_{i=1}^{N}(x_i - \bar{x}_N)^2}, \tag{11.1}$$

where $\hat{\sigma}^2$ is the sample variance of the residuals

$$\hat{\varepsilon}_i = y_i - \hat{\beta}_0 - \hat{\beta}_1 x_i. \tag{11.2}$$

To understand the issues involved, it is useful to look closer at the derivation of (11.1). Set

$$b_i = \frac{x_i - \bar{x}_N}{\sum_{i=1}^{N}(x_i - \bar{x}_N)^2}.$$

Then

$$\mathrm{Var}[\hat{\beta}_1] = \mathrm{Var}\left[\sum_{i=1}^{N} b_i(y_i - \bar{y}_N)\right] = \sum_{i,j=1}^{N} b_i b_j \mathrm{Cov}(y_i - \bar{y}_N, y_j - \bar{y}_N).$$

Since the x_i are fixed, we obtain

$$\mathrm{Var}[\hat{\beta}_1] = \sum_{i,j=1}^{N} b_i b_j \mathrm{Cov}(\varepsilon_i - \bar{\varepsilon}_N, \varepsilon_j - \bar{\varepsilon}_N). \tag{11.3}$$

If the ε_i are uncorrelated, all off–diagonal terms in (11.3) can be neglected, and we arrive at the estimator (11.1). However, if the ε_i are correlated, the off–diagonal terms in (11.3) contribute to the variance of $\hat{\beta}_1$.

To show how large the bias in the estimation of $\mathrm{Var}[\hat{\beta}_1]$ via (11.1) can be, we consider the following setting:

$$y_i = 2x_i + \varepsilon_i, \quad x_i = i/N, \; i = 1, 2, \ldots, N, \quad N = 100,$$

where the errors ε_i follows an AR(1) process

$$\varepsilon_i = \varphi \varepsilon_{i-1} + w_i, \quad w_i \sim iid\ N(0, 1).$$

Table 11.1 Approximate ratios of $\text{Var}[\hat{\beta}_1]$ to $E\,\widehat{\text{Var}}(\hat{\beta}_1)$ as a function of the autoregressive coefficient φ.

φ	0.0	0.1	0.3	0.5	0.7	0.9
V/A	1.00	1.23	1.89	3.01	5.76	19.56

We generated $R = 10{,}000$ such regressions, for each of them we computed $\hat{\beta}_1$, and then found the sample variance, V_R of the R numbers $\hat{\beta}_1$; V_R is close to $\text{Var}[\hat{\beta}_1]$. For each simulated regression, we also computed the estimate (11.1), and found the average A_R of these R estimates; A_R is close to the expected variance estimate when formula (11.1) is used. Table 11.1 displays the ratios V_R/A_R for selected values of φ. We see that if the errors ε_i are positively correlated, the estimate of the variance of $\hat{\beta}_1$ produced by a standard procedure may be far too small. Consequently, the confidence intervals will be too narrow, and the empirical size of hypothesis tests on β_1 will be too small.

Analogous variance and size distortions will occur in the the functional setting. Few asymptotic inferential procedures procedures for the functional linear model have been developed so far, but the use of the residual bootstrap is common. If the errors are dependent, using their standard bootstrap distribution will lead to problems fully analogous to those illustrated above in the scalar setting. Testing for error correlation is thus an important initial step before further work with a functional linear model is undertaken.

The two methods we study start with two ways of defining the residuals. Method I uses projections of all curves on the functional principal components of the regressors X_n, and so is closer to the standard regression in that one common basis is used. Method II uses two bases: the eigenfunctions of the covariance operators of the regressors and of the responses. The complexity of the requisite asymptotic theory is due to the fact that in order to construct a computable test statistic, finite dimensional objects reflecting the relevant properties of the infinite dimensional unobservable errors $\varepsilon_n(t)$ must be constructed. In the standard regression setting, the explanatory variables live in a finite dimensional Euclidean space with a fixed (standard) basis, and the residuals reflect the effect of parameter estimation (cf. (11.2)). In the functional setting, before any estimation can be undertaken, the dimension of the data must be reduced, typically by projecting on an "optimal" finite dimensional subspace. This projection operation introduces an error. The "optimal subspace" must be estimated, and this introduces another error. Finally, estimation of the kernel $\psi(\cdot,\cdot)$, conditional on the optimal subspace, introduces still another error. Our asymptotic approach focuses on the impact of these errors. We do not consider the dimensions of the optimal projection spaces growing to infinity with the sample size. Such an asymptotic analysis is much more complex. In a simpler setting of testing the equality of covariance operators, discussed in Chapter 5, it is developed by Panaretos *et al.* (2010).

11.2 Least squares estimation

This section explains the three steps, discussed in Section 11.1, involved in the construction of the residuals in the setting of model (8.1). The idea is that the curves are represented by their coordinates with respect to the FPC's of the X_n, e.g. $Y_{nk} = \langle Y_n, v_k \rangle$ is the projection of the nth response onto the kth largest FPC. A formal linear model for these coordinates is constructed and estimated by least squares. This formal model does not however satisfy the usual assumptions due to the effect of the projection of infinite dimensional curves on a finite dimensional subspace.

Since the v_k form a basis in $L^2([0,1])$, the products $v_i(t)v_j(s)$ form a basis in $L^2([0,1] \times [0,1])$. Thus, if $\psi(\cdot,\cdot)$ is a Hilbert–Schmidt kernel, then

$$\psi(t,s) = \sum_{i,j=1}^{\infty} \psi_{ij} v_i(t) v_j(s), \tag{11.4}$$

where $\psi_{ij} = \iint \psi(t,s) v_i(t) v_j(s) dt\, ds$. Therefore,

$$\int \psi(t,s) X_n(s) ds = \sum_{i,j=1}^{\infty} \psi_{ij} v_i(t) \langle X_n, v_j \rangle.$$

Hence, for any $1 \le k \le p$, we have

$$Y_{nk} = \sum_{j=1}^{p} \psi_{kj} \xi_{nj} + e_{nk} + \eta_{nk}, \tag{11.5}$$

where

$$Y_{nk} = \langle Y_n, v_k \rangle, \quad \xi_{nj} = \langle X_n, v_j \rangle, \quad e_{nk} = \langle \varepsilon_n, v_k \rangle,$$

and where

$$\eta_{nk} = \sum_{j=p+1}^{\infty} \psi_{kj} \langle X_n, v_j \rangle.$$

We combine the errors e_{nk} and η_{nk} by setting

$$\delta_{nk} = e_{nk} + \eta_{nk}.$$

Note that the δ_{nk} are no longer iid.

Setting

$$\mathbf{X}_n = [\xi_{n1}, \dots, \xi_{np}]^T \quad \mathbf{Y}_n = [Y_{n1}, \dots, Y_{np}]^T, \quad \boldsymbol{\delta}_n = [\delta_{n1}, \dots, \delta_{np}]^T,$$
$$\boldsymbol{\psi} = [\psi_{11}, \dots, \psi_{1p}, \psi_{21}, \dots, \psi_{2p} \dots, \psi_{p1}, \dots, \psi_{pp}]^T,$$

we rewrite (11.5) as

$$\mathbf{Y}_n = \mathbf{Z}_n \boldsymbol{\psi} + \boldsymbol{\delta}_n, \quad n = 1, 2, \dots, N,$$

where each $\mathbf{Z}_n$ is a $p \times p^2$ matrix

$$\mathbf{Z}_n = \begin{bmatrix} \mathbf{X}_n^T & \mathbf{0}_p^T & \cdots & \mathbf{0}_p^T \\ \mathbf{0}_p^T & \mathbf{X}_n^T & \cdots & \mathbf{0}_p^T \\ \vdots & \vdots & \vdots & \vdots \\ \mathbf{0}_p^T & \mathbf{0}_p^T & \cdots & \mathbf{X}_n^T \end{bmatrix}$$

with $\mathbf{0}_p = [0, \ldots, 0]^T$.

Finally, defining the $Np \times 1$ vectors $\mathbf{Y}$ and $\boldsymbol{\delta}$ and the $Np \times p^2$ matrix $\mathbf{Z}$ by

$$\mathbf{Y} = \begin{bmatrix} \mathbf{Y}_1 \\ \mathbf{Y}_2 \\ \vdots \\ \mathbf{Y}_N \end{bmatrix}, \quad \boldsymbol{\delta} = \begin{bmatrix} \boldsymbol{\delta}_1 \\ \boldsymbol{\delta}_2 \\ \vdots \\ \boldsymbol{\delta}_N \end{bmatrix}, \quad \mathbf{Z} = \begin{bmatrix} \mathbf{Z}_1 \\ \mathbf{Z}_2 \\ \vdots \\ \mathbf{Z}_N \end{bmatrix},$$

we obtain the following linear model

$$\mathbf{Y} = \mathbf{Z}\boldsymbol{\psi} + \boldsymbol{\delta}. \tag{11.6}$$

Note that (11.6) is not a standard linear model. Firstly, the design matrix $\mathbf{Z}$ is random. Secondly, $\mathbf{Z}$ and $\boldsymbol{\delta}$ are not independent. The error term $\boldsymbol{\delta}$ in (11.6) consists of two parts: the projections of the ε_n, and the remainder of an infinite sum. Thus, while (11.6) looks like the standard linear model, the existing asymptotic results do not apply to it, and a new asymptotic analysis involving the interplay of the various approximation errors is needed. Representation (11.6) leads to the formal "least squares estimator" for $\boldsymbol{\psi}$ is

$$\hat{\boldsymbol{\psi}} = (\mathbf{Z}^T\mathbf{Z})^{-1}\mathbf{Z}^T\mathbf{Y} = \boldsymbol{\psi} + (\mathbf{Z}^T\mathbf{Z})^{-1}\mathbf{Z}^T\boldsymbol{\delta}. \tag{11.7}$$

which cannot be computed because the v_k must be replaced by the $\hat{v}_k$. Projecting onto the $\hat{v}_k$, we are "estimating" the *random* vector

$$\widetilde{\boldsymbol{\psi}} = [\hat{c}_1\psi_{11}\hat{c}_1, \ldots, \hat{c}_1\psi_{1p}\hat{c}_p, \ldots, \hat{c}_p\psi_{p1}\hat{c}_1, \ldots, \hat{c}_p\psi_{pp}\hat{c}_p]^T. \tag{11.8}$$

with the "estimator"

$$\widetilde{\boldsymbol{\psi}}^{\wedge} = (\hat{\mathbf{Z}}^T\hat{\mathbf{Z}})^{-1}\hat{\mathbf{Z}}^T\hat{\mathbf{Y}}$$

obtained by replacing the v_k by the $\hat{v}_k$ in (11.7). It will be convenient to associate this vector of dimension p^2 with the $p \times p$ matrix

$$\widetilde{\boldsymbol{\Psi}}_p^{\wedge} = \begin{bmatrix} \tilde{\psi}_{11}^{\wedge} & \tilde{\psi}_{12}^{\wedge} & \cdots & \tilde{\psi}_{1p}^{\wedge} \\ \tilde{\psi}_{21}^{\wedge} & \tilde{\psi}_{22}^{\wedge} & \cdots & \tilde{\psi}_{2p}^{\wedge} \\ \vdots & \vdots & \vdots & \vdots \\ \tilde{\psi}_{p1}^{\wedge} & \tilde{\psi}_{p2}^{\wedge} & \cdots & \tilde{\psi}_{pp}^{\wedge} \end{bmatrix}. \tag{11.9}$$

In Section 11.7, we will use Proposition 11.1, which can be conveniently stated here because we have just introduced the required notation. It holds under the assumptions listed in Section 11.6, and the following additional assumption.

Assumption 11.1. *The coefficients* ψ_{ij} *of the kernel* $\psi(\cdot,\cdot)$ *satisfy* $\sum_{i,j=1}^{\infty} |\psi_{ij}| < \infty$.

Proposition 11.1. *If Assumptions (A1)–(A5) and 11.1 hold, then* $\widetilde{\boldsymbol{\psi}}^{\wedge} - \widetilde{\boldsymbol{\psi}} = O_P(N^{-1/2})$.

The proof of Proposition 11.1 is fairly technical and is developed in Aue *et al.* (2010).

11.3 Description of the test procedures

We consider two test statistics, (11.14) and (11.17) which arise from two different ways of defining finite dimensional vectors of residuals. Method I builds on the ideas presented in Section 11.2, the residuals are derived using the estimator $\widetilde{\boldsymbol{\psi}}^{\wedge}$ obtained by projecting both the Y_n and the X_n on the $\hat{v}_i$, the functional principal components of the regressors. Method II uses two projections; the X_n are projected on the $\hat{v}_i$, but the Y_n are projected on the $\hat{u}_i$. Motivated by Lemma 8.1, in Method II, we approximate $\psi(\cdot,\cdot)$ by

$$\widehat{\psi}_{pq}(t,s) = \sum_{j=1}^{q}\sum_{i=1}^{p} \hat{\lambda}_i^{-1}\hat{\sigma}_{ij}\hat{u}_j(t)\hat{v}_i(s) \quad \hat{\sigma}_{ij} = N^{-1}\sum_{n=1}^{N}\langle X_n,\hat{v}_i\rangle\langle Y_n,\hat{u}_j\rangle. \tag{11.10}$$

Method I emphasizes the role of the regressors X_n, and is, in a very loose sense, analogous to the plot of the residuals against the independent variable in a straight line regression. Method II emphasizes the role of the responses, and is somewhat analogous to the plot of the residuals against the fitted values. Both statistics have the form $\sum_{h=1}^{H} \hat{\mathbf{r}}_h^T \hat{\boldsymbol{\Sigma}}^{-1} \hat{\mathbf{r}}_h$, where $\hat{\mathbf{r}}_h$ are vectorized covariance matrices of appropriately constructed residuals, and $\hat{\boldsymbol{\Sigma}}$ is a suitably constructed matrix which approximates the covariance matrix of the the $\hat{\mathbf{r}}_h$, which are asymptotically iid. As in all procedures of this type, the P-values are computed for a range of values of H, typically $H \le 5$ or $H \le 10$. The main difficulty lies in deriving explicit formulas for the $\hat{\mathbf{r}}_h$ and $\hat{\boldsymbol{\Sigma}}$ and showing that the test statistics converge to the χ^2 distribution.

We continue to use the notation

$$C(v_k) = \lambda_k v_k, \quad X_n = \sum_{i=1}^{\infty} \xi_{ni} v_i, \quad \xi_{ni} = \langle v_i, X_n\rangle;$$

$$\Gamma(u_k) = \gamma_k u_k, \quad Y_n = \sum_{j=1}^{\infty} \zeta_{nj} u_j, \quad \zeta_{nj} = \langle u_j, Y_n\rangle,$$

with the sample counterparts denoted by $\hat{C}, \hat{\lambda}_k, \hat{v}_k, \hat{\xi}_{ni}$ and $\hat{\Gamma}, \hat{\gamma}_k, \hat{u}_k, \hat{\zeta}_{nj}$.

Method I. Recall the definition of the matrix $\widetilde{\boldsymbol{\Psi}}_p^{\wedge}$ (11.9) whose (i, j) entry approximates $\hat{c}_i \psi_{ij} \hat{c}_j$, and define also $p \times 1$ vectors

$$\hat{\mathbf{Y}}_n = [\hat{Y}_{n1}, \hat{Y}_{n2}, \ldots, \hat{Y}_{np}]^T, \quad \hat{Y}_{nk} = \langle Y_n, \hat{v}_k \rangle;$$
$$\hat{\mathbf{X}}_n = [\hat{\xi}_{n1}, \hat{\xi}_{n2}, \ldots, \hat{\xi}_{np}]^T, \quad \hat{\xi}_{nk} = \langle X_n, \hat{v}_k \rangle.$$

The fitted vectors are then

$$\widetilde{\mathbf{Y}}_n^{\wedge} = \widetilde{\boldsymbol{\Psi}}_p^{\wedge} \hat{\mathbf{X}}_n, \quad n = 1, 2, \ldots, N, \tag{11.11}$$

and the residuals are $\mathbf{R}_n = \hat{\mathbf{Y}}_n - \widetilde{\mathbf{Y}}_n^{\wedge}$. For $0 \le h < N$, define the sample autocovariance matrices of these residuals as

$$\mathbf{V}_h = N^{-1} \sum_{n=1}^{N-h} \mathbf{R}_n \mathbf{R}_{n+h}^T. \tag{11.12}$$

Finally, by $\mathrm{vec}(\mathbf{V}_h)$ denote the column vectors of dimension p^2 obtained by stacking the columns of the matrices $\mathbf{V}_h$ on top of each other, starting from the left. Next, define

$$e_{nk}^{\wedge} = \langle Y_n, \hat{v}_k \rangle - \sum_{j=1}^{p} \tilde{\psi}_{kj}^{\wedge} \langle X_n, \hat{v}_j \rangle,$$

$$\widehat{\mathbf{M}}_0 = \left[\frac{1}{N} \sum_{n=1}^{N} e_{nk}^{\wedge} e_{nk'}^{\wedge}, \ 1 \le k, k' \le p \right]$$

and

$$\widehat{\mathbf{M}} = \widehat{\mathbf{M}}_0 \otimes \widehat{\mathbf{M}}_0. \tag{11.13}$$

With this notation in place, we can define the test statistic

$$Q_N^{\wedge} = N \sum_{h=1}^{H} [\mathrm{vec}(\mathbf{V}_h)]^T \widehat{\mathbf{M}}^{-1} \mathrm{vec}(\mathbf{V}_h). \tag{11.14}$$

Properties of the Kronecker product, $\otimes$, give simplified formulae for $Q_N^{\wedge}$. Since $\widehat{\mathbf{M}}^{-1} = \widehat{\mathbf{M}}_0^{-1} \otimes \widehat{\mathbf{M}}_0^{-1}$ (see Horn and Johnson (1991) p. 244), Problem 25 on p. 252 of Horn and Johnson (1991), yields

$$Q_N^{\wedge} = N \sum_{h=1}^{H} \mathrm{tr} \left[\widehat{\mathbf{M}}_0^{-1} \mathbf{V}_h^T \widehat{\mathbf{M}}_0^{-1} \mathbf{V}_h \right].$$

Denoting by $\hat{m}_{f,h}(i, j)$ and $\hat{m}_{b,h}(i, j)$ the (i, j) entries, respectively, of $\widehat{\mathbf{M}}^{-1}\mathbf{V}_h$ and $\mathbf{V}_h \widehat{\mathbf{M}}^{-1}$, we can write according to the definition of the trace

$$Q_N^{\wedge} = N \sum_{h=1}^{H} \sum_{i,j=1}^{p} \hat{m}_{f,h}(i, j) \hat{m}_{b,h}(i, j).$$

The null hypothesis is rejected if $Q_N^{\wedge}$ exceeds an upper quantile of the chi–square distribution with $p^2 H$ degrees of freedom, see Theorem 11.2.

Method II. Equation (8.1) can be rewritten as

$$\sum_{j=1}^{\infty} \zeta_{nj} u_j = \sum_{i=1}^{\infty} \xi_{ni} \Psi(v_i) + \varepsilon_n, \tag{11.15}$$

where Ψ is the Hilbert–Schmidt operator with kernel $\psi(\cdot,\cdot)$. To define the residuals, we replace the infinite sums in (11.15) by finite sums, the unobservable u_j, v_i with the $\hat{u}_j, \hat{v}_i$, and Ψ with the estimator $\widehat{\Psi}_{pq}$ with kernel (11.10). This leads to the equation

$$\sum_{j=1}^{q} \hat{\zeta}_{nj} \hat{u}_j = \sum_{i=1}^{p} \hat{\xi}_{ni} \widehat{\Psi}_{pq}(\hat{v}_i) + \hat{z}_n,$$

where, similarly as in Section 11.2, $\hat{z}_n$ contains the ε_n, the effect of replacing the infinite sums with finite ones, and the effect of the estimation of the eigenfunctions. Method II is based on the residuals defined by

$$\hat{z}_n = \hat{z}_n(p,q) = \sum_{j=1}^{q} \hat{\zeta}_{nj} \hat{u}_j - \sum_{i=1}^{p} \hat{\xi}_{ni} \widehat{\Psi}_{pq}(\hat{v}_i) \tag{11.16}$$

Since $\widehat{\Psi}_{pq}(\hat{v}_i) = \sum_{j=1}^{q} \hat{\lambda}_i^{-1} \hat{\sigma}_{ij} \hat{u}_j(t)$, we see that

$$\hat{z}_n = \sum_{j=1}^{q} \left(\hat{\zeta}_{nj} - \sum_{i=1}^{p} \hat{\xi}_{ni} \hat{\lambda}_i^{-1} \hat{\sigma}_{ij} \right) \hat{u}_j(t).$$

Next define

$$\hat{Z}_{nj} := \langle \hat{u}_j, \hat{z}_n \rangle = \hat{\zeta}_{nj} - \sum_{i=1}^{p} \hat{\xi}_{ni} \hat{\lambda}_i^{-1} \hat{\sigma}_{ij}.$$

and denote by $\widehat{\mathbf{C}}_h$ the $q \times q$ autocovariance matrix with entries

$$\hat{c}_h(k,\ell) = \frac{1}{N} \sum_{n=1}^{N-h} \left(\hat{Z}_{nk} - \hat{\mu}_Z(k) \right) \left(\hat{Z}_{n+h,\ell} - \hat{\mu}_Z(\ell) \right),$$

where $\hat{\mu}_Z(k) = N^{-1} \sum_{n=1}^{N} \hat{Z}_{nk}$. Finally denote by $\hat{r}_{f,h}(i,j)$ and $\hat{r}_{b,h}(i,j)$ the (i,j) entries, respectively, of $\widehat{\mathbf{C}}_0^{-1} \widehat{\mathbf{C}}_h$ and $\widehat{\mathbf{C}}_h \widehat{\mathbf{C}}_0^{-1}$.

The null hypothesis is rejected if the statistic

$$\hat{Q}_N = N \sum_{h=1}^{H} \sum_{i,j=1}^{q} \hat{r}_{f,h}(i,j) \hat{r}_{b,h}(i,j) \tag{11.17}$$

exceeds an upper quantile of the chi–square distribution with $q^2 H$ degrees of freedom, see Theorem 11.3.

Repeating the arguments in the discussion of Method I, we get the following equivalent expressions for $\hat{Q}_N$:

$$\hat{Q}_N = N \sum_{h=1}^{H} \operatorname{tr}\left[\widehat{\mathbf{C}}_0^{-1} \widehat{\mathbf{C}}_h^T \widehat{\mathbf{C}}_0^{-1} \widehat{\mathbf{C}}_h\right]$$

and

$$\hat{Q}_N = N \sum_{h=1}^{H} [\operatorname{vec}(\widehat{\mathbf{C}}_h)]^T [\widehat{\mathbf{C}}_0 \otimes \widehat{\mathbf{C}}_0]^{-1} [\operatorname{vec}(\widehat{\mathbf{C}}_h)].$$

Both methods require the selection of p and q (Method I, only of p). We recommend the popular method based on the cumulative percentage of total variability (CPV) calculated as

$$CPV(p) = \frac{\sum_{k=1}^{p} \hat{\lambda}_k}{\sum_{k=1}^{\infty} \hat{\lambda}_k},$$

with a corresponding formula for the q. The numbers of eigenfunctions, p and q, are chosen as the smallest numbers, p and q, such that $CPV(p) \geq 0.85$ and $CPV(q) \geq 0.85$. Other ways of selecting p (and q) are discussed in Section 3.3.

As p and q increase, the normalized statistics $Q_N^{\wedge}$ and $\hat{Q}_N$ converge to the standard normal distribution. The normal approximation works very well even for small p or q (in the range 3-5 if $N \geq 100$) because the number of the degrees of freedom increases like p^2 or q^2. For Method I, which turns out to be conservative in small samples, the normal approximation brings the size closer to the nominal size. It also improves the power of Method I by up to 10%

Finally, we note that the methods of this chapter are suitable for testing the correlation of errors in model (8.1), but not in its special case known as the historical functional model of Malfait and Ramsay (2003). The latter is model (8.1) with $\psi(t,s) = \beta(s,t) I_H(s,t)$, where $\beta(\cdot,\cdot)$ is an arbitrary Hilbert–Schmidt kernel and $I_H(\cdot,\cdot)$ is the indicator function of the set $H = \{(s,t) : 0 \leq s \leq t \leq 1\}$. This model requires that $Y_n(t)$ depends only on the values of $Y_n(s)$ for $s \leq t$, i.e. it postulates temporal causality within the pairs of curves. Our approach cannot be readily extended to test for error correlation in the historical model because it uses series expansions of a general kernel $\psi(t,s)$, and the restriction that the kernel vanishes in the complement of H does not translate to any obvious restrictions on the coefficients of these expansions.

11.4 A simulation study

In this section we report the results of a simulation study performed to asses the empirical size and power of the proposed tests (Method I and Method II) for small to moderate sample sizes. The sample size N ranges from 50 to 500. Both independent and dependent regressor functions X_i are considered. The simulation runs have $1,000$ replications each. We used the `R` package `fda`.

To model the ε_i under H_0, independent trajectories of the Brownian bridge (BB) and the Brownian motion (BM) are generated by transforming cumulative sums of independent normal random variables computed on a grid of $1,000$ equispaced points in $[0, 1]$. In order to evaluate the effect of non Gaussian errors on the finite sample performance, we also simulated t_5 and uniform BB and BM (BB_{t_5}, BB_U, BM_{t_5} and BM_U) by generating t_5 and uniform, instead of normal increments. We also generate errors as

$$\varepsilon_n(t) = \sum_{j=1}^{5} \vartheta_{nj} j^{-1/2} \sin(j\pi t),$$

with the iid ϑ_{nj} distributed according to the normal, t_5 and uniform distributions.

We report simulation results obtained using by converting the curves into functional objects using B-splines with 20 basis functions. We also performed the simulations using the Fourier basis, and found that the results are not significantly different. To determine the number of principal components (p for X_n and q for Y_n), the cumulative percentage of total variability (CPV) is used as described in Section 11.3.

Three different kernel functions in (8.1) are considered: the Gaussian kernel $\psi(t,s) = \exp\{t^2 + s^2/2\}$, the Wiener kernel $\psi(t,s) = \min(t,s)$, and the Parabolic kernel $\psi(t,s) = -4\left[(t+1/2)^2 + (s+1/2)^2\right] + 2$. The regressors X_i in (8.1) are either iid BB or BM, or follow the functional autoregressive FAR(1) model studied in detail in Chapter 13. To simulate the FAR(1) X_n we used the kernels of the three types above, but multiplied by a constant K, so that their Hilbert–Schmidt norm is 0.5. Thus, the dependent regressors follow the model

$$X_n(t) = K \int \psi_X(t,s) X_{n-1}(s) ds + \alpha_n(t),$$

where the α_n are iid BB, BM, BB_{t_5}, BB_U, BM_{t_5} or BM_U.

We present here only a small selection of the results of our numerical experiments, and state general conclusions based on all simulations.

Starting with the empirical size, Tables 11.2 and 11.3 show that Method I is more conservative and slightly underestimates the nominal levels while Method II tends to overestimate them. The empirical sizes do not depend on whether the BB or the BM is used, nor whether regressors are iid or dependent, nor on the shape of the kernel. These sizes do not deteriorate if errors are not Gaussian either. The empirical size of both methods is thus robust to the form of the kernel, to moderate dependence in the regressors, and to departures from normality in the errors.

For the power simulations, we consider model (8.1) with the Gaussian kernel and $\varepsilon_n \sim$ ARH(1), i.e.

$$\varepsilon_n(t) = K \int \psi_\varepsilon(t,s) \varepsilon_{n-1}(s) ds + u_n(t),$$

where $\psi_\varepsilon(t,s)$ is Gaussian, Wiener or Parabolic and K is chosen so that the Hilbert-Schmidt norm of the above ARH(1) operator is 0.5 and the $u_n(t)$ are iid BB, BM, BB_{t_5}, BB_U, BM_{t_5} or BM_U.

Table 11.2 Empirical size for independent predictors: $X = BB$, $\varepsilon = BB$, ψ =Gaussian, Wiener and Parabolic, $p = 3$.

	Method I						Method II					
Sample	Gaussian		Wiener		Parabolic		Gaussian		Wiener		Parabolic	
size	10%	5%	10%	5%	10%	5%	10%	5%	10%	5%	10%	5%
						$H = 1$						
50	6.7	2.5	5.8	3.2	7.4	3.7	7.9	3.7	7.8	3.3	8.2	3.6
100	7.4	3.7	9.5	4.4	8.9	3.8	10.6	5.2	9.9	4.2	9.8	4.7
200	9.8	4.6	8.9	4.2	9.0	4.1	8.9	4.4	10.0	4.0	9.6	4.0
300	9.3	4.8	10.0	5.1	8.1	3.5	8.7	4.4	8.8	4.7	10.3	5.5
500	8.8	5.2	9.8	5.3	9.6	4.9	8.8	4.2	8.9	4.3	8.7	4.0
						$H = 3$						
50	4.3	2.5	5.6	2.1	6.0	3.4	10.7	5.3	8.9	4.7	9.0	4.2
100	7.6	3.7	6.9	3.6	6.4	3.3	9.9	4.5	10.2	4.0	10.1	4.9
200	8.7	4.6	6.4	3.2	8.0	3.3	9.6	4.8	10.1	5.1	9.6	5.0
300	7.6	3.5	9.5	4.2	9.5	4.8	11.0	5.1	8.9	4.0	8.1	4.6
500	9.8	4.6	9.1	3.9	9.2	4.9	11.1	6.8	9.1	4.4	10.0	5.1
						$H = 5$						
50	2.6	0.9	3.5	1.1	4.1	1.4	10.4	5.7	11.2	5.7	10.0	5.1
100	6.5	3.7	5.9	3.0	4.8	1.9	11.3	5.3	10.5	5.2	8.9	4.6
200	8.5	4.4	7.5	3.7	7.4	3.3	11.3	5.7	9.7	4.5	9.7	4.4
300	7.6	4.0	9.9	4.7	7.6	2.8	9.4	4.9	9.8	5.1	10.6	5.5
500	10.1	4.6	9.8	4.4	7.9	3.6	12.1	6.8	9.7	4.7	10.4	5.8

Typical power results are shown in Table 11.4. Just as for size, power is not affected by the dependence of the regressors. As expected from the results for the empirical size, power is uniformly higher for method II, but this difference is visible only for $N < 200$ (in our numerical experiments). The power is highest for $H = 1$, especially for smaller samples, because the errors follow the ARH(1) process.

11.5 Application to space physics and high–frequency financial data

We now illustrate the application of the tests on functional data sets arising in space physics and finance.

Application to Magnetometer data. We continue the study of the association between the auroral (high latitude) electrical currents and the currents flowing at mid– and low latitudes. This problem was introduced in Section 9.4. Maslova *et al.* (2010b) provide extensive references to the relevant space physics literature. The problem was cast into the setting of the functional linear model (8.1) in which the X_n are centered high-latitude records and Y_n are centered mid- or low-latitude magnetometer records. We consider two settings 1) consecutive days, 2) non-consecutive

Table 11.3 Empirical size for dependent predictors: $X \sim ARH(1)$ with the BB innovations and ψ_X =Gaussian, Wiener and Parabolic, ψ =Gaussian, $\varepsilon = BB$, $p = 3$.

Sample	Method I						Method II					
	Gaussian		Wiener		Parabolic		Gaussian		Wiener		Parabolic	
size	10%	5%	10%	5%	10%	5%	10%	5%	10%	5%	10%	5%
							$H = 1$					
50	8.4	3.9	5.9	2.1	7.3	2.9	9.2	4.6	7.2	2.7	8.6	3.8
100	8.9	4.4	8.8	3.7	8.4	3.7	10.4	4.6	10.2	4.9	9.9	4.8
200	10.2	4.7	9.7	4.6	10.1	4.7	9.5	4.8	8.9	4.0	9.8	5.2
300	9.2	4.9	8.9	4.4	8.6	4.6	10.1	4.1	8.5	3.4	12.0	5.3
500	10.5	5.2	9.3	4.5	9.0	4.7	9.0	4.2	9.5	4.8	11.5	5.6
							$H = 3$					
50	4.4	2.2	5.3	2.9	5.5	2.8	8.1	4.1	10.7	4.5	10.1	4.0
100	6.6	3.1	6.0	2.7	7.0	2.9	10.7	5.4	9.1	4.9	9.9	4.5
200	7.8	3.1	8.5	4.1	8.9	3.9	11.9	6.2	8.5	4.0	7.7	2.9
300	8.2	4.8	8.6	3.9	9.4	4.8	11.9	5.2	8.8	4.4	9.3	5.2
500	11.4	5.3	10.3	5.7	9.1	4.3	10.6	5.4	9.9	5.1	9.9	4.9
							$H = 5$					
50	4.2	1.8	3.2	1.5	4.0	1.9	9.9	5.2	11.1	6.6	11.9	6.7
100	7.2	3.2	4.9	2.4	5.2	2.1	10.5	5.5	10.2	5.5	11.2	6.0
200	7.6	2.8	8.1	3.7	8.8	4.4	11.4	4.6	10.3	4.6	11.6	7.3
300	8.3	4.2	8.3	3.4	7.3	3.9	10.7	5.5	9.3	5.2	9.7	4.7
500	10.7	5.8	10.4	4.9	7.9	4.2	9.0	4.1	9.2	4.0	10.4	5.3

days on which disturbances known as substorms occur. For consecutive days, we expect the rejection of the null hypothesis as there is a visible dependence of the responses from one day to the next, see the bottom panel of Figure 11.2. The low latitude curves, like those measured at Honolulu, exhibit changes on scales of several days. The high latitude curves exhibit much shorter dependence essentially confined to one day. This is because the auroral electrojects change on a scale of about 4 hours. In setting 2, the answer is less clear: the substorm days are chronologically arranged, but substorms may be separated by several days, and after each substorm the auroral current system resets itself to a quieter state.

To apply the tests, we converted the data to functional objects using 20 spline basis functions, and computed the EFPC's $\hat{v}_k$ and $\hat{u}_j$. For low latitude magnetometer data, 2 or 3 FPC's are needed to explain $87-89$, or $92-94$, percent of variability while for high latitude stations to explain $88-91$ percent of variability we need $8-9$ FPC's.

Setting 1 (consecutive days): We applied both methods to pairs (X_n, Y_n) in which the X_n are daily records at College, Alaska, and the Y_n are the corresponding records at six equatorial stations. Ten such pairs are shown in Figure 11.2. The samples consisted of all days in 2001, and of about 90 days corresponding to the four seasons. For all six stations and for the whole year the P–values were smaller than 10^{-12}. For the four seasons, all p-values, except two, were smaller than 2%. The higher P–values for the samples restricted to 90 days, are likely due to a smaller seasonal

Table 11.4 Method I: Empirical power for dependent predictor functions: $X \sim ARH(1)$ and $\varepsilon \sim ARH(1)$ with the BB innovations, $\psi_\varepsilon = \psi_X$ = Gaussian, Winer and Parabolic, ψ =Gaussian, $p = 3$.

Sample	Gaussian		Wiener		Parabolic	
size	10%	5%	10%	5%	10%	5%
			$H = 1$			
50	79.2	68.6	68.5	54.0	62.3	47.3
100	99.9	99.6	98.6	96.7	97.7	96.0
200	100	100	100	100	100	100
300	100	100	100	100	100	100
500	100	100	100	100	100	100
			$H = 3$			
50	53.8	40.7	45.4	32.8	40.0	29.0
100	98.0	95.7	93.6	89.5	87.5	81.3
200	100	100	100	99.9	100	99.8
300	100	100	100	100	100	100
500	100	100	100	100	100	100
			$H = 5$			
50	41.2	27.9	31.7	20.8	25.4	15.6
100	95.1	90.3	84.4	74.9	78.2	68.1
200	100	100	100	99.8	99.9	99.3
300	100	100	100	100	100	100
500	100	100	100	100	100	100

effect (the structure of the M-I system in the northern hemisphere changes with season). We conclude that it is not appropriate to use model (8.1) with iid errors to study the interaction of high– and low latitude currents when the data are derived from consecutive days.

Setting 2 (substorm days): We now focus on two samples studied in Maslova *et al.* (2010b). They are derived from 37 days on which isolated substorms were recorded at College, Alaska (CMO). A substorm is classified as an isolated substorm, if it is followed by 2 quiet days. There were only 37 isolated substorms in 2001, data for 10 such days are shown in Figure 11.3. The first sample consists of 37 pairs (X_n, Y_n), where X_n is the curve of the nth isolated storm recorded at CMO, and Y_n is the curve recorded on the same UT day at Honolulu, Hawaii, (HON). The second sample is constructed in the same way, except that Y_n is the curve recorded at Boulder, Colorado (BOU). The Boulder observatory is located in geomagnetic mid-latitude, i.e. roughly half way between the magnetic north pole and the magnetic equator. Honolulu is located very close to the magnetic equator.

The p-values for both methods and the two samples are listed in Table 11.5. For Honolulu, both tests indicate the suitability of model (8.1) with iid errors. For Boulder, the picture is less clear. The acceptance by Method I may be due to the small sample size ($N = 37$). The simulations in Section 11.4 show that for $N = 50$ this method has the power of about 50% at the nominal level of 5%. On the

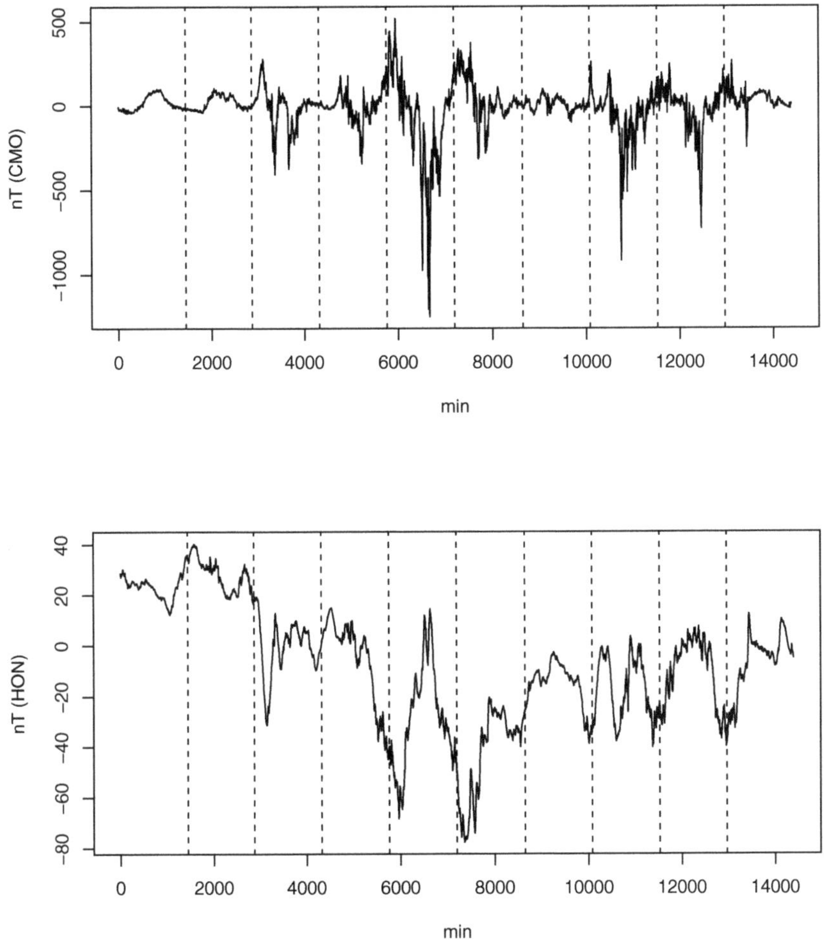

Fig. 11.2 Magnetometer data on **10** consecutive days (separated by vertical dashed lines) recorded at College, Alaska (CMO) and Honolulu, Hawaii, (HON).

other hand, Method II has the tendency to overreject. The sample with the Boulder records as responses confirms the general behavior of the two methods observed in Section 11.4, and emphasizes that it is useful to apply both of them to obtain more reliable conclusions. From the space physics perspective, midlatitude records are very difficult to interpret because they combine features of high latitude events (exceptionally strong auroras have been seen as far south as Virginia) and those of low latitude and field aligned currents.

We also applied the tests to samples in which the regressors are curves on days on which different types of substorms (according to a space physics classification)

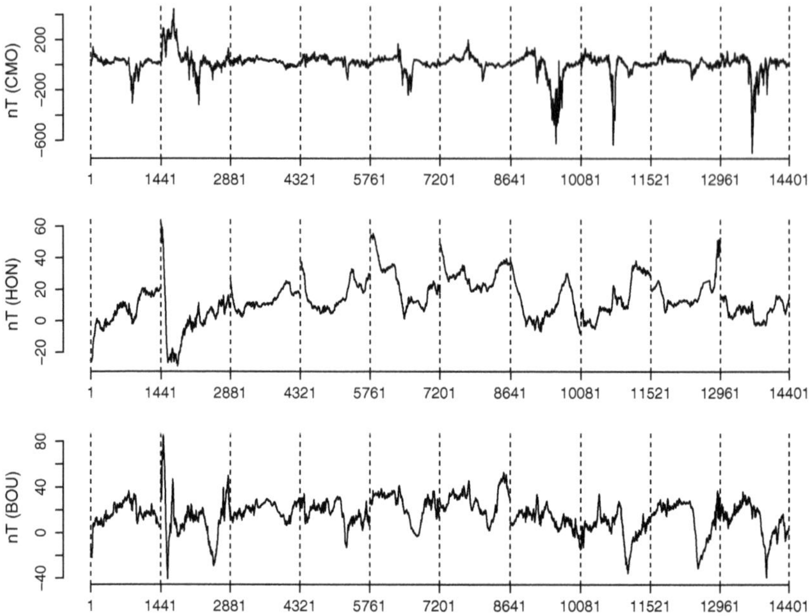

Fig. 11.3 Magnetometer data on 10 chronologically arranged isolated substorm days recorded at College, Alaska (CMO), Honolulu, Hawaii, (HON) and Boulder, Colorado (BOU).

Table 11.5 Isolated substorms data. P-values in percent.

	Response	
Method	HON	BOU
I	9.80	26.3
II	6.57	1.15

occurred. The broad conclusion remains that for substorm days, the errors in model (8.1) can be assumed iid if the period under consideration is not longer than a few months. For longer periods, seasonal trends apparently cause differences in distribution (possibly also of the X_n).

Application to intraday returns. Perhaps the best known application of linear regression to financial data is the celebrated Capital Asset Pricing Model (CAMP), see e.g. Chapter 5 of Campbell *et al.* (1997). In its simplest form, it is defined by

$$r_n = \alpha + \beta r_n^{(I)} + \varepsilon_n,$$

where

$$r_n = 100(\ln P_n - \ln P_{n-1}) \approx 100 \frac{P_n - P_{n-1}}{P_{n-1}}$$

is the return, in percent, over a unit of time on a specific asset, e.g. a stock of a corporation, and $r_n^{(I)}$ is the analogously defined return on a relevant market index. The unit of time can be can be day, month or year.

In this section we work with intra–daily price data, which are known to have properties quite different than those of daily or monthly closing prices, see e.g. Chapter 5 of Tsay (2005); Guillaume *et al.* (1997) and Andersen and Bollerslev (1997a, 1997b) also offer interesting perspectives. For these data, $P_n(t_j)$ is the price on day n at tick t_j (time of trade); we do not discuss issues related to the bid–ask spread, which are not relevant to what follows. For such data, it is not appropriate to define returns by looking at price movements between the ticks because that would lead to very noisy trajectories for which the methods based on the FPC's are not appropriate (Johnstone and Lu (2009) explain why principal components cannot be meaningfully estimated for noisy data). Instead, we adopt the following definition.

Definition 11.1. Suppose $P_n(t_j),\ n = 1,\dots,N,\ j = 1,\dots,m$, is the price of a financial asses at time t_j on day n. We call the functions

$$r_n(t_j) = 100[\ln P_n(t_j) - \ln P_n(t_1)], \quad j = 2,\dots,m,\ \ n = 1,\dots,N,$$

the *intra–day cumulative returns.*

Figure 11.4 shows intra-day cumulative returns on 10 consecutive days for the Standard & Poor's 100 index and the Exxon Mobil corporation. These returns have an appearance amenable to smoothing via FPC's.

We propose an extension of the CAPM to such return by postulating that

$$r_n(t) = \alpha(t) + \int \beta(t,s) r_n^{(I)}(s)ds + \varepsilon_n(t), \quad t \in [0,1], \tag{11.18}$$

where the interval [0, 1] is the rescaled trading period (in our examples, 9:30 to 16:00 EST). We refer to model (11.18) as the functional CAPM (FCAPM). As far as we know, this model has not been considered in the financial literature, but just as for the classical CAPM, it is designed to evaluate the extent to which intraday market returns determine the intraday returns on a specific asset. It is not our goal in this example to systematically estimate the parameters in (11.18) and compare them for various assets and markets, we merely want to use the methods developed in this paper to see if this model can be assumed to hold for some well–known assets. With this goal in mind, we considered FCAPM for S&P 100 and its major component, the Exxon Mobil Corporation (currently it contributes 6.78% to this index). The price processes over the period of about 8 years are shown in Figure 11.5. The functional observations are however not these processes, but the cumulative intra–daily returns, examples of which are shown in Figure 11.4.

After some initial data cleaning and preprocessing steps, we could compute the p-values for any period within the time stretch shown in Figure 11.5. The p-values for calendar years, the sample size N is equal to about 250, are reported in Table 11.6. In this example, both methods lead to the same conclusions, which match the well–known macroeconomic background. The tests do not indicate departures from

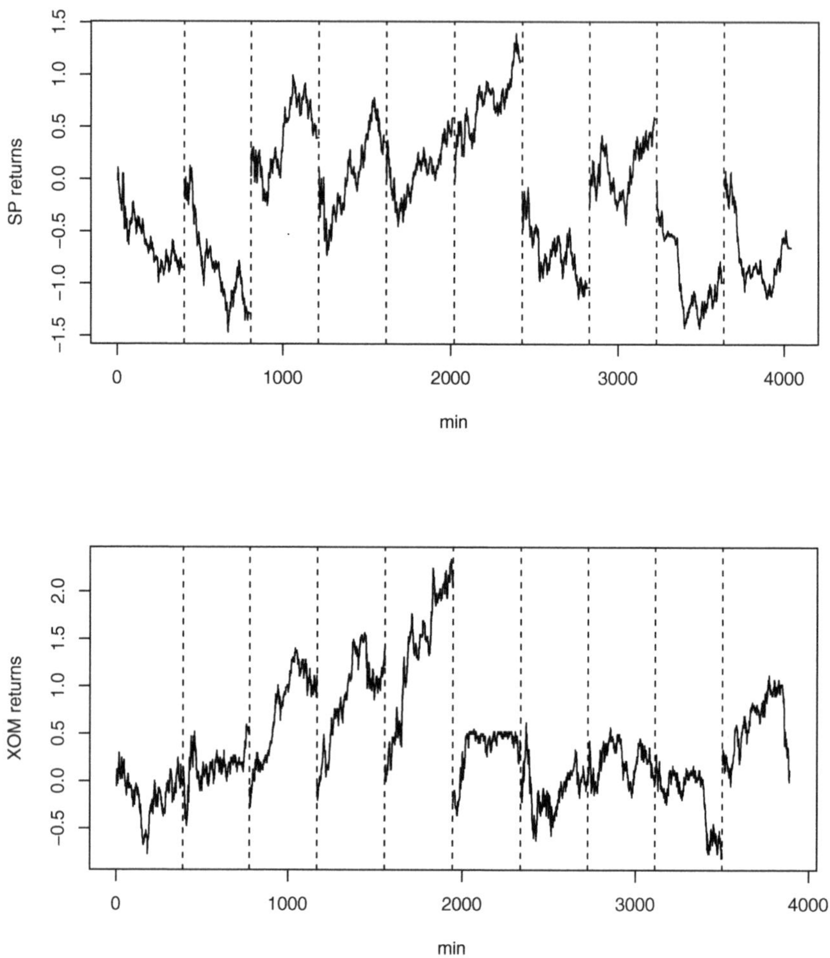

Fig. 11.4 Intra-day cumulative returns on 10 consecutive days for the Standard & Poor's 100 index (SP) and the Exxon–Mobil corporation (XOM).

the FCAMP model, except in 2002, the year between September 11 attacks and the invasion of Iraq, and in 2006 and 2007, the years preceding the collapse of 2008 in which oil prices were growing at a much faster rate than then the rest of the economy.

In the above examples we tested the correlation of errors in model (8.1), but not in the historical functional linear model defined at the end of Section 11.3. This is justified because the magnetometer data are obtained at locations with different

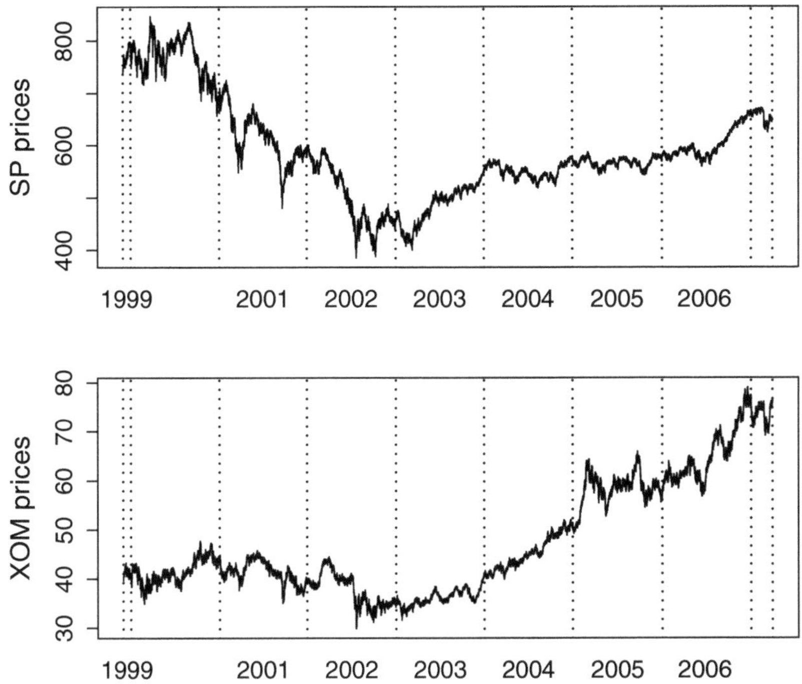

Fig. 11.5 Share prices of the Standard & Poor's 100 index (SP) and the Exxon–Mobil corporation (XOM). Dashed lines separate years.

Table 11.6 P–values, in percent, for the FCAPM (11.18) in which the regressors are the intra-daily cumulative returns on the Standard & Poor's 100 index, and the responses are such returns on the Exxon–Mobil stock.

Year	Method I	Method II
2000	46.30	55.65
2001	43.23	56.25
2002	0.72	0.59
2003	22.99	27.19
2004	83.05	68.52
2005	21.45	23.67
2006	2.91	3.04
2007	0.78	0.72

local times, and for space physics applications the dependence between the shapes of the daily curves is of importance. Temporal causality for financial data is often not assumed, as asset values reflect both historical returns and expectations of future market conditions.

11.6 Asymptotic theory

The exact asymptotic χ^2 distributions are obtained only under Assumption 11.2 which, in particular, requires that the X_n be iid. Under Assumption (A1)–(A5), these χ^2 distributions provide only approximations to the true limit distributions. The approximations are however very good, as the simulations in Section 11.4 show; size and power for dependent X_n are the same as for iid X_n, within the standard error. Thus, to understand the asymptotic properties of the tests, we first consider their behavior under Assumption 11.2. We begin the presentation of the asymptotic theory by stating the required assumptions.

Assumption 11.2. *The errors ε_n are independent identically distributed mean zero elements of L^2 satisfying $E\|\varepsilon_n\|^4 < \infty$. The regressors X_n are independent identically distributed mean zero elements of L^2 satisfying $E\|X_n\|^4 < \infty$. The sequences $\{X_n\}$ and $\{\varepsilon_n\}$ are independent.*

For data collected sequentially over time, the regressors X_n need not be independent. We formalize the notion of dependence in functional observations using the notion of L^4–m–approximability studied in detail in Chapter 16. For ease of reference, we repeat some conditions contained in Assumption 11.2; the weak dependence of the $\{X_n\}$ is quantified in Conditions (A2) and (A5).

(A1) The ε_n are independent, identically distributed with $E\varepsilon_n = 0$ and $E\|\varepsilon_n\|^4 < \infty$.

(A2) Each X_n admits the representation

$$X_n = g(\alpha_n, \alpha_{n-1}, \ldots),$$

in which the α_k are independent, identically distributed elements of a measurable space S, and $g : S^\infty \to L^2$ is a measurable function.

(A3) The sequences $\{\varepsilon_n\}$ and $\{\alpha_n\}$ are independent.

(A4) $EX_n = 0,\ E\|X_n\|^4 < \infty$.

(A5) There are $c_0 > 0$ and $\kappa > 2$ such that

$$\left(E\|X_n - X_n^{(k)}\|^4\right)^{1/4} \le c_0 k^{-\kappa},$$

where

$$X_n^{(k)} = g(\alpha_n, \alpha_{n-1}, \ldots, \alpha_{n-k+1}, \alpha_{n-k}^{(k)}, \alpha_{n-k-1}^{(k)}, \ldots),$$

and where the $\alpha_\ell^{(k)}$ are independent copies of α_0.

Condition (A2) means that the sequence $\{X_n\}$ admits a causal representation known as a Bernoulli shift. It follows from (A2) that $\{X_n\}$ is stationary and ergodic, see Section 3.5 of Stout (1974) or Sections 24 and 36 of Billingsley (1995). The structure of the function $g(\cdot)$ is not important, it can be a linear or a highly nonlinear function. What matters is that according to (A5), $\{X_n\}$ is weakly dependent, as it can be approximated with sequences of k–dependent variables, and the approximation improves as k increases. Several examples of functional sequences satisfying (A2), (A4) and (A5) are given in Chapter 16.

To state the alternative, we must impose dependence conditions on the ε_n. We use the same conditions that we imposed on the X_n, because then the asymptotic arguments under H_A can use the results derived for the X_n under H_0. Specifically, we introduce the following assumptions:

(B1) $E\varepsilon_n = 0$ and $E\|\varepsilon_n\|^4 < \infty$.
(B2) Each ε_n admits the representation

$$\varepsilon_n = h(u_n, u_{n-1}, \ldots),$$

in which the u_k are independent, identically distributed elements of a measurable space S, and $h : S^\infty \to L^2$ is a measurable function. (B3) The sequences $\{u_n\}$ and $\{\alpha_n\}$ are independent.
(B4) There are $c_0 > 0$ and $\kappa > 2$ such that

$$\left(E\|\varepsilon_n - \varepsilon_n^{(k)}\|^4\right)^{1/4} \leq c_0 k^{-\kappa},$$

where

$$\varepsilon_n^{(k)} = h(u_n, u_{n-1}, \ldots, u_{n-k+1}, u_{n-k}^{(k)}, u_{n-k-1}^{(k)}, \ldots),$$

and where the $u_\ell^{(k)}$ are independent copies of u_0.

The tests introduced in Section 11.3 detect dependence which manifests itself in a correlation between ε_n and ε_{n+h} for at least one h. Following Bosq (2000), we say that ε_n and ε_{n+h} are uncorrelated if $E[\langle\varepsilon_n, x\rangle \langle\varepsilon_{n+h}, y\rangle] = 0$ for all $x, y \in L^2$. If $\{e_j\}$ is any orthonormal basis in L^2, this is equivalent to $E[\langle\varepsilon_n, e_i\rangle \langle\varepsilon_{n+h}, e_j\rangle] = 0$ for all i, j. The two methods introduced in Section 11.3 detect the alternatives with $e_i = v_i$ (Method I) and $e_i = u_i$ (Method II). These methods test for correlation up to lag H, and use the FPC v_i, $i \leq p$, and u_i, $i \leq q$.

With this background, we can state the null and alternative hypotheses as follows.

H_0: Model (8.1) holds together with Assumptions (A1)–(A5).

The key assumption is (A1), i.e. the independence of the ε_n.

$H_{A,I}$: Model (8.1) holds together with Assumptions, (A2), (A4), (A5), (B1)–(B4), and $E[\langle\varepsilon_0, v_i\rangle \langle\varepsilon_h, v_j\rangle] \neq 0$ for some $1 \leq h \leq H$ and $1 \leq i, j \leq p$.

$H_{A,II}$: Model (8.1) holds together with Assumptions, (A2), (A4), (A5), (B1)–(B4), and $E[\langle\varepsilon_0, u_i\rangle \langle\varepsilon_h, u_j\rangle] \neq 0$ for some $1 \leq h \leq H$ and $1 \leq i, j \leq q$.

Note that the u_i are well defined under the alternative, because (A2), (A4), (A5) and (B1)–(B4) imply that the Y_n form a stationary sequence.

For ease of reference, we state the following Theorem, which follows immediately from Theorem 16.2.

Theorem 11.1. *If assumptions (A2), (A4), (A5) and* (2.12) *hold, then relations* (2.13) *hold.*

Method I is based on the following theorem.

Theorem 11.2. *Suppose Assumptions 11.2 and 11.1 and condition* (2.12) *hold. Then the statistics* $Q_N^{\wedge}$ *converges to the* χ^2*–distribution with* $p^2 H$ *degrees of freedom.*

Method II is based on Theorem 11.3. It is analogous to Theorem 7.1, but the observations are now replaced by residuals (11.16), so more delicate arguments are required.

Theorem 11.3. *Suppose Assumption 11.2 and condition* (2.12) *hold. Then statistic* (11.17) *converges in distribution to a chi–squared random variable with* $q^2 H$ *degrees of freedom.*

We now turn to the case of dependent regressors X_n. We focus on Method I. Similar results can be developed to justify the use of Method II, except that the u_j will also be involved. The case of dependent regressors involves the $p \times p$ matrices $\boldsymbol{D}_h$ with entries

$$D_h(i,j) = \sum_{\ell=p+1}^{\infty} \sum_{k=p+1}^{\infty} \iint v_\ell(s) e_h(s,t) v_k(t) ds\, dt, \quad 1 \le i,j \le p,$$

where

$$e_h(s,t) = E[X_0(s) X_h(t)].$$

Theorem 11.4. *Suppose Assumptions (A1)–(A5), Assumption 11.1 and condition* (2.12) *hold. Then, for any* $h > 0$,

$$N^{-1/2}\mathbf{V}_h = N^{-1/2}\left[\hat{c}_i \hat{c}_j V_h^*(i,j),\ 1 \le i,j \le p\right] + \mathbf{R}_{N,p}(h) + o_P(1).$$

The matrices $\mathbf{V}_h^* = \left[V_h^*(i,j),\ 1 \le i,j \le p\right]$, $1 \le h \le H$, *are jointly asymptotically normal. More precisely,*

$$N^{-1/2}\left\{\mathrm{vec}(\mathbf{V}_h^* - N\boldsymbol{D}_h),\ 1 \le h \le H\right\} \xrightarrow{d} \{\mathbf{Z}_1, \mathbf{Z}_2, \ldots, \mathbf{Z}_H\},$$

where the p^2*–dimensional vectors* $\mathbf{Z}_h$ *are iid normal, and coincide with the limits of* $N^{-1/2}\mathrm{vec}(\mathbf{V}_h)$, *if the* X_n *are independent.*

For any $r > 0$, *the terms* $\mathbf{R}_{N,p}(h)$ *satisfy,*

$$\lim_{p\to\infty} \limsup_{N\to\infty} P\left\{\left\|\mathbf{R}_{N,p}(h)\right\| > r\right\} = 0. \tag{11.19}$$

Theorem 11.4 justifies using Method I for weakly dependent X_n, provided p is so large that the first p FPC v_k explain a large percentage of variance of the X_n. To understand why, first notice that $|D_h(i,j)| \le (\lambda_\ell \lambda_k)^{1/2}$, and since $k, \ell > p$, the eigenvalues λ_ℓ, λ_k are negligible, as for functional data sets encountered in practice the graph of the λ_k approaches zero very rapidly. The exact form of $\mathbf{R}_{N,p}(h)$ is very complex. If $E[X_0(u)X_h(v)] = 0$, the $\mathbf{R}_{N,p}(h)$ and the matrices $\boldsymbol{D}_h$ vanish. If the X_n are dependent, these terms do not vanish, but are practically negligible because they all involve coefficients ψ_{jk} with at least one index greater than p multiplied by factors of order $O_P(N^{-1/2})$. In (11.19), the limit of p increasing to infinity should not be interpreted literally, but again merely indicates that p is so large that the first p FPC's v_k explain a large percentage of variance of the X_n.

Our last theorem states conditions under which the test is consistent. The interpretation of the limit as $p \to \infty$ is the same as above. Theorem 11.5 states that for such p and sufficiently large N the test will reject with large probability if ε_n and ε_{n+h} are correlated in the subspace spanned by $\{v_i,\ 1 \le i \le p\}$.

Theorem 11.5. *Suppose Assumptions (B1)–(B4), (A2), (A4), (A5), Assumption 11.1 and condition* (2.12) *hold. Then, for all* $R > 0$,

$$\lim_{p\to\infty} \liminf_{N\to\infty} P\left\{Q_N^{\wedge} > R\right\} = 1,$$

provided $E[\langle\varepsilon_0, v_i\rangle \langle\varepsilon_h, v_j\rangle] \neq 0$, *for some* $1 \le h \le H$ *and* $1 \le i, j \le p$.

To illustrate the arguments, we present in Section 11.7 the proof of Theorem 11.2. The proof of Theorem 11.3 follows the general outline of the proof of Theorem 7.1. The proof of Theorem 11.4 is very long, but the general idea is like that used in the proof of Theorem 11.2. Similarly, the proof of Theorem 11.5 is a modification and extension of the proof of Theorem 11.2.

11.7 Proof of Theorem 11.2

Relation (11.5) can be rewritten as

$$\mathbf{Y}_n = \boldsymbol{\Psi}_p \mathbf{X}_n + \boldsymbol{\delta}_n, \tag{11.20}$$

where

$$\boldsymbol{\Psi}_p = \begin{bmatrix} \psi_{11} & \psi_{12} & \cdots & \psi_{1p} \\ \psi_{21} & \psi_{22} & \cdots & \psi_{2p} \\ \vdots & \vdots & \vdots & \vdots \\ \psi_{p1} & \psi_{p2} & \cdots & \psi_{pp} \end{bmatrix}.$$

The vectors $\mathbf{Y}_n, \mathbf{X}_n, \boldsymbol{\delta}_n$ are defined in Section 11.2 as the projections on the FPC's $v_1, v_2, \ldots v_p$. Proposition 11.2 establishes an analog of (11.20) if these FPC's are replaced by the EFPC's $\hat{v}_1, \hat{v}_2, \ldots \hat{v}_p$. These replacement introduces additional terms generically denoted with the letter γ. First we prove Lemma 11.1 which leads to a decomposition analogous to (11.5).

Lemma 11.1. *If relation* (16.39) *holds with a Hilbert–Schmidt kernel* $\psi(\cdot,\cdot)$, *then*

$$Y_n(t) = \int \left(\sum_{i,j=1}^{p} \hat{c}_i \psi_{ij} \hat{c}_j \hat{v}_i(t) \hat{v}_j(s) \right) X_n(s) ds + \Delta_n(t),$$

where

$$\Delta_n(t) = \varepsilon_n(t) + \eta_n(t) + \gamma_n(t).$$

The terms $\eta_n(t)$ *and* $\gamma_n(t)$ *are defined as follows:*

$$\eta_n(t) = \eta_{n1}(t) + \eta_{n2}(t);$$

$$\eta_{n1}(t) = \int \left(\sum_{i=p+1}^{\infty} \sum_{j=1}^{\infty} \psi_{ij} v_i(t) v_j(s) \right) X_n(s) ds,$$

$$\eta_{n2}(t) = \int \left(\sum_{i=1}^{p} \sum_{j=p+1}^{\infty} \psi_{ij} v_i(t) v_j(s) \right) X_n(s) ds.$$

$$\gamma_n(t) = \gamma_{n1}(t) + \gamma_{n2}(t);$$

$$\gamma_{n1}(t) = \int \sum_{i,j=1}^{p} \hat{c}_i \psi_{ij} [\hat{c}_i v_i(t) - \hat{v}_i(t)] v_j(s) X_n(s) ds,$$

$$\gamma_{n2}(t) = \int \sum_{i,j=1}^{p} \hat{c}_i \psi_{ij} \hat{c}_j \hat{v}_i(t) [\hat{c}_j v_j(s) - \hat{v}_j(s)] X_n(s) ds.$$

Proof of Lemma 11.1. Observe that by (11.4),

$$\begin{aligned}\int \psi(t,s) X_n(s) ds &= \int \left(\sum_{i,j=1}^{\infty} \psi_{ij} v_i(t) v_j(s) \right) X_n(s) ds \\ &= \int \left(\sum_{i,j=1}^{p} \psi_{ij} v_i(t) v_j(s) \right) X_n(s) ds + \delta_n(t),\end{aligned}$$

where $\eta_n(t) = \eta_{n1}(t) + \eta_{n2}(t)$. Thus model (16.39) can be written as

$$Y_n(t) = \int \left(\sum_{i,j=1}^{p} \psi_{ij} v_i(t) v_j(s) \right) X_n(s) ds + \eta_n(t) + \varepsilon_n(t)$$

To take into account the effect of the estimation of the v_k, we will use the decomposition

$$\begin{aligned}\psi_{ij} v_i(t) v_j(s) &= \hat{c}_i \psi_{ij} \hat{c}_j (\hat{c}_i v_i(t))(\hat{c}_j v_j(s)) \\ &= \hat{c}_i \psi_{ij} \hat{c}_j \hat{v}_i(t) \hat{v}_j(s) \\ &\quad + \hat{c}_i \psi_{ij} \hat{c}_j [\hat{c}_i v_i(t) - \hat{v}_i(t)] \hat{c}_j v_j(s) \\ &\quad + \hat{c}_i \psi_{ij} \hat{c}_j \hat{v}_i(t) [\hat{c}_j v_j(s) - \hat{v}_j(s)],\end{aligned}$$

which allows us to rewrite (16.39) as

$$Y_n(t) = \int \left(\sum_{i,j=1}^{p} \hat{c}_i \psi_{ij} \hat{c}_j \hat{v}_i(t) \hat{v}_j(s) \right) X_n(s) ds + \Delta_n(t),$$

where $\Delta_n(t) = \varepsilon_n(t) + \eta_n(t) + \gamma_n(t)$ and $\gamma_n(t) = \gamma_{n1}(t) + \gamma_{n2}(t)$. □

To state Proposition 11.2, we introduce the vectors

$$\hat{\mathbf{Y}}_n = [\hat{Y}_{n1}, \hat{Y}_{n2}, \ldots, \hat{Y}_{np}]^T, \quad \hat{Y}_{nk} = \langle Y_n, \hat{v}_k \rangle;$$
$$\hat{\mathbf{X}}_n = [\hat{\xi}_{n1}, \hat{\xi}_{n2}, \ldots, \hat{\xi}_{np}]^T, \quad \hat{\xi}_{nk} = \langle X_n, \hat{v}_k \rangle;$$
$$\hat{\boldsymbol{\Delta}}_n = [\hat{\Delta}_{n1}, \hat{\Delta}_{n2}, \ldots, \hat{\Delta}_{np}]^T, \quad \hat{\Delta}_{nk} = \langle \Delta_n, \hat{v}_k \rangle.$$

Projecting relation (16.39) onto $\hat{v}_k$, we obtain by Lemma 11.1,

$$\langle Y_n, \hat{v}_k \rangle = \sum_{j=1}^{p} \hat{c}_k \psi_{kj} \hat{c}_j \langle X_n, \hat{v}_j \rangle + \langle \Delta_n, \hat{v}_k \rangle, \quad 1 \le k \le p,$$

from which the following proposition follows.

Proposition 11.2. *If relation* (16.39) *holds with a Hilbert–Schmidt kernel* $\psi(\cdot,\cdot)$, *then*

$$\hat{\mathbf{Y}}_n = \widetilde{\boldsymbol{\Psi}}_p \hat{\mathbf{X}}_n + \hat{\boldsymbol{\Delta}}_n, \quad n = 1, 2, \ldots N,$$

where $\widetilde{\boldsymbol{\Psi}}_p$ *is the* $p \times p$ *matrix with entries* $\hat{c}_k \psi_{kj} \hat{c}_j$, $k, j = 1, 2, \ldots p$.

To find the asymptotic distribution of the matrices $\mathbf{V}_h$, we establish several lemmas. Each of them removes terms which are asymptotically negligible, and in the process the leading terms are identified. Our first lemma shows that, asymptotically, in the definition of $\mathbf{V}_h$, the residuals

$$\mathbf{R}_n = \hat{\mathbf{Y}}_n - \widetilde{\mathbf{Y}}_n^{\wedge} = (\widetilde{\boldsymbol{\Psi}}_p - \widetilde{\boldsymbol{\Psi}}_p^{\wedge}) \hat{\mathbf{X}}_n + \hat{\boldsymbol{\Delta}}_n \tag{11.21}$$

can be replaced by the "errors" $\hat{\boldsymbol{\Delta}}_n$. The essential element of the proof is the relation $\widetilde{\boldsymbol{\Psi}}_p - \widetilde{\boldsymbol{\Psi}}_p^{\wedge} = O_P(N^{-1/2})$ stated in Proposition 11.1.

Lemma 11.2. *Suppose Assumptions 11.2 and 11.1 and condition* (2.12) *hold. Then, for any fixed* $h > 0$,

$$\left\| \mathbf{V}_h - N^{-1} \sum_{n=1}^{N-h} \hat{\boldsymbol{\Delta}}_n \hat{\boldsymbol{\Delta}}_{n+h}^T \right\| = O_P(N^{-1}).$$

Proof of Lemma 11.2. By (11.21) and (11.12),

$$\mathbf{V}_h = N^{-1} \sum_{n=1}^{N-h} [(\widetilde{\boldsymbol{\Psi}}_p - \widetilde{\boldsymbol{\Psi}}_p^{\wedge}) \hat{\mathbf{X}}_n + \hat{\boldsymbol{\Delta}}_n][(\widetilde{\boldsymbol{\Psi}}_p - \widetilde{\boldsymbol{\Psi}}_p^{\wedge}) \hat{\mathbf{X}}_{n+h} + \hat{\boldsymbol{\Delta}}_{n+h}]^T.$$

Denoting, $\hat{\mathbf{C}}_h = N^{-1} \sum_{n=1}^{N-h} \hat{\mathbf{X}}_n \hat{\mathbf{X}}_{n+h}^T$, we thus obtain

$$\mathbf{V}_h = (\widetilde{\boldsymbol{\Psi}}_p - \widetilde{\boldsymbol{\Psi}}_p^{\wedge}) \hat{\mathbf{C}}_h (\widetilde{\boldsymbol{\Psi}}_p - \widetilde{\boldsymbol{\Psi}}_p^{\wedge})^T + (\widetilde{\boldsymbol{\Psi}}_p - \widetilde{\boldsymbol{\Psi}}_p^{\wedge}) N^{-1} \sum_{n=1}^{N-h} \hat{\mathbf{X}}_n \hat{\boldsymbol{\Delta}}_{n+h}^T$$
$$+ N^{-1} \sum_{n=1}^{N-h} \hat{\boldsymbol{\Delta}}_n \hat{\mathbf{X}}_{n+h}^T (\widetilde{\boldsymbol{\Psi}}_p - \widetilde{\boldsymbol{\Psi}}_p^{\wedge}) + N^{-1} \sum_{n=1}^{N-h} \hat{\boldsymbol{\Delta}}_n \hat{\boldsymbol{\Delta}}_{n+h}^T.$$

By the CLT for h–dependent vectors, $\hat{\mathbf{C}}_h = O_P(1)$, so the first term satisfies

$$(\widetilde{\boldsymbol{\Psi}}_p - \widetilde{\boldsymbol{\Psi}}^{\wedge}_p)\hat{\mathbf{C}}_h(\widetilde{\boldsymbol{\Psi}}_p - \widetilde{\boldsymbol{\Psi}}^{\wedge}_p)^T = O_P(N^{-1/2}N^{-1/2}) = O_P(N^{-1}).$$

To deal with the remaining three terms, we use the decomposition of Lemma 11.1. It is enough to bound the coordinates of each of the resulting terms. Since $\Delta_n = \varepsilon_n + \eta_{n1} + \eta_{n2} + \gamma_{n1} + \gamma_{n2}$, we need to establish bounds for $2 \times 5 = 10$ terms, but these bounds fall only to a few categories, so we will only deal with some typical cases.

Starting with the decomposition of $\hat{\mathbf{X}}_n \hat{\boldsymbol{\Delta}}^T_{n+h}$, observe that

$$N^{-\frac{1}{2}} \sum_{n=1}^{N-h} \langle X_n, \hat{v}_i \rangle \langle \varepsilon_{n+h}, \hat{v}_j \rangle = \iint \left(N^{-1/2} \sum_{n=1}^{N-h} X_n(t)\varepsilon_{n+h}(s) \right) \hat{v}_i(t)\hat{v}_j(s)dtds.$$

The terms $X_n(t)\varepsilon_{n+h}(s)$ are iid elements of the Hilbert space $L^2([0,1] \times [0,1])$, so by the CLT in a Hilbert space, see Chapter 2,

$$\iint \left(N^{-1/2} \sum_{n=1}^{N-h} X_n(t)\varepsilon_{n+h}(s)dt\, ds \right)^2 = O_P(1).$$

Since the $\hat{v}_j$ have unit norm, $\iint (\hat{v}_i(t)\hat{v}_j(s))^2 dt\, ds = 1$. It therefore follows from the Cauchy–Schwarz inequality that

$$\sum_{n=1}^{N-h} \langle X_n, \hat{v}_i \rangle \langle \varepsilon_{n+h}, \hat{v}_j \rangle = O_P(N^{1/2}).$$

Thus, the ε_n contribute to $(\widetilde{\boldsymbol{\Psi}}_p - \widetilde{\boldsymbol{\Psi}}^{\wedge}_p)N^{-1}\sum_{n=1}^{N-h} \hat{\mathbf{X}}_n \hat{\boldsymbol{\Delta}}^T_{n+h}$ a term of the order $O_P(N^{-1/2}N^{-1}N^{1/2}) = O_P(N^{-1})$, as required.

We now turn to the contribution of the $\eta_{n,1}$. As above, we have

$$\begin{aligned}
&N^{-1/2} \sum_{n=1}^{N-h} \langle X_n, \hat{v}_i \rangle \langle \eta_{n+h,1}, \hat{v}_j \rangle \\
&= \iint \left(N^{-1/2} \sum_{n=1}^{N-h} X_n(t)\eta_{n+h,1}(s) \right) \hat{v}_i(t)\hat{v}_j(s)dt\, ds \\
&= \iint \left(N^{-1/2} \sum_{n=1}^{N-h} X_n(t) \int \left(\sum_{k=p+1}^{\infty} \sum_{\ell=1}^{\infty} \psi_{k\ell} v_k(s) v_\ell(u) \right) X_{n+h}(u)du \right) \\
&\qquad \times \hat{v}_i(t)\hat{v}_j(s)dt\, ds \\
&= \int \left[\iint N_h(t,u)R_p(t,u)dt\, du \right] v_k(s)\hat{v}_j(s)ds,
\end{aligned}$$

where

$$N_h(t,u) = N^{-1/2} \sum_{n=1}^{N-h} X_n(t) X_{n+h}(u)$$

and

$$R_p(t,u) = \sum_{\ell=1}^{\infty} \sum_{k=p+1}^{\infty} \psi_{k\ell} v_\ell(u) \hat{v}_i(t).$$

By the CLT for m–dependent elements in a Hilbert space, (follows e.g. from Theorem 2.17 of Bosq (2000)), $N_h(\cdot,\cdot)$ is $O_P(1)$ in $L^2([0,1]\times[0,1])$, so

$$\iint N_h^2(t,u)dt\,du = O_P(1).$$

A direct verification using Assumption 11.1 shows that also

$$\iint R_p^2(t,u)dt\,du = O_P(1).$$

Thus, by the Cauchy–Schwarz inequality, we obtain that

$$\sum_{n=1}^{N-h} \langle X_n, \hat{v}_i \rangle \langle \eta_{n+h,1}, \hat{v}_j \rangle = O_P(N^{1/2}),$$

and this again implies that the η_{n1} make a contribution of the same order as the ε_n. The same argument applies to the η_{n2}.

We now turn to the contribution of the γ_{n1}, the same argument applies to the γ_{n2}. Observe that, similarly as for the η_{n1},

$$\begin{aligned}
&N^{-1/2} \sum_{n=1}^{N-h} \langle X_n, \hat{v}_i \rangle \langle \gamma_{n+h,1}, \hat{v}_j \rangle \\
&= \iint \left(N^{-1/2} \sum_{n=1}^{N-h} X_n(t) \gamma_{n+h,1}(s) \right) \hat{v}_i(t) \hat{v}_j(s) dt\,ds \\
&= \int \left[\iint N_h(t,u) \sum_{k,\ell=1}^{p} \hat{c}_k \psi_{k\ell} v_\ell(u) \hat{v}_i(t) dt\,du \right] [\hat{c}_k v_k(s) - \hat{v}_k(s)] \hat{v}_j(s) ds
\end{aligned} \tag{11.22}$$

Clearly,

$$\iint \left(\sum_{k,\ell=1}^{p} \hat{c}_k \psi_{k\ell} v_\ell(u) \hat{v}_i(t) \right)^2 dt\,du = O_P(1),$$

By Theorem 2.7,

$$\left\{ \int [\hat{c}_k v_k(s) - \hat{v}_k(s)]^2 ds \right\}^{1/2} = O_P(N^{-1/2}). \tag{11.23}$$

We thus obtain

$$\sum_{n=1}^{N-h} \langle X_n, \hat{v}_i \rangle \left\langle \gamma_{n+h,1}, \hat{v}_j \right\rangle = O_P(1), \tag{11.24}$$

so the contribution of γ_n is smaller than that of ε_n and η_n.

To summarize, we have proven that

$$(\widetilde{\boldsymbol{\Psi}}_p - \widetilde{\boldsymbol{\Psi}}_p^{\wedge}) N^{-1} \sum_{n=1}^{N-h} \hat{\mathbf{X}}_n \hat{\boldsymbol{\Delta}}_{n+h}^T = O_P(N^{-1}).$$

The term $N^{-1} \sum_{n=1}^{N-h} \hat{\boldsymbol{\Delta}}_n \hat{\mathbf{X}}_{n+h}^T (\widetilde{\boldsymbol{\Psi}}_p - \widetilde{\boldsymbol{\Psi}}_p^{\wedge})$ can be dealt with in a fully analogous way. □

By Lemma 11.1, the errors $\hat{\boldsymbol{\Delta}}_n$ can be decomposed as follows

$$\hat{\boldsymbol{\Delta}}_n = \hat{\boldsymbol{\varepsilon}}_n + \hat{\boldsymbol{\eta}}_n + \hat{\boldsymbol{\gamma}}_n,$$

with the coordinates obtained by projecting the functions $\varepsilon_n, \eta_n, \gamma_n$ onto the EFPC's $\hat{v}_j$. For example,

$$\hat{\boldsymbol{\eta}}_n = [\langle \eta_n, \hat{v}_1 \rangle, \langle \eta_n, \hat{v}_2 \rangle, \ldots, \langle \eta_n, \hat{v}_p \rangle]^T.$$

Lemma 11.3 shows that the vectors $\hat{\boldsymbol{\gamma}}_n$ do not contribute to the asymptotic distribution of the $\mathbf{V}_h$. This is essentially due to the fact that by Theorem 2.7, the difference between $\hat{v}_j$ and $\hat{c}_j v_j$ is of the order $O_P(N^{-1/2})$. For the same reason, in the definition of $\hat{\beta}_n$ and $\hat{\boldsymbol{\eta}}_n$, the $\hat{v}_j$ can be replaced by the $\hat{c}_j v_j$, as stated in Lemma 11.4. Lemma 11.4 can be proven in a similar way as Lemma 11.3, so we present only the proof of Lemma 11.3.

Lemma 11.3. *Suppose Assumptions 11.2 and 11.1 and condition* (2.12) *hold. Then, for any fixed* $h > 0$,

$$\left\| \mathbf{V}_h - N^{-1} \sum_{n=1}^{N-h} [\hat{\boldsymbol{\varepsilon}}_n + \hat{\boldsymbol{\eta}}_n][\hat{\boldsymbol{\varepsilon}}_{n+h} + \hat{\boldsymbol{\eta}}_{n+h}]^T \right\| = O_P(N^{-1}).$$

Lemma 11.4. *Suppose Assumptions 11.2 and 11.1 and condition* (2.12) *hold. Then, for any fixed* $h > 0$,

$$\left\| \mathbf{V}_h - N^{-1} \sum_{n=1}^{N-h} [\tilde{\boldsymbol{\varepsilon}}_n + \tilde{\boldsymbol{\eta}}_n][\tilde{\boldsymbol{\varepsilon}}_{n+h} + \tilde{\boldsymbol{\eta}}_{n+h}]^T \right\| = O_P(N^{-1}),$$

where

$$\tilde{\boldsymbol{\varepsilon}}_n = \left[\hat{c}_1 \langle \varepsilon_n, v_1 \rangle, \hat{c}_2 \langle \varepsilon_n, v_2 \rangle, \ldots, \hat{c}_p \langle \varepsilon_n, v_p \rangle\right]^T$$

and

$$\tilde{\boldsymbol{\eta}}_n = \left[\hat{c}_1 \langle \eta_n, v_1 \rangle, \hat{c}_2 \langle \eta_n, v_2 \rangle, \ldots, \hat{c}_p \langle \eta_n, v_p \rangle\right]^T.$$

Proof of Lemma 11.3. In light of Lemma 11.2, we must show that the norm of difference between

$$N^{-1} \sum_{n=1}^{N-h} [\hat{\boldsymbol{\varepsilon}}_n + \hat{\boldsymbol{\eta}}_n][\hat{\boldsymbol{\varepsilon}}_n + \hat{\boldsymbol{\eta}}_n]^T$$

and

$$N^{-1} \sum_{n=1}^{N-h} [\hat{\boldsymbol{\varepsilon}}_n + \hat{\boldsymbol{\eta}}_n + \hat{\boldsymbol{\gamma}}_n][\hat{\boldsymbol{\varepsilon}}_n + \hat{\boldsymbol{\eta}}_n + \hat{\boldsymbol{\gamma}}_n]^T$$

is $O_P(N^{-1})$.

Writing $\hat{\boldsymbol{\eta}}_n = \hat{\boldsymbol{\eta}}_{n1} + \hat{\boldsymbol{\eta}}_{n2}$ and $\hat{\boldsymbol{\gamma}}_n = \hat{\boldsymbol{\gamma}}_{n1} + \hat{\boldsymbol{\gamma}}_{n2}$, we see that this difference consists of 20 terms which involve multiplication by $\hat{\boldsymbol{\gamma}}_{n1}$ or $\hat{\boldsymbol{\gamma}}_{n2}$. For example, analogously to (11.22), the term involving ε_n and and $\gamma_{n+h,1}$ has coordinates

$$\begin{aligned} N^{-1} \sum_{n=1}^{N-h} \langle \varepsilon_n, \hat{v}_i \rangle \langle \gamma_{n+h,1}, \hat{v}_j \rangle &= N^{-1/2} \int \left[\iint N_{\varepsilon,h}(t,u) \sum_{k,\ell=1}^{p} \hat{c}_k \psi_{k\ell} v_\ell(u) \hat{v}_i(t) dt\, du \right] \\ &\quad \times [\hat{c}_k v_k(s) - \hat{v}_k(s)] \hat{v}_j(s) ds, \end{aligned}$$

where

$$N_{\varepsilon,h}(t,u) = N^{-1/2} \sum_{n=1}^{N-h} \varepsilon_n(t) X_{n+h}(u).$$

By the argument leading to (11.24) (in particular by (11.23)),

$$N^{-1} \sum_{n=1}^{N-h} \langle \varepsilon_n, \hat{v}_i \rangle \langle \gamma_{n+h,1}, \hat{v}_j \rangle = O_P(N^{-1}).$$

The other terms can be bounded using similar arguments. The key point is that by (11.23), all these terms are $N^{1/2}$ times smaller than the other terms appearing in the decomposition of $N^{-1} \sum_{n=1}^{N-h} \hat{\boldsymbol{\Delta}}_n \hat{\boldsymbol{\Delta}}_n^T$. □

No more terms can be dropped. The asymptotic approximation to $\mathbf{V}_h$ thus involves linear functionals of the following processes.

$$R_{N,h}^{(1)} = N^{-1/2} \sum_{n=1}^{N} \varepsilon_n(t)\varepsilon_{n+h}(s),$$

$$R_{N,h}^{(2)} = N^{-1/2} \sum_{n=1}^{N} \varepsilon_n(t)X_{n+h}(s),$$

$$R_{N,h}^{(3)} = N^{-1/2} \sum_{n=1}^{N} \varepsilon_{n+h}(t)X_n(s),$$

$$R_{N,h}^{(4)} = N^{-1/2} \sum_{n=1}^{N} X_n(t)X_{n+h}(s).$$

Lemma 11.5, which follows directly for the CLT in the space $L^2([0,1] \times [0,1])$ and the calculation of the covariances, summarizes the asymptotic behavior of the processes $R_{N,h}^{(i)}$.

Lemma 11.5. *Suppose Assumptions 11.2 and 11.1 and condition* (2.12) *hold. Then*

$$\left\{R_{N,h}^{(i)},\ 1 \le i \le 4,\ 1 \le h \le H\right\} \xrightarrow{d} \left\{\Gamma_h^{(i)},\ 1 \le i \le 4,\ 1 \le h \le H\right\},$$

where the $\Gamma_h^{(i)}$ are $L^2([0,1] \times [0,1])$–valued jointly Gaussian process such that the processes $\left\{\Gamma_h^{(i)},\ 1 \le i \le 4\right\}$ are independent and identically distributed.

According to Lemmas 11.4 and 11.5, if

$$\hat{c}_1 = \hat{c}_2 = \ldots = \hat{c}_p = 1, \tag{11.25}$$

then

$$N^{1/2} \{\mathbf{V}_h,\ 1 \le h \le H\} \xrightarrow{d} \{\mathbf{T}_h,\ 1 \le h \le H\},$$

where the $\mathbf{T}_h$, $1 \le h \le H$, are independent identically distributed normal random matrices. Their covariances can be computed using Lemma 11.1. After lengthy but straightforward calculations, the following lemma is established

Lemma 11.6. *Suppose Assumptions 11.2 and 11.1 and condition* (2.12) *hold. If* (11.25) *holds, then for any fixed $h > 0$,*

$$N \operatorname{Cov}(\mathbf{V}_h(k,\ell), \mathbf{V}_h(k',\ell')) \to a(k,\ell;k',\ell'),$$

where

$$a(k,\ell;k',\ell') \\ = r_2(k,k')r_2(\ell,\ell') + r_2(k,k')r_1(\ell,\ell') + r_2(\ell,\ell')r_1(k,k') + r_1(k,k')r_1(\ell,\ell'),$$

with

$$r_1(\ell,\ell') = \sum_{j=p+1}^{\infty} \lambda_j \psi_{\ell j} \psi_{\ell' j}$$

and

$$r_2(k,k') = \iint E[\varepsilon_1(t)\varepsilon_1(s)] v_k(t) v_{k'}(s) dt\, ds.$$

While assumption (11.25) is needed to obtain the asymptotic distribution of the autocovariance matrices $\mathbf{V}_h$, we will now show that it is possible to construct a test statistic which does not require assumption (11.25). The arguments presented below use a heuristic derivation, and the approximate equalities are denoted with "$\approx$". The arguments could be formalized as in the proofs of Lemmas 11.3 and 11.4, but the details are not presented to conserve space.

We estimate $\langle \varepsilon_n, v_k \rangle$ by

$$\begin{aligned}
\varepsilon_{nk}^{\wedge} &= \langle Y_n, \hat{v}_k \rangle - \sum_{j=1}^{p} \tilde{\psi}_{kj}^{\wedge} \langle X_n, \hat{v}_j \rangle \\
&\approx \hat{c}_k \langle Y_n, v_k \rangle - \sum_{j=1}^{p} \hat{c}_k \psi_{kj} \hat{c}_j \hat{c}_j \langle X_n, v_j \rangle \\
&= \hat{c}_k \left(\langle Y_n, v_k \rangle - \sum_{j=1}^{p} \psi_{kj} \langle X_n, v_j \rangle \right) \\
&= \hat{c}_k \left(\langle \varepsilon_n, v_k \rangle + \sum_{j=p+1}^{\infty} \psi_{kj} \langle X_n, v_j \rangle \right).
\end{aligned}$$

By the strong law of large numbers

$$\begin{aligned}
&\frac{1}{N} \sum_{n=1}^{N} \left(\langle \varepsilon_n, v_k \rangle + \sum_{j=p+1}^{\infty} \psi_{kj} \langle X_n, v_j \rangle \right) \left(\langle \varepsilon_n, v_{k'} \rangle + \sum_{j=p+1}^{\infty} \psi_{k'j} \langle X_n, v_j \rangle \right) \\
&\overset{a.s.}{\to} E \left[\left(\langle \varepsilon_n, v_k \rangle + \sum_{j=p+1}^{\infty} \psi_{kj} \langle X_n, v_j \rangle \right) \left(\langle \varepsilon_n, v_{k'} \rangle + \sum_{j=p+1}^{\infty} \psi_{k'j} \langle X_n, v_j \rangle \right) \right] \\
&= r_1(k,k') + r_2(k,k').
\end{aligned}$$

Therefore, defining,

$$\hat{a}(k,k',\ell,\ell') = \left(\frac{1}{N} \sum_{n=1}^{N} \varepsilon_{nk}^{\wedge} \varepsilon_{nk'}^{\wedge} \right) \left(\frac{1}{N} \sum_{n=1}^{N} \varepsilon_{n\ell}^{\wedge} \varepsilon_{n\ell'}^{\wedge} \right),$$

we see that

$$\hat{a}(k,k',\ell,\ell') \approx \hat{c}_k \hat{c}_{k'} \hat{c}_\ell \hat{c}_{\ell'} a(k,k',\ell,\ell'). \tag{11.26}$$

By Lemma 11.6, under (11.25), the asymptotic covariance matrix of $N^{1/2}\text{vec}(\mathbf{V}_h)$ is a $p^2 \times p^2$ matrix

$$\mathbf{M} = [\,\mathbf{A}(i,j),\ 1 \le i,j \le p\,],$$

where

$$\mathbf{A}(i,j) = [\,a(\ell,i,k,j),\ 1 \le \ell,k \le p\,].$$

By (11.26), an estimator of $\mathbf{M}$ is

$$\widehat{\mathbf{M}} = \left[\,\widehat{\mathbf{M}}(i,j),\ 1 \le i,j \le p\,\right],$$

where

$$\widehat{\mathbf{M}}(i,j) = [\,\hat{a}(\ell,i,k,j),\ 1 \le \ell,k \le p\,].$$

Direct verification shows that $\widehat{\mathbf{M}}$ can be written in the form (11.13), which is convenient for coding.

As seen from (11.26), it cannot be guaranteed that the matrix $\widehat{\mathbf{M}}$ will be close to the matrix $\mathbf{M}$ because of the unknown signs $\hat{c}_i$. However, as will be seen in the proof of Theorem 11.2, statistic (11.14) does not depend on these signs.

Proof of Theorem 11.2. By Lemmas 11.2 and 11.3,

$$\text{vec}(\mathbf{V}_h) = \text{vec}\left(N^{-1}\sum_{n=1}^{N-h}[\hat{\boldsymbol{\beta}}_n + \hat{\boldsymbol{\eta}}_n][\hat{\boldsymbol{\beta}}_{n+h} + \hat{\boldsymbol{\eta}}_{n+h}]^T\right) + O_P(N^{-1}).$$

The arguments used in the proof of Lemma 11.2 show that

$$\begin{aligned}
&\text{vec}\left(N^{-1}\sum_{n=1}^{N-h}[\hat{\boldsymbol{\beta}}_n + \hat{\boldsymbol{\eta}}_n][\hat{\boldsymbol{\beta}}_{n+h} + \hat{\boldsymbol{\eta}}_{n+h}]^T\right)\\
&= [\widehat{\mathbf{C}} \otimes \widehat{\mathbf{C}}]\ \text{vec}\left(N^{-1}\sum_{n=1}^{N-h}[\boldsymbol{\varepsilon}_n + \boldsymbol{\eta}_n][\boldsymbol{\varepsilon}_{n+h} + \boldsymbol{\eta}_{n+h}]^T\right) + o_P(1),
\end{aligned}$$

where the matrix $\widehat{\mathbf{C}}$ is defined by

$$\widehat{\mathbf{C}} = \begin{bmatrix} \hat{c}_1 & 0 & \cdots & 0 \\ 0 & \hat{c}_2 & \cdots & 0 \\ \vdots & \vdots & \vdots & \vdots \\ 0 & 0 & \cdots & \hat{c}_p \end{bmatrix},$$

and where

$$\begin{aligned}
\boldsymbol{\varepsilon}_n &= [\langle \varepsilon_n, v_1\rangle, \langle \varepsilon_n, v_2\rangle, \ldots, \langle \varepsilon_n, v_p\rangle]^T;\\
\boldsymbol{\eta}_n &= [\langle \eta_n, v_1\rangle, \langle \eta_n, v_2\rangle, \ldots, \langle \eta_n, v_p\rangle]^T.
\end{aligned}$$

Similar arguments also show that

$$\widehat{\mathbf{M}} = [\widehat{\mathbf{C}} \otimes \widehat{\mathbf{C}}]\mathbf{M}[\widehat{\mathbf{C}} \otimes \widehat{\mathbf{C}}] + o_P(1).$$

Since $[\widehat{\mathbf{C}} \otimes \widehat{\mathbf{C}}]^T [\widehat{\mathbf{C}} \otimes \widehat{\mathbf{C}}]$ is the $p^2 \times p^2$ identity matrix, we obtain by Lemma 11.5 that

$$Q_N^{\wedge} = N \sum_{h=1}^{H} \left\{ \mathrm{vec}\left(N^{-1} \sum_{n=1}^{N-h} [\boldsymbol{\varepsilon}_n + \boldsymbol{\eta}_n][\boldsymbol{\varepsilon}_{n+h} + \boldsymbol{\eta}_{n+h}]^T \right) \right.$$

$$\left. \mathbf{M} \left[\mathrm{vec}\left(N^{-1} \sum_{n=1}^{N-h} [\boldsymbol{\varepsilon}_n + \boldsymbol{\eta}_n][\boldsymbol{\varepsilon}_{n+h} + \boldsymbol{\eta}_{n+h}]^T \right) \right]^T \right\} + o_P(1).$$

In particular, we see that the asymptotic distribution of $Q_N^{\wedge}$ does not depend on the signs $\hat{c}_1, \hat{c}_2, \ldots, \hat{c}_p$ (the same argument shows that $Q_N^{\wedge}$ itself does not depend on these signs), so we may assume that they are all equal to 1. The claim then follows form Lemmas 11.5 and 11.6. □

11.8 Bibliographical notes

There are relatively few papers dealing with goodness-of fit testing in the functional linear model, see Section 8.6. We have often used the the methodology of Chiou and Müller (2007) who emphasize the role of the functional residuals $\hat{\varepsilon}_i(t) = \hat{Y}_i(t) - Y_i(t)$, where the $Y_i(t)$ are the response curves, and the $\hat{Y}_i(t)$ are the fitted curves, and propose a number of graphical tools, akin to the usual residual plots. They also propose a test statistic based on Cook's distance, Cook (1977) or Cook and Weisberg (1982), whose null distribution can be computed by randomizing a binning scheme.

In the context of scalar data, Cochrane and Orcutt (1949) drew attention to the presence of serial correlation in the errors of models for economic time series, and investigated the effect of this correlation by means of simulations. Their paper is one of the first contributions advocating the use of simulation to study the behavior of statistical procedures. In the absence of a computer, they used tables of uniformly distributed random integers from 1 to 99 to construct a large number of tables similar to Table 11.1, but for more complex regression and dependence settings.

Tests for serial correlation in the standard scalar linear regression were developed by Durbin and Watson (1950, 1951, 1971), see also Chatfield (1998) and Section 10.4.4 of Seber and Lee (2003). Their statistics are functions of sample autocorrelations of the residuals, but their asymptotic distributions depend on the distribution of the regressors, and so various additional steps and rough approximations are required, see Thiel and Nagar (1961) and Thiel (1965), among others. To overcome these difficulties, Schmoyer (1994) proposed permutation tests based on quadratic forms of the residuals.

Textbook treatments addressing correlation in regression errors are available in Chapters 9 and 10 of Seber and Lee (2003), a good summary is given in Section

5.5 of Shumway and Stoffer (2006). The general idea is that when dependence in errors is detected, it must be modeled, and inference must be suitably adjusted. The relevant research is very extensive, so we mention only the influential papers of Sacks and Ylvisaker (1966) and Rao and Griliches (1969). Opsomer *et al.* (2001) and Xiao *et al.* (2003) consider a nonparametric regression $Y_t = m(X_t) + \varepsilon_t$.

Several other variants of the Functional CAPM and their predictive power are examined in Kokoszka and Zhang (2011).

As briefly discussed in Chapter 8, there are many possible departures from the specification of a scalar regression model. In addition to error autocorrelation, one may test the specification of the error distribution function or the parametric form of the regression function. Koul (2002) provides an exhaustive theoretical treatment of such issues.

Chapter 12
A test of significance in functional quadratic regression

The functional quadratic model in which a scalar response, Y_n, is paired with a functional predictor, $X_n(t)$, is defined as

$$Y_n = \mu + \int k(t)X_n^c(t)\,dt + \iint h(s,t)X_n^c(s)X_n^c(t)\,dt\,ds + \varepsilon_n, \qquad (12.1)$$

where $X_n^c(t) = X_n(t) - E\left(X_n(t)\right)$ is the centered predictor process. If $h(s,t) = 0$, then $\mu = E(Y_n)$ and (12.1) reduces to the functional linear model

$$Y_n = \mu + \int k(t)X_n^c(t)\,dt + \varepsilon_n. \qquad (12.2)$$

In this section we develop a test to determine if the use of a quadratic model is justified when a simpler linear model could be used.

12.1 Testing procedure

To test the significance of the quadratic term in (12.1), we test the null hypothesis,

$$H_0 : h(s,t) = 0, \qquad (12.3)$$

against the alternative

$$H_A : h(s,t) \neq 0.$$

It is clear that we can assume that h is symmetric, and we also impose the condition that the kernels are in L^2:

$$h(s,t) = h(t,s) \;\text{ and } \iint h^2(s,t)dt\,ds < \infty, \qquad (12.4)$$

$$\int k^2(t)dt < \infty. \qquad (12.5)$$

Thus we have the expansions

$$\begin{aligned} h(s,t) &= \sum_{i=1}^{\infty}\sum_{j=1}^{\infty} a_{i,j} v_j(s) v_i(t) \\ &= \sum_{i=1}^{\infty} a_{i,i} v_i(s) v_i(t) + \sum_{i=1}^{\infty}\sum_{j=i+1}^{\infty} a_{i,j}\left(v_j(s)v_i(t) + v_i(s)v_j(t)\right) \end{aligned} \tag{12.6}$$

and

$$k(t) = \sum_{i=1}^{\infty} b_i v_i(t). \tag{12.7}$$

We estimate the mean, $\mu_X(t)$, of the predictor process and the associated covariance function, $C(t,s)$, with the corresponding empiricals

$$\bar{X}(t) = \frac{1}{N}\sum_{n=1}^{N} X_n(t)$$

and

$$\hat{C}(t,s) = \frac{1}{N}\sum_{n=1}^{N}\left(X_n(t) - \bar{X}(t)\right)\left(X_n(s) - \bar{X}(s)\right).$$

The eigenvalues and the corresponding eigenfunctions of $\hat{C}(t,s)$ are denoted by $\hat{\lambda}_1 \geq \hat{\lambda}_2 \geq \cdots$ and $\hat{v}_1, \hat{v}_2, \ldots$.

Note that throughout this chapter the arguments of functions are sometimes omitted to make equations somewhat less cumbersome. Thus we use $v_j(t)$ and v_j interchangeably.

After projection the model is

$$\begin{aligned} Y_n &= \mu + \sum_{i=1}^{p} b_i \langle X_n - \bar{X}, \hat{c}_i \hat{v}_i\rangle \\ &\quad + \sum_{i=1}^{p}\sum_{j=i}^{p}(2 - 1\{i=j\}) a_{i,j}\langle X_n - \bar{X}, \hat{c}_i\hat{v}_i\rangle\langle X_n - \bar{X}, \hat{c}_j\hat{v}_j\rangle + \varepsilon_n^{**}, \end{aligned} \tag{12.8}$$

where

$$\begin{aligned} \varepsilon_n^{**} &= \varepsilon_n + \sum_{i=p+1}^{\infty} b_i\langle X_n^c, v_i\rangle + \sum_{i=p+1}^{\infty}\sum_{j=i}^{\infty}(2 - 1\{i=j\}) a_{i,j}\langle X_n^c, v_i\rangle\langle X_n^c, v_j\rangle \\ &\quad + \sum_{i=1}^{p}\sum_{j=p+1}^{\infty} 2a_{i,j}\langle X_n^c, v_i\rangle\langle X_n^c, v_j\rangle + \sum_{i=1}^{p} b_i\langle X_n^c, v_i - \hat{c}_i\hat{v}_i\rangle \\ &\quad + \sum_{i=1}^{p} b_i\langle \bar{X} - \mu_X, \hat{c}_i\hat{v}_i\rangle - \sum_{i=1}^{p}\sum_{j=i}^{p}(2 - 1\{i=j\}) a_{i,j} \\ &\qquad \times \left(\langle X_n - \bar{X}, \hat{c}_i\hat{v}_i\rangle\langle X_n - \bar{X}, \hat{c}_j\hat{v}_j\rangle - \langle X_n^c, v_i\rangle\langle X_n^c, v_j\rangle\right). \end{aligned}$$

Then

$$\mathbf{Y} = \hat{\mathbf{Z}} \begin{bmatrix} \tilde{\mathbf{A}} \\ \tilde{\mathbf{B}} \\ \mu \end{bmatrix} + \boldsymbol{\varepsilon}^{**}, \tag{12.9}$$

where

$$\begin{aligned}
\mathbf{Y} &= \left[Y_1, Y_2, \ldots, Y_N\right]^T, \\
\tilde{\mathbf{A}} &= \operatorname{vech}\left(\{\hat{c}_i \hat{c}_j a_{i,j}\,(2 - 1\{i = j\}),\ 1 \le i \le j \le p\}^T\right), \\
\tilde{\mathbf{B}} &= \left[\hat{c}_1 b_1, \hat{c}_2 b_2, \ldots, \hat{c}_p b_p\right]^T, \\
\boldsymbol{\varepsilon}^{**} &= \left[\varepsilon_1^{**}, \varepsilon_2^{**}, \ldots, \varepsilon_N^{**}\right]^T,
\end{aligned}$$

and

$$\hat{\mathbf{Z}} = \begin{bmatrix} \hat{\mathbf{D}}_1^T & \hat{\mathbf{F}}_1^T & 1 \\ \hat{\mathbf{D}}_2^T & \hat{\mathbf{F}}_2^T & 1 \\ \vdots & \vdots & \vdots \\ \hat{\mathbf{D}}_N^T & \hat{\mathbf{F}}_N^T & 1 \end{bmatrix}$$

with

$$\begin{aligned}
\hat{\mathbf{D}}_n &= \operatorname{vech}\left(\{\langle \hat{v}_i, X_n - \bar{X}\rangle\langle \hat{v}_j, X_n - \bar{X}\rangle,\ 1 \le i \le j \le p\}^T\right), \\
\hat{\mathbf{F}}_n &= \left[\langle X_n - \bar{X}, \hat{v}_1\rangle, \langle X_n - \bar{X}, \hat{v}_2\rangle, \ldots, \langle X_n - \bar{X}, \hat{v}_p\rangle\right]^T.
\end{aligned}$$

We estimate $\tilde{\mathbf{A}}$, $\tilde{\mathbf{B}}$, and μ using the least squares estimator:

$$\begin{bmatrix} \hat{\mathbf{A}} \\ \hat{\mathbf{B}} \\ \hat{\mu} \end{bmatrix} = \left(\hat{\mathbf{Z}}^T \hat{\mathbf{Z}}\right)^{-1} \hat{\mathbf{Z}}^T \mathbf{Y}. \tag{12.10}$$

To represent elements of $\hat{\mathbf{A}}$ and $\hat{\mathbf{B}}$, we will use the notation that $\hat{\mathbf{A}} = \operatorname{vech}(\{\hat{a}_{i,j}(2 - 1\{i = j\}), 1 \le i \le j \le p\}^T)$ and $\hat{\mathbf{B}} = \left[\hat{b}_1, \hat{b}_2, \ldots, \hat{b}_p\right]^T$.

We expect, under H_0, that $\hat{\mathbf{A}}$ will be close to zero since $\tilde{\mathbf{A}}$ is zero. If H_0 is not correct, we expect the magnitude of $\hat{\mathbf{A}}$ to be relatively large. Let

$$\hat{\mathbf{G}} = \frac{1}{N} \sum_{n=1}^{N} \hat{\mathbf{D}}_n \hat{\mathbf{D}}_n^T,$$

$$\hat{\mathbf{M}} = \frac{1}{N} \sum_{n=1}^{N} \hat{\mathbf{D}}_n,$$

and

$$\hat{\tau}^2 = \frac{1}{N} \sum_{n=1}^{N} \hat{\varepsilon}_n^2,$$

where

$$\hat{\varepsilon}_n = Y_n - \hat{\mu} - \sum_{i=1}^{p} \hat{b}_i \langle X_n - \bar{X}, \hat{v}_i \rangle$$
$$- \sum_{i=1}^{p} \sum_{j=i}^{p} (2 - 1\{i = j\}) \hat{a}_{i,j} \langle X_n - \bar{X}, \hat{v}_i \rangle \langle X_n - \bar{X}, \hat{v}_j \rangle$$

are the residuals under H_0. We reject the null hypothesis if

$$U_N = \frac{N}{\hat{\tau}^2} \hat{\mathbf{A}}^T (\hat{\mathbf{G}} - \hat{\mathbf{M}}\hat{\mathbf{M}}^T) \hat{\mathbf{A}}$$

is large. The main result of this paper is the asymptotic distribution of U_N under the null hypothesis. First, we discuss the assumptions needed to establish asymptotics for U_N:

Assumption 12.1. $\{X_n(t), n \geq 1\}$ *is a sequence of independent, identically distributed Gaussian processes.*

Assumption 12.2.

$$E\left(\int X_n^2(t)\,dt \right)^4 < \infty.$$

Assumption 12.3. $\{\varepsilon_n\}$ *is a sequence of independent, identically distributed random variables satisfying* $E\varepsilon_n = 0$ *and* $E\varepsilon_n^4 < \infty$,

and

Assumption 12.4. *the sequences* $\{\varepsilon_n\}$ *and* $\{X_n(t)\}$ *are independent.*

The last condition is standard in functional data analysis. It implies that the eigenfunctions $v_1, v_2, \ldots, v_p$ are unique up to a sign.

Assumption 12.5.

$$\lambda_1 > \lambda_2 > \cdots > \lambda_{p+1}.$$

Theorem 12.1. *If* H_0, *(12.5) and Assumptions 12.1–12.5 are satisfied, then*

$$U_N \xrightarrow{\mathcal{D}} \chi^2(r),$$

where $r = p(p+1)/2$ *is the dimension of the vector* $\hat{\mathbf{A}}$.

The proof of Theorem 12.1 is given in Section 12.3.

Remark: By the Karhunen-Loève expansion, every centered, square integrable process, $X_n^c(t)$, can be written as

$$X_n^c(t) = \sum_{\ell=1}^{\infty} \xi_{n,\ell} \varphi_\ell(t),$$

where φ_ℓ are orthonormal functions. Assumption 12.1 can be replaced with the requirement that $\xi_{n,1}, \xi_{n,2}, \ldots, \xi_{n,p}$ are independent with $E\xi_{n,\ell}^3 = 0$ and $E\xi_{n,\ell} = 0$ for all $1 \leq \ell \leq p$.

Our last result provides a simple condition for the consistency of the test based on U_N. Let $\mathbf{A} = \text{vech}(\{a_{i,j}(2 - 1\{i = j\}),\ 1 \leq i \leq j \leq p\}^T)$, i.e. the first $r = p(p+1)/2$ coefficients in the expansion of h in (12.6).

Theorem 12.2. *If* (12.4), (12.5), *Assumptions 12.1–12.5 are satisfied and* $\mathbf{A} \neq \mathbf{0}$, *then we have that*

$$U_N \xrightarrow{P} \infty.$$

The condition $\mathbf{A} \neq \mathbf{0}$ means that h is not the 0 function in the space spanned by the functions $v_i(t)v_j(s), 1 \leq i, j \leq p$.

12.2 Application to spectral data

In this section we apply our test to the Tecator data set available at `http://lib.stat.cmu.edu/datasets/tecator`. This data set is studied in Ferraty and Vieu (2006) Tecator Infratec food used 240 samples of finely chopped pure meat with different fat contents. For each sample of meat, a 100 channel spectrum of absorbances was recorded. These absorbances can be thought of as a discrete approximation to the continuous record, $X_n(t)$. Also, for each sample of meat, the fat content, Y_n was measured by analytic chemistry.

The absorbance curve measured from the nth meat sample is given by $X_n(t) = \log_{10}(I_0/I)$, where t is the wavelength of the light, I_0 is the intensity of the light before passing through the meat sample, and I is the intensity of the light after it passes through the meat sample. The Tecator Infratec food and feed analyzer measured absorbance at 100 different wavelengths between 850 and 1050 nanometers. This gives the values of $X_n(t)$ on a discrete grid from which we can use cubic splines to interpolate the values anywhere within the interval.

Yao and Müller (2010) proposed using a functional quadratic model to predict the fat content, Y_n, of a meat sample based on its absorbance spectrum, $X_n(t)$. We are interested in determining whether the quadratic term in (12.1) is needed by testing its significance for this data set. From the data, we calculate U_{240}. The p-value is then $P\left(\chi^2(r) > U_{240}\right)$. The test statistic and hence the p-value are influenced by the number of principal components that we choose to keep. If we select p according to the advice of Ramsay and Silverman (2005), we will keep only $p = 1$ principal component because this explains more than 85% of the variation between absorbance curves in the sample. Table 12.1 gives p-values obtained using $p = 1, 2,$ and 3 principal components, which strongly supports that the quadratic regression provides a better model for the Tecator data.

Since Theorem 12.1 assumes that the X_n's are Gaussian, we now check the plausibility of this assumption for the absorbance spectra. If the X_n's are Gaussian processes, then the projections $\langle X_n, v_i \rangle$ would be normally distributed. Using the

Table 12.1 p-values (in %) obtained by applying our testing procedure to the Tecator data set with $p = 1, 2$, and 3 principal components.

p	1	2	3
p-value	1.25	13.15	0.00

Shapiro-Wilks test for normality, we conclude that the first projection, $\langle X_n, \hat{v}_1 \rangle$, is not normally distributed (p-value$= 3.15 \times 10^{-7}$). The Box-Cox family of transformations is commonly employed to transform data to be more like the realizations of a normal random variable:

$$f(X(t)) = \frac{(X(t) + \omega_2)^{\omega_1} - 1}{\omega_1}. \tag{12.11}$$

We apply the Box-Cox transformation with $\omega_1 = -.0204$ and $\omega_2 = -1.6539$. We can now verify the plausibility of the Gaussian assumption for the transformed spectra by testing the first projection (p-value$= 0.38$). If we now apply our test of the significance of the quadratic term for the transformed data, we get a p-value which is essentially zero using $p = 1, 2$, or 3 principal components. This strongly supports that the quadratic regression provides a better model for the transformed Tecator data.

12.3 Outline for the Proof of Theorem 12.1

We have from (12.9) and (12.10) that

$$\begin{aligned} \begin{bmatrix} \hat{\mathbf{A}} \\ \hat{\mathbf{B}} \\ \hat{\mu} \end{bmatrix} &= \left(\hat{\mathbf{Z}}^T \hat{\mathbf{Z}}\right)^{-1} \hat{\mathbf{Z}}^T \left(\hat{\mathbf{Z}} \begin{bmatrix} \tilde{\mathbf{A}} \\ \tilde{\mathbf{B}} \\ \mu \end{bmatrix} + \boldsymbol{\varepsilon}^{**} \right) \\ &= \begin{bmatrix} \tilde{\mathbf{A}} \\ \tilde{\mathbf{B}} \\ \mu \end{bmatrix} + \left(\hat{\mathbf{Z}}^T \hat{\mathbf{Z}}\right)^{-1} \hat{\mathbf{Z}}^T \boldsymbol{\varepsilon}^{**}. \end{aligned} \tag{12.12}$$

We also note that, under the null hypothesis, $a_{i,j} = 0$ for all i and j and therefore ε_n^{**} reduces to

$$\varepsilon_n^{**} = \varepsilon_n + \sum_{i=p+1}^{\infty} b_i \langle X_n^c, v_i \rangle + \sum_{i=1}^{p} b_i \langle X_n^c, v_i - \hat{c}_i \hat{v}_i \rangle + \sum_{i=1}^{p} b_i \langle \bar{X} - \mu_X, \hat{c}_i \hat{v}_i \rangle.$$

To obtain the limiting distribution of $\sqrt{N}\hat{\mathbf{A}}$, we need to consider the vector $\sqrt{N}\left(\hat{\mathbf{Z}}^T \hat{\mathbf{Z}}\right)^{-1} \hat{\mathbf{Z}}^T \boldsymbol{\varepsilon}^{**}$. First, we need to show that

$$\left(\left(\frac{\hat{\mathbf{Z}}^T \hat{\mathbf{Z}}}{N} \right) - \begin{bmatrix} \boldsymbol{\zeta} \mathbf{G} \boldsymbol{\zeta} & \mathbf{0}_{r \times p} & \mathbf{M} \\ \mathbf{0}_{p \times r} & \boldsymbol{\Lambda} & \mathbf{0}_{p \times 1} \\ \mathbf{M}^T & \mathbf{0}_{1 \times p} & 1 \end{bmatrix} \right) = o_P(1), \tag{12.13}$$

where $\boldsymbol{\zeta}$ is an unobservable matrix of random signs, $\boldsymbol{\Lambda} = \mathrm{diag}(\lambda_1, \lambda_2, \ldots, \lambda_p)$, $\mathbf{M} = E(\mathbf{D}_n)$, and $\mathbf{G} = E\left(\mathbf{D}_n \mathbf{D}_n^T\right)$, where

$$\mathbf{D}_n = \mathrm{vech}\left(\{\langle v_i, X_n^c\rangle\langle v_j, X_n^c\rangle, \ 1 \le i \le j \le p\}^T\right).$$

We see from (12.13) that the vector $\sqrt{N}\left(\hat{\mathbf{Z}}^T\hat{\mathbf{Z}}\right)^{-1}\hat{\mathbf{Z}}^T\boldsymbol{\varepsilon}^{**}$ has the same limiting distribution as

$$\frac{1}{\sqrt{N}}\sum_{n=1}^{N}\varepsilon_n^{**}\begin{bmatrix} \zeta\left(\mathbf{G}-\mathbf{M}\mathbf{M}^T\right)^{-1}\zeta & \mathbf{0}_{r\times p} & -\zeta\left(\mathbf{G}-\mathbf{M}\mathbf{M}^T\right)^{-1}\zeta\mathbf{M} \\ \mathbf{0}_{p\times r} & \boldsymbol{\Lambda}^{-1} & \mathbf{0}_{p\times 1} \\ -\mathbf{M}^T\left(\mathbf{G}-\mathbf{M}\mathbf{M}^T\right)^{-1} & \mathbf{0}_{1\times p} & 1+\mathbf{M}^T\left(\mathbf{G}-\mathbf{M}\mathbf{M}^T\right)^{-1}\mathbf{M} \end{bmatrix}\begin{bmatrix}\hat{\mathbf{D}}_n \\ \hat{\mathbf{F}}_n \\ 1\end{bmatrix}. \tag{12.14}$$

Since we are only interested in $\sqrt{N}\hat{\mathbf{A}}$ we need only consider the first $r = p(p+1)/2$ elements of the vector in (12.14). One can show that these are given by

$$\begin{aligned}
&\frac{1}{\sqrt{N}}\sum_{n=1}^{N}\varepsilon_n^{**}\left[\zeta\left(\mathbf{G}-\mathbf{M}\mathbf{M}^T\right)^{-1}\zeta \quad \mathbf{0}_{r\times p} \quad -\zeta\left(\mathbf{G}-\mathbf{M}\mathbf{M}^T\right)^{-1}\zeta\mathbf{M}\right]\begin{bmatrix}\hat{\mathbf{D}}_n \\ \hat{\mathbf{F}}_n \\ 1\end{bmatrix} \\
&= \frac{1}{\sqrt{N}}\sum_{n=1}^{N}\varepsilon_n^{**}\left(\zeta\left(\mathbf{G}-\mathbf{M}\mathbf{M}^T\right)^{-1}\zeta\hat{\mathbf{D}}_n - \zeta\left(\mathbf{G}-\mathbf{M}\mathbf{M}^T\right)^{-1}\zeta\mathbf{M}\right) \\
&= \frac{1}{\sqrt{N}}\sum_{n=1}^{N}\varepsilon_n^{**}\zeta\left(\mathbf{G}-\mathbf{M}\mathbf{M}^T\right)^{-1}\zeta\left(\hat{\mathbf{D}}_n - \mathbf{M}\right).
\end{aligned}$$

Applying Theorem 2.1, one can show that

$$\frac{1}{\sqrt{N}}\sum_{n=1}^{N}\varepsilon_n^{**}\left(\mathbf{G}-\mathbf{M}\mathbf{M}^T\right)^{-1}\zeta\left(\hat{\mathbf{D}}_n - \mathbf{M}\right) \xrightarrow{\mathcal{D}} N\left(0, \tau^2\left(\mathbf{G}-\mathbf{M}\mathbf{M}^T\right)^{-1}\right),$$

where

$$\begin{aligned}
\tau^2 = \mathrm{Var}\Bigg(\varepsilon_n &+ \sum_{i=p+1}^{\infty} b_i\langle X_n^c, v_i\rangle \\
&+ \sum_{i=p+1}^{\infty}\sum_{j=i}^{\infty}(2-1\{i=j\})a_{i,j}\langle X_n^c, v_i\rangle\langle X_n^c, v_j\rangle \\
&+ \sum_{i=1}^{p}\sum_{j=p+1}^{\infty} 2a_{i,j}\langle X_n^c, v_i\rangle\langle X_n^c, v_j\rangle\Bigg).
\end{aligned}$$

The next step is to verify that $\hat{\tau}^2 - \tau^2 = o_P(1)$. As a consequence of (12.13), we see that $\left(\hat{\mathbf{G}} - \hat{\mathbf{M}}\hat{\mathbf{M}}^T\right) - \zeta\left(\mathbf{G}-\mathbf{M}\mathbf{M}^T\right)\zeta = o_P(1)$, completing the proof of the theorem. The details of this proof are given in Horváth and Reeder (2011b) and Reeder (2011).

12.4 Outline for the Proof of Theorem 12.2

We establish the weak law

$$\hat{\mathbf{A}}^T(\hat{\mathbf{G}} - \hat{\mathbf{M}}\hat{\mathbf{M}}^T)\hat{\mathbf{A}} \xrightarrow{P} \mathbf{A}^T(\mathbf{G} - \mathbf{M}\mathbf{M}^T)\mathbf{A}, \tag{12.15}$$

where $\mathbf{A} = \operatorname{vech}\left(\{a_{i,j}\,(2 - 1\{i = j\}),\ 1 \le i \le j \le p\}^T\right)$ is like the vector $\tilde{\mathbf{A}}$ except without the random signs.

The estimation of $v_1, \dots, v_p$ by $\hat{v}_1, \dots, \hat{v}_p$ causes only the introduction of the random signs $\hat{c}_1, \dots, \hat{c}_p$. As in the proof of Theorem 12.1 one can verify that

$$\hat{\mathbf{A}} - \boldsymbol{\zeta}\mathbf{A} \xrightarrow{P} \mathbf{0}.$$

Under H_0 or H_A, one can establish that

$$\hat{\mathbf{G}} - \boldsymbol{\zeta}\mathbf{G}\boldsymbol{\zeta} = o_P(1)$$

and

$$\hat{\mathbf{M}}\hat{\mathbf{M}}^T - \boldsymbol{\zeta}\mathbf{M}\mathbf{M}^T\boldsymbol{\zeta} = o_P(1),$$

completing the proof of (12.15).

Part III
Dependent functional data

Chapter 13
Functional autoregressive model

This chapter studies the functional autoregressive (FAR) process which has found many applications. The theory of autoregressive and more general linear processes in Hilbert and Banach spaces is developed in the monograph of Bosq (2000), on which Sections 13.1 and 13.2 are based. We present only a few selected results which provide an introduction to the central ideas, and are needed in the sequel. Section 13.3 is devoted to prediction by means of the FAR process; some theoretical background is given in Section 13.5.

We say that a sequence $\{X_n, -\infty < n < \infty\}$ of mean zero elements of L^2 follows a functional AR(1) model if

$$X_n = \Psi(X_{n-1}) + \varepsilon_n, \tag{13.1}$$

where $\Psi \in \mathcal{L}$ and $\{\varepsilon_n, -\infty < n < \infty\}$ is a sequence of iid mean zero errors in L^2 satisfying $E\|\varepsilon_n\|^2 < \infty$.

The above definition defines a somewhat narrower class of processes than that considered by Bosq (2000) who does not assume that the ε_n are iid, but rather that they are uncorrelated in an appropriate Hilbert space sense, see his Definitions 3.1 and 3.2. The theory of estimation for the process (13.1) is however developed only under the assumption that the errors are iid.

To lighten the notation, we set in this chapter, $\|\cdot\|_{\mathcal{L}} = \|\cdot\|$.

13.1 Existence

A scalar AR(1) process $\{X_n, -\infty < n < \infty\}$ is said to be causal if it admits the expansion

$$X_n = \sum_{j=0}^{\infty} c_j \varepsilon_{n-j}.$$

The X_n depends then only on the present and past errors, but not on the future ones. If $|\psi| < 1$, scalar AR(1) equations have a unique solution of this form, in which $c_j = \psi^j$. A detailed treatment of these issues is presented in Chapter 3 of Brockwell and Davis (1991).

Our goal in this section is to establish a condition analogous to $|\psi| < 1$ for functional AR(1) equations (13.1). We begin with the following lemma:

Lemma 13.1. *For any $\Psi \in \mathcal{L}$, the following two conditions are equivalent:*

C0: There exists an integer j_0 such that $\|\Psi^{j_0}\| < 1$.
C1: There exist $a > 0$ and $0 < b < 1$ such that for every $j \geq 0$, $\|\Psi^j\| \leq ab^j$.

Proof. Since C1 clearly implies C0, we must only show that C0 implies C1.

Write $j = j_0 q + r$ for some $q \geq 0$ and $0 \leq r < j_0$. Therefore,

$$\|\Psi^j\| = \|\Psi^{j_0 q}\Psi^r\| \leq \|\Psi^{j_0}\|^q \|\Psi^r\|.$$

If $\|\Psi^{j_0}\| = 0$, then $C1$ holds for any $a > 0$ and $0 < b < 1$, so we assume in the following that $\|\Psi^{j_0}\| > 0$. Since $q > j/j_0 - 1$ and $\|\Psi^{j_0}\| < 1$, we obtain

$$\|\Psi^j\| \leq \|\Psi^{j_0}\|^{j/j_0 - 1}\|\Psi^r\| \leq \left(\|\Psi^{j_0}\|^{1/j_0}\right)^j \|\Psi^{j_0}\|^{-1} \max_{0 \leq r < j_0} \|\Psi^r\|,$$

so C1 holds with $a = \|\Psi^{j_0}\|^{-1} \max_{0 \leq r < j_0} \|\Psi^r\|$, $b = \|\Psi^{j_0}\|^{1/j_0}$. □

Note that condition C0 is weaker than the condition $\|\Psi\| < 1$; in the scalar case these two conditions are clearly equivalent. Nevertheless, C1 is a sufficiently strong condition to ensure the convergence of the series $\sum_j \Psi^j(\varepsilon_{n-j})$, and the existence of a stationary causal solution to functional AR(1) equations, as stated in the following theorem.

Theorem 13.1. *If condition C0 holds, then there is a unique strictly stationary causal solution to* (13.1). *This solution is given by*

$$X_n = \sum_{j=0}^{\infty} \Psi^j(\varepsilon_{n-j}). \tag{13.2}$$

The series converges almost surely, and in the L^2 norm, i.e.

$$E\left\|X_n - \sum_{j=0}^{m} \Psi^j(\varepsilon_{n-j})\right\|^2 \to 0, \quad \text{as } m \to \infty.$$

Proof. To establish the existence of the limit of the infinite series, we work with the space of square integrable random functions in $L^2 = L^2([0,1])$, see Section 2.3. If the random functions are defined on a probability space Ω, then we work with $L^2(\Omega, L^2([0,1]))$, which is a Hilbert space with the inner product $E\langle X, Y\rangle$, $X, Y \in L^2([0,1])$. Thus, to show that the sequence $X_n^{(m)} = \sum_{j=0}^{m} \Psi^j(\varepsilon_{n-j})$ has a limit in $L^2(\Omega, L^2([0,1]))$, it suffices to check that it is a Cauchy sequence in m for every fixed n.

Observe that by Lemma 2.1,

$$E\left\|\sum_{j=m}^{m'} \Psi^j(\varepsilon_{n-j})\right\|^2 = \sum_{j=m}^{m'}\sum_{k=m}^{m'} E\left\langle \Psi^j(\varepsilon_{n-j}), \Psi^k(\varepsilon_{n-k})\right\rangle$$
$$= \sum_{j=m}^{m'} E\|\Psi^j(\varepsilon_{n-j})\|^2.$$

Note that $E\Psi^j(\varepsilon_{n-j}) = 0$ because the expectation commutes with bounded operators. Therefore, by Lemma 13.1,

$$E\left\|\sum_{j=m}^{m'} \Psi^j(\varepsilon_{n-j})\right\|^2 \le \left(\sum_{j=m}^{m'} \|\Psi^j\|^2\right) E\|\varepsilon_0\|^2 \le E\|\varepsilon_0\|^2 a^2 \sum_{j=m}^{m'} b^{2j}.$$

Thus $X_n^{(m)}$ converges in $L^2(\Omega, L^2([0,1]))$.

To show the a.s. convergence, it is enough to verify that

$$\sum_{j=0}^{\infty} \|\Psi^j(\varepsilon_{n-j})\| < \infty \quad a.s.$$

This holds because by condition C1

$$E\left(\sum_{j=0}^{\infty} \|\Psi^j\|\|\varepsilon_{n-j}\|\right)^2 \le \sum_{j,k=0}^{\infty} \|\Psi^j\|\|\Psi^k\| E\|\varepsilon_0\|^2$$
$$\le E\|\varepsilon_0\|^2 \left(\sum_{j=0}^{\infty} ab^j\right)^2 < \infty,$$

and so $\sum_{j=0}^{\infty} \|\Psi^j\|\|\varepsilon_{n-j}\| < \infty$ a.s.

The series (13.2) is clearly strictly stationary, and it satisfies equation (13.1). Suppose $\{X_n'\}$ is another strictly stationary causal sequence satisfying (13.1). Then, iterating (13.1), we obtain, for any $m \ge 1$,

$$X_n' = \sum_{j=1}^{m} \Psi^j(\varepsilon_{n-j}) + \Psi^{m+1}(X'_{n-m+1}).$$

Therefore,

$$E\|X_n' - X_n^{(m)}\| \le \|\Psi^{m+1}\| E\|X'_{n-m+1}\| \le E\|X_0\| ab^{m+1}.$$

Thus X_n' is equal a.s. to the limit of $X_n^{(m)}$ i.e. to X_n. □

Example 13.1. As in Section 2.2, consider an integral Hilbert–Schmidt operator on L^2 defined by

$$\Psi(x)(t) = \int \psi(t,s)x(s)ds, \quad x \in L^2, \tag{13.3}$$

which satisfies

$$\iint \psi^2(t,s)dt\,ds < 1. \tag{13.4}$$

Recall from section 2.2 that the left–hand side of (13.4) is equal to $\|\Psi\|_{\mathcal{S}}^2$. Since $\|\Psi\| \le \|\Psi\|_{\mathcal{S}}$, we see that (13.4) implies condition C0 of Lemma 13.1 with $j_0 = 1$.

13.2 Estimation

This section is devoted to the estimation of the autoregressive operator Ψ, but first we state Theorem 13.2 on the convergence of the EFPC's and the corresponding eigenvalues, which states that the expected distances between the population and the sample eigenelements are $O(N^{-1/2})$, just as for independent functional observations. Theorem 13.2 follows from Example 16.1, Theorem 16.2 and Lemma 13.1.

Theorem 13.2. *Suppose the operator* Ψ *in* (13.1) *satisfies condition C0 of Lemma 13.1, and the process* $\{X_n\}$ *satisfies* $E\|X_0\|^4 < \infty$. *If* (16.12) *holds, then, for each* $1 \le j \le d$, *relations* (2.13) *hold.*

We now turn to the estimation of the autoregressive operator Ψ. It is instructive to focus first on the univariate case $X_n = \psi X_{n-1} + \varepsilon_n$, in which all quantities are scalars. We assume that $|\psi| < 1$, so that there is a stationary solution such that ε_n is independent of X_{n-1}. Then, multiplying the AR(1) equation by X_{n-1} and taking the expectation, we obtain $\gamma_1 = \psi\gamma_0$, where $\gamma_k = E[X_n X_{n+k}] = \mathrm{Cov}(X_n, X_{n+k})$. The autocovariances γ_k are estimated by the sample autocovariances $\hat{\gamma}_k$, so the usual estimator of ψ is $\hat{\psi} = \hat{\gamma}_1/\hat{\gamma}_0$. This estimator is optimal in many ways, see Chapter 8 of Brockwell and Davis (1991), and the approach outlined above, known as the Yule–Walker estimation, works for higher order and multivariate autoregressive processes. To apply this technique to the functional model, note that by (13.1), under condition C0 of Lemma 13.1,

$$E\left[\langle X_n, x\rangle X_{n-1}\right] = E\left[\langle \Psi(X_{n-1}), x\rangle X_{n-1}\right], \quad x \in L^2.$$

Define the lag–1 autocovariance operator by

$$C_1(x) = E[\langle X_n, x\rangle X_{n+1}]$$

and denote with superscript $\cdot^T$ the adjoint operator. Then, $C_1^T = C\Psi^T$ because, by a direct verification, $C_1^T = E\left[\langle X_n, x\rangle X_{n-1}\right]$, i.e.

$$C_1 = \Psi C. \tag{13.5}$$

The above identity is analogous to the scalar case, so we would like to obtain an estimate of Ψ by using a finite sample version of the relation $\Psi = C_1 C^{-1}$. The operator C does not however have a bounded inverse on the whole of H. To see it, recall that C admits representation (2.4), which implies that $C^{-1}(C(x)) = x$, where

$$C^{-1}(y) = \sum_{j=1}^{\infty} \lambda_j^{-1} \langle y, v_j \rangle v_j.$$

The operator C^{-1} is defined if all λ_j are positive. (If $\lambda_1 \geq \lambda_2 \geq \cdots \geq \lambda_p > \lambda_{p+1} = 0$, then $\{X_n\}$ is in the space spanned by $\{v_1, \ldots, v_p\}$. On this subspace, we can define C^{-1} by $C^{-1}(y) = \sum_{j=1}^{p} \lambda_j^{-1} \langle y, v_i \rangle v_i$.) Since $\|C^{-1}(v_n)\| = \lambda_n^{-1} \to \infty$, as $n \to \infty$, it is unbounded. This makes it difficult to estimate the bounded operator Ψ using the relation $\Psi = C_1 C^{-1}$. A practical solution is to use only the first p most important EFPC's $\hat{v}_j$, and to define

$$\widehat{IC}_p(x) = \sum_{j=1}^{p} \hat{\lambda}_j^{-1} \langle x, \hat{v}_j \rangle \hat{v}_j.$$

The operator $\widehat{IC}_p$ is defined on the whole of L^2, and it is bounded if $\hat{\lambda}_j > 0$ for $j \leq p$. By judiciously choosing p we find a balance between retaining the relevant information in the sample, and the danger of working with the reciprocals of small eigenvalues $\hat{\lambda}_j$. To derive a computable estimator of Ψ, we use an empirical version of (13.5). Since C_1 is estimated by

$$\widehat{C}_1(x) = \frac{1}{N-1} \sum_{k=1}^{N-1} \langle X_k, x \rangle X_{k+1},$$

we obtain, for any $x \in L^2$,

$$\begin{aligned}
\widehat{C}_1 \widehat{IC}_p(x) &= \widehat{C}_1 \left(\sum_{j=1}^{p} \hat{\lambda}_j^{-1} \langle x, \hat{v}_j \rangle \hat{v}_j \right) \\
&= \frac{1}{N-1} \sum_{k=1}^{N-1} \left\langle X_k, \sum_{j=1}^{p} \hat{\lambda}_j^{-1} \langle x, \hat{v}_j \rangle \hat{v}_j \right\rangle X_{k+1} \\
&= \frac{1}{N-1} \sum_{k=1}^{N-1} \sum_{j=1}^{p} \hat{\lambda}_j^{-1} \langle x, \hat{v}_j \rangle \langle X_k, \hat{v}_j \rangle X_{k+1}.
\end{aligned}$$

The estimator $\widehat{C}_1 \widehat{IC}_p$ can be used in principle, but typically an additional smoothing step is introduced by using the approximation $X_{k+1} \approx \sum_{i=1}^{p} \langle X_{k+1}, \hat{v}_i \rangle \hat{v}_i$. This leads to the estimator

$$\widehat{\Psi}_p(x) = \frac{1}{N-1} \sum_{k=1}^{N-1} \sum_{j=1}^{p} \sum_{i=1}^{p} \hat{\lambda}_j^{-1} \langle x, \hat{v}_j \rangle \langle X_k, \hat{v}_j \rangle \langle X_{k+1}, \hat{v}_i \rangle \hat{v}_i. \tag{13.6}$$

To establish the consistency of this estimator, it must be assumed that $p = p_N$ is a function of the sample size N. Theorem 8.7 of Bosq (2000) then establishes sufficient conditions for $\|\widehat{\Psi}_p - \Psi\|$ to tend to zero. They are technical, but, intuitively, they mean that the λ_j and the distances between them cannot tend to zero too fast.

The estimator (13.6) is a kernel operator with the kernel

$$\hat{\psi}_p(t,s) = \frac{1}{N-1}\sum_{k=1}^{N-1}\sum_{j=1}^{p}\sum_{i=1}^{p}\hat{\lambda}_j^{-1}\langle X_k, \hat{v}_j\rangle\langle X_{k+1}, \hat{v}_i\rangle\,\hat{v}_j(s)\hat{v}_i(t). \tag{13.7}$$

This is verified by noting that

$$\widehat{\Psi}_p(x)(t) = \int \hat{\psi}_p(t,s)x(s)ds.$$

All quantities at the right–hand side of (13.7) are available as output of the R function `pca.fd`, so this estimator is very easy to compute. Kokoszka and Zhang (2010) conducted a number of numerical experiments to determine how close the estimated surface $\hat{\psi}_p(t,s)$ is to the surface $\psi(t,s)$ used to simulate an FAR(1) process. Broadly speaking, for $N \le 100$, the discrepancies are very large, both in magnitude and in shape. This is illustrated in Figure 13.1, which shows the Gaussian kernel $\psi(t,s) = \alpha \exp\{-(t^2+s^2)/2\}$, with α chosen so that the Hilbert–Schmidt norm of ψ is $1/2$, and three estimates which use $p = 2,3,4$. The innovations ε_n were generated as Brownian bridges. Such discrepancies are observed for other kernels and other innovation processes as well. Moreover, by any reasonable measure of a distance between two surfaces, the distance between ψ and $\hat{\psi}_p$ increases as p increases. This is counterintuitive because by using more EFPC' $\hat{v}_j$, we would expect the approximation (13.7) to improve. For the FAR(1) used to produce Figure 13.1, the sums $\sum_{j=1}^{p}\hat{\lambda}_j$ explain, respectively, 74, 83 and 87 percent of the variance for $p = 2,3$ and 4, but (for the series length $N = 100$), the absolute deviation distances between ψ and $\hat{\psi}_p$ are $0.40, 0.44$ and 0.55. The same pattern is observed for the RMSE distance $\|\hat{\psi} - \psi\|_S$ and the relative absolute distance. As N increases, these distances decrease, but their tendency to increase with p remains. This problem is partially due to the fact that for many FAR(1) models, the estimated eigenvalues $\hat{\lambda}_j$ are very small, except $\hat{\lambda}_1$ and $\hat{\lambda}_2$, and so a small error in their estimation translates to a large error in the reciprocals $\hat{\lambda}_j^{-1}$ appearing in (13.7). Kokoszka and Zhang (2010) show that this problem can be alleviated to some extent by adding a positive baseline to the $\hat{\lambda}_j$. However, as we will see in Section 13.3, precise estimation of the kernel ψ is not necessary to obtain satisfactory predictions.

13.3 Prediction

In this section, we discuss finite sample properties of forecasts with the FAR(1) model. Besse *et al.* (2000) compare several prediction methods for functional time

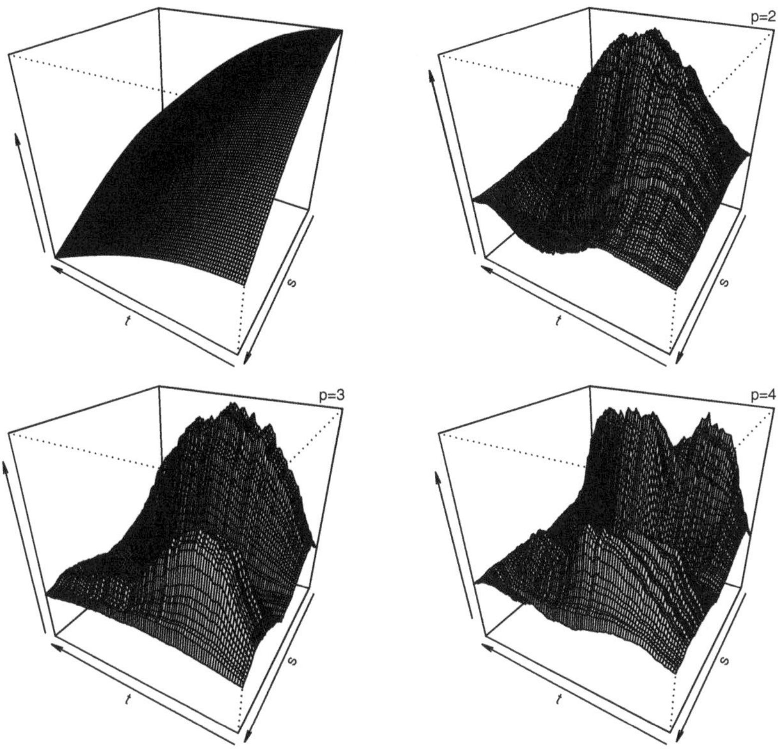

Fig. 13.1 The kernel surface $\psi(t, s)$ (top left) and its estimates $\hat{\psi}_p(t, s)$ for $p = 2, 3, 4$.

series by application to real geophysical data. Their conclusion is that the method which we call below *Estimated Kernel* performs better than the "non–functional" methods rooted in classical time series analysis. A different approach to prediction of functional data was proposed by Antoniadis *et al.* (2006). In this section, we mostly report the findings of Didericksen *et al.* (2011), whose simulation study includes a new method proposed by Kargin and Onatski (2008), which we call below *Predictive Factors*, and which seeks to replace the FPC's by directions which are most relevant for predictions.

We begin by describing the prediction methods we compare. This is followed by the discussion of their finite sample properties.

Estimated Kernel (EK). This method uses estimator (13.7). The predictions are calculated as

$$\hat{X}_{n+1}(t) = \int \hat{\psi}_p(t, s) X_n(s) ds = \sum_{k=1}^{p} \left(\sum_{\ell=1}^{p} \hat{\psi}_{k\ell} \langle X_n, \hat{v}_\ell \rangle \right) \hat{v}_k(t), \tag{13.8}$$

where

$$\hat{\psi}_{ji} = \hat{\lambda}_i^{-1}(N-1)^{-1}\sum_{n=1}^{N-1}\langle X_n, \hat{v}_i\rangle\langle X_{n+1}, \hat{v}_j\rangle. \tag{13.9}$$

There are several variants of this method which depend on where and what kind of smoothing is applied. In our implementation, all curves are converted to functional objects in R using 99 Fourier basis functions. The same minimal smoothing is used for the Predictive Factors method.

Predictive Factors (PF). Estimator (13.7) is not directly justified by the problem of prediction, it is based on FPC's, which may focus on the features of the data that are not relevant to prediction. An approach known as predictive factors may (potentially) be better suited for forecasting. It finds directions most relevant to prediction, rather than explaining the variability, as the FPC's do. Roughly speaking, it focuses on the optimal expansion of $\Psi(X_n)$, which is, theoretically, the best predictor of X_{n+1}, rather than the optimal expansion of X_n. Since Ψ is unknown, Kargin and Onatski (2008) developed a way of approximating such an expansion in finite samples. We describe this approach in Section 13.4. It's practical implementation depends on choosing an integer k and a positive number α. We used $k = p$ (the same as the number of the EFPC's), and $\alpha = 0.75$, as recommended by Kargin and Onatski (2008).

We selected five prediction methods for comparison, two of which do not use the autoregressive structure. To obtain further insights, we also included the errors obtained by assuming perfect knowledge of the operator Ψ. For ease of reference, we now describe these methods, and introduce some convenient notation.

MP (Mean Prediction) We set $\hat{X}_{n+1}(t) = 0$. Since the simulated curves have mean zero at every t, this corresponds to using the mean function as a predictor. This predictor is optimal if the data are uncorrelated.

NP (Naive Prediction) We set $\hat{X}_{n+1} = X_n$. This method does not attempt to model temporal dependence. It is included to see how much can be gained by utilizing the autoregressive structure of the data.

EX (Exact) We set $\hat{X}_{n+1} = \Psi(X_n)$. This is not really a prediction method because the autoregressive operator Ψ is unknown. It is included to see if poor predictions might be due to poor estimation of Ψ (cf. Section 13.2).

EK (Estimated Kernel) This method is described above.

EKI (Estimated Kernel Improved) This is method EK, but the $\hat{\lambda}_i$ in (13.9) are replaced by $\hat{\lambda}_i + \hat{b}$, as described in Section 13.2.

PF (Predictive Factors) This method is introduced above and described in detail in Section 13.4.

Didericksen *et al.* (2011) studied the errors E_n and R_n, $N-50 < n < N$, defined by

$$E_n = \sqrt{\int_0^1 \left(X_n(t) - \hat{X}_n(t)\right)^2 dt} \quad \text{and} \quad R_n = \int_0^1 \left|X_n(t) - \hat{X}_n(t)\right| dt.$$

for $N = 50, 100, 200$, and $\|\Psi\|_{\mathcal{S}} = 0.5, 0.8$. They considered several kernels and innovation processes, including smooth errors obtained as sum of two trigonometric function, irregular errors generated as Brownian bridges, an intermediate errors. Examples of boxplots are shown in Figures 13.2 and 13.3. In addition to boxplots, Didericksen *et al.* (2011) reported the averages of the E_n and R_n, $N-50 < n < N$, and the standard errors of these averages, which allow to assess if the differences in the performance of the predictors are statistically significant. Their conclusions can be summarized as follows:

1. Taking the autoregressive structure into account reduces prediction errors, but, in some settings, this reduction is not be statistically significant relative to method MP, especially if $\|\Psi\| = 0.5$. Generally if $\|\Psi\| = 0.8$, using the autoregressive structure significantly and visibly improves the predictions.
2. None of the Methods EX, EK, EKI uniformly dominates the other. In most cases method EK is the best, or at least as good at the others.
3. In some cases, method PF performs visibly worse than the other methods, but always better than NP.
4. Using the improved estimation described in Section 13.2 does not generally reduce prediction errors.

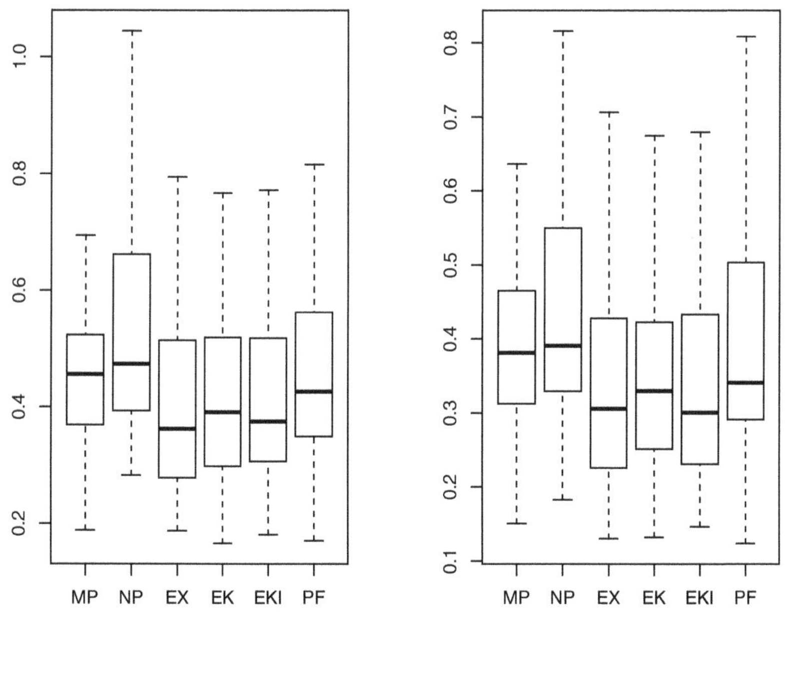

Fig. 13.2 Boxplots of the prediction errors E_n (left) and R_n (right); Brownian bridge innovations, $\psi(t,s) = Ct$, $N = 100$, $p = 3$, $\|\Psi\| = 0.5$.

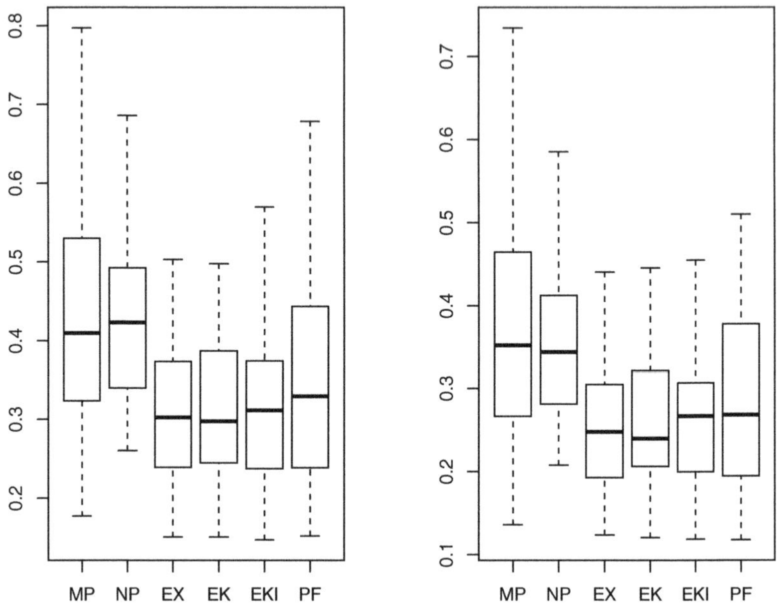

Fig. 13.3 Boxplots of the prediction errors E_n (left) and R_n (right); Brownian bridge innovations, $\psi(t,s) = Ct$, $N = 100$, $p = 3$, $\|\Psi\| = 0.8$.

Didericksen *et al.* (2011) also applied all prediction methods to mean corrected precipitation data studied in Besse *et al.* (2000). For this data set, the averages of the E_n and the R_n are not significantly different between the first five methods, method PF performs significantly worse than the others. We should point out that method PF depends on the choice of the parameters α and k. It is possible that its performance can be improved by better tuning these parameters. On the other hand, our simulations show that method EK essentially reaches the limit of what is possible, it is comparable to the theoretically perfect method EX. While taking into account the autoregressive structure of the observations does reduce prediction errors, many prediction errors are comparable to those of the trivial MP method. To analyze this observation further, we present in Figure 13.4 six consecutive trajectories of a FAR(1) process with $\|\Psi\| = 0.5$, and Brownian bridge innovations, together with EK predictions. Predictions obtained with other nontrivial methods look similar. We see that the predictions look much smoother than the observations, and their range is much smaller. If the innovations ε_n are smooth, the observations are also smooth, but the predicted curves have a visibly smaller range than the observations. This is also true for smooth real data, as shown in Figure 13.5.

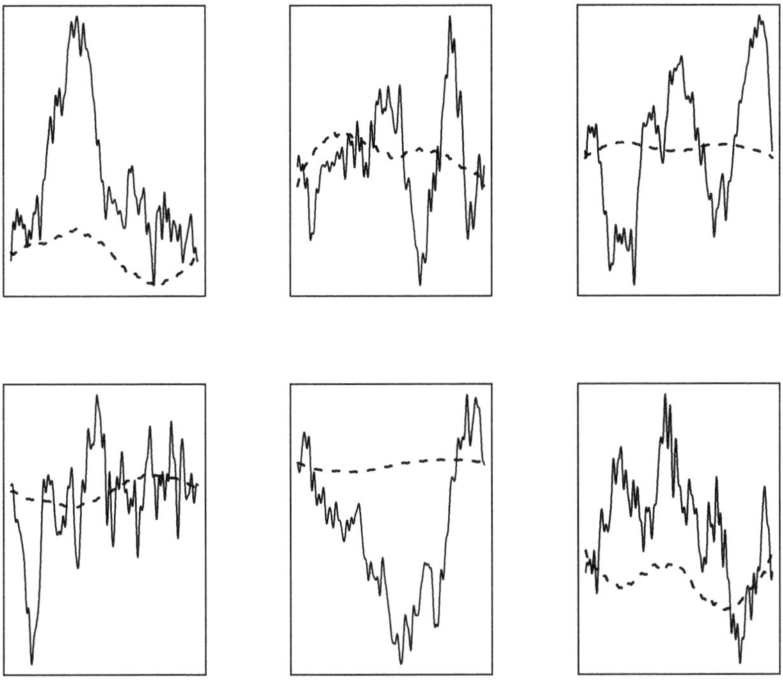

Fig. 13.4 Six consecutive trajectories of the FAR(1) process with Gaussian kernel, $\|\Psi\| = 0.5$, and Brownian bridge innovations. Dashed lines show EK predictions with $p = 3$.

The smoothness of the predicted curves follows from representation (13.8), which shows that each predictor is a linear combination of a few EFPC's, which are smooth curves themselves. The smaller range of the the predictors is not peculiar to functional data, but is enhanced in the functional setting. For a mean zero scalar AR(1) process $X_n = \psi X_{n-1} + \varepsilon_n$, we have $\mathrm{Var}(X_n) = \psi^2 \mathrm{Var}(X_{n-1}) + \mathrm{Var}(\varepsilon_n)$, so the variance of the predictor $\hat{\psi} X_{n-1}$ is about ψ^{-2} times smaller than the variance of X_n. In the functional setting, the variance of $\hat{X}_n(t)$ is close to $\mathrm{Var}[\int \psi(t,s) X_n(s) ds]$. If the kernel ψ admits the decomposition $\psi(t,s) = \psi_1(t)\psi_2(s)$, as all the kernels we use do, then

$$\mathrm{Var}\left[\hat{X}_n(t)\right] \approx \psi_1^2(t) \mathrm{Var}\left[\int_0^1 \psi_2(s) X_{n-1}(s) ds\right].$$

If the function ψ_1 is small for some values of $t \in [0,1]$, it will automatically drive down the predictions. If ψ_2 is small for some $s \in [0,1]$, it will reduce the integral $\int_0^1 \psi_2(s) X_{n-1}(s) ds$. The estimated kernels do not admit a factorization of this type, but are always weighted sums of products of orthonormal functions (the EFPC's $\hat{v}_k$). A conclusion of this discussion is that the predicted curves will in general look

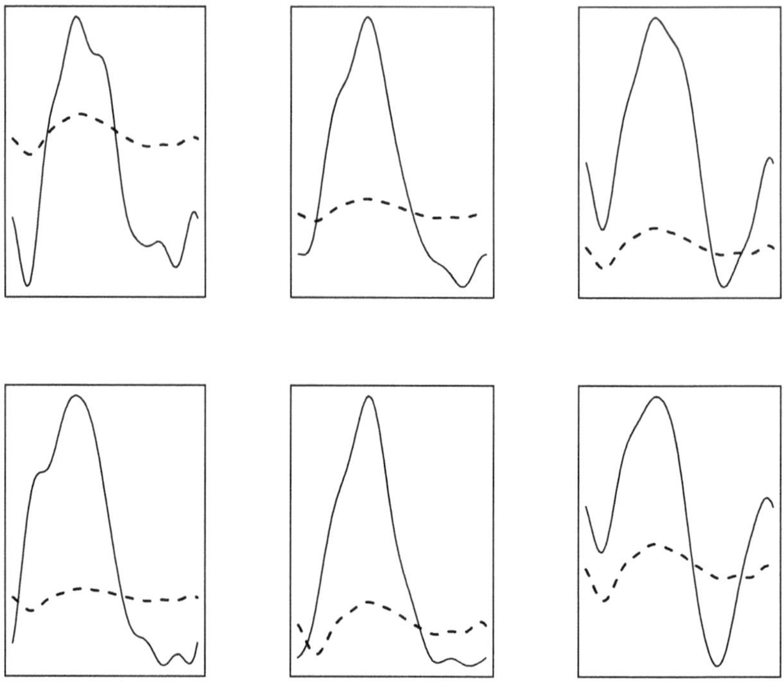

Fig. 13.5 Six consecutive trajectories (1989–1994) of centered pacific precipitation curves (solid) with their EK predictions (dashed).

smoother and "smaller" than the data. This somewhat disappointing performance is however not due to poor prediction methods, but to a natural limit of predictive power of the FAR(1) model; the curves $\Psi(X_n)$ share the general properties of the curves $\hat{\Psi}(X_n)$, no matter how Ψ is estimated.

13.4 Predictive factors

As we have seen in Section 13.3, it is difficult to improve on the predictor $\hat{\Psi}(X_n)$, and, in particular, the method of predictive factors, does not do it. It is however a very interesting approach because it focusses on directions different than the FPC's, which are a main focus of this book, and uses ideas which are central to the theory of operators in Hilbert spaces. For these reasons, we describe in this section the method of predictive factors in some in some detail. Section 13.5 contains some required theoretical background, which may also be of independent interest.

We continue to assume that $X_1, \ldots, X_N$ follow the AR(1) model (13.1). We denote by $\mathcal{R}_k$ the set of all rank k operators in $\mathcal{L}$, i.e. those operators which map L^2 into a subspace of dimension k. Rank k operators are clearly compact, and admit representation (2.1) with at most k nonzero λ_j. We first want to find $A \in \mathcal{R}_k$ which minimizes

$$E\|X_{n+1} - A(X_n)\|^2 = E\|(\Psi - A)(X_n)\|^2 + E\|\varepsilon_{n+1}\|^2.$$

To solve this problem, Kargin and Onatski (2008) introduce the polar decomposition of $\Psi C^{1/2}$, see Section 13.5,

$$\Psi C^{1/2} = U\,\Phi^{1/2}, \quad \Phi = C^{1/2}\Psi^T \Psi C^{1/2}.$$

To lighten the notation, suppose that k is smaller than the rank of Φ, and denote by $\sigma_1^2 > \cdots > \sigma_k^2$ the largest k eigenvalues of Φ, and by $x_1, \ldots, x_k$ the corresponding eigenfunctions.

Theorem 13.3. *If Condition C0 of Lemma 13.1 holds, and* $\sigma_1^2 > \cdots > \sigma_k^2 > 0$, *then*

$$\min\{E\|X_{n+1} - A(X_n)\|^2; \quad A \in \mathcal{R}_k\} = E\|X_{n+1} - \Psi_k(X_n)\|^2,$$

where Ψ_k *is defined by*

$$\Psi_k(y) = \sum_{i=1}^{k} \sigma_i^{-1} \left\langle y, \Psi^T \Psi C^{1/2}(x_i)\right\rangle U(x_i). \tag{13.10}$$

Proof. For any square integrable $Y \in L^2$ with covariance operator C_Y,

$$E\|Y\|^2 = \mathrm{tr}(C_Y), \tag{13.11}$$

with the trace tr defined in (13.19). Equality (13.11) follows from (2.7), which states that $E\|Y\|^2$ is equal to the sum of the eigenvalues of C_Y. This sum is clearly equal to $\mathrm{tr}(C_Y)$ (take the eigenfunction of C_Y as the e_n in (13.19)). Setting $Y = (\Psi - A)(X_n)$, it is easy to see that $C_Y = (\Psi - A)C(\Psi - A)^T$, and so by (13.11),

$$E\|X_{n+1} - A(X_n)\|^2 = \mathrm{tr}[(\Psi - A)C(\Psi - A)^T], \tag{13.12}$$

where C is the covariance operator of X_1.

Using property P2 of Section 13.5 and identity (13.23), we obtain

$$\begin{aligned} \mathrm{tr}[(\Psi - A)C(\Psi - A)^T] &= \mathrm{tr}[(\Psi - A)C^{1/2}C^{1/2}(\Psi - A)^T] \\ &= \mathrm{tr}\left[(\Psi - A)C^{1/2}[(\Psi - A)C^{1/2}]^T\right] \\ &= \mathrm{tr}\left[[(\Psi - A)C^{1/2}]^T(\Psi - A)C^{1/2}\right] \\ &= \|(\Psi - A)C^{1/2}\|_{\mathcal{S}}^2 = \|\Psi C^{1/2} - R\|_{\mathcal{S}}^2, \quad R = AC^{1/2}. \end{aligned} \tag{13.13}$$

If $A \in \mathcal{R}_k$, then $R = AC^{1/2} \in \mathcal{R}_k$, so our problem will be solved if we can find $R_k \in \mathcal{R}_k$ of the form $R_k = \Psi_k C^{1/2}$, $\Psi_k \in \mathcal{R}_k$, such that

$$\min\{\|\Psi C^{1/2} - R\|_{\mathcal{S}} : R \in \mathcal{R}_k\} = \|\Psi C^{1/2} - R_k\|_{\mathcal{S}}$$

The problem of finding R_k can be solved using the results of Section 13.5. To apply them, we notice that $\Psi C^{1/2} \in \mathcal{S}$. This is because the operator $C^{1/2}$ is Hilbert–Schmidt as C is trace class, so $\Psi C^{1/2} \in \mathcal{S}$, see Section 13.5. By (13.24), R_k is given by

$$R_k(y) = \sum_{i=1}^{k} \sigma_i \langle y, x_i \rangle U(x_i), \quad y \in L^2, \tag{13.14}$$

and it remains to verify that $R_k = \Psi_k C^{1/2}$. Observe that

$$\begin{aligned} \Psi_k C^{1/2}(x) &= \sum_{i=1}^{k} \sigma_i^{-1} \left\langle C^{1/2}(x), \Psi^T \Psi C^{1/2}(x_i)\right\rangle U(x_i) \\ &= \sum_{i=1}^{k} \sigma_i^{-1} \left\langle x, C^{1/2}\Psi^T \Psi C^{1/2}(x_i)\right\rangle U(x_i) \\ &= \sum_{i=1}^{k} \sigma_i^{-1} \langle x, \Phi(x_i)\rangle U(x_i) \\ &= \sum_{i=1}^{k} \sigma_i^{-1} \left\langle x, \sigma_i^2 x_i\right\rangle U(x_i) = R_k(x). \end{aligned}$$ □

Before moving on to the construction of a feasible predictor, we state a proposition which quantifies the population prediction error.

Proposition 13.1. *If Condition C0 of Lemma 13.1 holds, and* $\sigma_1^2 > \cdots > \sigma_k^2 > 0$, *then*

$$E\|X_{n+1} - \Psi_k(X_n)\|^2 = \sum_{i=k+1}^{\infty} \sigma_i^2,$$

where the σ_i^2 *are the eigenvalues of* $\Phi = C^{1/2}\Psi^T \Psi C^{1/2}$ *in decreasing order.*

Proof. The proposition follows by combining (13.12), (13.13) and (13.25). □

By (2.7), $C \in \mathcal{T}$. Thus $C^{1/2} \in \mathcal{S}$, so $C^{1/2}\Psi^T \in \mathcal{S}$ and $\Psi C^{1/2} \in \mathcal{S}$. Consequently $\Phi \in \mathcal{T}$, and so $\sum_{i=1}^{\infty} \sigma_i^2 < \infty$. Hence $\sum_{i=k+1}^{\infty} \sigma_i^2$ tends to zero, as $k \to \infty$.

Representation (13.10) cannot be used directly to compute predictions because it contains unknown quantities, and the operator U is not specified. The following lemma is a step towards the construction of a feasible predictor.

Lemma 13.2. *The operator* (13.10) *admits the representation*

$$\Psi_k(y) = \sum_{i=1}^{k} \langle y, b_i \rangle C_1(b_i), \quad b_i = C^{-1/2}(x_i), \tag{13.15}$$

where C_1 is the lag–1 autocovariance operator defined in Section 13.2.

Proof. We first verify that x_i is in the range of $C^{1/2}$, so that $C^{-1/2}(x_i)$ is well–defined, see Section 4.5. Since the x_i are the eigenfunctions of Φ, we have

$$\sigma_i^2 x_i = \Phi(x_i) = C^{1/2}(\Psi^T \Psi C^{1/2}(x_i)),$$

so x_i is in the range of $C^{1/2}$, and

$$\sigma_i^2 C^{-1/2}(x_i) = \Psi^T \Psi C^{1/2}(x_i). \tag{13.16}$$

Inserting (13.16) into (13.10), we obtain

$$\Psi_k(y) = \sum_{i=1}^{k} \sigma_i \langle y, b_i \rangle U(x_i), \tag{13.17}$$

so it remains to identify $U(x_i)$.

Using (13.5) and the the identity $\Phi^{1/2}(x_i) = \sigma_i x_i$, see Section 13.5, we have

$$C_1 C^{-1/2}(x_i) = \Psi C C^{-1/2}(x_i) = \Psi C^{1/2}(x_i) = U \Phi^{1/2}(x_i) = \sigma_i U(x_i).$$

Consequently,

$$U(x_i) = \sigma_i^{-1} C_1 C^{-1/2}(x_i) = \sigma_i^{-1} C_1(b_i).$$

Inserting into (13.17) yields (13.15). □

The random sequences $\{\langle X_n, b_i \rangle, \ -\infty < n < \infty\}$ are called the *predictive factors*. The functions $C_1(b_i)$ are called the *predictive loadings*. The loadings $C_1(b_i)$, $i = 1, 2, \ldots k$, are the "directions" in L^2 most relevant for prediction. Since the x_i are defined only up to a sign, the same is true for the predictive factors and loadings. However the operator Ψ_k (13.15) is uniquely defined.

To implement the prediction strategy suggested by Theorem 13.3 and Lemma 13.2, we need to estimate the eigenfunctions x_i and the eigenvalues σ_i^2 of $\Phi = C^{1/2}\Psi^T \Psi C^{1/2} = C^{-1/2} C_1^T C_1 C^{-1/2}$, cf. (13.5), and then approximate the $b_i = C^{-1/2}(x_i)$. Similarly as in the problem of the estimation of Ψ, the difficulty arises from the fact $C^{-1/2}$ is not a bounded estimator. This introduces an instability to the estimation of the eigenfunctions and eignevalues of $\hat{C}^{-1/2}\hat{C}_1^T \hat{C}_1 \hat{C}^{-1/2}$, and it cannot be ensured that these estimates converge to their population counterparts. To deal with these difficulties, Kargin and Onatski (2008) propose the following approach. To facilitate the inversion, introduce

$$\hat{\Phi}_\alpha = \hat{C}_\alpha^{-1/2} \hat{C}_1^T \hat{C}_1 \hat{C}_\alpha^{-1/2}, \quad \hat{C}_\alpha = \hat{C} + \alpha I,$$

where α is a small positive number and I is the identity operator. Denote by $\hat{\sigma}^2_{\alpha,1} \geq \cdots \geq \hat{\sigma}^2_{\alpha,k}$ the largest k eigenvalues of $\hat{\Phi}_\alpha$, and by $\hat{x}_{\alpha,1}, \ldots, \hat{x}_{\alpha,k}$ the corresponding eigenfunctions. Then define the predictor by

$$\hat{\Psi}_{\alpha,k}(y) = \sum_{i=1}^{k} \left\langle y, \hat{b}_{\alpha,i} \right\rangle \hat{C}_1(\hat{b}_{\alpha,i}), \quad \hat{b}_{\alpha,i} = \hat{C}_\alpha^{-1/2}(\hat{x}_{\alpha,i}). \tag{13.18}$$

Finding a bound on the prediction error requires a long technical argument. Kargin and Onatski (2008) established the following result.

Theorem 13.4. *Suppose Assumptions of Theorem 13.2 hold, and α and k are functions of the sample size N such that*

$$N^{1/6}\alpha \to A > 0 \quad \text{and} \quad N \geq N^{-1/4}k \geq K > 0,$$

for some constants A and K. Then

$$E\|X_{n+1} - \hat{\Psi}_{\alpha,k}(X_n)\|^2 = O\left(N^{-1/6}\log^2(N)\right).$$

Proof. The claim follows from Theorem 4 of Kargin and Onatski (2008). □

13.5 The trace class and the polar and singular decompositions

We present in this section several useful properties of operators in a Hilbert space. They were used in Section 13.4, but they applicability is much broader.

We first review some properties of *trace class* operators. Detailed proofs are presented in Section VI.6 of Reed and Simon (1972).

The trace of any positive–definite operator $A \in \mathcal{L}$ is defined by

$$\operatorname{tr}(A) = \sum_{n=1}^{\infty} \langle e_n, A(e_n) \rangle, \tag{13.19}$$

where $\{e_n\}$ is any orthonormal basis.

The value of the trace does not depend on the choice of the basis $\{e_n\}$. If $\{f_m\}$ is another orthonormal basis, then

$$\begin{aligned}\sum_{m=1}^{\infty} \langle f_m, A(f_m) \rangle &= \sum_{m=1}^{\infty} \left\langle A^{1/2}(f_m), A^{1/2}(f_m) \right\rangle \\ &= \sum_{m=1}^{\infty} \|A^{1/2}(f_m)\|^2 = \sum_{m=1}^{\infty}\sum_{n=1}^{\infty} \left\langle A^{1/2}(f_m), e_n \right\rangle^2 \\ &= \sum_{n=1}^{\infty}\sum_{m=1}^{\infty} \left\langle f_m, A^{1/2}(e_n) \right\rangle^2 = \sum_{n=1}^{\infty} \|A^{1/2}(e_n)\|^2\end{aligned}$$

$$\sum_{n=1}^{\infty} \left\langle A^{1/2}(e_n), A^{1/2}(e_n) \right\rangle = \sum_{n=1}^{\infty} \langle e_n, A(e_n) \rangle,$$

where we used the fact that $A^{1/2}$ is symmetric, see Section 4.5.

The trace has the following properties:

$$\mathrm{tr}(A+B) = \mathrm{tr}(A) + \mathrm{tr}(B). \tag{13.20}$$

$$\mathrm{tr}(\alpha A) = \alpha \mathrm{tr}(A), \quad \alpha \geq 0. \tag{13.21}$$

For any unitary operator U,

$$\mathrm{tr}(UAU^{-1}) = \mathrm{tr}(A). \tag{13.22}$$

Properties (13.20) and (13.21) are trivial, (13.22) follows from a simple verification which uses the fact that the trace can be computed using $\{e_n\}$ or $\{U(e_n)\}$.

Definition 13.1. An operator $A \in \mathcal{L}$ is called *trace class* or *nuclear* if $\mathrm{tr}[(A^T A)^{1/2}] < \infty$. The set of all trace class operators is denoted $\mathcal{T}$.

The class $\mathcal{T}$ has the following properties:

P1 $\mathcal{T}$ is a vector space.
P2 If $A \in \mathcal{T}$ and $B \in \mathcal{L}$, then $AB \in \mathcal{T}$ and $BA \in \mathcal{T}$, and $\mathrm{tr}(AB) = \mathrm{tr}(BA)$.
P3 If $A \in \mathcal{T}$, then $A^T \in \mathcal{T}$, and $\mathrm{tr}(A^T) = \mathrm{tr}(A)$.

The verification of these properties, especially the first one, requires some background in the theory of decompositions of linear operators.

It can be shown that

$$\|A\|_{\mathcal{T}} = \mathrm{tr}[(A^T A)^{1/2}]$$

defines a norm on $\mathcal{T}$, and that finite rank operators are dense in $\mathcal{T}$ equipped with this norm.

We have the following class inclusions:

Trace class $\subset$ Hilbert–Schmidt $\subset$ Compact,

with the corresponding norm inequalities:

$$\|A\|_{\mathcal{L}} \leq \|A\|_{\mathcal{S}} \leq \|A\|_{\mathcal{T}}.$$

In terms of expansion (2.1), trace class operators are those with $\sum_j |\lambda_j| < \infty$, Hilbert–Schmidt those with $\sum_j \lambda_j^2 < \infty$, and compact with $\lambda_j \to 0$.

An operator S is Hilbert–Schmidt if and only if $S^T S$ is trace class, i.e. if and only if $\mathrm{tr}[S^T S] < \infty$. In that case

$$\|S\|_{\mathcal{S}}^2 = \mathrm{tr}[S^T S]. \tag{13.23}$$

If $S \in \mathcal{S}$ and $A \in \mathcal{L}$, then $AS \in \mathcal{S}$ and $SA \in \mathcal{S}$.

We now turn the the polar and singular decompositions. These are extensions to operators in a Hilbert space of fundamental decompositions of matrices, see e.g. Chapter 7 of Horn and Johnson (1985). The proofs of the following results, which go back to Schmidt (1907), can be found in Gohberg and Krein (1969).

An operator $U \in \mathcal{L}$ is called a *partial isometry* if $\|U(x)\| = \|x\|$ for all x in the orthogonal complement of $\ker(U) = \{x \in L^2 : U(x) = 0\}$. Every operator $A \in \mathcal{L}$ admits the *polar decomposition*:

$$A = U\,(A^T A)^{1/2}$$

in which U is a partial isometry uniquely defined by the requirement that $\ker(U) = \ker(A)$.

Suppose now that $S \in \mathcal{S}$, and we want to find $S_k \in \mathcal{R}_k$ such that $\|S - S_k\|_{\mathcal{S}}$ is minimum. Using the polar decomposition, we can write

$$S = U\,\Phi^{1/2}, \quad \Phi = S^T S.$$

Set $r = \mathrm{rank}(\Phi)$, which is the dimension of the range of Φ, and may be infinity. Denote by σ_i^2 the eigenvalues of Φ, and by x_i the corresponding eigenfunctions, $i = 1, 2, \ldots, r$. One can show that $\mathrm{rank}(\Phi^{1/2}) = \mathrm{rank}(\Phi)$ and that the x_i are are the eigenvectors of $\Phi^{1/2}$ with eignevalues σ_i.

The *singular value decomposition* of S is

$$S(y) = \sum_{i=1}^{r} \sigma_i \,\langle y, x_i\rangle\, U(x_i), \quad y \in L^2.$$

The approximation S_k is then given by

$$S_k(y) = \sum_{i=1}^{k \wedge r} \sigma_i \,\langle y, x_i\rangle\, U(x_i), \quad y \in L^2, \tag{13.24}$$

and satisfies

$$\|S - S_k\|_{\mathcal{S}}^2 = \sum_{i=k+1}^{\infty} \sigma_i^2. \tag{13.25}$$

Chapter 14
Change point detection in the functional autoregressive process

In this chapter, we develop a change point test for the FAR(1) model introduced in Chapter 13. The importance of change point testing was discussed in Chapter 6. Failure to take change points into account leads to spurious inference. This chapter is based on the work of Horváth *et al.* (2010). Zhang *et al.* (2011) proposed a self–normalized statistic to solve the problem discussed in this chapter. Self–normalized statistics are discussed in Section 16.6.

The remainder of this chapter is organized as follows. The testing problem and the assumptions are stated in Section 14.1. The testing procedure is described and heuristically justified in Section 14.2. Its application and finite sample performance are examined in Section 14.3. Asymptotic justification is presented in Section 14.4, with the proofs developed in Sections 14.5 and 14.6.

14.1 Introduction

The problem can be stated as follows. We observe the random functions $\{X_n(t),\ t \in [0,1],\ n = 1, 2, \ldots N\}$ and assume that they follow the model

$$X_{n+1} = \Psi_n(X_n) + \varepsilon_{n+1}, \quad n = 1, 2, \ldots, N, \tag{14.1}$$

with independent identically distributed (iid) mean zero innovations $\varepsilon_n \in L^2$.

We want to test

$$H_0: \Psi_1 = \Psi_2 = \cdots = \Psi_N$$

against the alternative

$$H_A: \text{there is } 1 \le k^* < N: \ \Psi_1 = \cdots = \Psi_{k^*} \ne \Psi_{k^*+1} = \cdots = \Psi_N.$$

Under H_0, the common operator is denoted by Ψ.

The test statistic is based on the differences of the sample autocovariances of projections of the X_n on the EFPC's. The limit distribution can be derived by replacing the EFPC's by their population counterparts and using a functional central limit theorem for ergodic sequences. But in the functional setting, this replacement introduces asymptotically nonnegligible terms, see Section 14.6, which cancel because of the special form of the test statistic. To show that the remaining terms due to the estimation of the FPC's are asymptotically negligible, we develop a new technique which involves the truncation at lag $O(\log N)$ of the moving average representation of the FAR(1) process (Lemma 14.3), a blocking technique that utilizes this truncation (Lemma 14.4) and Mensov's inequality (Lemma 14.8).

The following assumption formalizes the structure of the observations under the null hypothesis.

Assumption 14.1. *The functional observations $X_n \in L^2$ satisfy*

$$X_{n+1} = \Psi X_n + \varepsilon_{n+1}, \quad n = 0, 1, \ldots, N-1, \tag{14.2}$$

where Ψ is an integral operator with the kernel $\psi(t,s)$ satisfying

$$\iint \psi^2(s,t)ds\,dt < 1, \tag{14.3}$$

and the iid mean zero innovations $\varepsilon_n \in L^2$ satisfy

$$E\|\varepsilon_n\|^4 = E\left[\int \varepsilon_n^2(t)dt\right]^2 < \infty. \tag{14.4}$$

Equation (14.2) can then be written more explicitelly as

$$X_{n+1}(t) = \int \psi(t,s)X_n(s)ds + \varepsilon_{n+1}(t), \quad t,s \in [0,1]. \tag{14.5}$$

Assumption 14.1 ensures that (14.5) has a unique strictly stationary solution $\{X_n(t),\ t \in [0,1]\}$ with finite fourth moment in L^2 such that ε_{n+1} is independent of $X_n, X_{n-1}, \ldots$, see Chapter 13.

If Assumption 14.1 holds, we can define the covariance operator

$$C(x) = E[\langle X_n, x\rangle\, X_n], \quad x \in L^2,$$

and its eigenfunctions v_j and eigenvalues λ_j. Since the X_n are assumed to have mean zero, it is convenient to work with the sample covariance operator defined by

$$\hat{C}(x) = \frac{1}{N}\sum_{n=1}^{N} \langle X_n, x\rangle\, X_n, \quad x \in L^2.$$

14.2 Testing procedure

In this section, we describe the idea of the test and explain its practical application. The requisite asymptotic theory is presented in Section 14.4.

The idea is to check if the action of Ψ on the span of the p most important principal components of the observations $X_1, X_2, \ldots, X_N$ changes at some unknown time point k. If there is no change in the autoregressive operator Ψ, the functions $\Psi v_j,\ j = 1, 2, \ldots, p$, remain constant. Since $\Psi v_j = \sum_\ell \langle \Psi v_j, v_\ell \rangle v_\ell$, this is the case, to a good approximation, if the coefficients $\langle \Psi v_j, v_\ell \rangle, \ell \le p$ remain constant. Direct verification shows that under H_0, $\langle \Psi v_j, v_\ell \rangle = \lambda_j^{-1} \langle R v_j, v_\ell \rangle$ where

$$Rx = E[\langle X_n, x \rangle X_{n+1}], \quad x \in L^2,$$

is the lag–1 autocovariance operator. Thus, the constancy of Ψ is approximately equivalent to the constancy of the products $\langle R v_j, v_\ell \rangle,\ j, \ell = 1, 2, \ldots, p$.

The restriction to the action of Ψ on the span of $v_j,\ j, = 1, 2, \ldots, p$, means that the test will not detect changes on the orthogonal complement of this space. Typically p is chosen so that the empirical counterparts $\hat{v}_j,\ j, = 1, 2, \ldots, p$, explain a large percentage of the variability of the data and their linear combinations approximate the data very closely. We therefore view a change in the action of Ψ on $v_j, j > p$, as not relevant. This restriction quantifies the intuition that very small changes cannot be detected. Another point to note is that since $\langle R v_j, v_\ell \rangle = \lambda_j \langle \Psi v_j, v_\ell \rangle$, a change in Ψ may be obscured by a change in the eigenfunctions λ_j, thus potentially reducing power. Nevertheless, the test introduced below is effective in practical settings, and its large sample properties are tractable.

To devise a test against the alternative of a change–point, we must first estimate these products from observations $X_1, X_2, \ldots, X_k$, then from observations $X_{k+1}, X_{k+2}, \ldots, X_N$, and compare the resulting estimates. To achieve it, we define p–dimensional projections

$$\mathbf{X}_i = [X_{i1}, \ldots, X_{ip}]^T, \quad \hat{\mathbf{X}}_i = [\hat{X}_{i1}, \ldots, \hat{X}_{ip}]^T,$$

where

$$X_{ij} = \langle X_i, v_j \rangle = \int X_i(t) v_j(t) dt, \quad \hat{X}_{ij} = \langle X_i, \hat{v}_j \rangle = \int X_i(t) \hat{v}_j(t) dt.$$

We also define the $p \times p$ lag-1 autocovariance matrices:

$$\mathbf{R}_k = \frac{1}{k} \sum_{2 \le i \le k} \mathbf{X}_{i-1} \mathbf{X}_i^T, \quad \mathbf{R}^*_{N-k} = \frac{1}{N-k} \sum_{k < i \le N} \mathbf{X}_{i-1} \mathbf{X}_i^T;$$

$$\hat{\mathbf{R}}_k = \frac{1}{k} \sum_{2 \le i \le k} \hat{\mathbf{X}}_{i-1} \hat{\mathbf{X}}_i^T, \quad \hat{\mathbf{R}}^*_{N-k} = \frac{1}{N-k} \sum_{k < i \le N} \hat{\mathbf{X}}_{i-1} \hat{\mathbf{X}}_i^T.$$

Observe that by the ergodic theorem, as $k \to \infty$,

$$\mathbf{R}_k(j, \ell) = \frac{1}{k} \sum_{2 \le i \le k} \langle X_{i-1}, v_j \rangle \langle X_i, v_\ell \rangle \stackrel{a.s.}{\to} E[\langle X_{n-1}, v_j \rangle \langle X_n, v_\ell \rangle] = \langle R v_j, v_\ell \rangle.$$

Thus the matrices $\mathbf{R}_k$ and $\mathbf{R}^*_{N-k}$ approximate the matrix $[\langle Rv_j, v_\ell\rangle,\ j,\ell = 1,2,\dots,p]$ based, correspondingly, on the observations before and after time k, and so it is appealing to base the test on their difference. The matrices $\mathbf{R}_k$ and $\mathbf{R}^*_{N-k}$ cannot however be computed from the data because the population principal components v_j are unknown. Thus, we must replace them by their empirical counterparts $\hat{\mathbf{R}}_k$ and $\hat{\mathbf{R}}^*_{N-k}$. Relation (2.13) means that $\hat{c}_j\hat{v}_j$ is close to v_j. Consequently, the (j,ℓ) entry of $\hat{\mathbf{R}}_k$ must be multiplied by $\hat{c}_j\hat{c}_\ell$ in order to approximate the (j,ℓ) entry of $\mathbf{R}_k$. The random signs $\hat{c}_j$ and $\hat{c}_\ell$ are unknown, so a test statistic must be constructed in such a way that they do not appear in it. This is not a mere technical point; changing just a few observations can flip the curves $\hat{v}_j$, sometimes the sign changes in another estimation run, even if the data do not change. Another important point is that using the EFPC's $\hat{v}_j$ introduces a bias. Roughly speaking, details are presented in Section 14.6,

$$k\hat{\mathbf{R}}_k = k\mathbf{R}_k + k\boldsymbol{\eta}_N + o_P(N^{1/2}), \qquad 1 \le k \le N,$$

where the random matrix $\boldsymbol{\eta}_N$ depends on the differences $\hat{c}_j\hat{v}_j - v_j,\ 1 \le j \le p$. The order of $\boldsymbol{\eta}_N$ is thus $O_P(N^{-1/2})$, so $k\boldsymbol{\eta}_N,\ 2 \le k \le N$, is of the same order, $O_P(N^{1/2})$, as $k\mathbf{R}_k, 2 \le k \le N$. However if a procedure is based on a CUSUM statistic, the contribution of $\boldsymbol{\eta}_N$ cancels out. Under H_0, we expect the partial sum $\sum_{2\le i\le k}\hat{\mathbf{X}}_{i-1}\hat{\mathbf{X}}_i^T = k\hat{\mathbf{R}}_k$ to be close to $k\hat{\mathbf{R}}_N$, so CUSUM test statistics are functionals of the the differences $k\hat{\mathbf{R}}_k - k\hat{\mathbf{R}}_N, 2 \le k \le N$. Notice that

$$\begin{aligned} k\hat{\mathbf{R}}_k - k\hat{\mathbf{R}}_N &= k\mathbf{R}_k + k\boldsymbol{\eta}_N + o_P(N^{1/2} - k\left(\mathbf{R}_N + \boldsymbol{\eta}_N + o_P(N^{-1/2})\right) \\ &= k\mathbf{R}_k - k\mathbf{R}_N + o_P(N^{1/2}) + k o_P(N^{-1/2}) \\ &= k\mathbf{R}_k - k\mathbf{R}_N + o_P(N^{1/2}). \end{aligned}$$

We now describe how to construct the test statistics.
Denote

$$Y_i(j,\ell) = \langle X_{i-1}, v_j\rangle\langle X_i, v_\ell\rangle, \quad \hat{Y}_i(j,\ell) = \langle X_{i-1}, \hat{v}_j\rangle\langle X_i, \hat{v}_\ell\rangle \tag{14.6}$$

and consider the column vectors of length p^2:

$$\mathbf{Y}_i = [Y_i(1,1),\dots,Y_i(1,p),\ Y_i(2,1),\dots,Y_i(2,p),\dots,Y_i(p,1),\dots,Y_i(p,p)]^T;$$
$$\hat{\mathbf{Y}}_i = [\hat{Y}_i(1,1),\dots,\hat{Y}_i(1,p),\ \hat{Y}_i(2,1),\dots,\hat{Y}_i(2,p),\dots,\hat{Y}_i(p,1),\dots,\hat{Y}_i(p,p)]^T.$$

Define further

$$\mathbf{Z}_k = \sum_{2\le i\le k}\mathbf{Y}_i, \quad \mathbf{Z}^*_{N-k} = \sum_{k<i\le N}\mathbf{Y}_i;$$
$$\hat{\mathbf{Z}}_k = \sum_{2\le i\le k}\hat{\mathbf{Y}}_i, \quad \hat{\mathbf{Z}}^*_{N-k} = \sum_{k<i\le N}\hat{\mathbf{Y}}_i.$$

Since the X_i follow a functional AR(1) model, the vectors $\mathbf{Y}_i$ form a weakly dependent stationary sequence, and so, as $k \to \infty$,

$$\sqrt{k}\left[\frac{1}{k}\sum_{2\le i\le k}\mathbf{Y}_i - E\,\mathbf{Y}_k\right] \xrightarrow{d} N(\mathbf{0}, \boldsymbol{D}), \tag{14.7}$$

where $\boldsymbol{D}$ is the $p^2 \times p^2$ long run covariance matrix defined by

$$\boldsymbol{D} = \sum_{h=-\infty}^{\infty} E\left[(\mathbf{Y}_0 - E\mathbf{Y}_0)(\mathbf{Y}_h - E\mathbf{Y}_h)^T\right]. \tag{14.8}$$

Relation (14.7), and the corresponding relation for the sum over $k < i \le N$, can be rewritten as

$$\mathbf{Z}_k - kE\mathbf{Y}_N \approx N(0, k\boldsymbol{D}), \quad \mathbf{Z}^*_{N-k} - (N-k)E\mathbf{Y}_N \approx N(0, (N-k)\boldsymbol{D}).$$

Denoting by $\{\mathbf{W}_D(t),\ t \ge 0\}$ a p^2–dimensional Brownian motion with covariance matrix $\boldsymbol{D}$, we have, in fact,

$$\mathbf{Z}_k - kE\mathbf{Y}_N \approx \mathbf{W}_D(k), \quad \mathbf{Z}^*_{N-k} - (N-k)E\mathbf{Y}_N \approx \mathbf{W}_D(N) - \mathbf{W}_D(k). \tag{14.9}$$

By (14.9), under H_0 we have,

$$\begin{aligned}\frac{1}{k}\mathbf{Z}_k - \frac{1}{N-k}\mathbf{Z}^*_{N-k} &\approx \frac{1}{k}\mathbf{W}_D(k) - \frac{1}{N-k}(\mathbf{W}_D(N) - \mathbf{W}_D(k))\\ &= \frac{1}{k(N-k)}[N\mathbf{W}_D(k) - k\mathbf{W}_D(k)\\ &\quad -k\mathbf{W}_D(N) + k\mathbf{W}_D(k)]\\ &= \frac{N}{k(N-k)}\left[\mathbf{W}_D(k) - \frac{k}{N}\mathbf{W}_D(N)\right].\end{aligned}$$

Denote

$$\mathbf{U}_N(k) = \frac{k(N-k)}{N}\left(\frac{1}{k}\mathbf{Z}_k - \frac{1}{N-k}\mathbf{Z}^*_{N-k}\right). \tag{14.10}$$

The above calculation shows that

$$\mathbf{U}_N(k) \approx \mathbf{W}_D(k) - \frac{k}{N}\mathbf{W}_D(N).$$

Comparing covariances, we see that

$$\frac{1}{N}\left[\mathbf{W}_D(k) - \frac{k}{N}\mathbf{W}_D(N)\right]^T \boldsymbol{D}^{-1}\left[\mathbf{W}_D(k) - \frac{k}{N}\mathbf{W}_D(N)\right], \quad 1 \le k \le N,$$

has the same distribution as

$$\sum_{1\le m\le p^2} B_m^2(k/N), \quad 1 \le k \le N, \tag{14.11}$$

where the $B_m(\cdot)$ are independent Brownian bridges on $[0, 1]$. Consequently, any functional of

$$G_N(k) = \frac{1}{N}\mathbf{U}_N(k)^T \boldsymbol{D}^{-1}\mathbf{U}_N(k), \quad 1 \le k \le N, \tag{14.12}$$

can be approximated by the corresponding functional of (14.11).

Asymptotic theory for functionals of the process $\sum_{1\le m\le d} B_m^2(u),\ u \in [0,1]$, including weighted sums and maximally selected statistics, is well–known, see e.g. Csörgő and Horváth (1993, 1997), and goes back to Kiefer (1959). A Cramér–von–Mises type functional $K_d := \int_0^1 \sum_{1\le m\le d} B_m^2(u)du$ leads to tests with good finite sample properties, and so we focus on it in the following, but clearly other functionals can be used as well, see e.g. Horváth *et al.* (1999) for more examples.

To implement the test, we need to estimate the matrix $\boldsymbol{D}$ (14.8). The estimation of the long run covariance matrix is one of the most extensively studied topics in time series analysis and econometrics, see e.g. Andrews (1991), Andrews and Monahan (1992) and Robinson (1998) for recent approaches and references. Any reasonable method can be used, but for concreteness, we focus on the popular and simple Bartlett estimator, and explain how to adapt it to the change point problem.

Denote by

$$\widehat{\boldsymbol{\gamma}}_h(k) = \frac{1}{k}\sum_{1\le i\le k-h}\left(\hat{\mathbf{Y}}_i - \frac{1}{k}\sum_{1\le i\le k}\hat{\mathbf{Y}}_i\right)\left(\hat{\mathbf{Y}}_{i+h} - \frac{1}{k}\sum_{1\le i\le k}\hat{\mathbf{Y}}_i\right)^T$$

and

$$\widehat{\boldsymbol{\gamma}}_h^*(N-k) = \frac{1}{N-k}\sum_{k<i\le N-h}\left(\hat{\mathbf{Y}}_i - \frac{1}{N-k}\sum_{k<i\le N-h}\hat{\mathbf{Y}}_i\right) \times\left(\hat{\mathbf{Y}}_{i+h} - \frac{1}{N-k}\sum_{k<i\le N-h}\hat{\mathbf{Y}}_i\right)^T$$

the lag h $p^2 \times p^2$ autocovariance matrices computed, respectively, from the first k and the last $N-k$ observations. The corresponding Bartlett estimators of $\boldsymbol{D}$ are then

$$\widehat{\boldsymbol{D}}_k = \sum_{|h|\le q}\left(1 - \frac{h}{q+1}\right)\widehat{\boldsymbol{\gamma}}_h(k) \tag{14.13}$$

and

$$\widehat{\boldsymbol{D}}^*_{N-k} = \sum_{|h|\le q}\left(1 - \frac{h}{q+1}\right)\widehat{\boldsymbol{\gamma}}_h^*(N-k). \tag{14.14}$$

The sequence $G_N(k)$ (14.12) is approximated by the sequence

$$\hat{G}_N(k) = \frac{1}{N}\widehat{\mathbf{U}}_N(k)^T\left[\frac{k}{N}\widehat{\boldsymbol{D}}_k + \left(1 - \frac{k}{N}\right)\widehat{\boldsymbol{D}}^*_{N-k}\right]^{-1}\widehat{\mathbf{U}}_N(k), \tag{14.15}$$

where

$$\widehat{\mathbf{U}}_N(k) = \frac{k(N-k)}{N}\left(\frac{1}{k}\hat{\mathbf{Z}}_k - \frac{1}{N-k}\hat{\mathbf{Z}}^*_{N-k}\right). \tag{14.16}$$

Using the weighted sum of the estimators $\widehat{\boldsymbol{D}}_k$ and $\widehat{\boldsymbol{D}}^*_{N-k}$ in (14.15) has been shown in different settings to lead to better power than using just $\widehat{\boldsymbol{D}}_N$, see Antoch *et al.* (1997) and Hušková *et al.* (2007).

Defining the critical value $c(\alpha, d)$ by $P(K_d > c(\alpha, d)) = \alpha$, and

$$\hat{I}_N = \frac{1}{N}\sum_{k=1}^{N} \hat{G}_N(k), \tag{14.17}$$

the test rejects if $\hat{I}_N > c(\alpha, p^2)$ The critical values $c(\alpha, d)$ can be computed using an analytic formula derived by Kiefer (1959), but the simulated critical values in Table 6.1 give better results in finite samples.

It is possible to develop a rigorous theory for the behavior of the test under the alternative, but the analysis becomes even more technical and would take up space. We therefore outline only the essential arguments which explain why and when the test is consistent.

First we introduce the following notation: Let $k^* = [n\theta],\ 0 < \theta < 1$, be the time of change. The kernel changes from ψ to ψ^* which satisfies $\iint (\psi^*(s,t))^2 ds\, dt < 1$.

Following the proof of Theorem 2.6, one can show that as $N \to \infty$,

$$\iint (\hat{C}_N(x,y) - \bar{C}(x,y))^2 dx\, dy \stackrel{P}{\to} 0,$$

where

$$\hat{C}_N(x,y) = \frac{1}{N}\sum_{1\le i\le n} X_i(x)X_i(y)$$

and

$$\bar{C}(x,y) = \theta E[X_0(x)X_0(y)] + (1-\theta)\lim_{N\to\infty} E[X_N(x)X_N(y)].$$

The kernel $\bar{C}(x,y)$ is symmetric, positive–definite and Hilbert–Schmidt with eigenvalues and eigenfunctions $\bar{\lambda}_i$ and $\bar{v}_i$. It follows from Lemmas 2.2 and 2.3 that as $N \to \infty$, $\|\hat{v}_i - \bar{v}_i\|$ and $|\hat{\lambda}_i - \bar{\lambda}_i|$ tend to 0 in probability.

An application of the ergodic theorem yields that for all $0 \le u \le \theta$,

$$\frac{1}{N}\sum_{1\le i\le Nu} \langle X_{i-1}, \hat{v}_j\rangle\langle X_i, \hat{v}_\ell\rangle \ \to u \iint R(t,s)\bar{v}_j(t)\bar{v}_\ell(s)dt\, ds \quad \text{a.s.},$$

where $R(t,s) = E[X_1(t)X_2(s)]$.

Under the alternative, $X_{k^*+1}, X_{k^*+2}, \ldots, X_N, X_{N+1}, \ldots$ is not stationary (X_{k^*} is not the stationary initial value), but because $\iint (\psi^*(s,t))^2 dsdt < 1$ the effect of X_{k^*} is dying out exponentially fast and the elements of X_{k^*+m} are very close to a stationary solution if m is large. So carefully applying the ergodic theorem again, we obtain for all $\theta \le u \le 1$,

$$\frac{1}{N} \sum_{Nu \le i \le N} \langle X_{i-1}, \hat{v}_j \rangle \langle X_i, \hat{v}_\ell \rangle \xrightarrow{P} (1-u) \iint R^*(t,s) \bar{v}_j(t) \bar{v}_\ell(s) dt\, ds,$$

where $R^*(t,s) = \lim_{N\to\infty} EX_N(t)X_{N+1}(s)$. This means that we have consistency if for at least one (j, ℓ)

$$\iint R(t,s)\bar{v}_j(t)\bar{v}_\ell(s)dt\, ds \ne \iint R^*(t,s)\bar{v}_j(t)\bar{v}_\ell(s)dt\, ds,$$

i.e. if R and R^* are different on the space spanned by $\{\bar{v}_j(t)\bar{v}_\ell(s), 1 \le j, \ell \le p\}$.

We conclude this section with a summary of the practical implementation of the test procedure:

1) Find p so large that $\sum_{j=1}^{p} \hat{\lambda}_j / \sum_{j=1}^{N} \hat{\lambda}_j > 0.8$, but not greater than 5.
2) Compute $\hat{I}_N$ (14.17).
3) Choose a significance level α and find the critical value $c(\alpha, d)$ with $d = p^2$ from Table 6.1.
4) Reject H_0 if $\hat{I}_N > c(\alpha, p^2)$.

In step 1), p cannot be too large because it is then difficult to estimate $\boldsymbol{D}$. In step 2) good results are also obtained if in (14.15) $\frac{k}{N}\widehat{\boldsymbol{D}}_k + \left(1 - \frac{k}{N}\right)\widehat{\boldsymbol{D}}^*_{N-k}$ is replaced by $\widehat{\boldsymbol{D}}_N$, the computations are then much faster.

14.3 Application to credit card transactions and Eurodollar futures

In this section we report the results of a small simulation study that examined the finite sample performance of our test. Calculations were performed using the R package `fda`. We used the functional time series X_n of differenced counts of credit card transactions described in Section 7.3. The first three weeks (21 functional observations) are shown in Figure 1.7. The whole data set contains $N = 200$ curves. Applied to these data, our test does not reject the null hypothesis, indicating that a functional AR(1) model is appropriate for all 200 X_n. This is in agreement with the conclusions of Laukaitis and Račkauskas (2002) and of Section 7.3. The long run covariance matrix was estimated using the code Hansen (1995) (with some modifications).

In the following, we use the curves X_n to generate functional AR(1) processes which will allow us to assess the finite sample performance of our test in a realistic setting. To do it, we estimate the kernel $\psi(\cdot,\cdot)$ using the function `linmod`, see Malfait and Ramsay (2003) (we omit the details of regularization). Then, residual functions are computed as $\hat{\varepsilon}_n(t) = X_{n+1}(t) - \hat{\Psi} X_{n+1}(t),\ n = 1, \ldots, 193$. Drawing these residuals with replacement, we can simulate functional AR(1) series of any length via

$$Z_m(t) = \int \hat{\psi}(t,s) Z_{m-1}(s) ds + \varepsilon_m^*(t), \quad m = 1, 2, \ldots, N,$$

where the $\varepsilon_m^*(\cdot)$ are the bootstrap draws of the $\hat{\varepsilon}_n(\cdot)$. If we change the kernel $\psi(\cdot,\cdot)$ at some point, we can assess the power of the test. To remove the initialization effect, the first "burn-in" 100 simulated functional observations were removed. The empirical rejection rates reported below are based on one thousand replications.

Table 14.1 shows empirical sizes for several values of p and N. The test becomes conservative as p increases. This is because the critical values increase in proportion to p^2, but only the first few principal components explain most of the variance. The same phenomenon was observed in Chapter 7. To save space, we report the empirical power only for $p = 2$ and $p = 3$; for $p = 4$ the power is about 30% lower than for $p = 3$. We introduced a change at half length by multiplying $\hat{\psi}(\cdot,\cdot)$ by $c = 0.1,\ 0.3,\ 0.6$, sample realizations for $N = 200$ are shown in Figure 14.1. The change is not readily seen by eye, especially for $c = 0.6$. For $c = 0.1$, the second half of the series looks more like white noise, and the power is correspondingly very close to 100%, and so is not reported. Table 14.2 shows that the power increases with the sample size N, and is satisfactory for $N = 200$, supporting the claim the the functional AR(1) model is suitable for the whole credit card transactions record.

We now turn to the application of the change point test to the data set consisting of Eurodollar futures contract prices studied by Kargin and Onatski (2008). The seller of a Eurodollar futures contract takes on an obligation to deliver a 3 month deposit of one million US dollars to a bank account outside the United States i months from today. The price the buyer is willing to pay for this contract depends on the prevailing interest rate. These contracts are traded at the Chicago Mercantile Exchange, and provide a way to lock in an interest rate. They are liquid assets responsive to Federal Reserve policy, inflation, and economic indicators.

The data we study consist of 114 points per day; point i corresponds to the price of a contract with closing date i months from today. We work with centered data, i.e. the mean function has been subtracted from all observations. Examples of these

Table 14.1 Empirical size (in percent)

p=2			p=3			p=4		
10%	5%	1%	10%	5%	1%	10%	5%	1%
			N=50					
9.4	3.4	0.3	11.9	5.9	0.4	6.2	1.9	0.0
			N=100					
9.7	3.6	0.6	9.9	5.0	1.0	7.2	2.3	0.5
			N=200					
8.1	3.8	0.5	10.3	4.8	0.8	6.3	2.8	0.3

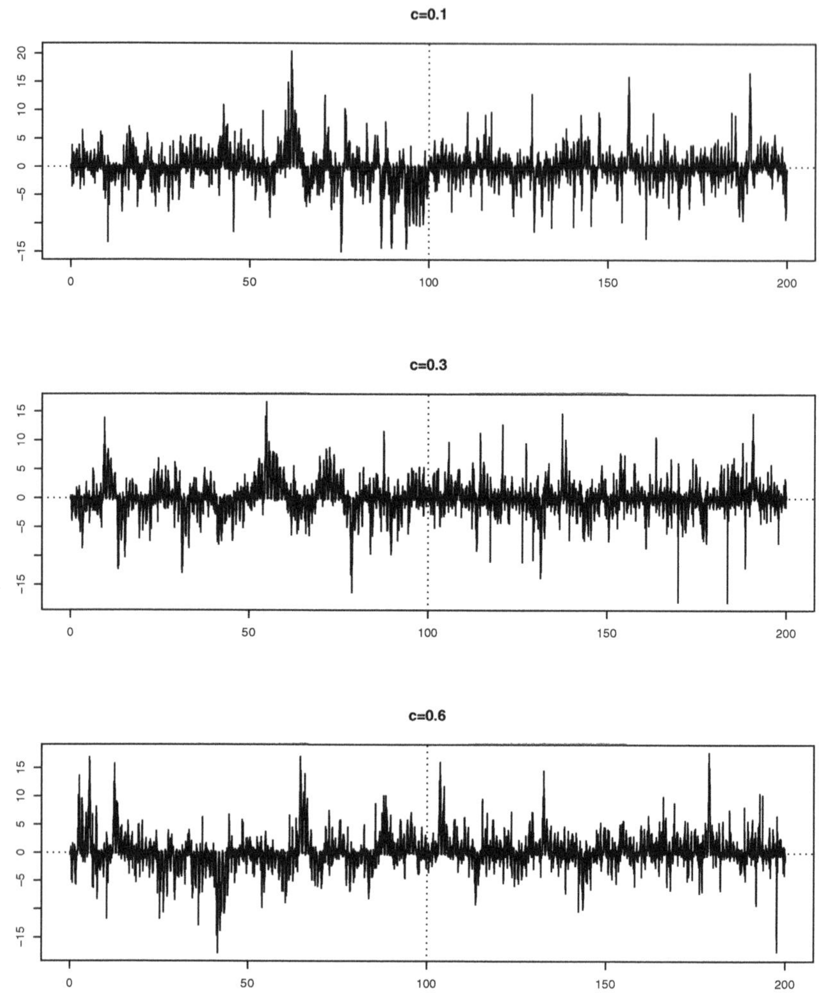

Fig. 14.1 Bootstrap realizations under alternatives

centered functions are shown in Figure 14.2, the middle panel reflects a change in expectations of future interest rates following the September 11, 2001 terrorist attacks.

The test rejects the null hypothesis of a constant operator Ψ for some periods and accepts for others. Figure 14.3 shows a period of 50 days for which the null hypothesis is accepted. Even though the prices of the contract fluctuate, these fluctuation can be modeled by assuming a single FAR(1) model. By contrast, the curves shown in Figure 14.4 cannot be assumed to follow an FAR(1) model, according to the change point test. No single change point is apparent, but the range of the data increases in

Table 14.2 Empirical power (in percent) for a change occurring at $k^* = N/2$, and $\hat{\psi}$ changing to $c\hat{\psi}$ for $c = 0.3$ (in parentheses $c = 0.6$).

p=2			p=3		
10%	5%	1%	10%	5%	1%
N=50					
46.1 (30.9)	28.3 (16.5)	6.3 (1.7)	23.1 (15.9)	10.4 (5.6)	0.3 (0.1)
N=100					
82.5 (58.1)	67.7 (44.3)	33.5 (16.7)	64.4 (46.9)	46.9 (28.8)	18.2 (7.8)
N=200					
98.7 (91.6)	95.8 (81.6)	82.3 (52.8)	96.3 (82.8)	90.4 (67.4)	65.6 (34.9)

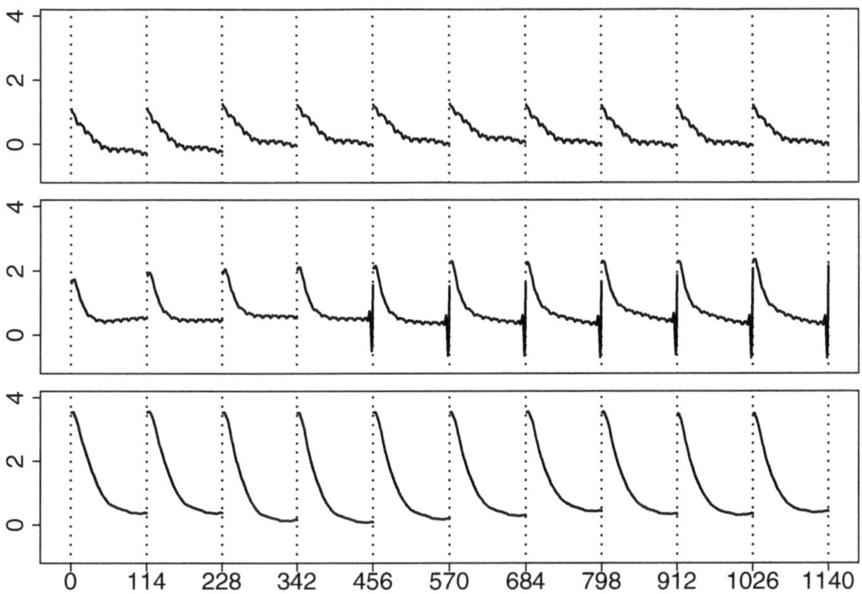

Fig. 14.2 Eurodollar futures curves over three disjoint 10 day long periods.

a systematic way, making modeling with a stationary FAR(1) model inappropriate. In general, the test rejects H_0 for longer series and accepts for shorter series. This is illustrated in Figure 14.5. For example, out of 8 consecutive periods of $N = 300$ trading days, only one can be modeled as a stationary FAR(1) process. By contrast, out of 49 periods of length $N = 50$, 37 can be assumed to follow an FAR(1) model.

The Eurodollar futures data set shows that while the test of this chapter is intended to detect a single change point on the autoregressive operator, it can also be used more generally to assess the suitability of a single FAR(1) model for the whole functional time series.

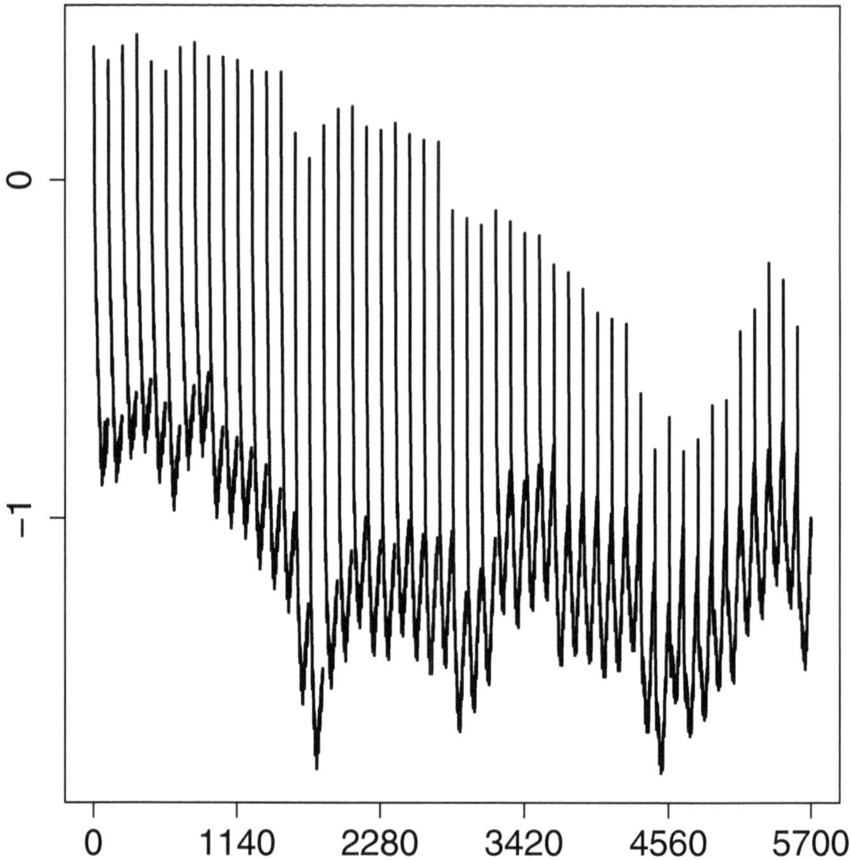

Fig. 14.3 Eurodollar futures curves on fifty consecutive days. The stability of the autoregressive operator is not rejected; P–value 0.291.

14.4 Asymptotic results

In order to develop an asymptotic theory, we must verify that the test statistic does not change if the principal components $\hat{v}_j$ are replaced by $\hat{c}_j \hat{v}_j$, as only the latter converge to the population principal components v_j. For this purpose, it is convenient to introduce a $p \times p$ diagonal matrix $\boldsymbol{C}_p$ and a $p^2 \times p^2$ diagonal matrix $\mathbf{M}$ defined by

$$\boldsymbol{C}_p = \begin{bmatrix} \hat{c}_1 & & & \\ & \hat{c}_2 & & \\ & & \ddots & \\ & & & \hat{c}_p \end{bmatrix}, \quad \mathbf{M} = \boldsymbol{C}_p \otimes \boldsymbol{C}_p ,$$

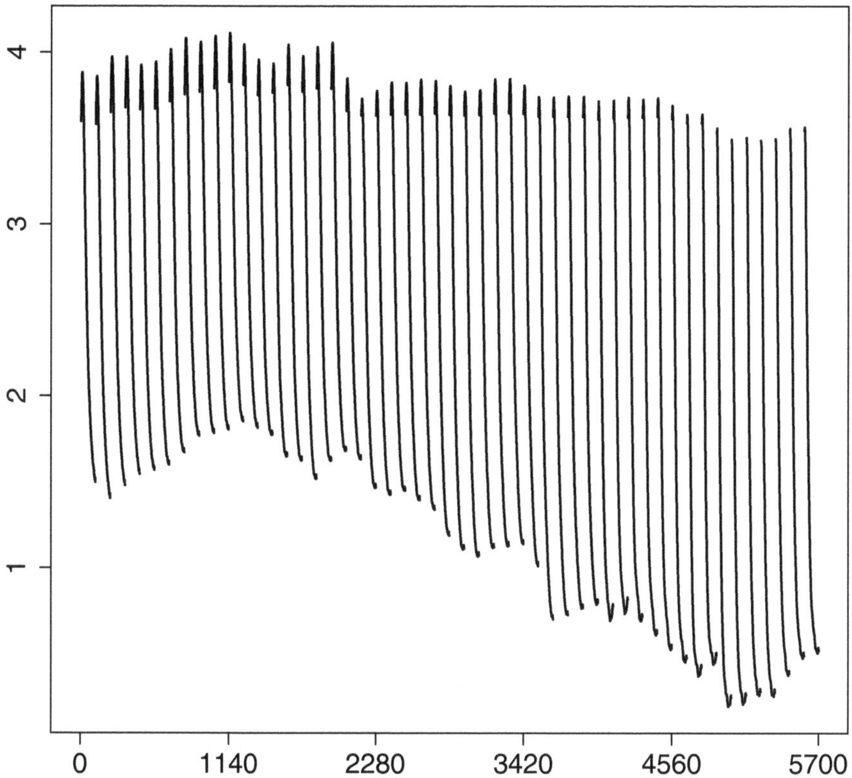

Fig. 14.4 Eurodollar futures curves on fifty consecutive days. The stability of the autoregressive operator is rejected; P–value 0.019.

where $\otimes$ denotes the Kronecker product, see e.g. Graham (1981). For example, if $p = 2$

$$\mathbf{M} = \begin{bmatrix} \hat{c}_1\hat{c}_1 & & & \\ & \hat{c}_1\hat{c}_2 & & \\ & & \hat{c}_2\hat{c}_1 & \\ & & & \hat{c}_2\hat{c}_2 \end{bmatrix}.$$

Replacing $\hat{v}_j$ by $\hat{c}_j\hat{v}_j$ implies replacing the vectors $\hat{\mathbf{Y}}_i$ by $\mathbf{M}\hat{\mathbf{Y}}_i$, which in turn implies replacing $\widehat{\mathbf{U}}_N(k)$ by $\mathbf{M}\widehat{\mathbf{U}}_N(k)$, while $\widehat{\boldsymbol{D}}_k$ and $\widehat{\boldsymbol{D}}^*_{N-k}$ are replaced, respectively, by $\mathbf{M}\widehat{\boldsymbol{D}}_k\mathbf{M}^T$ and $\mathbf{M}\widehat{\boldsymbol{D}}^*_{N-k}\mathbf{M}^T$. Since $\mathbf{M}^2$ is a $p^2 \times p^2$ identity matrix, it follows that the $\hat{G}(k)$ (14.15) are invariant to the signs of the $\hat{v}_j$. To develop asymptotic arguments, we can thus work with quantities $\hat{c}_j \langle X_i, \hat{v}_j \rangle$ in place of the actual scores $\langle X_i, \hat{v}_j \rangle$.

Recall the definition (14.15) of $\hat{G}(k)$ and introduce the process

$$\hat{Q}_N(u) = \hat{G}_N([Nu]), \quad u \in [0, 1].$$

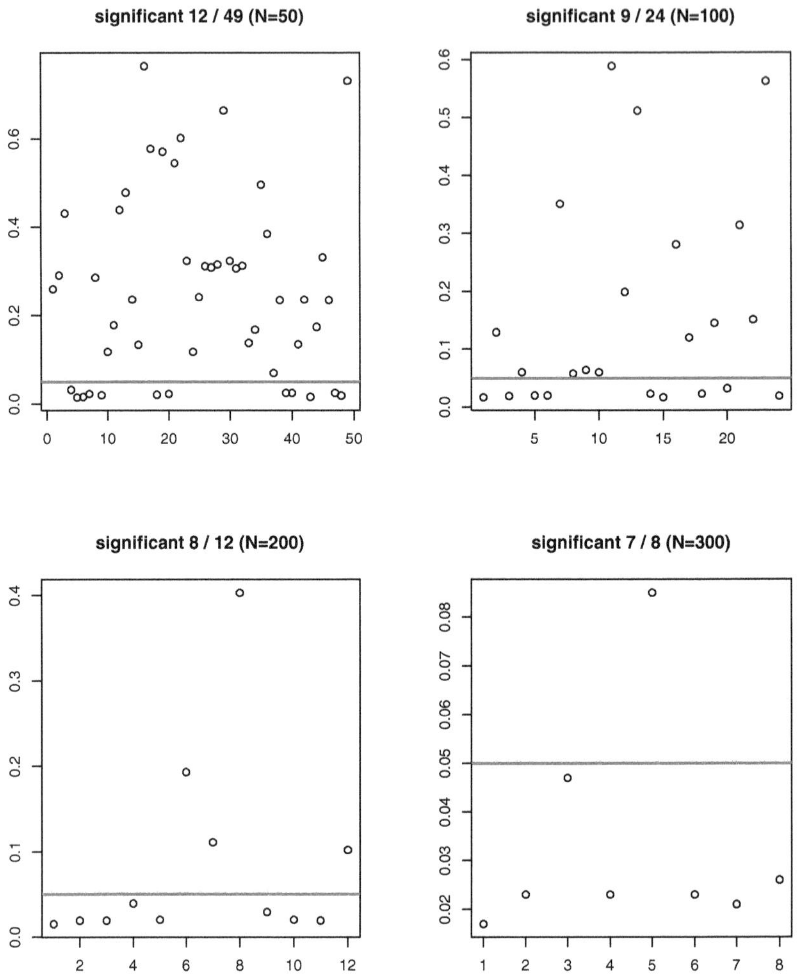

Fig. 14.5 P-values for consecutive segments. The continuous line indicates the five percent threshold.

Recall also the definition of the Bartlett estimators (14.13) and (14.14), and introduce the following assumption of the rate of growth of the bandwidth $q = q(N)$.

Assumption 14.2. *Suppose $q(N)$ is nondecreasing and satisfies*

$$\sup_{k \geq 0} \frac{q(2^{k+1})}{q(2^k)} < \infty \tag{14.18}$$

and

$$q(N) \to \infty \quad \text{and} \quad q(N)(\log N)^4 = O(N). \tag{14.19}$$

The following theorem shows that the test procedure described in Section 14.2 has asymptotically correct size.

Theorem 14.1. *Suppose Assumptions 14.1, 14.2 and condition* (2.12) *hold. Then*

$$\hat{Q}_N(u) \to \sum_{1 \le m \le p^2} B_m^2(u) \quad \text{in } D([0,1]),$$

where $\{B_m(u),\ u \in [0,1]\},\ 1 \le m \le p^2$*, are iid Brownian bridges.*

As we discussed in the previous section, the proof of theorem 14.1 is split into two steps. The first step, Proposition 14.1 is the weak convergence of the process $Q_N(u) = G_N([Nu]),\ u \in [0,1]$, where G_N is defined by (14.12). This is the CUSUM process based on the projections on population eigenfunctions of the covariance operator. In the second step, Proposition 14.2, it is shown that the estimation of the eigenfunctions and eigenvalues has only asymptotically negligible effect. The second step is more delicate, relies on the special structure of the process Q_N, a truncation and blocking technique, and an application of Mensov's inequality.

Proposition 14.1. *Under Assumption 14.1,*

$$Q_N(u) \to \sum_{1 \le m \le p^2} B_m^2(u) \quad \text{in } D([0,1]),$$

where $Q_N(u) = G_N([Nu]),\ u \in [0,1]$*, and* G_N *is defined by* (14.12).

Proposition 14.2. *Under Assumption 14.1 and condition* (2.12)*,*

$$N^{-1/2} \max_{2 \le k \le N} \left\| \mathbf{M}\widehat{\mathbf{U}}_N(k) - \mathbf{U}_N(k) \right\| \xrightarrow{P} 0.$$

Propositions 14.1 and 14.2 are proven, respectively, in Sections 14.5 and 14.6. Using them, it is easy to prove Theorem 14.1.

Proof of Theorem 14.1.. Recall that $Q_N(u) = G_N([Nu]),\ u \in [0,1]$, where G_N is defined by (14.12). By Proposition 14.1, $Q_N(u) \to \sum_{1 \le m \le p^2} B_m^2(u)$ in $D([0,1])$. To complete the proof, we must show that

$$\max_{2 \le k \le N} |\hat{G}_N(k) - G_N(k)| \xrightarrow{P} 0. \tag{14.20}$$

Relation (14.20) will follow once we have verified that

$$N^{-1/2} \max_{2 \le k \le N} \left\| \mathbf{M}\widehat{\mathbf{U}}_N(k) - \mathbf{U}_N(k) \right\| \xrightarrow{P} 0 \tag{14.21}$$

and

$$\max_{2 \le k \le N} \left\| \left[\frac{k}{N} \widehat{\boldsymbol{D}}_k + \left(1 - \frac{k}{N}\right) \widehat{\boldsymbol{D}}^*_{N-k} \right]^{-1} - \boldsymbol{D}^{-1} \right\| \xrightarrow{P} 0. \tag{14.22}$$

Relation (14.21) is stated as Proposition 14.2. To prove (14.22), we use Theorem A.1 and Remark A.1 of Berkes *et al.* (2006) which imply that under Assumption 14.2, $\widehat{\boldsymbol{D}}_k$ and $\widehat{\boldsymbol{D}}^*_{N-k}$ converge almost surely to $\boldsymbol{D}$. Recall that if a sequence ζ_n converges to zero a.s., then $\max_{1\le n\le N} |\zeta_n| \xrightarrow{P} 0$, as $N \to \infty$. Therefore, $\sup_{1<u<1} \|u\widehat{\boldsymbol{D}}_{[Nu]} - u\boldsymbol{D}\| \xrightarrow{P} 0$ and $\sup_{1<u<1} \|(1-u)\widehat{\boldsymbol{D}}^*_{N-[Nu]} - (1-u)\boldsymbol{D}\| \xrightarrow{P} 0$, and so

$$\sup_{1<u<1} \|u\widehat{\boldsymbol{D}}_{[Nu]} + (1-u)\widehat{\boldsymbol{D}}^*_{N-[Nu]} - \boldsymbol{D}\| \xrightarrow{P} 0.$$

Since the inverse is a continuous map, (14.22) follows. □

14.5 Proof of Proposition 14.1

Proposition 14.1 follows from Proposition 14.3 because by (14.23),

$$Q_N(u) \to [\mathbf{W}_D(u) - u\mathbf{W}_D(1)]^T \boldsymbol{D}^{-1}[\mathbf{W}_D(u) - u\mathbf{W}_D(1)] \quad \text{in } D([0,1])$$

and a direct computation shows that the Gaussian vectors–valued processes $\{\boldsymbol{D}^{-1/2}[\mathbf{W}_D(u) - u\mathbf{W}(u)],\ u \in [0,1]\}$ and $\{[B_1(u), \ldots, B_{p^2}(u)]^T,\ u \in [0,1]\}$ have equal covariance functions. Recall that $\mathbf{W}_D(\cdot)$, introduced in Section 14.2, is a Gaussian process with $E\mathbf{W}_D(u) = 0$ and $E\left[\mathbf{W}_D(u)\mathbf{W}_D^T(s)\right] = \boldsymbol{D}\min(u,s),\ u,s \in [0,1]$.

Proposition 14.3. *If Assumption 14.1 holds, then*

$$N^{-1/2}\left(\mathbf{Z}_{[Nu]} - E\mathbf{Z}_{[Nu]}\right) \to \mathbf{W}_D(u), \quad \text{in } D^{p^2}([0,1]). \tag{14.23}$$

Proof. Denote $Z_k(j,\ell) = \sum_{2\le i\le k} Y_i(j,\ell)$. To prove the proposition, it is enough to establish the convergence in $D([0,1])$ of all linear combinations, namely

$$N^{-1/2}\sum_{j,\ell=1}^{p} \theta(j,\ell)\left(Z_{[Nu]}(j,\ell) - EZ_{[Nu]}(j,\ell)\right) \xrightarrow{d} W_{\theta,D}(u),$$

where $\{W_{\theta,D}(u),\ u \in [0,1]\}$ is a Brownian motion with variance

$$E\left[W^2_{\theta,D}(u)\right] = u\sum_{j,\ell=1}^{p}\sum_{j',\ell'=1}^{p} \theta(j,\ell)\theta(j',\ell')D(j,\ell;j',\ell').$$

To reduce the notational burden, we focus on just one component, i.e. we want to show that

$$N^{-1/2}\sum_{2\le i\le [Nu]} [Y_i(j,\ell) - EY_i(j,\ell)] \xrightarrow{d} W_{D(i,j)}(u). \tag{14.24}$$

($W_{D(i,j)}(u)$ is defined by setting $\theta(i',j') = \delta_{i'i}\delta_{j'j}$ where $\delta_{..}$ is the Kronecker delta.) Convergence (14.24) (in $D([0,1])$) follows essentially from Theorem 19.1

of Billingsley (1999); we must verify that the sequence $\{Y_i(j,\ell)\}$ is stationary and ergodic and that

$$\sum_{i=1}^{\infty} |\mathrm{Cov}(Y_0(j,\ell), Y_i(j,\ell))| < \infty. \tag{14.25}$$

Relation (14.25) is established in Lemma 14.1. Ergodicity follows from the representation

$$\begin{aligned} Y_i(j,\ell) &= \langle X_{i-1}, v_j\rangle [\langle \Psi X_{i-1}, v_\ell\rangle + \langle \varepsilon_i, v_\ell\rangle] \\ &= \langle X_{i-1}, v_j\rangle \langle X_{i-1}, \Psi^T v_\ell\rangle + \langle X_{i-1}, v_j\rangle \langle \varepsilon_i, \Psi^T v_\ell\rangle \end{aligned}$$

and Theorem 13.1 (moving average representation of X_k) and Theorem 36.4 of Billingsley (1995) (a function of shifts of an iid sequence forms an ergodic sequence). □

Now we establish (14.25).

Lemma 14.1. *Under Assumption 14.1, the $Y_i(j,\ell)$ defined by* (14.6) *satisfy,*

$$\sum_{1\le i<\infty} |\mathrm{Cov}(Y_1(j,\ell), Y_i(m,n))| < \infty.$$

Proof. Since

$$\begin{aligned} Y_i(j,\ell) &= \langle X_{i-1}, v_j\rangle \langle X_{i-1}, \Psi^T v_\ell\rangle + \langle X_{i-1}, v_j\rangle \langle \varepsilon_i, \Psi^T v_\ell\rangle, \\ \mathrm{Cov}(Y_1(j,\ell), Y_i(m,n)) &= C_1(i) + C_2(i) + C_3(i) + C_4(i), \end{aligned}$$

where

$$\begin{aligned} C_1(i) &= \mathrm{Cov}\left(\langle X_0, v_j\rangle \langle X_0, \Psi^T v_\ell\rangle, \langle X_{i-1}, v_m\rangle \langle X_{i-1}, \Psi^T v_n\rangle\right); \\ C_2(i) &= \mathrm{Cov}\left(\langle X_0, v_j\rangle \langle X_0, \Psi^T v_\ell\rangle, \langle X_{i-1}, v_m\rangle \langle \varepsilon_i, \Psi^T v_n\rangle\right); \\ C_3(i) &= \mathrm{Cov}\left(\langle X_0, v_j\rangle \langle \varepsilon_1, \Psi^T v_\ell\rangle, \langle X_{i-1}, v_m\rangle \langle X_{i-1}, \Psi^T v_n\rangle\right); \\ C_4(i) &= \mathrm{Cov}\left(\langle X_0, v_j\rangle \langle \varepsilon_1, \Psi^T v_\ell\rangle, \langle X_{i-1}, v_m\rangle \langle \varepsilon_i, \Psi^T v_n\rangle\right). \end{aligned}$$

It is easy to see that $C_2(i) = C_4(i) = 0$, for $i > 1$, so it remains to find an absolutely convergent bounds on $C_1(i)$ and $C_3(i)$. We focus on the term $C_1(i)$, the argument for $C_3(i)$ being similar. Consider arbitrary $x, y, u, v \in L^2([0,1])$. Since $X_k = \Psi^k X_0 + \sum_{j=0}^{k-1} \Psi^j \varepsilon_{k-j}$,

$$\begin{aligned} &\mathrm{Cov}\left(\langle X_0, x\rangle \langle X_0, y\rangle, \langle X_k, u\rangle \langle X_k, v\rangle\right) \\ &\quad = \mathrm{Cov}\left(\langle X_0, x\rangle \langle X_0, y\rangle, \langle \Psi^k X_0, u\rangle \langle \Psi^k X_0, v\rangle\right). \end{aligned}$$

Consequently

$$\begin{aligned} &|\mathrm{Cov}\left(\langle X_0, x\rangle \langle X_0, y\rangle, \langle X_k, u\rangle \langle X_k, v\rangle\right)| \\ &\quad \le E\left|\langle X_0, x\rangle \langle X_0, y\rangle \langle \Psi^k X_0, u\rangle \langle \Psi^k X_0, v\rangle\right| \\ &\qquad + E\,|\langle X_0, x\rangle \langle X_0, y\rangle|\, E\,|\langle X_k, u\rangle \langle X_k, v\rangle| \\ &\quad \le \|\Psi\|^{2k} \left\{E\|X_0\|^4 + \left[E\|X_0\|^2\right]^2\right\} \|x\|\,\|y\|\,\|u\|\,\|v\|. \end{aligned}$$

Therefore

$$|C_1(i)| \le \|\Psi\|^{2(i-1)} \left\{ E\|X_0\|^4 + \left[E\|X_0\|^2\right]^2 \right\} \|v_j\| \, \|v_\ell\| \, \|\Psi^T v_m\| \, \|\Psi^T v_n\| \\ \le 2\|\Psi\|^{2i} E\|X_0\|^4. \qquad \square$$

14.6 Proof of Proposition 14.2

Denote $r(t,s) = E[X_1(t)X_2(s)]$ and

$$\hat{R}(j,\ell) = \iint r(t,s)\hat{u}(t,s)dt\,ds,$$

where

$$\hat{u}(t,s) = v_j(t)v_\ell(s) - \hat{c}_j\hat{v}_j(t)\hat{c}_\ell\hat{v}_\ell(s), \quad 0 < s,t < 1. \tag{14.26}$$

Proof of Proposition 14.2.. The component of $\mathbf{M}\widehat{\mathbf{U}}_N(k) - \mathbf{U}_N(k)$ corresponding to the product of the jth and the ℓth score is equal to

$$\frac{k(N-k)}{N}\left\{\frac{1}{k}\left[\hat{c}_j\hat{c}_\ell\hat{Z}_k(j,\ell) - Z_k(j,\ell) - k\hat{R}(j,\ell)\right] \right. \\ \left. -\frac{1}{N-k}\left[\hat{c}_j\hat{c}_\ell\hat{Z}^*_{N-k}(j,\ell) - Z^*_{N-k}(j,\ell) - (N-k)\hat{R}(j,\ell)\right]\right\}.$$

Thus the claim will follow once we have verified that

$$N^{-1/2}\max_{2\le k\le N}\left[\hat{c}_j\hat{c}_\ell\hat{Z}_k(j,\ell) - Z_k(j,\ell) - k\hat{R}(j,\ell)\right] \xrightarrow{P} 0 \tag{14.27}$$

and

$$N^{-1/2}\max_{2\le k\le N}\left[\hat{c}_j\hat{c}_\ell\hat{Z}^*_{N-k}(j,\ell) - Z^*_{N-k}(j,\ell) - (N-k)\hat{R}(j,\ell)\right] \xrightarrow{P} 0. \tag{14.28}$$

Since the above two relations are verified in the same way, we will show only the verification of (14.27).

Observe that

$$Z_k(j,\ell) = \iint \sum_{2\le i\le k} X_{i-1}(t)X_i(s)v_j(t)v_\ell(s)dt\,ds$$

and

$$\hat{c}_j\hat{c}_\ell\hat{Z}_k(j,\ell) = \iint \sum_{2\le i\le k} X_{i-1}(t)X_i(s)\hat{c}_j\hat{v}_j(t)\hat{c}_\ell\hat{v}_\ell(s)dt\,ds.$$

Therefore

$$
\begin{aligned}
&Z_k(j,\ell) - \hat{c}_j\hat{c}_\ell\hat{Z}_k(j,\ell)\\
&\quad = \iint \sum_{2\le i\le k} [X_{i-1}(t)X_i(s) - r(t,s)]v_j(t)v_\ell(s)dt\,ds\\
&\qquad - \iint \sum_{2\le i\le k} [X_{i-1}(t)X_i(s) - r(t,s)]\hat{c}_j\hat{v}_j(t)\hat{c}_\ell\hat{v}_\ell(s)dt\,ds\\
&\qquad + (k-1)\iint r(t,s)[v_j(t)v_\ell(s) - \hat{c}_j\hat{v}_j(t)\hat{c}_\ell\hat{v}_\ell(s)]dt\,ds\\
&\quad = \iint \sum_{2\le i\le k} [X_{i-1}(t)X_i(s) - r(t,s)]\hat{u}(t,s)dt\,ds + (k-1)\hat{R}(j,\ell).
\end{aligned}
$$

As $\hat{R}(j,\ell) = O_P(1)$, to prove (14.27), it thus remains to show that

$$
\max_{2\le k\le N}\left|\iint \sum_{2\le i\le k} [X_{i-1}(t)X_i(s) - r(t,s)]\hat{u}(t,s)dt\,ds\right| = o_P(N^{1/2}). \quad (14.29)
$$

Since

$$
\begin{aligned}
&\iint \left|\sum_{1\le i\le k} [X_{i-1}(t)X_i(s) - r(t,s)]\right| |\hat{u}(t,s)|\,dt\,ds\\
&\le \left(\iint \left|\sum_{1\le i\le k} [X_{i-1}(t)X_i(s) - r(t,s)]\right|^2 dt\,ds\right)^{1/2} \left(\iint |\hat{u}(t,s)|^2\,dt\,ds\right)^{1/2},
\end{aligned}
$$

(14.29) follows from Lemmas 14.2 and 14.3. □

Lemma 14.2. *The function* $\hat{u} \in L^2([0,1]^2)$ *defined by* (14.26) *satisfies*

$$
\|\hat{u}\| = \left(\iint [\hat{u}(t,s)]^2 dt\,ds\right)^{1/2} = O_P(N^{-1/2}).
$$

Proof. Since

$$
\begin{aligned}
\left|v_j(t)v_\ell(s) - \hat{c}_j\hat{v}_j(t)\hat{c}_\ell\hat{v}_\ell(s)\right|^2 &\le 2v_j^2(t)[v_\ell(s) - \hat{c}_j\hat{v}_\ell(s)]^2\\
&\quad + 2\hat{v}_\ell^2(s)[v_j(t) - \hat{c}_\ell\hat{v}_j(t)]^2
\end{aligned}
$$

and v_j and $\hat{v}_\ell$ have unit norm in $L^2([0,1])$, $\|\hat{u}\|^2 \le 2\left\{\|v_\ell - \hat{c}_\ell\hat{v}_\ell\|^2 + \|v_j - \hat{c}_j\hat{v}_j\|^2\right\}$. Consequently, by (2.13), there is a constant K such that $E\|\hat{u}\|^2 \le KN^{-1}$. □

Lemma 14.3. *Under Assumption 14.1,*

$$N^{-1} \max_{2\le k\le N} \left(\iint \left[\sum_{2\le i\le k} [X_{i-1}(t)X_i(s) - r(t,s)] \right]^2 dt\, ds \right)^{1/2} \xrightarrow{P} 0.$$

Proof. By Theorem 13.1,

$$X_k = \sum_{j=0}^{\infty} \Psi^j \varepsilon_{k-j}, \tag{14.30}$$

where the series converges in the L^2 norm and almost surely. For $c > 0$ to be determined later, introduce the truncated series

$$X_{k,N} = \sum_{j=0}^{c \log N} \Psi^j \varepsilon_{k-j}. \tag{14.31}$$

We will use the decomposition

$$\begin{aligned} X_{i-1}(t)X_i(s) - r(t,s) &= X_{i-1}(t)X_i(s) - X_{i-1,N}(t)X_{i,N}(s) \\ &\quad + [X_{i-1,N}(t)X_{i,N}(s) - r_N(t,s)] \\ &\quad + [r_N(t,s) - r(t,s)], \end{aligned}$$

where

$$r_N(t,s) = E[X_{i-1,N}(t)X_{i,N}(s)]. \tag{14.32}$$

Introduce also the functions

$$V_{i,N}(t,s) = X_{i-1}(t)X_i(s) - X_{i-1,N}(t)X_{i,N}(s) \tag{14.33}$$

and

$$U_{i,N}(t,s) = X_{i-1,N}(t)X_{i,N}(s). \tag{14.34}$$

To prove the lemma, it suffices to show that

$$N^{-1} E \max_{2\le k\le N} \left\| \sum_{2\le i\le k} V_{i,N} \right\| \to 0, \tag{14.35}$$

$$N^{-1} E \max_{2\le k\le N} \left\| \sum_{2\le i\le k} [U_{i,N} - r_N] \right\| \to 0 \tag{14.36}$$

and

$$\|r_N - r\| \to 0. \tag{14.37}$$

In (14.35), (14.36), (14.37), the norm is taken in the space $L^2([0,1]^2)$.

By Lemma 14.4,

$$E \max_{2\le k\le N} \left\| \sum_{2\le i\le k} V_{i,N} \right\| \le KN^{2-\kappa},$$

for some K and any $\kappa > 0$, provided c is sufficiently large, so (14.35) follows. Relations (14.36) and (14.37) follow, respectively, from Lemmas 14.5 and 14.6. □

Lemma 14.4. *For $c > 0$ define $X_{k,N} = X_{k,N,c}$ by* (14.31)*. Consider the function $V_{i,N}(t,s)$ defined by* (14.33)*. Then for any $\kappa > 0$, there is c so large that*

$$E\|V_{i,N}\| = E\left\{\iint V_{i,N}^2(t,s)dt\,ds\right\}^{1/2} \le KN^{-\kappa}$$

for some constant K.

Proof. Observe that

$$\begin{aligned}
\|V_{i,N}\|^2 &= \iint [X_{i-1}(t)X_i(s) - X_{i-1,N}(t)X_{i,N}(s)]^2 dt\,ds \\
&= \iint [X_{i-1}(t)(X_i(s) - X_{i,N}(s)) \\
&\quad + X_{i,N}(s)(X_{i-1}(t) - X_{i-1,N}(t))]^2 dt\,ds \\
&\le 2\left\{\int X_{i-1}^2(t)dt \int (X_i(s) - X_{i,N}(s))^2 ds \right.\\
&\quad \left. + \int X_{i,N}^2(s)ds \int (X_{i-1}(t) - X_{i-1,N}(t))^2 dt\right\} \\
&= 2\left\{\|X_{i-1}\|^2\|X_i - X_{i,N}\|^2 + \|X_{i,N}\|^2\|X_{i-1} - X_{i-1,N}\|^2\right\}
\end{aligned}$$

Define $r > 0$ by $\|\Psi\| = e^{-r}$. Then

$$E\|X_k - X_{k,N}\| \le \sum_{j>c\log N} \|\Psi\|^j E\|\varepsilon_0\| \le (1-e^{-r})^{-1}N^{-cr}E\|\varepsilon_0\|,$$

and the claim follows. □

Lemma 14.5. *The functions $U_{i,N} \in L^2([0,1]^2)$ defined by* (14.34) *satisfy*

$$E \max_{2\le k\le N} \left\| \sum_{2\le i\le k} [U_{i,N} - EU_{i,N}] \right\| \le KN^{1/2}(\log N)^{3/2},$$

where K is a constant and the norm is in the space $L^2([0,1]^2)$.

Proof. Set

$$U_{i,N}^*(t,s) = U_{i,N}(t,s) - EU_{i,N}(t,s).$$

Let $m = c \log N$ and assume without loss of generality that m is an integer. We will work with the decomposition

$$\sum_{1 \le i \le k} U^*_{i,N} = S_1(k) + S_2(k) + \ldots + S_m(k).$$

The idea is that $S_1(k)$ is the sum of (available) $U^*_{1,N}$, $U^*_{1+m,N}, \ldots$, $S_2(k)$ of $U^*_{2,N}$, $U^*_{2+m,N}, \ldots$, etc. Formally, for $1 \le k \le N$ and $1 \le j \le m$, define

$$S_j(k) = \begin{cases} \sum_{\ell=1}^{[k/m]} U^*_{(\ell-1)m+j,N} + U^*_{m[k/m]+j,N}, & \text{if } k/m \text{ is not an integer} \\ \sum_{\ell=1}^{k/m} U^*_{(\ell-1)m+j,N}, & \text{if } k/m \text{ is an integer.} \end{cases} \tag{14.38}$$

By (14.31) and (14.34), for any fixed j, $S_j(k)$ is a sum of independent identically distributed random functions in $L^2([0,1]^2)$. Since $\left\|\sum_{1 \le i \le k} U^*_{i,N}\right\| \le \sum_{j=1}^m \left\|S_j(k)\right\|$,

$$\left\| \sum_{1 \le i \le k} U^*_{i,N} \right\|^2 \le m \sum_{j=1}^m \left\|S_j(k)\right\|^2 . \tag{14.39}$$

By (14.39) and Lemma 14.7, we obtain $E\left\|\sum_{1 \le i \le k} U^*_{i,N}\right\|^2 \le Cmk$, where C is a constant which does not depend on N. Since $U^*_{i,N}$ is a stationary sequence, this bound implies that for all $K < L$,

$$E \left\| \sum_{K \le i \le L} U^*_{i,N} \right\|^2 \le Cm(L-K). \tag{14.40}$$

Relation (14.40) together with the Mensov inequality (Lemma 14.8) imply that

$$E \max_{1 \le k \le N} \left\| \sum_{1 \le i \le k} U^*_{i,N} \right\|^2 \le Cm(\log N)^2 N. \tag{14.41}$$

Recall that $m = O(\log N)$, to obtain the claim of the lemma. □

Lemma 14.6. *Recall the functions* $r(t,s) = E[X_{i-1}(t) X_i(s)]$ *and* $r_N(t,s)$ (14.32). *Then*

$$\|r - r_N\|^2 = \iint |r(t,s) - r_N(t,s)|^2 \, dt \, ds = O(N^{-2rc}),$$

where $r > 0$ *is defined by* $\|\Psi\| = e^{-r}$.

Proof. For ease of notation set $m = c \log N$ and observe that

$$r_N(t,s) = E\left[\sum_{j=0}^{m} \Psi^j \varepsilon_{i-1-j}(t) \sum_{\ell=0}^{m} \Psi^\ell \varepsilon_{i-\ell}(t)\right]$$
$$= \sum_{j=0}^{m} E\left[\Psi^j \varepsilon_{i-1-j}(t)\Psi^{j+1}\varepsilon_{i-1-j}(t)\right].$$

using an analogous expansion of $r(t,s)$, we obtain

$$\|r - r_N\|^2 = \iint \left|\sum_{j>m} E[\Psi^j \varepsilon_{-j}(t)\Psi^{j+1}\varepsilon_{-j}(s)]\right|^2 dt\, ds$$
$$= \sum_{j,\ell>m} E \iint [\Psi^j \varepsilon_{-j}(t)\Psi^{j+1}\varepsilon_{-j}(s)\Psi^\ell \varepsilon_{-l}(t)\Psi^{\ell+1}\varepsilon_{-\ell}(s)] dt\, ds$$
$$= \sum_{j,\ell>m} E\left[\int \Psi^j \varepsilon_{-j}(t)\Psi^l \varepsilon_{-\ell}(t) dt \int \Psi^{j+1}\varepsilon_{-j}(s)\Psi^{\ell+1}\varepsilon_{-\ell}(s) ds\right]$$
$$\le \sum_{j,\ell>m} E\left[\|\Psi^j \varepsilon_{-j}\| \, \|\Psi^\ell \varepsilon_{-\ell}\| \, \|\Psi^{j+1}\varepsilon_{-j}\| \, \|\Psi^{\ell+1}\varepsilon_{-\ell}\|\right]$$
$$\le \sum_{j,\ell>m} \|\Psi\|^{2j+1} \|\Psi\|^{2\ell+1} E\|\varepsilon_0\|^4 \le K\|\Psi\|^{4m}. \qquad \square$$

The following two lemmas are used in the proof of Lemma 14.5.

Lemma 14.7. *The functions $S_j(k) \in L^2([0,1]^2)$ defined by* (14.38) *satisfy*

$$E\|S_j(k)\|^2 \le Ck/m, \quad 1 \le j \le m,$$

where C is a constant which does not depend on N.

Proof. To lighten the notation, suppose $k/m = n$ is an integer. By stationarity of the $X_{i,N}$,

$$E\|S_j(k)\|^2 = E\iint \left|\sum_{\ell=1}^{n} U^*_{(\ell-1)m+j,N}(t,s)\right|^2 dt\, ds$$
$$= E\iint \left|\sum_{\ell=1}^{n} U^*_{(\ell-1)m,N}(t,s)\right|^2 dt\, ds$$

does not depend on j. By construction, the $U^*_{(\ell-1)m,N}(t,s)$ are mean zero and $U^*_{(\ell-1)m,N}(t,s)$ is independent of $U^*_{(\ell'-1)m,N}(t,s)$ if $l' \neq \ell$. Therefore

$$E\|S_j(k)\|^2 = \iint \sum_{\ell=1}^{n} E\left[U^{*2}_{(\ell-1)m,N}(t,s)\right] dt\, ds = n \iint E\left[U^{*2}_{1,N}(t,s)\right] dt\, ds.$$

It thus remains to show that $\iint E\left[U_{1,N}^{*2}(t,s)\right] dt\, ds$ is bounded by a constant which does not depend on N.

Observe that

$$\begin{aligned}\iint E\left[U_{1,N}^{*2}(t,s)\right] dt\, ds &\le \iint E\left[X_{0,N}^2(t) X_{1,N}^2(s)\right] dt\, ds \\ &= E\left(\int X_{0,N}^2(t)dt\right)\left(\int X_{1,N}^2(s)ds\right) \\ &\le E\left(\int X_{0,N}^2(t)dt\right)^2 = E\|X_{0,N}\|^4.\end{aligned}$$

Setting $m = c\log N$, we get

$$\begin{aligned}E\|X_{0,N}\|^4 &= E\left\|\sum_{j=0}^{m}\Psi^j\varepsilon_{-j}\right\|^4 \le E\left(\sum_{j=0}^{m}\|\Psi\|^j\|\varepsilon_{-j}\|\right)^4 \\ &= \sum_{j_1=0}^{m}\sum_{j_2=0}^{m}\sum_{j_3=0}^{m}\sum_{j_4=0}^{m}\|\Psi\|^{j_1}\|\Psi\|^{j_2}\|\Psi\|^{j_3}\|\Psi\|^{j_4} \\ &\quad\times E\left[\|\varepsilon_{-j_1}\|\,\|\varepsilon_{-j_2}\|\,\|\varepsilon_{-j_3}\|\,\|\varepsilon_{-j_4}\|\right] \\ &\le \left(\sum_{j=0}^{m}\|\Psi\|^j\right)^4 E\|\varepsilon_0\|^4 \le (1-\|\Psi\|)^{-4}E\|\varepsilon_0\|^4.\end{aligned}$$

□

Lemma 14.8. *(Mensov inequality) Let $\xi_1, \xi_2, \dots$ be arbitrary Hilbert space valued random variables. If for any $K < L$*

$$E\left\|\sum_{i=K+1}^{L}\xi_i\right\|^2 \le C(L-K) \tag{14.42}$$

then, for any b,

$$E\max_{1\le k\le N}\left\|\sum_{i=1+b}^{k+b}\xi_i\right\|^2 \le C[\log(2N)]^2 N. \tag{14.43}$$

Proof. The proof is practically the same as for real–valued random variables ξ_i, see Móricz (1976), and so is omitted. □

Chapter 15
Determining the order of the functional autoregressive model

This chapter is concerned with determining the order p in the FAR(p) model

$$Z_i = \sum_{j=1}^{p} \Phi_j(Z_{i-j}) + \varepsilon_i. \tag{15.1}$$

We describe a testing procedure proposed by Kokoszka and Reimherr (2011). At its core is the representation of the FAR(p) process as a fully functional linear model with dependent regressors. Estimating the kernel function in this linear model allows us to construct a test statistic which has, approximately, a chi–square distribution with the number of degrees of freedom determined by the number of functional principal components used to represent the data. The procedure enjoys very good finite sample properties, as confirmed by a simulation study and applications to functional time series derived from credit card transactions and Eurodollar futures data.

Order selection has been a central problem in time series analysis, and the resulting research has had a transforming impact on the application of time series models. The literature is very extensive, we mention only the pioneering work of Akaike (1978), Hannan and Quinn (1979), Hannan (1980), Shibata (1980) and Hannan and Rissannen (1982). A comprehensive review is provided by Bhansali (1993), and a brief introduction in Section 9.3 of Brockwell and Davis (1991). In contrast to scalar autoregression, only very small values of p, say 0,1,2, are of interest in the functional setting because the curves Z_k already consist of a large number of scalar observations, often hundreds, and the goal of functional data analysis is to replace all of these observations by a single functional object. Consequently, we do not attempt to develop analogues of the well known information or prediction error based order selection criteria, but propose an approach based on hypothesis testing. Our approach is specifically designed for functional data and fundamentally differs from the approaches in common use for scalar and vector valued time series. It relies on the observation that the union of intervals, $[0, 1] \cup [1, 2] \cup \ldots \cup [(p-1), p]$, is again an interval (which can be treated as a unit interval after rescaling), and on a multistage testing procedure rather than penalized likelihood.

The issue of determining an optimal order p can be approached in a problem specific manner by checking if using the FAR(p) produces better results than using FAR($p-1$), in a sense defined by a statistical problem at hand. Nevertheless, an appropriate criterion may not be obvious, and we believe that a universal approach that can be applied to any such situation is useful. In this paper we propose a suitable testing procedure. We focus on practical applicability, but we also provide a large sample justification, which is based on the theory presented in Chapter 16. An asymptotic justification of the procedure described in this paper is presented in Kokoszka and Reimherr (2011). It is of independent interest because it concerns kernel estimation in the extensively used fully functional linear model without assuming the independence of the regressor/response pairs.

This Chapter is organized as follows. In Section 15.1, we state model assumptions and develop a representation and an estimation technique for the FAR(p) process suitable for the testing problem. Section 15.2 describes the testing procedure whose performance is assessed in Section 15.3 by a simulation study and application to credit card transaction and Eurodollar futures data.

15.1 Representation of an FAR(p) process as a functional linear model

In this Chapter, we will work with the direct products

$$(x \otimes y)(t,s) = x(s)y(t), \quad x, y \in L^2,$$

which are elements of the space $L^2([0,1] \times [0,1])$. The inner product in the latter space will also be denoted by $\langle \cdot, \cdot \rangle$, as it will always be clear from the context what space the product is in.

We observe a sample of curves $Z_1(t), Z_2(t), \ldots, Z_N(t),\ t \in [0,1]$. We assume that these curves are a part–realization of an infinite sequence $\{Z_j\}$ which satisfies (15.1) and the following assumptions.

Assumption 15.1. *The operators Φ_j in* (15.1) *are Hilbert–Schmidt integral operators in L^2, i.e.*

$$\Phi_j(x)(t) = \int \phi_j(t,s)x(s)ds, \quad \iint \phi_j^2(t,s)dtds < \infty. \tag{15.2}$$

The operator

$$\Phi' = \begin{bmatrix} \Phi_1 & \Phi_2 & \ldots & \Phi_{p-1} & \Phi_p \\ I & 0 & \ldots & 0 & 0 \\ 0 & I & \ldots & 0 & 0 \\ \vdots & \vdots & \vdots & \vdots & \vdots \\ 0 & 0 & \ldots & I & 0 \end{bmatrix} \tag{15.3}$$

acting on the cartesian product $(L^2)^p$ *satisfies*

$$\|\Phi'\|_{\mathcal{L}} < 1, \tag{15.4}$$

where $\|\cdot\|_{\mathcal{L}}$ *is the operator norm in the cartesian product* $(L^2)^p$.

Assumption 15.2. *The* $\varepsilon_i \in L^2$ *in* (15.1) *are independent and identically distributed.*

Condition (15.4) and Assumption 15.2 imply that the Z_i form a stationary and ergodic sequence in L^2 such that ε_i is independent of $Z_{i-1}, Z_{i-2}, \ldots$, see Section 5.1 of Bosq (2000). For ease of reference, we state the following definition.

Definition 15.1. We say that the functional observations $Z_1, Z_2, \ldots, Z_N$ follow an FAR(p) process if Φ_p is not the zero operator, and Assumptions 15.1 and 15.2 hold.

Sufficient conditions for (15.4) to hold are established in Chapter 5 of Bosq (2000). A condition analogous to the usual condition for the existence of a scalar AR(p) process is the following: if the operator

$$Q_p(z) = z^p I - \sum_{j=1}^{p} z^{p-j} \Phi_j$$

does not have a bounded inverse, then $|z| < 1$. A stronger condition is $\sum_{j=1}^{p} \|\Phi_j\| < 1$. These conditions are derived using a Markovian representation of the process (15.1) as a FAR(1) process in the cartesian product $(L^2)^p$. For the task of testing FAR($p-1$) against FAR(p), a different representation is useful. It directly uses the structure of the observations as curves, and of the kernels ϕ_j as surfaces, rather than treating them as elements of abstract Hilbert spaces.

We start by expressing $\Phi_j(Z_{i-j})$ as an integral over the interval $((j-1)/p, j/p]$. Setting $x := (s + j - 1)/p$, a change of variables yields

$$\begin{aligned}
[\Phi_j(Z_{i-j})](t) &= \int_0^1 \phi_j(t,s) Z_{i-j}(s)\, ds \\
&= \int_{(j-1)/p}^{j/p} \phi_j(t, xp - (j-1)) Z_{i-j}(xp - (j-1))p\, dx.
\end{aligned}$$

Denoting by I_j the indicator function of the interval $((j-1)/p, j/p]$, we obtain

$$\sum_{j=1}^{p} [\Phi_j(Z_{i-j})](t) = \int_0^1 \sum_{j=1}^{p} I_j(x) \phi_j(t, xp - (j-1)) Z_{i-j}(xp - (j-1))p\, dx.$$

Next we define

$$X_i(s) = \sum_{j=1}^{p} Z_{i-j}(sp - (j-1)) I_j(s) \tag{15.5}$$

and

$$\psi(t,s) = p\sum_{j=1}^{p}\phi_j(t, sp-(j-1))I_j(s). \tag{15.6}$$

Setting $Y_i = Z_i$, we have

$$Y_i = \Psi(X_i) + \varepsilon_i, \tag{15.7}$$

where Ψ is an integral Hilbert–Schmidt operator with the kernel ψ, i.e.

$$Y_i(t) = \int \psi(t,s)X_i(s)ds + \varepsilon_i(t). \tag{15.8}$$

Thus, if we can estimate Ψ, then we can estimate each of the Φ_j. The FAR($p-1$) model will be rejected in favor of FAR(p) if the resulting estimate of $\hat{\Phi}_p$ is large in a sense established in Section 15.2. We now turn to the estimation of the operator Ψ.

Let $\{\hat{v}_k, 1 \le k \le N\}$ be an orthonormal basis of L^2 (for each N), constructed from the eigenfunctions of the covariance operator

$$\widehat{C}_X(t,s) = \frac{1}{N}\sum_{i=1}^{N}(X_i(t) - \bar{X}_N(t))(X_i(s) - \bar{X}_N(s)),$$

ordered by the corresponding eigenvalues $\hat{\lambda}_k$. To construct the test statistic, we will use only the first q_x eigenfunction/eigenvalue pairs $(\hat{v}_k, \hat{\lambda}_k)$. While we will use $\{\hat{v}_k\}$ in projecting the regressors, we allow for a separate basis in projecting the response variables. Define $\{\hat{u}_j\}_{j=1}^N$ and q_y analogously to $\{\hat{v}_k\}$ and q_x for the response functions. The tuning parameter q_y can be chosen to explain about 80–90 percent of the variance of the Y_i. Due to the nature of the X_i, q_x can either be chosen analogously to q_y or it can be taken to be $q_x = q_y p$. While the latter results in a much larger q_x, our procedure will involve a truncation step that will bring it back in line with q_y. In our experience, taking $q_x = q_y p$ results in a slightly more powerful procedure, though both approaches are valid. We take $q_x = q_y p$ for the simulations and applications presented in this chapter.

We estimate ψ projected onto the random subspace

$$\hat{H}_{q_x,q_y} := \text{span}\{\hat{v}_1, \ldots, \hat{v}_{q_x}\} \times \text{span}\{\hat{u}_1, \ldots, \hat{u}_{q_y}\}.$$

Let $\hat{\pi}_{q_x,q_y}$ denote the projection operator onto $\hat{H}_{q_x,q_y}$. Then we wish to estimate $\hat{\pi}_{q_x,q_y}(\psi)$. We should mention that this differs sharply from an analogous multivariate problem. While we wish to estimate ψ, we can only estimate ψ projected onto a finite dimensional subspace. Furthermore, that subspace is actually random since the space we choose depends on the random operators $\widehat{C}_X$ and $\widehat{C}_Y$. An asymptotic framework that handles these issues is developed in Kokoszka and Reimherr (2011).

To construct a least squares estimator, we define for $i = 1, \ldots, N$, $j = 1, \ldots, q_y$, and $k = 1, \ldots, q_x$

$$\begin{gathered}\mathbf{Y}(i,j) = \langle Y_i, \hat{u}_j\rangle, \quad \mathbf{X}(i,k) = \langle X_i, \hat{v}_k\rangle, \\ \boldsymbol{\psi}(k,j) = \langle \psi, \hat{v}_k \otimes \hat{u}_j\rangle = \iint \psi(t,s)\hat{v}_k(s)\hat{u}_j(t)dt\, ds.\end{gathered} \tag{15.9}$$

For ease of reference, we list the dimensions of the matrices introduced above

$$\mathbf{Y}\,(N \times q_y), \quad \mathbf{X}\,(N \times q_x), \quad \boldsymbol{\psi}\,(q_x \times q_y).$$

Using these matrices, we now reduce model (15.7) to a finite dimensional linear model. The precision of this finite dimensional approximation will be reflected in the structure of its random errors. Observe that

$$\mathbf{Y}(i,j) = \langle Y_i, \hat{u}_j \rangle = \langle \Psi(X_i) + \varepsilon_i, \hat{u}_j \rangle = \langle \Psi(X_i), \hat{u}_j \rangle + \langle \varepsilon_i, \hat{u}_j \rangle.$$

Since Ψ has a kernel ψ and $\{\hat{v}_k\}$ forms a basis for L^2, we have

$$\begin{aligned}
\langle \Psi(X_i), \hat{u}_j \rangle + \langle \varepsilon_i, \hat{u}_j \rangle &= \langle \psi, X_i \otimes \hat{u}_j \rangle + \langle \varepsilon_i, \hat{u}_j \rangle \\
&= \left\langle \psi, \sum_{k=1}^{\infty} \langle X_i, \hat{v}_k \rangle \hat{v}_k \otimes \hat{u}_j \right\rangle + \langle \varepsilon_i, \hat{u}_j \rangle \\
&= \sum_{k=1}^{\infty} \langle X_i, \hat{v}_k \rangle \langle \psi, \hat{v}_k \otimes \hat{u}_j \rangle + \langle \varepsilon_i, \hat{u}_j \rangle \\
&= \sum_{k=1}^{q_x} \langle X_i, \hat{v}_k \rangle \langle \psi, \hat{v}_k \otimes \hat{u}_j \rangle + \langle \varepsilon_i, \hat{u}_j \rangle + \sum_{k=q_x+1}^{\infty} \langle X_i, \hat{v}_k \rangle \langle \psi, \hat{v}_k \otimes \hat{u}_j \rangle \\
&= \sum_{k=1}^{q_x} \mathbf{X}(i,k) \boldsymbol{\psi}(k,j) + \langle \varepsilon_i, \hat{u}_j \rangle + \sum_{k=q_x+1}^{\infty} \langle X_i, \hat{v}_k \rangle \langle \psi, \hat{v}_k \otimes \hat{u}_j \rangle.
\end{aligned}$$

Therefore, the projections lead to the multivariate relation

$$\mathbf{Y} = \mathbf{X}\boldsymbol{\psi} + \boldsymbol{\varepsilon}',$$

The $N \times q_y$ matrix $\boldsymbol{\varepsilon}'$ has absorbed the error we made in projecting onto a finite dimensional space, and is given by

$$\boldsymbol{\varepsilon}'(i,j) = \langle \varepsilon_i, \hat{u}_j \rangle + \sum_{l > q_x} \langle X_i, \hat{v}_l \rangle \langle \psi,\ \hat{v}_l \otimes \hat{u}_j \rangle.$$

Observe also that the matrix $\boldsymbol{\psi}$ is not a population parameter, it is a projection of an unknown kernel function ψ onto a random subspace. It is therefore a random matrix. We can nevertheless compute the usual least squares estimator

$$\hat{\boldsymbol{\psi}} = (\mathbf{X}^T\mathbf{X})^{-1}\mathbf{X}^T\mathbf{Y} \tag{15.10}$$

and use its entries to $\hat{\boldsymbol{\psi}}(k,j)$, $k \le q_x, j \le q_y$, to construct a test statistic, as described in Section 15.2. The asymptotic properties of the estimator $\hat{\boldsymbol{\psi}}$ are established in Kokoszka and Reimherr (2011).

15.2 Order determination procedure

The testing procedure to determine the order of an FAR process consists of a sequence of tests of the hypotheses

$$H_p : \{Z_i\} \text{ are FAR}(p).$$

We start by testing

$$\begin{aligned}\text{Null Hypothesis} &:= H_0 : \{Z_i\} \text{ are iid vs}\\ \text{Alternative Hypothesis} &:= H_1 : \{Z_i\} \text{ are FAR}(1).\end{aligned}$$

If we accept H_0, then we conclude that the observations can be assumed to be iid. If we reject H_0, then we make H_1 our new null hypothesis and H_2 our new alternative. We continue until we accept a null hypothesis. We then conclude that the process is of the corresponding order. As explained in the introduction, the number of the individual tests will, in practice, be very small, one or two, so we are not concerned with problems arising in testing a large number of hypotheses. Our goal is consequently to construct a statistic to test the null hypothesis H_{p-1} against the alternative H_p.

We now describe how such a test statistic is constructed. As will be clear from the exposition that follows, some other variants are possible, but we focus on only one that seems most direct to us and leads to a test with very good finite sample properties. The test algorithm is summarized at the end of this section.

Using the estimator (15.10), we obtain an estimator of the kernel ψ given by

$$\hat{\psi}(t,s) = \sum_{k\le q_x,\ j\le q_y} \hat{\boldsymbol{\psi}}(k,j)\hat{v}_k(s)\hat{u}_j(t). \tag{15.11}$$

By (15.6), we can estimate the kernel ϕ_p by

$$\hat{\phi}_p(t,s) = \frac{1}{p}\hat{\psi}\left(t, \frac{s+p-1}{p}\right) = \frac{1}{p}\sum_{k\le q_x,\ j\le q_y} \hat{\boldsymbol{\psi}}(k,j)\hat{v}_k\left(\frac{s+p-1}{p}\right)\hat{u}_j(t).$$

Testing the nullity of ϕ_p is thus equivalent to checking if the sum

$$\sum_{k\le q_x,\ j\le q_y} \hat{\boldsymbol{\psi}}(k,j)\hat{v}_k(x)\hat{u}_j(t), \quad \frac{p-1}{p} \le x \le 1,\ 0 \le t \le 1 \tag{15.12}$$

is close to zero. The key element is the range of the argument x of $\hat{v}_k$, which reflects the part of ψ whose nullity we want to test. Based on the above representation, we want to find linear combinations of the $\hat{\boldsymbol{\psi}}(k,j)$ which make the sum (15.12) small. Clearly, we do not want to test if all $\hat{\boldsymbol{\psi}}(k,j)$ are small because that would mean that the whole kernel ψ and so all of the ϕ_j, $1 \le j \le p$, vanish. For further discussion, it is convenient to set

$$\hat{v}_{k,p}(s) = \hat{v}_k\left(\frac{s+p-1}{p}\right), \quad 0 \le s \le 1,$$

so that

$$\hat{\phi}_p(t,s) = \frac{1}{p} \sum_{k \le q_x,\ j \le q_y} \hat{\boldsymbol{\psi}}(k,j)\hat{v}_{k,p}(s)\hat{u}_j(t), \quad 0 \le s, \ t \le 1.$$

The idea behind the construction of the test statistic is to replace the $\hat{v}_{k,p}$ by a smaller set of functions that optimally describe the space spanned by them, and so, in a sense, by the $\hat{v}_k(x), x \ge (p-1)/p$. In other words, we test the nullity of ϕ_p only in the most significant orthogonal directions of the $\hat{v}_{k,p}$. We orthogonalize them as

$$\hat{w}_{k,p}(s) = \sum_{i=1}^{q_x} \hat{\alpha}_{i,k}\hat{v}_{i,p}(s)$$

with the vectors

$$\hat{\boldsymbol{\alpha}}_k = [\hat{\alpha}_{1,k}, \hat{\alpha}_{2,k}, \ldots, \hat{\alpha}_{q_x,k}]^T$$

such that $\|\hat{\boldsymbol{\alpha}}_k\| = 1$. To accomplish this, we construct the $q_x \times q_x$ matrix $\hat{\mathbf{V}}$ whose entries are the inner products

$$\hat{V}(k,k') = \langle \hat{v}_{k,p}, \hat{v}_{k',p} \rangle. \tag{15.13}$$

Since the matrix $\hat{\mathbf{V}}$ is positive–definite and symmetric, we define the $\hat{\boldsymbol{\alpha}}_k$ as its orthonormal eigenvectors ordered by their eigenvalues, i.e. we have

$$\hat{\mathbf{V}}\hat{\boldsymbol{\alpha}}_k = \hat{\nu}_k\hat{\boldsymbol{\alpha}}_k, \quad 1 \le k \le q_x, \tag{15.14}$$

where

$$\hat{\nu}_1 \ge \hat{\nu}_2 \ge \cdots \ge \hat{\nu}_{q_x}.$$

A direct verification shows that

$$\left\langle \hat{w}_{k,p}, \hat{w}_{k',p} \right\rangle = \hat{\nu}_k \delta_{k,k'},$$

where $\delta_{k,k'}$ is Dirac's delta.

Next we project $\hat{\phi}_p$ onto the functions $\{\hat{w}_{k,p} \otimes \hat{u}_j\}$. However, we will only include $\hat{w}_{k,p}$ whose norms are above a certain threshold, as the larger the value of $\|\hat{w}_{k,p}\|$ the greater its role in estimating ϕ_p. We obtained very good empirical performance by setting

$$q_\star = \max\{k \in \{1,\ldots,q_x\} : \|\hat{w}_{k,p}\|^2 \ge 0.9p\}.$$

What happens for both simulated and real data is that a few $\hat{w}_{k,p}$ have norms close to p, and the remaining norms are significantly smaller. An approximate upper bound of p, holds because

$$\|\hat{w}_{k,p}\| \le \sum_{i=1}^{q_x} |\hat{\alpha}_{i,k}|\|\hat{v}_{i,p}\| \le p\|\boldsymbol{\alpha}_k\|_1,$$

where we see $\|\hat{v}_{k,p}\| \le p$ by the change of variables

$$\|\hat{v}_{k,p}\| = p \int_{(p-1)/p}^{1} \hat{v}_k^2(x)dx.$$

Since $\int_0^1 \hat{v}_k^2(x)dx = 1$, $\|\hat{v}_{k,p}\|$ will generally not be very close to p, unless most of the mass of $\hat{v}_k$ is concentrated on the interval $[(p-1)/p, 1]$.

We thus want to determine if the coefficients

$$\langle \hat{\phi}_p, \hat{w}_{k,p} \otimes \hat{u}_j \rangle, \quad k = 1, \ldots, q_\star, \quad j = 1, \ldots, q_y$$

are collectively small. Observe that

$$\begin{aligned}
p\langle \hat{\phi}_p, \hat{w}_{k,p} \otimes \hat{u}_j \rangle &= p \iint \hat{\phi}_p(t,s)\hat{w}_{k,p}(s)\hat{u}_j(t)dsdt \\
&= \iint \left(\sum_{k',j'} \hat{\boldsymbol{\psi}}(k',j')\hat{v}_{k',p}(s)\hat{u}_{j'}(t) \right) \hat{w}_{k,p}(s)\hat{u}_j(t)dsdt \\
&= \int \sum_{k',j} \hat{\boldsymbol{\psi}}(k',j)\hat{v}_{k',p}(s)\hat{w}_{k,p}(s)ds \\
&= \int \sum_{k',j} \hat{\boldsymbol{\psi}}(k',j)\hat{v}_{k',p}(s) \left(\sum_i \hat{\alpha}_{i,k}\hat{v}_{i,p}(s) \right) ds \\
&= \sum_{k',i} \hat{\boldsymbol{\psi}}(k',j)\hat{V}(k',i)\hat{\alpha}_{i,k} \\
&= \sum_{k'} \hat{\boldsymbol{\psi}}(k',j)[\hat{\mathbf{V}}\hat{\boldsymbol{\alpha}}_k](k') \\
&= \sum_{k'} \hat{\boldsymbol{\psi}}(k',j)\hat{\delta}_k\hat{\alpha}_{k',k} = \hat{v}_k[\boldsymbol{\alpha}_k^T\hat{\boldsymbol{\psi}}](j).
\end{aligned}$$

The above calculation shows that the coefficients $\langle \hat{\phi}_p, \hat{w}_{k,p} \otimes \hat{u}_j \rangle$ are small if the matrices $\hat{v}_k\boldsymbol{\alpha}_k^T\hat{\boldsymbol{\psi}}$ have small entries. As explained above, $\hat{v}_k = \|\hat{w}_{k,p}\|^2 \ge 0.9p$, so we reject H_p if the entries of the matrices $\boldsymbol{\alpha}_k^T\hat{\boldsymbol{\psi}}$ are collectively large. To derive a test statistic, consider the following matrices (with their dimensions in parentheses)

$$\hat{\mathbf{A}}_\star = [\hat{\boldsymbol{\alpha}}_1, \ldots, \hat{\boldsymbol{\alpha}}_{q_\star}] \quad (q_x \times q_\star), \qquad \hat{\mathbf{A}}_\star^T\hat{\boldsymbol{\psi}} \quad (q_\star \times q_y). \tag{15.15}$$

We want to construct a quadratic form which is large when some entries of $\hat{\mathbf{A}}_\star^T\hat{\boldsymbol{\psi}}$ are large, and which has an approximately parameter free distribution. We will exploit the approximation $\mathbf{Z}^T(\text{Var}\mathbf{Z})^{-1}\mathbf{Z} \xrightarrow{d} \chi^2_{\dim(\mathbf{Z})}$, which holds for an asymptotically normal vector $\mathbf{Z}$. To this end, we form the column vector $\text{vec}(\hat{\mathbf{A}}_\star^T\hat{\boldsymbol{\psi}})$ by stacking the columns of $\hat{\mathbf{A}}_\star^T\hat{\boldsymbol{\psi}}$, a process known as *vectorization*. Using the property,

$\text{vec}(ABC) = (C^T \otimes A)\text{vec}(B)$, where $\otimes$ now denotes the Kronecker product of matrices, see e.g. Chapter 4 of Horn and Johnson (1985), we obtain

$$\text{vec}(\hat{\mathbf{A}}_\star^T \hat{\boldsymbol{\psi}}) = (\mathbf{I}(q_y) \otimes \hat{\mathbf{A}}_\star^T)\text{vec}(\hat{\boldsymbol{\psi}}),$$

where $\mathbf{I}(q_y)$ is the $q_y \times q_y$ identity matrix. We use the above identity to determine the approximate covariance matrix of $\text{vec}(\hat{\mathbf{A}}_\star^T \hat{\boldsymbol{\psi}})$. Applying the formula $\text{Var}(\mathbf{QZ}) = \mathbf{Q}\text{Var}(\mathbf{Z})\mathbf{Q}^T$, and treating the matrix $\hat{\mathbf{A}}_\star$ as deterministic, we obtain

$$\text{Var}\left[\text{vec}(\hat{\mathbf{A}}_\star^T \hat{\boldsymbol{\psi}})\right] \approx \left((\mathbf{I}(q_y) \otimes \hat{\mathbf{A}}_\star^T\right) \text{Var}(\text{vec}(\hat{\boldsymbol{\psi}})) \left((\mathbf{I}(q_y) \otimes \hat{\mathbf{A}}_\star^T\right),$$

where we used the property $(A \otimes B)^T = A^T \otimes B^T$. One can show that, see Kokoszka and Reimherr (2011),

$$N\,\text{Var}(\text{vec}(\hat{\boldsymbol{\psi}})) \approx \widehat{\mathbf{C}}_\varepsilon \otimes \hat{\boldsymbol{\Lambda}},$$

where

$$\hat{\boldsymbol{\Lambda}} = \text{diag}\{\hat{\lambda}_1, \ldots, \hat{\lambda}_{q_x}\}, \quad \widehat{\mathbf{C}}_\varepsilon = N^{-1}(\mathbf{Y} - \mathbf{X}\hat{\boldsymbol{\psi}})^T(\mathbf{Y} - \mathbf{X}\hat{\boldsymbol{\psi}}).$$

Combining these results, we arrive at the test statistic

$$\hat{\Delta}_p := N \left(\text{vec}\left[\hat{\mathbf{A}}_\star \hat{\boldsymbol{\psi}}\right]\right)^T \left[(\mathbf{I}(q_y) \otimes \hat{\mathbf{A}}_\star)(\widehat{\mathbf{C}}_\varepsilon \otimes \hat{\boldsymbol{\Lambda}})(\mathbf{I}(q_y) \otimes \hat{\mathbf{A}}_\star)\right]^{-1} \text{vec}(\hat{\mathbf{A}}_\star \hat{\boldsymbol{\psi}}). \tag{15.16}$$

The statistic $\hat{\Delta}_p$ has an approximately chi–square distribution with $q_y q_\star$ degrees of freedom. In Section 15.3 we evaluate the quality of this approximation. We conclude this section with an algorithmic description of the test procedure.

Test algorithm (H_{p-1} against H_p).

1. Subtract the sample mean from the functional observations. Continue to work with the centered data.
2. Construct the regressors X_i according to (15.5), and set $Y_i = Z_i$.
3. Determine q_y such that the first q_y eigenfunctions of the covariance operator $\widehat{C}_Y$ explain between 80 and 90 percent of the variance.

   a. Set $q_y = q_x p$ or
   b. take q_x analogous to q_y.

4. Construct the matrices $\mathbf{Y}$ and $\mathbf{X}$ according to (15.9).
5. Calculate the $q_x \times q_y$ matrix $\hat{\boldsymbol{\psi}}$ according to (15.10).
6. Calculate the $q_x \times q_x$ matrix $\hat{\mathbf{V}}$ according to (15.13), and its eigenvectors $\hat{\boldsymbol{\alpha}}_k$ and eigenfunctions $\hat{v}_k$ defined in (15.14).
7. Determine $q_\star$ such that the first $q_\star$ eigenvalues $\hat{v}_k$ are greater than 0.9p. (The procedure is not sensitive to the cut–off value of 0.9, taking 0.5 produced the same conclusions in data examples and simulations.)
8. Construct the matrices $\hat{\mathbf{A}}_\star$ and $\hat{\mathbf{A}}_\star^T \hat{\boldsymbol{\psi}}$ defined in (15.15) and compute the test statistic $\hat{\Delta}_p$ defined in (15.16)
9. Compute the P–value using the chi–square density with $q_y q_\star$ degrees of freedom.

15.3 Finite sample performance and application to financial data

We first evaluate the performance of the test using simulated data, then we turn to the application to two financial data sets.

Simulated data. The data are generated according to an FAR model, the choice of the autoregressive operators specifies the order. We consider two models

$$Z_i = c_1 Z_{i-1} + c_2 Z_{i-2} + \varepsilon_i, \tag{15.17}$$

where the $c_j \in [0, 1)$ are scalars, and

$$Z_i = \Phi_1(Z_{i-1}) + \Phi_2(Z_{i-2}) + \varepsilon_i, \tag{15.18}$$

where the kernel of Φ_i is given by

$$\phi_i(t,s) = \frac{c_i}{.7468} e^{-(t^2+s^2)/2}.$$

The L^2 norm of ϕ_i is approximately c_i.

The ε_i in both models are standard Brownian bridges. We used a burn–in period of 200 functional observations. The rejection rates are based on one thousand replications, so the standard errors for the empirical size are about 0.009, 0.006 and 0.003, respectively for the nominal sizes of 0.10, 0.05 and 0.01. To speed up the simulations, we used fixed values $q_y = 3$, which explain about 85 percent of the variance of the Y_i, and $q_x = 3p$.

The results of the simulation study are displayed in Tables 15.1 and 15.2. For $N \geq 100$, the sample sizes are generally within two standard errors off the nominal sizes. The power is practically 100% for testing the null hypothesis of the iid model against the alternative of an FAR(p) model with some $p = 1$ or $p = 2$. The power is also very high when testing the null hypothesis of the FAR(1) model against FAR(2) model, but lower than for testing the iid hypothesis. For $N = 300$, the power is 100% for all cases we considered.

Table 15.1 Empirical size and power for model (15.17).

Null Hyp	$p=0$	$p \leq 1$	$p=0$	$p \leq 1$	$p=0$	$p \leq 1$
Alt Hyp	$p \geq 1$	$p \geq 2$	$p \geq 1$	$p \geq 2$	$p \geq 1$	$p \geq 2$
	$c_1=0$	$c_1=0$	$c_1=0.5$	$c_1=0.5$	$c_1=0.5$	$c_1=0.5$
Sig. Level	$c_2=0$	$c_2=0$	$c_2=0$	$c_2=0$	$c_2=0.3$	$c_2=0.3$
			$N=100$			
0.10	0.115	0.122	1	0.112	1	0.831
0.05	0.070	0.068	1	0.060	1	0.753
0.01	0.022	0.015	1	0.016	1	0.558
			$N=200$			
0.10	0.117	0.120	1	0.105	1	0.986
0.05	0.054	0.062	1	0.058	1	0.968
0.01	0.012	0.013	1	0.010	1	0.925

Table 15.2 Empirical size and power for model (15.18).

Null Hyp	$p=0$	$p\le 1$	$p=0$	$p\le 1$	$p=0$	$p\le 1$
Alt Hyp	$p\ge 1$	$p\ge 2$	$p\ge 1$	$p\ge 2$	$p\ge 1$	$p\ge 2$
	$c_1=0$	$c_1=0$	$c_1=0.5$	$c_1=0.5$	$c_1=0.5$	$c_1=0.5$
Sig. Level	$c_2=0$	$c_2=0$	$c_2=0$	$c_2=0$	$c_2=0.3$	$c_2=0.3$
			$N=100$			
0.10	0.108	0.105	0.996	0.112	1	0.807
0.05	0.059	0.054	0.995	0.066	1	0.724
0.01	0.014	0.012	0.987	0.019	0.999	0.549
			$N=200$			
0.10	0.107	0.105	1	0.116	1	0.979
0.05	0.057	0.051	1	0.063	1	0.961
0.01	0.016	0.012	1	0.009	1	0.925

Table 15.3 P–values for the test applied to credit card data transformed by *differencing*.

Null Hyp	$p=0$	$p\le 1$
Alt Hyp	$p\ge 1$	$p\ge 2$
P–Value	0.000	0.427

Table 15.4 P–values for the test applied to credit card data transformed by *centering*.

Null Hyp	$p=0$	$p\le 1$	$p\le 2$
Alt Hyp	$p\ge 1$	$p\ge 2$	$p\ge 3$
P–Value	0.000	0.00	0.161

We now apply our multistage test procedure to two financial data sets we have already introduced in previous chapters: the daily credit card transactions and the curves of Eurodollar futures prices.

Credit Card Transactions. This data set is introduced in Section 1.3. Recall that we denote by $D_n(t_i)$ the number of credit card transactions in day n, $n=1,\dots,200$, between times t_{i-1} and t_i, where $t_i-t_{i-1}=8$ min, $i=1,\dots,128$. We thus have $N=200$ daily curves. The transactions are normalized to have time stamps in the interval $[0,1]$, which thus corresponds to one day. Some smoothing is applied to construct the functional objects, as explained in Section 1.3.

The curves thus obtained have non–zero mean and exhibit strong weekly periodicity. By computing the differences $Z_n(t)=Y_n(t)-Y_{n-7}(t)$, $n=8,9,\dots,200$, we can remove both. We refer to this method of obtaining the Z_i for further analysis as *differencing*. Another way to remove the weekly periodicity and the mean is to center the observations according to their day of the week. We refer to this method as *centering*.

The P–values are displayed in Tables 15.3 and 15.4. The stationary process obtained by differencing can be modeled as FAR(1). This agrees with the conclusions we reached in Chapters 7 and 14, where we tested the suitability of the FAR(1) model using significance tests against error correlations and change points.

Table 15.5 P–values of the test applied to Eurodollar futures curves.

Null Hyp	$p = 0$	$p \leq 1$	$p \leq 2$
Alt Hyp	$p \geq 1$	$p \geq 2$	$p \geq 3$
P–Value	0.000	0.000	0.731

Centering by week days leads to a more complex structure, which can be captured by the FAR(2) model.

Eurodollar Futures. We now turn to the application of our procedure to the data set consisting of Eurodollar futures contract prices studied in Section 14.3. Recall that each daily curve consists of 114 points per day; point i corresponds to the price of a contract with closing date i months from today. We work with centered data, i.e. the sample mean function has been subtracted from all observations.

The P–values displayed in Table 15.5 indicate that the FAR(1) model is not suitable for modelling the whole data set, but the FAR(2) model is acceptable. This conclusion agrees with the analysis presented in Section 14.3 where a change point test was applied to these data. We saw that the FAR(1) model is not suitable for the whole data set, merely for shorter subintervals. The present analysis shows that a slightly more complex FAR(2) model captures the stochastic structure of the whole data set.

Chapter 16
Functional time series

Functional data often arise from measurements obtained by separating an almost continuous time record into natural consecutive intervals, for example days. The functions thus obtained form a functional time series, and the central issue in the analysis of such data is to take into account the temporal dependence of these functional observations. In the previous chapters we have seen many examples, which include daily curves of financial transaction data and daily patterns of geophysical and environmental data. In Chapter 13, we introduced the functional autoregressive model which can approximate the temporal dependence in many such data sets. For many functional time series it is however not clear what specific model they follow, and for many statistical procedures it is not necessary to assume a specific model. In such cases, it is important to know what the effect of the dependence on a given procedure is. Is it robust to temporal dependence, or does this type of dependence introduce a serious, broadly understood, bias? To answer questions of this type, it is essential to quantify the notion of temporal dependence. For scalar and vector valued stochastic processes, a large number of dependence notions have been proposed, mostly involving mixing type distances between σ–algebras. In time series analysis, measures of dependence based on moments have proven most useful (autocovariances and cumulants). In this chapter, we introduce a moment based notion of dependence for functional time series which is an extension of m–dependence. We show that it is applicable to linear as well as nonlinear functional time series. Then we investigate the impact of dependence thus quantified on several important statistical procedures for functional data. We study the estimation of the functional principal components, the long-run covariance matrix, change point detection and the functional linear model. We explain when temporal dependence affects the results obtained for iid functional observations, and when these results are robust to weak dependence. Our examples are chosen to show that some statistical procedures for functional data are robust to temporal dependence, as quantified in this paper, while other require modifications that take this dependence into account.

While we focus here on a general theoretical framework, this research has been motivated by our work with functional data arising in space physics. For such data, no validated time series models are currently available, so to justify any inference

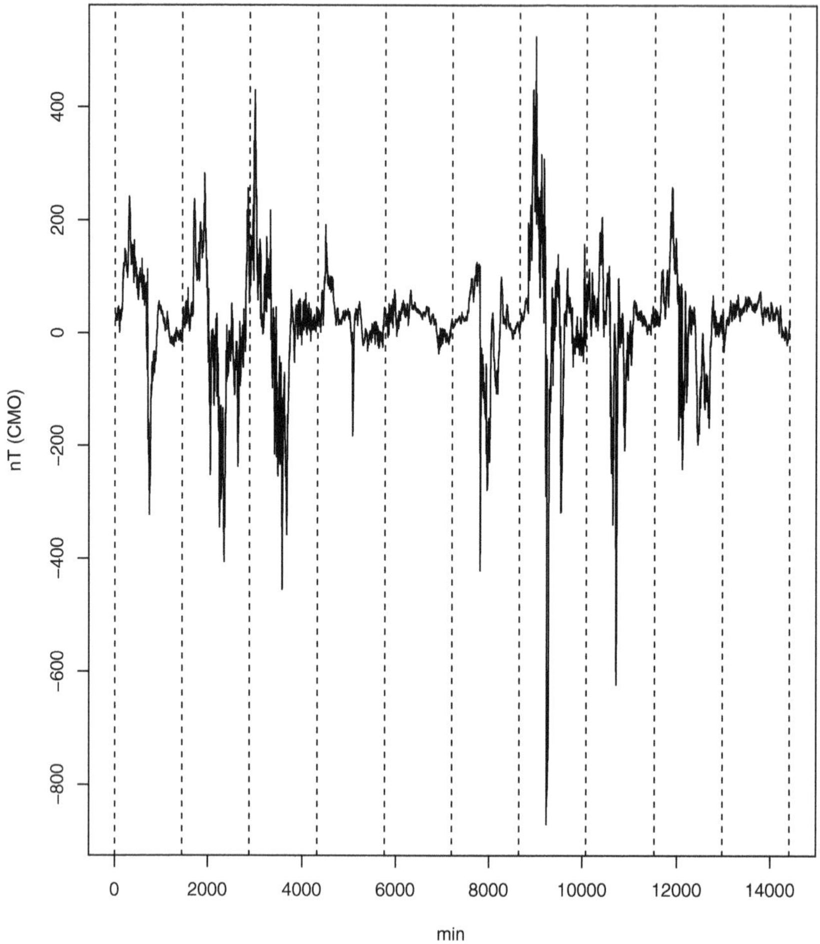

Fig. 16.1 Ten consecutive functional observations of a component of the magnetic field recorded at College, Alaska. The vertical lines separate days. Long negative spikes lasting a few hours correspond to the *aurora borealis*.

drawn from them, they must fit into a general, one might say nonparametric, dependence scheme. An example of space physics data is shown in Figure 16.1. Temporal dependence from day to day can be discerned, but has not been modeled yet.

The Chapter is organized as follows. In Section 16.1, we introduce the dependence condition and illustrate it with several examples. In particular, we show that the linear functional processes fall into this framework, and present some nonlinear models that also do. In Section 16.2 we show how the consistency of the estimators for the eigenvalues and eigenfunctions of the covariance operator extends to dependent functional data. Next, in Sections 16.3 and 16.4, we turn to the estimation of an appropriately defined long run variance matrix for functional data. For most time series procedures, the long run variance plays a role analogous to the variance–covariance matrix for independent observations. Its estimation is therefore of fundamental importance, and has been a subject of research for many decades, Andrews (1991), Anderson (1994) and Hamilton (1994) provide the background and numerous references. In Sections 16.5 and 16.7, we illustrate the application of the results of Sections 16.2 and 16.3 on two problems: change point detection for functional mean, and the estimation of kernel in the functional linear model. We show that the detection procedure introduced in Chapter 6 must be modified if the data exhibit dependence, but the kernel estimation procedure is robust to mild dependence. Section 16.5 also contains a small simulation study and a data example. The proofs are collected in the remaining sections. This chapter is partially based on the paper of Hörmann and Kokoszka (2010).

16.1 Approximable functional time series

The notion of weak dependence has over the past decades been formalized in many ways. Perhaps the most popular are various mixing conditions, see Doukhan (1994), Bradley (2007), but in recent years several other approaches have also been introduced, see Doukhan and Louhichi (1999) and Wu (2005, 2007), among others. In time series analysis, moment based measures of dependence, most notably autocorrelations and cumulants, have gained broad acceptance. The measure we consider below is a moment type quantity, but it is also related to the mixing conditions as it considers σ–algebras m time units apart, with m tending to infinity.

A most direct relaxation of independence is m–dependence. Suppose $\{X_n\}$ is a sequence of random elements taking values in a measurable space S. Denote by $\mathcal{F}_k^- = \sigma\{\dots X_{k-2}, X_{k-1}, X_k\}$ and $\mathcal{F}_k^+ = \sigma\{X_k, X_{k+1}, X_{k+2}, \dots\}$, the σ–algebras generated by the observations up to time k and after time k, respectively. Then the sequence $\{X_n\}$ is said to be m-dependent if for any k, the σ–algebras $\mathcal{F}_k^-$ and $\mathcal{F}_{k+m}^+$ are independent.

Most time series models are not m–dependent. Rather, various measures of dependence decay sufficiently fast, as the distance m between the σ–algebras $\mathcal{F}_k^-$ and $\mathcal{F}_{k+m}^+$ increases. However, m–dependence can be used as a tool to study properties of many nonlinear sequences, see e.g. Berkes and Horváth (2001), Berkes, Horváth and Kokoszka (2003, 2005), Berkes and Horváth (2003a, 2003b), Hörmann (2008), Berkes, Hörmann and Schauer (2008, 2009). The general idea is to approximate $\{X_n, n \in \mathbb{Z}\}$ by m–dependent processes $\{X_n^{(m)}, n \in \mathbb{Z}\}, m \geq 1$. The goal is to establish that for every n the sequence $\{X_n^{(m)}, m \geq 1\}$ converges in some sense to X_n, if we let $m \to \infty$. If the convergence is fast enough, then one can obtain the limiting behavior of the original process from corresponding results for m–dependent sequences. Definition 16.1 formalizes this idea and sets up the necessary framework for the construction of such m–dependent approximation sequences. The idea of approximating scalar sequences by m–dependent nonlinear moving averages appears already in Section 21 of Billingsley (1968), and it was developed in several direction by Pötscher and Prucha (1997). A version of Definition 16.1 for vector valued processes was used in Aue *et al.* (2009).

For $p \geq 1$, we denote by $L^p = L^p(\Omega, \mathcal{A}, P)$ the space of (classes of) real valued random variables such that $\|X\|_p = (E|X|^p)^{1/p} < \infty$. Further, we let $L^p_H = L^p_H(\Omega, \mathcal{A}, P)$ be the space of $H = L^2$ valued random functions X such that

$$\nu_p(X) = \left(E\|X\|^p\right)^{1/p} = \left(E\left\{\int X^2(t)dt\right\}^{p/2}\right)^{1/p} < \infty. \tag{16.1}$$

In this chapter we use H to denote the function space $L^2 = L^2([0,1])$ to avoid confusion with the space L^p of scalar random variables.

Definition 16.1. A sequence $\{X_n\} \in L^p_H$ is called L^p*–m–approximable* if each X_n admits the representation

$$X_n = f(\varepsilon_n, \varepsilon_{n-1}, \ldots), \tag{16.2}$$

where the ε_i are iid elements taking values in a measurable space S, and f is a measurable function $f : S^\infty \to H$. Moreover we assume that if $\{\varepsilon'_i\}$ is an independent copy of $\{\varepsilon_i\}$ defined on the same probability space, then letting

$$X_n^{(m)} = f(\varepsilon_n, \varepsilon_{n-1}, \ldots, \varepsilon_{n-m+1}, \varepsilon'_{n-m}, \varepsilon'_{n-m-1}, \ldots) \tag{16.3}$$

we have

$$\sum_{m=1}^{\infty} \nu_p\big(X_n - X_n^{(m)}\big) < \infty. \tag{16.4}$$

For our applications, choosing $p = 4$ will be convenient, but any $p \geq 1$ can be used, depending on what is needed. (Our definition makes even sense if $p < 1$, but then ν_p is no longer a norm.) Definition 16.1 implies that $\{X_n\}$ is strictly stationary. It is clear from the representation of X_n and $X_n^{(m)}$ that $E\|X_m - X_m^{(m)}\|^p = E\|X_1 - X_1^{(m)}\|^p$, so that condition (16.4) could be formulated solely in terms of X_1 and the approximations $X_1^{(m)}$. Obviously the sequence $\{X_n^{(m)}, n \in \mathbb{Z}\}$ as defined in (16.3) is *not* m–dependent. To this end we need to define for each n an independent copy $\{\varepsilon_k^{(n)}\}$ of $\{\varepsilon_k\}$ (this can always be achieved by enlarging the probability space) which is then used instead of $\{\varepsilon'_k\}$ to construct $X_n^{(m)}$, i.e. we set

$$X_n^{(m)} = f(\varepsilon_n, \varepsilon_{n-1}, \ldots, \varepsilon_{n-m+1}, \varepsilon^{(n)}_{n-m}, \varepsilon^{(n)}_{n-m-1}, \ldots). \tag{16.5}$$

We call this method the *coupling construction*. Since this modification lets condition (16.4) unchanged, we will assume from now on that the $X_n^{(m)}$ are defined by (16.5).

Then, for each $m \geq 1$, the sequences $\{X_n^{(m)}, n \in \mathbb{Z}\}$ are strictly stationary and m–dependent, and each $X_n^{(m)}$ is equal in distribution to X_n.

One can also define $X_n^{(m)}$ by

$$X_n^{(m)} = f(\varepsilon_n, \varepsilon_{n-1}, \ldots, \varepsilon_{n-m+1}, \varepsilon_{n,n-m}^{(m)}, \varepsilon_{n,n-m-1}^{(m)}, \ldots), \tag{16.6}$$

where $\left\{\varepsilon_{n,\ell}^{(m)}, m \geq 1, -\infty < n, \ell < \infty\right\}$ are iid copies of ε_0. We require (16.4), but now $X_n^{(m)}$ defined by (16.6) is used. To establish (16.4) with $X_n^{(m)}$ defined by (16.5) or (16.6) the same arguments are used.

L^p–m–approximability is related to L^p–approximability studied by Pötscher and Prucha (1997) for scalar– and vector–valued processes. Since our definition applies with an obvious modification to sequences with values in any normed vector spaces H (in particular, $\mathbb{R}$ or $\mathbb{R}^n$), it can been seen as a generalization of L^p–approximability. There are, however, important differences. By definition, L^p–approximability only allows for approximations that are, like the truncation construction, measurable with respect to a finite selection of basis vectors $\varepsilon_n, \ldots, \varepsilon_{n-m}$, whereas the coupling construction does not impose this condition. On the other hand, L^p–approximability is not based on independence of the innovation process. Instead independence is relaxed to certain mixing conditions.

Finally, we point out that only a straightforward modification is necessary in order to generalize the theory of this paper to non-causal processes $X_n = f(\ldots, \varepsilon_{n+1}, \varepsilon_n, \varepsilon_{n-1}, \ldots)$. At the expense of additional technical assumptions, our framework can also be extended to non-stationary sequences, e.g. those of the form (16.2) where $\{\varepsilon_k\}$ is a sequence of independent, but not necessarily identically distributed, random variables.

We now illustrate the applicability of Definition 16.1 with several examples. Let $\mathcal{L} = \mathcal{L}(H, H)$ be the set of bounded linear operators from H to H. Recall that for $A \in \mathcal{L}$ the operator norm is $\|A\|_{\mathcal{L}} = \sup_{\|x\| \leq 1} \|Ax\|$.

Example 16.1 (Functional autoregressive process). Suppose $\Psi \in \mathcal{L}$ satisfies $\|\Psi\|_{\mathcal{L}} < 1$. Let $\varepsilon_n \in L_H^2$ be iid with mean zero. Then there is a unique stationary sequence of random elements $X_n \in L_H^2$ such that

$$X_n(t) = \Psi(X_{n-1})(t) + \varepsilon_n(t). \tag{16.7}$$

For details see Chapter 13. The AR(1) sequence (16.7) admits the expansion $X_n = \sum_{j=0}^{\infty} \Psi^j(\varepsilon_{n-j})$, where Ψ^j is the j-th iterate of the operator Ψ. We thus set $X_n^{(m)} = \sum_{j=0}^{m-1} \Psi^j(\varepsilon_{n-j}) + \sum_{j=m}^{\infty} \Psi^j(\varepsilon_{n-j}^{(n)})$. It is easy to verify that for every A in $\mathcal{L}$, $\nu_p(A(Y)) \leq \|A\|_{\mathcal{L}} \nu_p(Y)$. Since $X_m - X_m^{(m)} = \sum_{j=m}^{\infty} \left(\Psi^j(\varepsilon_{m-j}) - \Psi^j(\varepsilon_{m-j}^{(m)})\right)$, it follows that $\nu_p(X_m - X_m^{(m)}) \leq 2 \sum_{j=m}^{\infty} \|\Psi\|_{\mathcal{L}}^j \nu_p(\varepsilon_0) = O(1) \nu_p(\varepsilon_0) \|\Psi\|_{\mathcal{L}}^m$. By assumption $\nu_2(\varepsilon_0) < \infty$ and therefore $\sum_{m=1}^{\infty} \nu_2(X_m - X_m^{(m)}) < \infty$, so condition (16.4) holds with $p \geq 2$, as long as $\nu_p(\varepsilon_0) < \infty$.

Proposition 16.1 establishes sufficient conditions for a general linear process to be L^p–m–approximable. Its verification follows the lines of Example 16.1, and so is omitted. A sequence $\{X_n\}$ is said to be a *linear process in* H if $X_n = \sum_{j=0}^{\infty} \Psi_j(\varepsilon_{n-j})$, where the errors $\varepsilon_n \in L_H^2$ are iid and zero mean, and each Ψ_j is a bounded operator. If $\sum_{j=1}^{\infty} \|\Psi_j\|_{\mathcal{L}}^2 < \infty$, then the series defining X_n converges a.s. and in L_H^2, see Section 7.1 of Bosq (2000).

Proposition 16.1. *Suppose $\{X_n\} \in L_H^2$ is a linear process whose errors satisfy $\nu_p(\varepsilon_0) < \infty$, $p \geq 2$. The operator coefficients satisfy $\sum_{m=1}^{\infty} \sum_{j=m}^{\infty} \|\Psi_j\| < \infty$. Then $\{X_n\}$ is L^p–m–approximable.*

We next give a simple example of a nonlinear L^p–m–approximable sequence. It is based on the model used by Maslova *et al.* (2010a) to simulate the so called solar quiet (Sq) variation in magnetometer records. In that model, $X_n(t) = U_n(S(t) + Z_n(t))$ represents the part of the magnetometer record on day n which reflects the magnetic field generated by ionospheric winds of charged particles driven by solar heating. These winds flow in two elliptic cells, one on each day–side of the equator. Their position changes from day to day, causing a different appearance of the curves $X_n(t)$, with changes in the amplitude being most pronounced. To simulate this behavior, $S(t)$ is introduced as the typical pattern for a specific magnetic observatory, $Z_n(t)$ as the change in shape on day n, and the scalar random variable U_n as the amplitude on day n. With this motivation, we formulate the following example.

Example 16.2. (Product model) Suppose $\{Y_n\} \in L_H^p$ and $\{U_n\} \in L^p$ are both L^p–m–approximable sequences, independent of each other. The respective representations are $Y_n = g(\eta_1, \eta_2, \ldots)$ and $U_n = h(\gamma_1, \gamma_2, \ldots)$. Each of these sequences could be a linear sequence satisfying the assumptions of Proposition 16.1, but they need not be. The sequence $X_n(t) = U_n Y_n(t)$ is then a nonlinear L^p–m–approximable sequence with the underlying iid variables $\varepsilon_n = (\eta_n, \gamma_n)$. To see this, set $X_n^{(m)}(t) = U_n^{(m)} Y_n^{(m)}(t)$ and observe that $\nu_p(X_m - X_m^{(m)}) \leq \nu_p\left((U_m - U_m^{(m)})Y_m\right) + \nu_p\left(U_m^{(m)}(Y_m - Y_m^{(m)})\right)$. Using the independence of $\{Y_n\}$ and $\{U_n\}$ it can be easily shown that $\nu_p\left((U_m - U_m^{(m)})Y_m\right) = \|U_m - U_m^{(m)}\|_p \, \nu_p(Y_0)$ and $\nu_p\left(U_m^{(m)}(Y_m - Y_m^{(m)})\right) = \|U_0\|_p \, \nu_p\left(Y_m - Y_m^{(m)}\right)$.

Example 16.2 illustrates the principle that in order for products of L^p–m–approximable sequences to be L^p–m–approximable, independence must be assumed. It does not have to be assumed as directly as in Example 16.2, the important point being that appropriately defined functional Volterra expansions should not contain diagonal terms, so that moments do not pile up. Such expansions exist, see e.g. Giraitis *et al.* (2000), for all nonlinear scalar processes used to model financial data. The model $X_n(t) = Y_n(t)U_n$ is similar to the popular scalar stochastic volatility model $r_n = v_n \varepsilon_n$ used to model returns r_n on a speculative asset. The dependent sequence $\{v_n\}$ models volatility, and the iid errors ε_n, independent of the v_n, generate unpredictability in returns. Our final example, focuses on a functional extension of the celebrated ARCH model of Engle (1982) which has a more complex Volterra expansion. The proof of Proposition 16.2 is more involved than those presented in Examples 16.1 and 16.2, and is not presented. The curves $y_k(t)$ appearing in Example 16.3 correspond to appropriately defined intradaily returns.

Example 16.3 (Functional ARCH). Let $\delta \in H$ be a positive function and let $\{\varepsilon_k\}$ an i.i.d. sequence in L^4_H. Further, let $\beta(s,t)$ be a non-negative kernel function in $L^2([0,1]^2, \mathcal{B}^2_{[0,1]}, \lambda^2)$. Then we call the process

$$y_k(t) = \varepsilon_k(t)\sigma_k(t), \quad t \in [0,1], \tag{16.8}$$

where

$$\sigma_k^2(t) = \delta(t) + \int_0^1 \beta(t,s) y_{k-1}^2(s) ds, \tag{16.9}$$

the *functional ARCH(1) process.*

Proposition 16.2 establishes conditions for the existence of a strictly stationary solution to equations (16.8) and (16.9) and its L^p–m–approximability.

Proposition 16.2. *Assume that there is a* $p > 0$ *such that*

$$E\left(\|\beta\|_{\mathcal{L}} \sup_{0\le s\le 1} |\varepsilon(s)|^2\right)^{p/2} < 1.$$

Then equations (16.8) *and* (16.9) *have a unique strictly stationary and causal solution and the sequence* $\{y_k\}$ *is* L^p*–m–approximable.*

Example 16.3 and Proposition 16.2 are taken from Hörmann *et al.* (2010), where further properties of the functional ARCH sequence are discussed.

We conclude this section we a simple but useful Lemma which shows that L^p–m–approximability is unaffected by linear transformations, whereas independence assumptions are needed for product type operations.

Lemma 16.1. *Let* $\{X_n\}$ *and* $\{Y_n\}$ *be two* L^p*–m–approximability sequences in* L^p_H*. Define*

- $Z_n^{(1)} = A(X_n)$, *where* $A \in \mathcal{L}$;
- $Z_n^{(2)} = X_n + Y_n$;
- $Z_n^{(3)} = X_n \circ Y_n$ $(X_n \circ Y_n(t) = X_n(t)Y_n(t))$;
- $Z_n^{(4)} = \langle X_n, Y_n \rangle$;
- $Z_n^{(5)} = X_n \otimes Y_n$ $(X_n \otimes Y_n(t,s) = X_n(s)Y_n(t))$.

Then $\{Z_n^{(1)}\}$ *and* $\{Z_n^{(2)}\}$ *are* L^p*–m–approximable sequences in* L^p_H*. If* X_n *and* Y_n *are independent then* $\{Z_n^{(4)}\}$ *and* $\{Z_n^{(5)}\}$ *are* L^p*–m–approximable sequences in the respective spaces. If* $E\sup_{t\in[0,1]}|X_n(t)|^p + E\sup_{t\in[0,1]}|Y_n(t)|^p < \infty$*, then* $\{Z_n^{(3)}\}$ *is* L^p*–m–approximable in* L^p_H*.*

Proof. The first two relations are immediate. We exemplify the proofs of the remaining claims by focusing on $Z_n = Z_n^{(5)}$. For this we set $Z_m^{(m)} = X_m^{(m)} \otimes Y_m^{(m)}$ and note that Z_m and $Z_m^{(m)}$ are (random) kernel operators, and thus Hilbert-Schmidt operators. Since

$$\begin{aligned}\|Z_m - Z_m^{(m)}\|_{\mathcal{L}} &\le \|Z_m - Z_m^{(m)}\|_{\mathcal{S}} \\ &\le \left(\iint \left(X_m(s)Y_m(t) - X_m^{(m)}(s)Y_m^{(m)}(t)\right)^2 dt\, ds\right)^{1/2} \\ &\le \sqrt{2}\left(\|X_m\|\|Y_m - Y_m^{(m)}\| + \|Y_m^{(m)}\|\|X_m - X_m^{(m)}\|\right),\end{aligned}$$

the proof follows from the independence of X_n and Y_n. □

The proof shows that for product type operations our assumptions can be modified and independence is not required. However, if X, Y are not independent, then we have then to use the Cauchy-Schwarz inequality and obviously need $2p$ moments.

16.2 Convergence of sample eigenfunctions and a central limit theorem

In this section we extend the results of Section 2.5 to weakly dependent functional time series and establish a central limit theorem for functional time series.

Let $\{X_n\} \in L_H^2$ be a stationary sequence with covariance operator C. We assume that C is an integral operator with kernel $c(t,s) = \mathrm{Cov}(X_1(t), X_1(s))$ whose estimator is

$$\hat{c}(t,s) = \frac{1}{N}\sum_{n=1}^{N}(X_n(t) - \bar{X}_N(t))(X_n(s) - \bar{X}_N(s)). \tag{16.10}$$

Then natural estimators of the eigenvalues λ_j and eigenfunctions v_j of C are the eigenvalues $\hat{\lambda}_j$ and eigenfunctions $\hat{v}_j$ of $\hat{C}$, the operator with the kernel (16.10). By Lemmas 2.2 and 2.3 we can bound the estimation errors for eigenvalues and eigenfunctions by $\|C - \hat{C}\|_{\mathcal{S}}^2$, where $\|\cdot\|_{\mathcal{S}}$ denotes the Hilbert–Schmidt norm. This motivates the next result.

Theorem 16.1. *Suppose $\{X_n\} \in L_H^4$ is an L^4–m–approximable sequence with covariance operator C. Then there is some constant $U_X < \infty$, which does not depend on N, such that*

$$E\|\hat{C} - C\|_{\mathcal{S}}^2 \le U_X N^{-1}. \tag{16.11}$$

The proof of Theorem 16.1 is given in Section 16.8. Let us note that by Lemma 2.2 and Theorem 16.1,

$$NE\left[|\lambda_j - \hat{\lambda}_j|^2\right] \le NE\|\hat{C} - C\|_{\mathcal{L}}^2 \le NE\|\hat{C} - C\|_{\mathcal{S}}^2 \le U_X.$$

Assuming

$$\lambda_1 > \lambda_2 > \cdots > \lambda_d > \lambda_{d+1}, \tag{16.12}$$

Lemma 2.3 and Theorem 16.1 yield for $j \geq d$ ($\hat{c}_j = \text{sign}(\langle \hat{v}_j, v_j \rangle)$),

$$NE\left[\|\hat{c}_j\hat{v}_j - v_j\|^2\right] \leq \left(\frac{2\sqrt{2}}{\alpha_j}\right)^2 NE\|\hat{C} - C\|_{\mathcal{L}}^2 \leq \frac{8}{\alpha_j^2} NE\|\hat{C} - C\|_{\mathcal{S}}^2 \leq \frac{8U_X}{\alpha_j^2},$$

with the α_j defined in Lemma 2.3.

These inequalities establish the following result.

Theorem 16.2. *Suppose* $\{X_n\} \in L_H^4$ *is an* L^4*–m–approximable sequence and assumption* (16.12) *holds. Then, for* $1 \leq j \leq d$,

$$\limsup_{N\to\infty} NE\left[|\lambda_j - \hat{\lambda}_j|^2\right] < \infty, \quad \limsup_{N\to\infty} NE\left[\|\hat{c}_j\hat{v}_j - v_j\|^2\right] < \infty. \tag{16.13}$$

Relations (16.13) are a fundamental tool for establishing asymptotic properties of procedures for functional simple random samples which are based on the functional principal components. Theorem 16.2 shows that in many cases one can expect that these properties will remain the same under weak dependence, an important example is discussed in Section 16.7.

The following theorem was established by Horváth *et al.* (2011).Its proof is presented in Section 16.8.

Theorem 16.3. *If* $\{X_i\}$ *is a zero mean* L^2*–m–approximable sequence, then*

$$N^{-1/2}\sum_{i=1}^{N} X_i \xrightarrow{d} G \ \ in \ \ L^2,$$

where G *is a Gaussian process with*

$$EG(t) = 0 \quad \text{and} \quad E[G(t)G(s)] = c(t,s);$$

$$c(t,s) = E[X_0(t)X_0(s)] + \sum_{i\geq 1} E[X_0(t)X_i(s)] + \sum_{i\geq 1} E[X_0(s)X_i(t)]. \tag{16.14}$$

16.3 The long–run variance

The concept of the long run variance, while fundamental in time series analysis, has not been studied for functional data, and not even for scalar approximable sequences. It is therefore necessary to start with some preliminaries, which lead to the main results and illustrate the role of the L^p–m–approximability.

Let $\{X_n\}$ be a scalar (weakly) stationary sequence. Its long run variance is defined as $\sigma^2 = \sum_{j\in\mathbb{Z}} \gamma_j$, where $\gamma_j = \text{Cov}(X_0, X_j)$, provided this series is absolutely convergent.Our first lemma shows that this is the case for L^2–m–approximable sequences.

Lemma 16.2. *Suppose $\{X_n\}$ is a scalar L^2–m–approximable sequence. Then its autocovariance function $\gamma_j = \mathrm{Cov}(X_0, X_j)$ is absolutely summable, i.e. $\sum_{j=-\infty}^{\infty} |\gamma_j| < \infty$.*

Proof. Observe that for $j > 0$,

$$\mathrm{Cov}(X_0, X_j) = \mathrm{Cov}(X_0, X_j - X_j^{(j)}) + \mathrm{Cov}(X_0, X_j^{(j)}).$$

Since

$$X_0 = f(\varepsilon_0, \varepsilon_{-1}, \ldots), \quad X_j^{(j)} = f^{(j)}(\varepsilon_j, \varepsilon_{j-1}, \ldots, \varepsilon_1, \varepsilon_0^{(j)}, \varepsilon_{-1}^{(j)}, \ldots),$$

the random variables X_0 and $X_j^{(j)}$ are independent, so $\mathrm{Cov}(X_0, X_j^{(j)}) = 0$, and

$$|\gamma_j| \le [EX_0^2]^{1/2}[E(X_j - X_j^{(j)})^2]^{1/2}.$$

The summability of the autocovariances is a fundamental property of weak dependence because then $N\mathrm{Var}[\bar{X}_N] \to \sum_{j=-\infty}^{\infty} \gamma_j$, i.e. the variance of the sample mean converges to zero at the rate N^{-1}, the same as for iid observations. A popular approach to the estimation of the long-run variance is to use the kernel estimator

$$\hat{\sigma}^2 = \sum_{|j| \le q} \omega_q(j)\hat{\gamma}_j, \quad \hat{\gamma}_j = \frac{1}{N}\sum_{i=1}^{N-|j|}(X_i - \bar{X}_N)(X_{i+|j|} - \bar{X}_N).$$

Various weights $\omega_q(j)$ have been proposed and their optimality properties studied, see Andrews (1991) and Anderson (1994), among others. In theoretical work, it is typically assumed that the bandwidth q is a deterministic function of the sample size such that $q = q(N) \to \infty$ and $q = o(N^r)$, for some $0 < r \le 1$. We will use the following assumption:

Assumption 16.1. *The bandwidth $q = q(N)$ satisfies $q \to \infty$, $q^2/N \to 0$ and the weights satisfy $\omega_q(j) = \omega_q(-j)$ and*

$$|\omega_q(j)| \le b \tag{16.15}$$

and, for every fixed j,

$$\omega_q(j) \to 1. \tag{16.16}$$

All kernels used in practice have symmetric weights and satisfy conditions (16.15) and (16.16).

The absolute summability of the autocovariances is not enough to establish the consistency of the kernel estimator $\hat{\sigma}^2$. Traditionally, summability of the cumulants has been assumed to control the fourth order structure of the data. Denoting $\mu = EX_0$, the fourth order cumulant of a stationary sequence is defined by

$$\kappa(h, r, s) = \mathrm{Cov}\left((X_0 - \mu)(X_h - \mu), (X_r - \mu)(X_s - \mu)\right) - \gamma_r\gamma_{h-s} - \gamma_s\gamma_{h-r}.$$

The usual sufficient condition for the consistency of $\hat{\sigma}$ is

$$\sum_{h=-\infty}^{\infty} \sum_{r=-\infty}^{\infty} \sum_{s=-\infty}^{\infty} |\kappa(h,r,s)| < \infty. \tag{16.17}$$

Recently, Giraitis *et al.* (2003) showed that condition (16.17) can be replaced by a weaker condition

$$\sup_h \sum_{r=-\infty}^{\infty} \sum_{s=-\infty}^{\infty} |\kappa(h,r,s)| < \infty. \tag{16.18}$$

A technical condition we need is

$$N^{-1} \sum_{k,l=0}^{q(N)} \sum_{r=1}^{N-1} \left| \mathrm{Cov}\left(X_0(X_k - X_k^{(k)}), X_r^{(r)} X_{r+\ell}^{(r+\ell)} \right) \right| \to 0. \tag{16.19}$$

By analogy to condition (16.18), it can be replaced by a much stronger, but a more transparent condition

$$\sup_{k,l\geq 0} \sum_{r=1}^{\infty} \left| \mathrm{Cov}\left(X_0(X_k - X_k^{(k)}), X_r^{(r)} X_{r+\ell}^{(r+\ell)} \right) \right| < \infty. \tag{16.20}$$

To explain the intuition behind conditions (16.19) and (16.20), consider the linear process $X_k = \sum_{j=0}^{\infty} c_j X_{k-j}$. For $k \geq 0$,

$$X_k - X_k^{(k)} = \sum_{j=k}^{\infty} c_j \varepsilon_{k-j} - \sum_{j=k}^{\infty} c_j \varepsilon_{k-j}^{(k)}.$$

Thus $X_0(X_k - X_k^{(k)})$ depends on

$$\varepsilon_0, \varepsilon_{-1}, \varepsilon_{-2}, \ldots \quad \text{and} \quad \varepsilon_0^{(k)}, \varepsilon_{-1}^{(k)}, \varepsilon_{-2}^{(k)}, \ldots \tag{16.21}$$

and $X_r^{(r)} X_{r+\ell}^{(r+\ell)}$ depends on

$$\varepsilon_{r+\ell}, \ldots, \varepsilon_1, \varepsilon_0^{(r)} \varepsilon_{-1}^{(r)}, \varepsilon_{-2}^{(r)}, \ldots \quad \text{and} \quad \varepsilon_0^{(r+\ell)} \varepsilon_{-1}^{(r+\ell)}, \varepsilon_{-2}^{(r+\ell)}, \ldots$$

Consequently, the covariances in (16.20) vanish except when $r = k$ or $r + \ell = k$, so condition (16.20) always holds for linear processes.

For general nonlinear sequences, the difference

$$X_k - X_k^{(k)} = f(\varepsilon_k, \ldots, \varepsilon_1, \varepsilon_0, \varepsilon_{-1}, \ldots) - f(\varepsilon_k, \ldots, \varepsilon_1, \varepsilon_0^{(k)}, \varepsilon_{-1}^{(k)}, \ldots)$$

cannot be expressed only in terms of the errors (16.21), but the errors $\varepsilon_k, \ldots, \varepsilon_1$ should approximately cancel, so that the difference $X_k - X_k^{(k)}$ is small, and very weakly correlated with $X_r^{(r)} X_{r+\ell}^{(r+\ell)}$.

With this background, we now formulate the following result.

Theorem 16.4. *Suppose $\{X_n\} \in L^4$ is a scalar L^4–m–approximable time series for which condition* (16.19) *holds. If Assumption 16.1 holds, then $\hat{\sigma}^2 \xrightarrow{P} \sum_{j=-\infty}^{\infty} \gamma_j$.*

Theorem 16.4 is proven in Section 16.8. The general plan of the proof is the same as that of the proof of Theorem 3.1 of Giraitis *et al.* (2003), but the verification of the crucial relation (16.49) uses a new approach based on L^4–m–approximability. The arguments preceding (16.49) show that replacing $\bar{X}_N$ by $\mu = EX_0$ does not change the limit. We note that the condition $q^2/N \to 0$ we assume is stronger than the condition $q/N \to 0$ assumed by Giraitis *et al.* (2003). This difference is of little practical consequence, as the optimal bandwidths for the kernels used in practice are typically of the order $O(N^{1/5})$. Finally, we notice that by further strengthening conditions on the behavior of the bandwidth function $q = q(N)$, the convergence in probability in Theorem 16.4 could be replaced by the almost sure convergence, but we do not pursue this research here. The corresponding result under condition (16.18) was established by Berkes *et al.* (2005), it is also stated without proof as part of Theorem A.1 of Berkes *et al.* (2006).

We now turn to the vector case in which the data are of the form

$$\mathbf{X}_n = [X_{1n}, X_{2n}, \ldots, X_{dn}]^T, \quad n = 1, 2, \ldots, N.$$

Just as in the scalar case, the estimation of the mean by the sample mean does not effect the limit of the kernel long–run variance estimators, so *we assume that* $EX_{in} = 0$ and define the autocovariances as

$$\gamma_r(i, j) = E[X_{i0} X_{jr}], \quad 1 \le i, j \le d.$$

If $r \ge 0$, $\gamma_r(i, j)$ is estimated by $N^{-1} \sum_{n=1}^{N-r} X_{in} X_{j,n+r}$ but if $r < 0$ it is estimated by $N^{-1} \sum_{n=1}^{N-|r|} X_{i,n+|r|} X_{j,n}$. We therefore define the autocovariance matrices

$$\hat{\boldsymbol{\Gamma}}_r = \begin{cases} N^{-1} \sum_{n=1}^{N-r} \mathbf{X}_n \mathbf{X}_{n+r}^T & \text{if } r \ge 0, \\ N^{-1} \sum_{n=1}^{N-|r|} \mathbf{X}_{n+|r|} \mathbf{X}_n^T & \text{if } r < 0. \end{cases}$$

The variance $\mathrm{Var}[N^{-1}\bar{\mathbf{X}}_n]$ has (i, j)–entry

$$N^{-2} \sum_{m,n=1}^{N} E[X_{im} X_{jn}] = N^{-1} \sum_{|r|<N} \left(1 - \frac{|r|}{N}\right) \gamma_r(i, j),$$

so the long–run variance is

$$\boldsymbol{\Sigma} = \sum_{r=-\infty}^{\infty} \boldsymbol{\Gamma}_r, \quad \boldsymbol{\Gamma}_r := [\gamma_r(i, j),\ 1 \le i, j \le d],$$

and its kernel estimator is

$$\hat{\boldsymbol{\Sigma}} = \sum_{|r| \le q} \omega_q(r) \hat{\boldsymbol{\Gamma}}_r. \tag{16.22}$$

The consistency of $\hat{\boldsymbol{\Sigma}}$ can be established by following the lines of the proof of Theorem 16.4 for every fixed entry of the matrix $\hat{\boldsymbol{\Sigma}}$. Condition (16.19) must be replaced by

$$N^{-1} \sum_{k,l=0}^{q(N)} \sum_{r=1}^{N-1} \max_{1 \le i,j \le d} \left| \mathrm{Cov}\left(X_{i0}(X_{jk} - X_{jk}^{(k)}), X_{ir}^{(r)} X_{j,r+\ell}^{(r+\ell)} \right) \right| \to 0. \tag{16.23}$$

Condition (16.23) is analogous to cumulant conditions for vector processes which require summability of fourth order cross–cumulants of all scalar components, see e.g. Assumption A on p. 823 of Andrews (1991).

For ease of reference we state these results as a theorem.

Theorem 16.5. *a) If* $\{\mathbf{X}_n\} \in L^2_{\mathbb{R}^d}$ *is an* L^2*–m–approximable sequence, then the series* $\sum_{r=-\infty}^{\infty} \boldsymbol{\Gamma}_r$ *converges absolutely. b) Suppose* $\{\mathbf{X}_n\} \in L^4_{\mathbb{R}^d}$ *an* L^4*–m–approximable sequence such that condition* (16.23) *holds. If Assumption 16.1 holds, then* $\hat{\boldsymbol{\Sigma}} \xrightarrow{P} \boldsymbol{\Sigma}$.

We are now able to turn to functional data. Suppose $\{X_n\} \in L^2_H$ is a zero mean sequence and $v_1, v_2, \ldots, v_d$ is any set of orthonormal functions in H. Define $X_{in} = \int X_n(t) v_i(t) dt$, $\mathbf{X}_n = [X_{1n}, X_{2n}, \ldots, X_{dn}]^T$ and $\boldsymbol{\Gamma}_r = \mathrm{Cov}(\mathbf{X}_0, \mathbf{X}_r)$. A direct verification shows that if $\{X_n\}$ is L^p–m–approximable, then so is the vector sequence $\{\mathbf{X}_n\}$. We thus obtain the following corollary.

Corollary 16.1. *a) If* $\{X_n\} \in L^2_H$ *is an* L^2*–m–approximable sequence, then the series* $\sum_{r=-\infty}^{\infty} \boldsymbol{\Gamma}_r$ *converges absolutely. b) If, in addition,* $\{X_n\}$ *is* L^4*–m–approximable and Assumption 16.1 and condition* (16.23) *hold, then* $\hat{\boldsymbol{\Sigma}} \xrightarrow{P} \boldsymbol{\Sigma}$.

The results of Section 16.4 show that the conclusions of parts $b)$ of Theorem 16.5 and Corollary 16.1 holds under L^2–m–approximability and mild additional assumptions; L^4–m–approximability and Condition (16.23) are not required.

In Corollary 16.1, the functions $v_1, v_2, \ldots, v_d$ form an arbitrary orthonormal deterministic basis. In many applications, a random basis consisting of the estimated principal components $\hat{v}_1, \hat{v}_2, \ldots, \hat{v}_d$ is used. The scores with respect to this basis are defined by

$$\hat{\eta}_{\ell i} = \int (X_i(t) - \bar{X}_N(t)) \hat{v}_\ell(t) dt, \quad 1 \le \ell \le d. \tag{16.24}$$

To use the results established so far, it is convenient to decompose the stationary sequence $\{X_n\}$ into its mean and a zero mean process, i.e. we set $X_n(t) = \mu(t) + Y_n(t)$, where $EY_n(t) = 0$. We introduce the unobservable quantities

$$\beta_{\ell n} = \int Y_n(t) v_\ell(t) dt, \quad \hat{\beta}_{\ell n} = \int Y_n(t) \hat{v}_\ell(t) dt. \quad 1 \le \ell \le d, \tag{16.25}$$

We then have the following proposition which will be useful in most statistical procedures for functional time series which, an application to change point detection is developed in Section 16.5.

Proposition 16.3. *Let* $\hat{\mathbf{C}} = \mathrm{diag}(\hat{c}_1, \ldots, \hat{c}_d)$*, with* $\hat{c}_i = \mathrm{sign}(\langle v_i, \hat{v}_i \rangle)$*. Suppose* $\{X_n\} \in L^4_H$ *is* L^4*–m–approximable and that* (16.12) *holds. Assume further that*

$$\kappa := \sup_{q \geq 1} \frac{1}{q} \sum_{j=-q}^{q} w_q(j) < \infty \tag{16.26}$$

and $q^4/N \to 0$*. Then*

$$|\hat{\boldsymbol{\Sigma}}(\boldsymbol{\beta}) - \hat{\boldsymbol{\Sigma}}(\hat{\mathbf{C}}\hat{\boldsymbol{\beta}})| = o_P(1) \quad and \quad |\hat{\boldsymbol{\Sigma}}(\hat{\boldsymbol{\eta}}) - \hat{\boldsymbol{\Sigma}}(\hat{\boldsymbol{\beta}})| = o_P(1). \tag{16.27}$$

Condition (16.26) holds for all weights used in practice. In particular, if $\omega_q(j) = K(j/q)$, as in Section 16.4, then $\kappa = 2\int K(x)dx$. Notice that Proposition 16.3 assumes a stronger condition $q^4/N \to 0$, which is common in the literature on the estimation of the long–run covariance matrix, see e.g. Newey and West (1987), but can be dropped, as we show in Section 16.4. We note that condition (16.23) does not appear in the statement of Proposition 16.3. Its point is that if $\hat{\boldsymbol{\Sigma}}(\boldsymbol{\beta})$ is consistent under some conditions, then so is $\hat{\boldsymbol{\Sigma}}(\hat{\boldsymbol{\eta}})$. The proof of Proposition 16.3 is presented in Section 16.8.

16.4 Estimation of the long–run covariance matrix under weak assumptions

The result of this section, Theorem 16.6, provides an elegant alternative to Theorem 16.5. It is a general consistency result for the kernel estimators of the long–run covariance matrix, which can be used in many problems of inference for vector–valued time series. Since its relevance goes beyond functional data, we restate some assumptions and definitions, to make this section as self–contained as possible. In this Section, $\mathbf{X}_\ell = [X_{1\ell}, \ldots, X_{d\ell}]^T$, is a sequence of zero mean L^2–m–approximable random vectors. For ease of reference, recall that this means that the following assumptions hold:

Assumption 16.2. *For a measurable function* $\mathbf{f}$ *taking values in* $\mathbb{R}^d$*,*

$$\mathbf{X}_\ell = \mathbf{f}(\varepsilon_\ell, \varepsilon_{\ell-1}, \ldots),$$

where ε_ℓ *is a sequence of independent identically distributed random elements as in Definition 16.1.*

Assumption 16.3.

$$E\mathbf{X}_\ell = \mathbf{0} \quad and \quad E\|\mathbf{X}_\ell\|^2 < \infty.$$

Assumption 16.4.

$$\max_{1\le j\le d}\sum_{m=1}^{\infty}\left(E(X_{j\ell}-X_{j\ell}^{(m)})^2\right)^{1/2}<\infty,$$

where $\mathbf{X}_\ell^{(m)}=[X_{1\ell}^{(m)},\dots,X_{d\ell}^{(m)}]^T$ *and*

$$\mathbf{X}_\ell^{(m)}=f(\varepsilon_\ell,\varepsilon_{\ell-1},\dots,\varepsilon_{\ell-m+1},\varepsilon_{\ell,\ell-m}^{(m)},\varepsilon_{\ell,\ell-m-1}^{(m)},\dots),$$

where $\{\varepsilon_{\ell,n}^{(m)},\ m\ge 1,-\infty<n,\ell<\infty\}$ *are iid copies of* ε_0.

Recall that the long–run variance matrix $\boldsymbol{\Sigma}$ introduced in Section 16.3 is defined by

$$\boldsymbol{\Sigma}=E\mathbf{X}_0\mathbf{X}_0^T+\sum_{l=1}^{\infty}E\mathbf{X}_0\mathbf{X}_\ell^T+\sum_{l=1}^{\infty}E\mathbf{X}_\ell\mathbf{X}_0^T.$$

Assumptions 16.3 and 16.4 yield that $\boldsymbol{\Sigma}$ is well–defined, and the infinite sums in the definition are (coordinate-wise) absolutely convergent. We consider the estimation of $\boldsymbol{\Sigma}$. The sample autocovariance matrices defined in Section 16.3 can be written as

$$\hat{\boldsymbol{\Gamma}}_k=\frac{1}{N}\sum_{\ell=\max(1,1-k)}^{\min(N,N-k)}\mathbf{X}_\ell\mathbf{X}_{\ell+k}^T$$

and the kernel estimator (16.22) as

$$\hat{\boldsymbol{\Sigma}}_N=\sum_{k=-(N-1)}^{N-1}K(k/B_N)\hat{\boldsymbol{\Gamma}}_k. \tag{16.28}$$

We write the weights $\omega_q(j)$ as $K(j/B_N)$ to emphasize the dependence of the bandwidth on the sample size N, and to facilitate the formulation of conditions in Assumption 16.5. If the support of the kernel K is the interval $[-1,1]$, then $q=B_N$; for a different compact support, q and B_N are proportional. The kernel K is assumed to satisfy the following conditions:

Assumption 16.5. *(i)* $K(0)=1$
(ii) K is a symmetric, Lipschitz function
(iii) K has a bounded support
(iv) $\hat{K}$, the Fourier transform of K, is also Lipschitz and integrable

The Fourier transform in Assumption 16.5 is defined as

$$\hat{K}(u)=\frac{1}{2\pi}\int_{-\infty}^{\infty}K(s)e^{-isu}ds.$$

Assumption 16.5 imposes smoothness conditions on the kernel K which are not required in Assumption 16.1, but these conditions are mild, and are satisfied by the most commonly used kernels, like the Bartlett (cf. Example 16.4) and Parzen (cf. Example 16.5). Assumption 16.5 has been used in other contexts, for example, Liu and Wu (2010) established consistency results for the estimation of spectral densities under Assumption 16.6. It does not specify the rate at which B_N tends to infinity. We formulate it as a separate assumption, namely,

Assumption 16.6.

$$B_N \to \infty \quad \textit{and} \quad B_N/N \to 0.$$

We can now state the following theorem, which is proven in Section 16.9

Theorem 16.6. *If Assumptions 16.2-16.6 hold, then*

$$\hat{\boldsymbol{\Sigma}}_N \xrightarrow{P} \boldsymbol{\Sigma}.$$

If Theorem 16.6 is used in the context of functional data, the vectors $\mathbf{X}_\ell$ are often projections onto the EFPC's $\hat{v}_1, \ldots, \hat{v}_d$. In this case, $\hat{\boldsymbol{\Sigma}}_N$ is close to $\hat{C}\boldsymbol{\Sigma}\hat{C}$, with the matrix $\hat{C}$ as in Proposition 16.3. For more applications of Theorem 16.6, see Horváth and Reeder (2011).

The main advantage of Theorem 16.6 over Theorem 16.4 is that the latter requires L^4–m–approximability, whereas only L^2–m–approximability is assumed in Theorem 16.6. This is of practical relevance as some data, most notably those arising in financial applications, may not have fourth moments. Moreover, Theorem 16.6 does not use the cumulant–like condition (16.23), which may be difficult to verify for some model classes. Finally, Theorem 16.6 uses a weaker and more standard assumption $B_N = o(N)$, rather than $B_N = o(N^{1/2})$ needed in Theorem 16.4. This is achieved at the expense of imposing smoothness condition on the kernel K and it its Fourier transform $\hat{K}$ (Assumption 16.5(iii) can be replaced with the requirement that $K(t)$ decays fast enough as $|t| \to \infty$). For all kernels and bandwidths used in practice, both the conditions on K and the rate $B_N = o(N^{1/2})$ hold, so these differences in assumptions are less important.

We conclude this section with some example illustrating Assumption 16.5.

Example 16.4. The Bartlett kernel is

$$K(s) = \begin{cases} 1 - |s|, & |s| \le 1, \\ 0, & \text{otherwise} \end{cases}$$

This kernel clearly satisfies parts (i)–(iii) of Assumption 16.5. Its Fourier transform is

$$\hat{K}(u) = \left\{ \frac{1}{\pi u} \sin\left(\frac{u}{2}\right) \right\}^2.$$

Thus, to verify part (iv), we must check that the function

$$F(t) = \left\{ \frac{\sin(t)}{t} \right\}^2$$

is integrable and Lipschitz. The integrability follows because $|F(t)| \leq t^{-2}$ and $F(t) \to 1$, as $t \to 0$.

The derivative of F for $t \neq 0$ is

$$F'(t) = \frac{2\sin(t)}{t}\left\{\frac{t\cos(t) - \sin(t)}{t^2}\right\}.$$

This function is clearly bounded outside any neighborhood of zero. Using the Taylor expansion of the sine and cosine functions, it is easy to verify that $F'(t) = o(t)$, as $t \to 0$. In a similar fashion, one can verify that $F(t) - F(0) = o(t^2)$, as $t \to 0$. Thus F is Lipschitz on the whole line.

Example 16.5. The Parzen kernel is given by

$$K(s) = \begin{cases} 1 - 6s^2 + 6|s|^3, & |s| \leq 1/2, \\ 2(1 - |s|^3), & 1/2 \leq |s| \leq 1, \\ 0, & |s| > 1. \end{cases}$$

Taniguchi and Kakizawa (2000), p. 391, show that

$$\hat{K}(u) = \frac{3}{8\pi}\left\{\frac{\sin(u/4)}{u/4}\right\}^4.$$

Following the arguments used in Example 16.4, one can verify that $\hat{K}(u)$ is also integrable and Lipschitz.

In Examples 16.4 and 16.5, the kernel is a scaled version of the convolution of uniform densities on $[-1, 1]$. The Bartlett kernel is the convolution of two, while the Parzen kernel is the convolution of four (this follows immediately from the form of $\hat{K}(u)$, cf. Example 16.6). Higher order convolutions can be used as well.

Example 16.6. Up to multiplicative constants, the Fourier transform of the rectangular kernel

$$K(s) = \begin{cases} 1, & |s| \leq 1, \\ 0, & \text{otherwise} \end{cases}$$

is

$$F(t) = \frac{\sin(t)}{t}.$$

This function is not absolutely integrable, so part (iv) of Assumption 16.5 does not hold.

The rectangular kernel is not used in practice due to its poor performance in finite samples, which can be theoretically explained by the slowly decaying Fourier transform. To some extend, this is also true of the Bartlett kernel, but it is more often used due to its simplicity. Optimal kernels are generally smoother in the time domain and "more compactly" supported in the frequency domain. In software implementations, these kernels are typically not defined directly through a function K, but through the weights $\omega_q(j)$ considered in Section 16.3. For example the modified Daniell kernel is obtained by repeated discrete convolutions of the weights $\omega(-1) = 1/3$, $\omega(0) = 1/3$, $\omega(1) = 1/3$, see e.g. Chapter 4 of Shumway and Stoffer (2006).

16.5 Change point detection

Functional time series are obtained from data collected sequentially over time, and it is natural to expect that conditions under which observations are made may change. If this is the case, procedures developed for stationary series will produce spurious results. In this section, we develop a procedure for the detection of a change in the mean function of a functional time series. In addition to its practical relevance, the requisite theory illustrates the application of the results developed in Sections 16.2 and 16.3. The main results of this Section, Theorems 16.7 and 16.8, are proven in Section 16.10. This Section is an extension of Chapter 6 to dependent curves. We thus consider testing the null hypothesis

$$H_0: \quad EX_1(t) = EX_2(t) = \cdots = EX_N(t),\ t \in [0,1].$$

(Note that under H_0, we do not specify the value of the common mean.) The test we construct, has a particularly good power against the alternative in which the data can be divided into several consecutive segments, and the mean is constant within each segment, but changes from segment to segment. The simplest case of only two segments (one change point) is specified in Assumption 16.8. First we note that under the null hypothesis, we can represent each functional observation as

$$X_i(t) = \mu(t) + Y_i(t), \quad EY_i(t) = 0. \tag{16.29}$$

The following assumption specifies conditions on $\mu(\cdot)$ and the errors $Y_i(\cdot)$ needed to establish the convergence of the test statistic under H_0.

Assumption 16.7. *The mean μ in* (16.29) *is in H. The error functions $Y_i \in L^4_H$ are L^4–m–approximable mean zero random elements such that the eigenvalues of their covariance operator satisfy* (16.12).

Recall that the L^4–m–approximability implies that the Y_i are identically distributed with $\nu_4(Y_i) < \infty$. In particular, their covariance function

$$c(t,s) = E[Y_i(t)Y_i(s)] \quad 0 \le t, s \le 1,$$

is square integrable, i.e. is in $L^2([0,1] \times [0,1])$.

We develop the theory under the alternative of exactly one change point, but the procedure is applicable to multiple change points by using a segmentation algorithm described in Chapter 6.

Assumption 16.8. *The observations follow the model*

$$X_i(t) = \begin{cases} \mu_1(t) + Y_i(t), & 1 \le i \le k^*, \\ \mu_2(t) + Y_i(t), & k^* < i \le N, \end{cases}$$

in which the Y_i satisfy Assumption 16.7, the mean functions μ_1 and μ_2 are in L^2 and

$$k^* = [N\theta] \quad \text{for some } 0 < \theta < 1.$$

The general idea of testing is similar to that developed in Chapter 6 for independent observations, the central difficulty is in accommodating the dependence. To define the test statistic, recall that bold symbols denote d–dimensional vectors, e.g. $\hat{\boldsymbol{\eta}}_i = [\hat{\eta}_{1i}, \hat{\eta}_{2i}, \dots, \hat{\eta}_{di}]^T$. Define the partial sums process

$$\mathbf{S}_N(x, \boldsymbol{\xi}) = \sum_{n=1}^{\lfloor Nx \rfloor} \boldsymbol{\xi}_n, \quad x \in [0,1],$$

and the bridge process

$$\mathbf{L}_N(x, \boldsymbol{\xi}) = \mathbf{S}_N(x, \boldsymbol{\xi}) - x\mathbf{S}_N(1, \boldsymbol{\xi}), \tag{16.30}$$

where $\{\boldsymbol{\xi}_n\}$ is a generic R^d–valued sequence. Denote by $\boldsymbol{\Sigma}(\boldsymbol{\xi})$ the long–run variance of the sequence $\{\boldsymbol{\xi}_n\}$, and by $\hat{\boldsymbol{\Sigma}}(\boldsymbol{\xi})$ its kernel estimator, see Section 16.3. The proposed test statistic is then

$$T_N(d) = \frac{1}{N} \int_0^1 \mathbf{L}_N(x, \hat{\boldsymbol{\eta}})^T \hat{\boldsymbol{\Sigma}}(\hat{\boldsymbol{\eta}})^{-1} \mathbf{L}_N(x, \hat{\boldsymbol{\eta}})\, dx, \tag{16.31}$$

with the scores $\hat{\eta}_{\ell i}$ given by (16.24).

Our first theorem establishes its asymptotic null distribution.

Theorem 16.7. *Suppose H_0 and Assumption 16.7 hold. If the estimator $\hat{\boldsymbol{\Sigma}}(\hat{\boldsymbol{\eta}})$ is consistent, then*

$$T_N(d) \xrightarrow{d} T(d) := \sum_{\ell=1}^{d} \int_0^1 B_\ell^2(x) dx, \tag{16.32}$$

where $\{B_\ell(x),\ x \in [0,1]\}$, $1 \le \ell \le d$, are independent Brownian bridges.

The distribution of the random variable $T(d)$ was derived by Kiefer (1959). The limit distribution is the same as in the case of independent observations, this is possible because the long–run variance estimator $\hat{\boldsymbol{\Sigma}}(\hat{\boldsymbol{\eta}})$ soaks up the dependence. Sufficient conditions for its consistency are stated in Section 16.3, and, in addition to the assumptions of Theorem 16.7, are: Assumption 16.1 with $q^4/N \to 0$, and condition (16.23).

The next result shows that our test has asymptotic power 1. Our proof requires the following condition:

$$\hat{\boldsymbol{\Sigma}}(\hat{\boldsymbol{\eta}}) \xrightarrow{a.s.} \boldsymbol{\Omega}, \quad \text{where } \boldsymbol{\Omega} \text{ is some positive definite matrix.} \tag{16.33}$$

Condition (16.33) could be replaced by weaker technical conditions, but we prefer it, as it leads to a transparent, short proof. Essentially, it states that the matrix $\hat{\boldsymbol{\Sigma}}(\hat{\boldsymbol{\eta}})$ does not become degenerate in the limit, the matrix $\boldsymbol{\Omega}$ has only positive eigenvalues. A condition like (16.33) is not needed for independent Y_i because that case does not require normalization with the long–run covariance matrix. To formulate our result, introduce vectors $\boldsymbol{\mu}_1, \boldsymbol{\mu}_2 \in \mathbb{R}^d$ with coordinates

$$\int \mu_1(t) v_\ell(t)dt \quad \text{and} \quad \int \mu_2(t) v_\ell(t)dt, \qquad 1 \le \ell \le d.$$

Theorem 16.8. *Suppose Assumption 16.8 and condition* (16.33) *hold. If the vectors* $\boldsymbol{\mu}_1$ *and* $\boldsymbol{\mu}_2$ *are not equal, then* $T_N(d) \xrightarrow{P} \infty$.

The behavior under the alternative of change point tests for dependent functional data is studied by Aston and Kirch (2011a). Their work addresses in detail the orthogonality conditions required for a test to have nontrivial power, and includes epidemic changes in which the mean is μ_2 at $k_1, k_1 + 1, \ldots, k_2$ with $k_1 > 1$ and $k_2 < n$, and μ_1 elsewhere.

We conclude this section with two numerical examples which illustrate the effect of dependence on our change point detection procedure. Example 16.7 uses synthetic data, while Example 16.8 focuses on particulate pollution data. Both show that using statistic (16.31) with $\hat{\boldsymbol{\Sigma}}(\hat{\boldsymbol{\eta}})$ being the estimate for just the covariance, not the long–run covariance matrix, leads to spurious rejections of H_0, a nonexistent change point can be detected with a large probability. An interesting example is presented in Aston and Kirch (2011b) who develop methodology for determining distributions of change points for 3D functional data from multiple subjects. They apply it to a large study on resting state functional magnetic resonance imaging.

Example 16.7. We simulate 200 observations of the functional AR(1) process of Example 16.1, when Ψ has the parabolic integral kernel $\psi(t,s) = \gamma \cdot \big(2-(2x-1)^2-(2y-1)^2\big)$. We chose the constant γ such that $\|\Psi\|_S = 0.6$ (the Hilbert–Schmidt norm). The innovations $\{\varepsilon_n\}$ are standard Brownian bridges. The first 3 principal components explain approximately 85% of the total variance, so we compute the test statistic $T_{200}(3)$ given in (16.31). For the estimation of the long–run covariance matrix $\boldsymbol{\Sigma}$ we use the Bartlett kernel

$$\omega_q^{(1)}(j) = \begin{cases} 1 - |j|/(1+q), & \text{if } |j| \le q; \\ 0, & \text{otherwise.} \end{cases}$$

We first let $q = 0$, which corresponds to using just the sample covariance of $\{\hat{\boldsymbol{\eta}}_n\}$ in the normalization for the test statistic (16.31) (dependence is ignored). We use 1000 replications and the 5% confidence level. The rejection rate is 23.9%, much higher than the nominal level of 5%. In contrast, using an appropriate estimate for the long–run variance, the reliability of the test improves dramatically. Choosing an optimal bandwidth q is a separate problem, which we do not pursue here. Here we adapt the formula $q \approx 1.1447\,(aN)^{1/3}$, $a = \frac{4\psi^2}{(1+\psi)^4}$ valid for a a scalar AR(1) process with the autoregressive coefficient ψ, Andrews (1991). Using this formula with $\psi = \|\Psi\|_{\mathcal{S}} = 0.6$ results in $q = 4$. This choice gives the empirical rejection rate of 3.7%, much closer to the nominal rate of 5%.

Example 16.8. This example, which uses `pm10` (particulate matter with diameter $< 10\mu$m, measured in μg/m^3) data, illustrates a similar phenomenon as Example 16.7. For the analysis we use `pm10` concentration data measured in the Austrian city of Graz during the winter of 2008/2009 (N=151). The data are given in 30 minutes resolution, yielding an intraday frequency of 48 observations. As in Stadtlober *et al.* (2008) we use a square root transformation to reduce heavy tails. Next we remove possible weekly periodicity by subtracting the corresponding mean vectors obtained from the different weekdays. A time series plot of this new sequence is given in Figure 16.2. The data look relatively stable, although a shift appears to be possible in the center of the time series. It should be emphasized however, that `pm10` data, like many geophysical time series, exhibit a strong, persistent positive autocorrelation structure. These series are stationary over long periods of time, with an appearance of local trends or shifts at various time scales (random self–similar or fractal structure).

The daily measurement vectors are transformed into smooth functional data using 15 B-splines functions of order 4. The functional principal component analysis yields that the first three principal components explain $\approx 84\%$ of the total variability, so we use statistic (16.31) with $d = 3$. A look at the `acf` and `pacf` of the first empirical PC scores (Figure 16.3) suggests an AR(1), maybe AR(3) behavior. The second and third empirical PC scores show no significant autocorrelation structure. We use the formula given in Example 16.7 with $\psi = 0.70$ (`acf` at lag 1) and $N = 151$ and obtain $q \approx 4$. This gives $T_{151}(3) = 0.94$, which is close to the critical value 1.00 when testing at a 95% confidence level, but does not support rejection

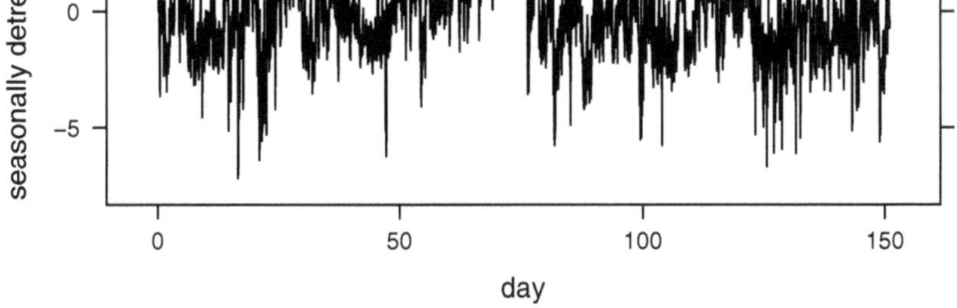

Fig. 16.2 Seasonally detrended $\sqrt{\texttt{pm10}}$, Nov 1, 2008 – Mar 31, 2009.

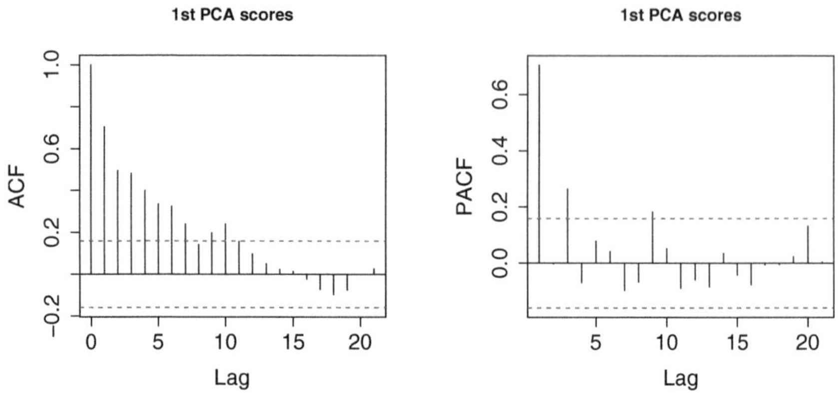

Fig. 16.3 Left panel: Sample autocorrelation function of the first empirical PC scores. Right panel: Sample partial autocorrelation function of the first empirical PC scores.

of the no-change hypothesis. In contrast, using only the sample covariance matrix in (16.32) gives $T_{151}(3) = 1.89$, and thus a clear and possibly spurious rejection of the null hypothesis.

16.6 Self–normalized statistics

We have seen in the previous sections of this chapter that in many inferential problems related to time series the long run variance plays a fundamental role. In particular, in Section 16.5 we used it to normalize the test statistic of Chapter 6 to obtain a limiting null distribution that is parameter free, see Theorem 16.7. It has been known in econometric research that using the long run variance in this way can lead to the so–called non–monotonic power. This phenomenon is illustrated in Figure 16.4 which shows the power of three tests of the null hypothesis of Section 16.5 under the alternative quantified in Assumption 16.8. The mean zero curves Y_i are Gaussian FAR(1) processes; $k^* = N/2$, $\mu_1 = 0$, and $\mu_2 = \delta f(t)$. Of central importance is the parameter δ which quantifies the magnitude of the change. The test called BGHK is the the test of Chapter 6 (it assumes independent Y_i), HK refers to the test of Section 16.5, and SN to the test based on a self–normalized statistic, which will be introduced later in this section. Focusing on the HK test , we see that if the change becomes very large, this test looses power; its power is non–monotonic. A heuristic explanation is that the "denominator" in statistic (16.31), an estimate of run variance matrix based on a data driven procedure discussed later in this section, becomes very large when δ increases. This is because the scores $\hat{\eta}_{\ell i}$, given by (16.24), are computed without adjusting for a possible change point. If the change point is very large, the sample autocorrelations of the scores decay

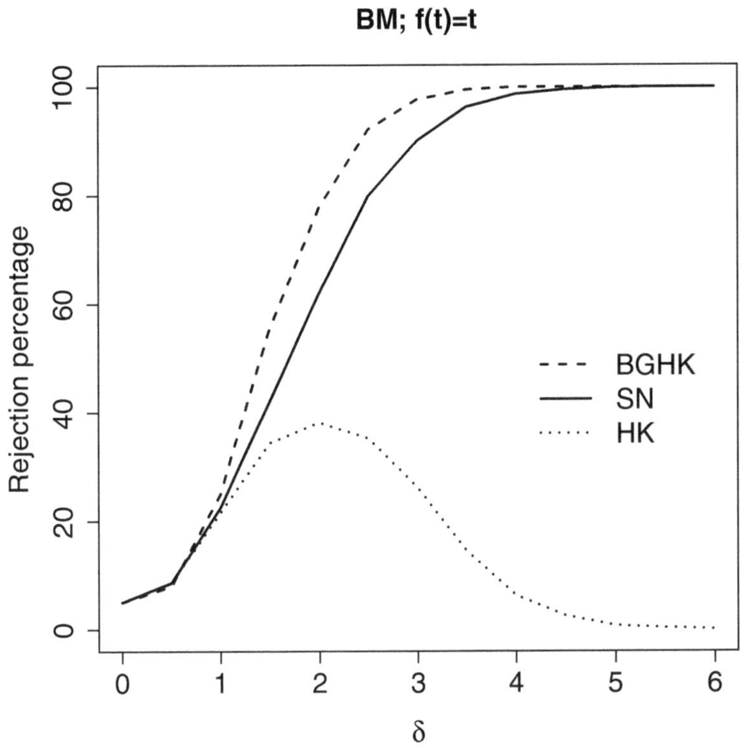

Fig. 16.4 Size-adjusted power for detecting the change in the mean function; δ measures the magnitude of change; sample size $N = 50$.

very slowly and it causes the data driven procedure to select large bandwidths and the estimator of the long run variance to behave like for a very strongly dependent sequence. This inflates its value so much that the test looses power. (We note that if the bandwidth is deterministic, the power is monotonic, but there is no universal formula for the kernel bandwidth that gives correct size.) A remedy is to adjust the definition of the scores $\hat{\eta}_{\ell i}$ to allow for a possible change point. Combined with the idea of self–normalization, this leads to a test that has monotonic power. Before proceeding further, we note that size of change corresponding to, say, $\delta = 2$ is very large relative to the Y_i, and a change point of this magnitude can be detected by eye. The tests based on self–normalized statistics correct however not only the problem of non–monotonic power, but perhaps more importantly eliminate the need to select the bandwidth parameter in the kernel estimators of the long run variance.

The remainder of this section is devoted to the discussion of these issues. It is based on the work of Shao (2010), Shao and Zhang (2010) and Zhang *et al.* (2011). These papers contain references to earlier work and to other applications of self–normalization. A different approach to change point detection which does not require bandwidth selection is proposed in Horváth *et al.* (2008). We focus only on the change point in the mean function. Zhang *et al.* (2011) show how self–normalization can be applied to the problem of change point detection in the functional AR(1) model studied in Chapter 14. Figure 16.4 and all numerical results presented in this section were made available to us by Xiaofeng Shao.

We first explain the idea of self–normalization for scalar time series. Suppose $\{X_n\}$ is a stationary time series such that

$$N^{-1/2} \sum_{1 \le k \le Nr} (X_k - \mu) \xrightarrow{d} \sigma W(r), \quad 0 \le r \le 1, \tag{16.34}$$

in the Skorokhod space. The parameter σ^2 is the long–run variance:

$$\sigma^2 = \lim_{N \to \infty} N \operatorname{Var}\left(\bar{X}_N\right) = \sum_h \gamma(h).$$

Set

$$D_N = N^{-2} \sum_{n=1}^{N} \left\{ \sum_{j=1}^{n} (X_j - \bar{X}_N) \right\}^2 .$$

Then, (16.34) implies

$$\frac{N(\bar{X}_N - \mu)^2}{D_N} \xrightarrow{d} \frac{W^2(1)}{\int_0^1 B^2(r)dr}. \tag{16.35}$$

To see why (16.35) holds, set

$$S_N(r) = N^{-1/2} \sum_{1 \le k \le Nr} (X_k - \mu), \quad 0 \le r \le 1,$$

and observe that

$$S_N(1) = N^{1/2} \left(\bar{X}_N - \mu\right),$$

so that

$$N(\bar{X}_N - \mu)^2 = S_N^2(1) \xrightarrow{d} \sigma^2 W^2(1). \tag{16.36}$$

Next, observe that

$$\sum_{j=1}^{n} (X_j - \bar{X}_N) = \sum_{j=1}^{n} (X_j - \mu) - n\left(\bar{X}_N - \mu\right);$$

$$N^{-1/2} \sum_{j=1}^{n} (X_j - \bar{X}_N) = S_N\left(\frac{n}{N}\right) - \frac{n}{N} S_N(1).$$

Consequently

$$D_N = N^{-1} \sum_{n=1}^{N} \left\{ S_N\left(\frac{n}{N}\right) - \frac{n}{N} S_N(1) \right\}^2 \xrightarrow{d} \sigma^2 \int_0^1 \{W(r) - rW(1)\}^2 dr. \tag{16.37}$$

The convergences in (16.36) and (16.37) are joint, so (16.35) follows.

The key point is the cancelation of σ^2 when (16.36) is divided by (16.37). Relation (16.37) shows that D_N is an inconsistent estimator of σ^2. The distribution of the right–hand side of (16.35) can however be simulated, and the critical values can be obtained with arbitrary precision. Relation (16.35) can be used to construct a confidence interval for μ without estimating the long run variance. Such a construction does not require the selection of a bandwidth parameter in the kernel estimates of σ^2.

The normalization analogous to D_N is however not suitable for the change point problem. Simulations reported in Shao and Zhang (2010) show that with such a normalization the power of change point tests tends to zero as δ increases. These authors propose a self–normalization that takes into account the behavior under the alternative. Their approach was extended to functional data by Zhang *et al.* (2011). To explain it, we extend and lighten the notation introduced in Section 16.5. Set

$$\mathbf{U}_N(n_1,n_2) = \sum_{j=n_1}^{n_2} \hat{\boldsymbol{\eta}}_j = \mathbf{S}_N\left(\frac{n_2}{N},\hat{\boldsymbol{\eta}}\right) - \mathbf{S}_N\left(\frac{n_1-1}{N},\hat{\boldsymbol{\eta}}\right),$$

with the scores defined by (16.24). Note the the sums $\mathbf{U}_N(n_1,n_2)$ depend also on the number d of the EFPC's to be used. Next, for each $1 \le k \le N$, introduce the $d \times d$ matrices

$$\boldsymbol{D}_N(n,k) = \left[\mathbf{U}_N(1,n) - \frac{n}{k}\mathbf{U}_N(1,k)\right]\left[\mathbf{U}_N(1,n) - \frac{n}{k}\mathbf{U}_N(1,k)\right]^T, \quad n \le k.$$

and

$$\boldsymbol{D}_N^*(n,k) = \left[\mathbf{U}_N(n,N) - \frac{N-n+1}{N-k}\mathbf{U}_N(k+1,N)\right] \times \left[\mathbf{U}_N(n,N) - \frac{N-n+1}{N-k}\mathbf{U}_N(k+1,N)\right]^T, \quad n > k.$$

Using these matrices, we can define the normalizing matrices as

$$\mathbf{V}_N(k) = \frac{1}{N}\left\{\sum_{n=1}^{k} \boldsymbol{D}_N(n,k) + \sum_{n=k+1}^{N} \boldsymbol{D}_N^*(n,k)\right\}.$$

A test statistic can be a functional of the process

$$\mathbf{L}_N(x,\hat{\boldsymbol{\eta}})^T\left[\mathbf{V}_N(\lfloor Nx \rfloor)\right]^{-1}\mathbf{L}_N(x,\hat{\boldsymbol{\eta}}), \quad x \in [0,1],$$

with the bridge process $\mathbf{L}_N(\cdot,\hat{\boldsymbol{\eta}})$ defined in (16.30). Zhang *et al.* (2011) focus on the Kolmogorov–Smirnov functional

$$G_N = \sup_{1 \le k < N} \mathbf{L}_N(k/N,\hat{\boldsymbol{\eta}})^T\left[\mathbf{V}_N(k)\right]^{-1}\mathbf{L}_N(k/N,\hat{\boldsymbol{\eta}}).$$

They show that under L^4–m–approximability, and additional technical assumptions, G_N converges in distribution to a random variable G which can be expressed as a

Table 16.1 Simulated critical values of G based on 10000 replications.

$\alpha\%$ d	1	2	3	4	5	6	7	8	9	10
90.0%	29.6	56.5	81.5	114.7	150.0	183.8	223.5	267.1	308.5	360.0
95.0%	40.1	73.7	103.6	141.5	182.7	218.8	267.3	317.9	360.7	420.5
97.5%	52.2	92.2	128.9	171.9	218.7	255.0	313.4	367.9	416.3	483.0
99.0%	68.6	117.7	160.0	209.7	265.8	318.3	368.0	432.5	483.6	567.2
99.5%	84.6	135.3	182.9	246.6	291.7	367.7	410.5	498.1	544.9	621.6
99.9%	121.9	192.5	246.8	319.2	358.1	464.9	530.6	614.1	649.0	751.1

functional of a d–dimensional Brownian motion whose components are independent standard Brownian motions. The formula for G is not difficult to derive heuristically from the form of G_N, but the main point is that the critical values of G can be obtained by simulation. The critical values are shown in Table 16.1.

We now compare the finite sample performance of the three tests described at the beginning of the Section. All simulation results are based on one thousand replications.

First, we consider independent functional observations. The mean function is zero under the null hypothesis. Two cases of the Y_i are considered, namely the standard Brownian motion (BM) and the Brownian Bridge (BB). Under the alternative, $\mu_1(t) = 0$ and $\mu_2(t) = t$ or $\mu_2(t) = \sin(t)$; $k^* = N/2$. We compare the SN test based on the Kolmogorov–Smirnov functional G_N to the BGHK test of Chapter 6 (based on the Cramér–von–Mises functional). The empirical size and size–adjusted power are summarized in Table 16.2. Size–adjusted power is computed using finite sample critical values based on the Monte Carlo simulation under the null hypothesis. When a test is conservative, size–adjusted power is higher than power; when it overrejects, size–adjusted power is smaller than power. In the latter scenario power is often very high just because the test has empirical size much higher than nominal. Size–adjusted power is often believed to offer a fairer comparison of several tests. The empirical size of the SN test is comparable with that of the BGHK test, but the SN test suffers from a small power loss.

To examine the effect of dependence, the functional sequence $\{Y_i(t)\}$ is generated according to the FAR(1) model (13.3). We consider the Gaussian kernel $\psi(t,s) = C\exp\left\{(t^2+s^2)/2\right\}$ and the Wiener kernel, $\psi(t,s) = C\min(t,s)$. The constant C is chosen so that the Hilbert–Schmidt norm of the kernels is 0.5. We now compare the SN test with the BGHK test and the HK test. To implement the HK test, we have to estimate the long run variance matrix of the first d scores. We use the kernel estimator (16.28) with the Bartlett kernel defined in Example 16.4. We use the popular data driven truncation lag of Andrews (1991), $B_N = 1.1447\{\hat{\alpha}(1)N\}^{1/3}$, where

$$\hat{\alpha}(1) = \left\{\sum_{\ell=1}^{p} \frac{4\hat{\sigma}_\ell^4\hat{\rho}_\ell^2}{(1-\hat{\rho}_\ell)^6(1+\hat{\rho}_\ell)^2}\right\}\left\{\sum_{\ell=1}^{p} \frac{\hat{\sigma}_\ell^4}{(1-\hat{\rho}_\ell)^4}\right\}^{-1}. \tag{16.38}$$

Table 16.2 Empirical size (upper panel) and size–adjusted power (lower panel) in percent for the SN test (i) and the BGHK test (ii) for independent functional data generated as BM or BB. The size-adjusted power is computed under the alternative with $\mu_2(t) = t$ or $\mu_2(t) = \sin(t)$, and $k^* = N/2$.

		$d = 1$			$d = 2$			$d = 3$		
		10%	5%	1%	10%	5%	1%	10%	5%	1%
$N = 50$										
BM	(i)	10.7	5.7	0.7	9.6	3.7	0.7	10.8	5.2	1.4
	(ii)	10.0	5.3	1.2	10.3	5.0	0.8	10.9	5.5	1.0
BB	(i)	7.5	3.8	0.8	8.2	4.6	1.1	10.7	6.0	1.3
	(ii)	10.6	5.4	0.8	10.9	5.1	1.1	10.5	5.2	1.2
$N = 100$										
BM	(i)	9.9	5.1	1.1	9.2	4.3	0.5	9.1	4.6	0.7
	(ii)	10.4	5.4	0.5	10.3	4.5	0.6	9.5	3.8	0.6
BB	(i)	10.0	5.1	1.3	8.4	3.5	0.7	9.9	4.7	0.7
	(ii)	9.6	5.2	0.9	9.3	4.9	0.6	9.1	4.1	0.9
$N = 50$										
BM, t	(i)	77.6	64.5	44.9	71.7	58.4	39.4	67.4	51.7	23.8
	(ii)	89.5	79.8	48.9	83.6	73.7	48.9	77.8	65.4	38.8
BB, t	(i)	99.8	99.4	95.6	100	100	99.6	100	100	99.9
	(ii)	100	100	99.7	100	100	100	100	100	100
BM, $sin(t)$	(i)	70.0	57.7	38.9	62.1	48.3	29.1	56.0	41.4	17.0
	(ii)	82.1	71.9	39.4	74.4	61.4	36.4	66.9	52.4	28.7
BB, $sin(t)$	(i)	99.3	98.1	89.7	100	99.6	96.9	100	99.9	99.4
	(ii)	99.9	99.7	97.6	100	100	100	100	100	100
$N = 100$										
BM, t	(i)	96.9	89.9	70.8	92.9	87.4	73.0	90.9	84.0	66.7
	(ii)	99.3	98.4	95.5	99.1	97.9	94.0	98.5	96.8	91.2
BB, t	(i)	100	99.9	99.6	100	100	100	100	100	100
	(ii)	100	100	100	100	100	100	100	100	100
BM, $sin(t)$	(i)	92.7	84.2	62.7	87.1	78.7	59.2	83.9	73.8	52.1
	(ii)	98.4	95.8	89.6	96.3	93.5	86.6	95.2	90.9	78.0
BB, $sin(t)$	(i)	99.9	99.7	98.8	100	100	100	100	100	100
	(ii)	100	100	100	100	100	100	100	100	100

Here $\hat{\rho}_\ell$ is the autoregressive coefficient estimate in the model $\hat{\eta}_{n,\ell} = \rho_\ell \hat{\eta}_{n-1,\ell} + \varepsilon_{n,\ell}$, and $\hat{\sigma}_\ell^2$ is the estimate of the innovation variance. Table 16.3 reports the empirical sizes. We see that the size distortion of the BGHK test is very large compared to the other two tests. This is due to the fact that it is designed only for independent functional data and is invalid in the temporally–dependent case. For the HK test, the size distortion is less severe but is sensitive to the choice of d. It tends to be oversized for small d but undersized for large d. For the SN test, size distortion is apparent for $N = 50$, but improves for $N = 100$. The size for the SN test is fairly robust to the choice of d. Based on the results reported in Zhang *et al.* (2011), the following comments can be made about the size–adjusted power. First, the BGHK test delivers the highest power among the three tests, which is largely due to its

Table 16.3 Empirical size in percent of the SN (i), the BGHK (ii) and the HK (iii) test for data following an FAR(1) process.

		$d = 1$			$d = 2$			$d = 3$		
		10%	5%	1%	10%	5%	1%	10%	5%	1%
$N = 50$										
Gaussian										
BM	(i)	15.2	10.3	3.9	15.2	8.4	2.4	14.5	8.0	2.2
	(ii)	44.1	32.2	16.2	37.5	25.0	12.4	32.7	23.0	11.4
	(iii)	17.7	9.2	0.6	11.1	3.2	0.3	4.9	1.1	0.0
BB	(i)	17.3	10.6	3.1	14.0	7.1	2.5	14.5	8.2	2.3
	(ii)	42.7	32.3	13.8	36.1	25.5	10.0	34.9	23.2	9.6
	(iii)	19.8	8.8	0.2	11.1	2.6	0.0	6.3	1.5	0.0
Wiener										
BM	(i)	16.0	10.4	4.0	16.1	9.2	3.0	16.0	9.8	2.9
	(ii)	46.4	33.6	16.6	40.2	26.9	12.6	36.6	25.0	10.2
	(iii)	17.5	8.4	0.5	10.9	2.8	0.1	6.1	0.7	0.0
BB	(i)	17.0	10.4	2.9	13.3	7.3	2.2	15.4	9.7	2.2
	(ii)	42.8	31.0	14.3	37.9	26.7	11.2	36.4	23.9	10.0
	(iii)	19.0	8.9	0.2	11.2	3.3	0.0	6.5	1.8	0.0
$N = 100$										
Gaussian										
BM	(i)	13.3	7.8	2.0	11.7	5.7	1.2	11.7	6.1	1.2
	(ii)	51.2	35.9	16.4	39.7	27.9	11.6	34.9	24.1	9.7
	(iii)	15.2	7.4	0.4	11.6	3.9	0.2	7.2	2.1	0.0
BB	(i)	11.6	6.7	1.6	10.9	4.9	1.1	11.5	7.1	1.2
	(ii)	46.7	33.0	13.9	35.9	25.1	10.2	36.4	25.8	11.4
	(iii)	16.1	8.0	1.4	12.4	5.3	0.3	10.0	3.5	0.1
Wiener										
BM	(i)	13.7	7.8	2.1	11.7	5.8	1.3	12.9	7.1	1.3
	(ii)	52.2	37.2	17.5	43.8	29.7	12.8	38.3	26.1	11.7
	(iii)	15.3	7.4	0.5	11.6	4.1	0.2	7.5	2.6	0.0
BB	(i)	11.9	6.4	1.9	10.4	5.6	1.2	12.0	7.8	1.3
	(ii)	45.1	32.0	13.5	38.5	27.5	12.8	37.9	27.3	11.9
	(iii)	16.4	7.6	1.5	13.3	5.9	0.5	10.8	3.7	0.2

severe upward size distortion. Second, the power of the SN test is comparable to that of the HK test for $N = 50$ and BM innovations, but the SN test tends to have moderate power loss when sample size increases to 100. In the case of the BB innovations, the SN test is superior to the HK test in power. Overall, the severe size distortion of the BGHK test under weak dependence suggests its inability to accommodate dependence and thus it is not recommended for testing for a change point for dependent functional data. The HK test is able to account for dependence but it is sensitive to the choice of bandwidth B_N and of d. As shown in Figure 16.4, the data driven bandwidth used in the HK test leads to non–monotonic power. Compared to the other two tests, the SN test tends to have more accurate size at the expense of some power loss.

16.7 Functional linear model with dependent regressors

We consider the fully functional model of the form

$$Y_n(t) = \int \psi(t,s)X_n(s) + \varepsilon_n(t), \quad n = 1, 2, \ldots, N, \tag{16.39}$$

in which both the regressors and the responses are functions. The results of this section can be easily specialized to the case of scalar responses.

In the existing theory, the X_n in (16.39) are assumed to be independent and identically distributed. For functional time series the assumption of the independence of the X_n is often questionable, so it is important to investigate if procedures developed and theoretically justified for independent regressors can still be used if the regressors are dependent.

We focus here on the estimation of the kernel $\psi(t,s)$. Our result is motivated by the work of Yao *et al.* (2005b) who considered functional regressors and responses obtained from sparse *independent* data measured with error. The data that motivates our work are measurements of physical quantities obtained with negligible errors or financial transaction data obtained without error. In both cases the data are available at fine time grids, and the main concern is the presence of temporal dependence between the curves X_n. We therefore merely assume that the sequence $\{X_n\} \in L^4_H$ is L^4–m–approximable, which, as can be easily seen, implies the L^4–m–approximability of $\{Y_n\}$. To formulate additional technical assumptions, we need to introduce some notation.

We assume that the errors ε_n are iid and independent of the X_n, and denote by X and Y random functions with the same distribution as X_n and Y_n, respectively. We work with their expansions

$$X(s) = \sum_{i=1}^{\infty} \xi_i v_i(s), \quad Y(t) = \sum_{j=1}^{\infty} \zeta_j u_j(t),$$

where the v_j are the FPC's of X and the u_j the FPC's of Y, and $\xi_i = \langle X, v_i \rangle$, $\zeta_j = \langle Y, u_j \rangle$. Indicating with the "hat" the corresponding empirical quantities, an estimator of $\psi(t,s)$ proposed by Yao *et al.* (2005b) is

$$\hat{\psi}_{KL}(t,s) = \sum_{k=1}^{K} \sum_{\ell=1}^{L} \hat{\lambda}_\ell^{-1} \hat{\sigma}_{\ell k} \hat{u}_k(t) \hat{v}_\ell(s),$$

where $\hat{\sigma}_{\ell k}$ is an estimator of $E[\xi_\ell \zeta_k]$. We will work with the simplest estimator

$$\hat{\sigma}_{\ell k} = \frac{1}{N} \sum_{i=1}^{N} \langle X_i, \hat{v}_\ell \rangle \langle Y_i, \hat{u}_k \rangle, \tag{16.40}$$

but any estimator for which Lemma 16.6 of Section 16.11 holds can be used without affecting the rates.

Let λ_j and γ_j be the eigenvalues corresponding to v_j and u_j. Define α_j as in Lemma 2.3 and define α'_j accordingly with γ_j instead of λ_j. Set

$$h_L = \min\{\alpha_j, 1 \le j \le L\}, \quad h'_L = \min\{\alpha'_j, 1 \le j \le L\}.$$

To establish the consistency of the estimator $\hat{\psi}_{KL}(t,s)$ we assume that

$$\sum_{k=1}^{\infty}\sum_{\ell=1}^{\infty} \frac{(E[\xi_\ell \zeta_k])^2}{\lambda_\ell^2} < \infty. \tag{16.41}$$

and that the following assumption holds:

Assumption 16.9. *(i) We have* $\lambda_1 > \lambda_2 > \cdots$ *and* $\gamma_1 > \gamma_2 > \cdots$.
(ii) We have $K = K(N)$, $L = L(N) \to \infty$ *and*

$$\frac{KL}{\lambda_L \min\{h_K, h'_L\}} = o\big(N^{1/2}\big).$$

For model (16.39), condition (16.41) is equivalent to the assumption that $\psi(t,s)$ is a Hilbert–Schmidt kernel, i.e. $\iint \psi^2(t,s)dt\,ds < \infty$. It is formulated in the same way as in Yao *et al.* (2005b), because this form is convenient in the theoretical arguments. Assumption 16.9 is much shorter than the corresponding assumptions of Yao *et al.* (2005b). This is because we do not deal with smoothing, and so can isolate the impact of the magnitude of the eigenvalues on the bandwidths K and L.

Theorem 16.9. *Suppose* $\{X_n\} \in L^4_H$ *is a zero mean* L^4*–m–approximable sequence independent of the sequence of iid errors* $\{\varepsilon_n\}$. *If* (16.41) *and Assumption 16.9 hold, then*

$$\iint \left[\hat{\psi}_{KL}(t,s) - \psi(t,s)\right]^2 dt\,ds \xrightarrow{P} 0, \quad (N \to \infty). \tag{16.42}$$

Remark 16.1. Horváth and Reeder (2011) showed, under more general conditions, that for fixed K and L

$$\iint \left[\hat{\psi}_{KL}(t,s) - \psi_{KL}(t,s)\right]^2 dt\,ds = O_P(N^{-1}),$$

where

$$\psi_{KL}(t,s) = \sum_{k=1}^{K}\sum_{\ell=1}^{L} \lambda_\ell^{-1} E[\xi_\ell \zeta_k] u_k(t) v_\ell(s).$$

The conclusion of Theorem 16.9 is comparable to the first part of Theorem 1 in Yao *et al.* (2005b). Both theorems are established under (16.41) and finite fourth moment conditions. Otherwise the settings are quite different. Yao *et al.* (2005b) work under the assumption that the subject (Y_i, X_i), $i = 1, 2, \ldots$ are independent and sparsely observed, whereas Theorem 16.9 admits dependence, but does not deal with curves measured with error at irregular points.

16.8 Proofs of the results of Sections 16.2 and 16.3

In this, and the following sections of this chapter, we use the following conventions: A generic X, which is assumed to be equal in distribution to X_1, will be used at some places. Any constants occurring will be denoted by $\kappa_1, \kappa_2, \ldots$ The κ_i may change their values from proof to proof.

Proof of Theorem 16.1. We assume for simplicity that $EX = 0$. For $k \in \mathbb{Z}$ define the operators $B_k(y) = \langle X_k, y\rangle X_k - C(y)$, $y \in H$. Then since B_k are iid Hilbert-Schmidt operators we have

$$E\|\hat{C}_N - C\|_{\mathcal{S}}^2 = E\left\|\frac{1}{N}\sum_{k=1}^{N} B_k\right\|_{\mathcal{S}}^2 = \frac{1}{N}\sum_{k=-(N-1)}^{N-1}\left(1-\frac{|k|}{N}\right)E\langle B_0, B_k\rangle_{\mathcal{S}}$$
$$\leq \frac{1}{N}\sum_{k\in\mathbb{Z}}|E\langle B_0, B_k\rangle_{\mathcal{S}}|,$$

and it remains to show that $|E\langle B_0, B_k\rangle_{\mathcal{S}}|$ decays sufficiently fast. We let $\lambda_1 \geq \lambda_2 \geq \cdots$ b e the eigenvalues of the operator C and we let $\{e_i\}$ be the corresponding eigenvectors. Then

$$E\left\langle X_0, X_k^{(k)}\right\rangle = \sum_{j\geq 1}\lambda_j^2, \quad k \geq 1. \tag{16.43}$$

This can be shown by using that X_0 and $X_k^{(k)}$ are independent. Furthermore it can be readily verified that

$$E\,\langle B_0, B_k\rangle_{\mathcal{S}} = E\,\langle X_0, X_k\rangle^2 - \sum_{j\geq 1}\lambda_j^2, \quad k \geq 1. \tag{16.44}$$

For ease of notation we set $X_k' = X_k^{(k)}$. Then we have

$$\left|\langle X_0, X_k - X_k'\rangle^2\right| = \langle X_0, X_k\rangle^2 + \langle X_0, X_k'\rangle^2 - 2\,\langle X_0, X_k\rangle\,\langle X_0, X_k'\rangle$$
$$= \langle X_0, X_k\rangle^2 - \langle X_0, X_k'\rangle^2 - 2\langle X_0, X_k - X_k'\rangle\langle X_0, X_k'\rangle.$$

Thus

$$\langle X_0, X_k\rangle^2 - \langle X_0, X_k'\rangle^2 = \langle X_0, X_k - X_k'\rangle^2 + 2\langle X_0, X_k - X_k'\rangle\langle X_0, X_k'\rangle.$$

and by repeated application of Cauchy-Schwarz it follows that

$$\left| E \langle X_0, X_k \rangle^2 - E \langle X_0, X_k' \rangle^2 \right| \leq \nu_4^2(X_0)\nu_4^2\left(X_k - X_k'\right) + 2\nu_4^3(X_0)\nu_4\left(X_k - X_k'\right). \tag{16.45}$$

Combining (16.43), (16.44), (16.45) and using the Definition of L^4–m–approximability yields the proof of our theorem, with U_X equal to the sum over $k \geq 1$ of the right hand side of (16.45). □

Proof of Theorem 16.3. The proof is split into two steps. First we show that $N^{-1/2}\sum_{i=1}^N X_i(t)$ is close to $N^{-1/2}\sum_{i=1}^N X_i^{(m)}(t)$, if m is sufficiently large. Then we establish the claim for m-dependent functions, for any $m \geq 1$.

As the first step, we show that

$$\limsup_{m\to\infty}\limsup_{N\to\infty} E\int\left[N^{-1/2}\sum_{i=1}^N\left(X_i(t) - X_i^{(m)}(t)\right)\right]^2 dt = 0, \tag{16.46}$$

where the variables $X_i^{(m)}$ are defined in (16.6). By stationarity,

$$\begin{aligned}
E&\left[\sum_{1\leq i\leq N}\left(X_i(t) - X_i^{(m)}(t)\right)\right]^2 \\
&= \sum_{1\leq i\leq N}\sum_{1\leq j\leq N} E\left(X_i(t) - X_i^{(m)}(t)\right)\left(X_j(t) - X_j^{(m)}(t)\right) \\
&= NE\left(X_0(t) - X_0^{(m)}(t)\right)^2 \\
&\quad + 2\sum_{1\leq i<j\leq N} E\left(X_i(t) - X_i^{(m)}(t)\right)\left(X_j(t) - X_j^{(m)}(t)\right).
\end{aligned}$$

In the proof, we will repeatedly use independence relations which follow from representation (16.6). First observe that if $j > i$, then $(X_i, X_i^{(m)})$ is independent of $X_j^{(j-i)}$ because

$$X_j^{(j-i)} = f(\varepsilon_j, \ldots, \varepsilon_{i+1}, \varepsilon_{j,i}^{(j-i)}, \varepsilon_{j,i-1}^{(j-i)}, \ldots).$$

Consequently, $E\left(X_i(t) - X_i^{(m)}(t)\right)X_j^{(j-i)}(t) = 0$, and so

$$\begin{aligned}
&\sum_{1\leq i<j\leq N} E\left(X_i(t) - X_i^{(m)}(t)\right)X_j(t) \\
&\quad = \sum_{1\leq i<j\leq N} E\left(X_i(t) - X_i^{(m)}(t)\right)\left(X_j(t) - X_j^{(j-i)}(t)\right).
\end{aligned}$$

Using the Cauchy-Schwarz inequality, we conclude that

$$\left| \int \sum_{1 \le i < j \le N} E\left(X_i(t) - X_i^{(m)}(t)\right)\left(X_j(t) - X_j^{(j-i)}(t)\right) dt \right|$$
$$\le \sum_{1 \le i < j \le N} \int \left[E\left(X_i(t) - X_i^{(m)}(t)\right)^2\right]^{\frac{1}{2}} \left[E\left(X_j(t) - X_j^{(j-i)}(t)\right)^2\right]^{\frac{1}{2}} dt$$
$$\le \sum_{1 \le i < j \le N} \left[\int E\left(X_i(t) - X_i^{(m)}(t)\right)^2 dt\right]^{\frac{1}{2}} \left[\int E\left(X_j(t) - X_j^{(j-i)}(t)\right)^2 dt\right]^{\frac{1}{2}}$$
$$= \sum_{1 \le i < j \le N} \left[\int E\left(X_0(t) - X_0^{(m)}(t)\right)^2 dt\right]^{\frac{1}{2}} \left[\int E\left(X_0(t) - X_0^{(j-i)}(t)\right)^2 dt\right]^{\frac{1}{2}}$$
$$\le N \left[\int E\left(X_0(t) - X_0^{(m)}(t)\right)^2 dt\right]^{\frac{1}{2}} \sum_{k \ge 1} \left[\int \left(X_0(t) - X_0^{(k)}(t)\right)^2 dt\right]^{\frac{1}{2}}$$
$$= N \nu_2\left(X_0 - X_0^{(m)}\right) \sum_{k \ge 1} \nu_2\left(X_0 - X_0^{(k)}\right).$$

Hence, by (16.4),

$$\limsup_{m \to \infty} \limsup_{N \to \infty} \frac{1}{N} \left| \int \sum_{1 \le i < j \le N} E\left[\left(X_i^{(m)}(t) - X_i(t)\right) X_j(t)\right] dt \right| = 0.$$

Similar arguments give

$$\limsup_{m \to \infty} \limsup_{N \to \infty} \frac{1}{N} \left| \int \sum_{1 \le i < j \le N} E\left[\left(X_i^{(m)}(t) - X_i(t)\right) X_j^{(m)}(t)\right] dt \right| = 0.$$

Completing the verification of (16.46).

The next the step is to show that $N^{-1/2} \sum_{1 \le i \le N} X_i^{(m)}$ converges to a Gaussian process Z_m with covariances defined analogously to (16.14). Recall that for every integer $m \ge 1$, $\{X_i^{(m)}\}$ is an m–dependent sequence of functions. To lighten the notation, in the remainder of the proof, we fix m and denote sequence $\{X_i^{(m)}\}$ by $\{X_i\}$, so $\{X_i\}$ is now m–dependent.

Let $K > 1$ be an integer and let the v_i be the orthonormal eigenfunctions of the integral operator with the kernel

$$E[X_0(t)X_0(s)] + \sum_{i=1}^{m} E[X_0(t)X_i(s)] + \sum_{i=1}^{m} E[X_0(s)X_i(t)].$$

The corresponding eigenvalues are denoted by λ_i. Then, by the Karhunen-Loéve expansion, we have

$$X_i(t) = \sum_{\ell \ge 1} \langle X_i, v_\ell \rangle v_\ell(t).$$

Next we define

$$X_i^{(K)}(t) = \sum_{1 \le \ell \le K} \langle X_i, v_\ell \rangle \, v_\ell(t).$$

By the triangle inequality we have that

$$\left\{ E \int \left[\sum_{1 \le i \le N} \left((X_i(t) - X_i^{(K)}(t) \right) \right]^2 dt \right\}^{1/2}$$
$$\le \left\{ E \int \left[\sum_{i \in V(0)} \left((X_i(t) - X_i^{(K)}(t) \right) \right]^2 dt \right\}^{1/2} + \cdots$$
$$+ \left\{ E \int \left[\sum_{i \in V(m-1)} \left((X_i(t) - X_i^{(K)}(t) \right) \right]^2 dt \right\}^{1/2},$$

where $V(k) = \{i : 1 \le i \le N, i = k \pmod{m}\}, 0 \le k \le m-1$. Due to the m dependence of the sequence $\{X_i\}$, $\sum_{i \in V(k)} (X_i(t) - X_i^{(K)}(t))$ is a sum of independent, identically distributed random variables, and thus we get

$$E \int \left[\sum_{i \in V(m-1)} \left((X_i(t) - X_i^{(K)}(t) \right) \right]^2 dt \le N \sum_{\ell \ge K} E \langle X_0, v_\ell \rangle^2 .$$

Utilizing

$$\lim_{K \to \infty} \sum_{\ell \ge K} E \langle X_0, v_\ell \rangle^2 = 0$$

we conclude that for any $r > 0$

$$\limsup_{K \to \infty} \limsup_{N \to \infty} P \left\{ \int \left[\frac{1}{N^{1/2}} \sum_{1 \le i \le N} \left((X_i(t) - X_i^{(K)}(t) \right) \right]^2 dt > r \right\} = 0.$$

The sum of the $X_i^{(K)}$'s can be written as

$$\frac{1}{N^{1/2}} \sum_{1 \le i \le N} X_i^{(K)}(t) = \sum_{1 \le \ell \le K} v_\ell(t) \frac{1}{N^{1/2}} \sum_{1 \le i \le N} \langle X_i, v_\ell \rangle .$$

Next, we use the central limit theorem for stationary m-dependent sequences of random vectors (see Lehmann (1999) and the Cramér-Wold theorems in DasGupta (2008), pages 9 and 120)) and get that

$$\left\{ \frac{1}{N^{1/2}} \sum_{1 \le i \le N} \langle X_i, v_\ell \rangle , 1 \le \ell \le K \right\}^T \xrightarrow{d} \mathbf{N}_K(\mathbf{0}, \boldsymbol{\Delta}_K),$$

where $\mathbf{N}_K(\mathbf{0}, \boldsymbol{\Delta}_K)$ is a K-dimensional normal random variable with zero mean and covariance matrix $\boldsymbol{\Delta}_K = \mathrm{diag}(\lambda_1, \ldots, \lambda_K)$. Thus we proved that for all $K > 1$

$$N^{-1/2} \sum_{1 \le i \le N} X_i^{(K)}(t) \xrightarrow{d} \sum_{1 \le \ell \le K} \lambda_\ell^{1/2} N_\ell v_\ell(t) \quad \text{in } L^2,$$

where $N_i, i \ge 1$ are independent standard normal random variables. It is easy to see that

$$\int \left(\sum_{K < \ell < \infty} \lambda_\ell^{1/2} N_\ell v_\ell(t) \right)^2 dt = \sum_{K < \ell < \infty} \lambda_\ell N_\ell^2 \xrightarrow{P} 0,$$

as $K \to \infty$. Thus we have the convergence of $N^{-1/2} \sum_{1 \le i \le N} X_i$ for any m and therefore the proof of the theorem is now complete. □

Proof of Theorem 16.4. As in Giraitis *et al.* (2003), set $\mu = EX_0$ and

$$\tilde{\gamma}_j = \frac{1}{N} \sum_{i=1}^{N-|j|} (X_i - \mu)(X_{i+|j|} - \mu),$$

$$S_{k,\ell} = \sum_{i=k}^{\ell} (X_i - \mu).$$

Observe that

$$\hat{\gamma}_j - \tilde{\gamma}_j = \left(1 - \frac{|j|}{N}\right)(\bar{X}_N - \mu)^2 + \frac{1}{N}(\bar{X}_N - \mu)(S_{1,N-|j|} + S_{|j|+1,N}) =: \delta_j.$$

We therefore have the decomposition

$$\hat{\sigma}^2 = \sum_{|j| \le q} \omega_q(j)\tilde{\gamma}_j + \sum_{|j| \le q} \omega_q(j)\delta_j =: \hat{\sigma}_1^2 + \hat{\sigma}_2^2.$$

The proof will be complete once we have shown that

$$\hat{\sigma}_1^2 \xrightarrow{P} \sum_{j=-\infty}^{\infty} \gamma_j \tag{16.47}$$

and

$$\hat{\sigma}_2^2 \xrightarrow{P} 0. \tag{16.48}$$

We begin with the verification of the easier relation (16.48). By (16.15),

$$E|\hat{\sigma}_2^2| \le b \sum_{|j| \le q} E|\delta_j| \le b \sum_{|j| \le q} E(\bar{X}_N - \mu)^2$$
$$+ \frac{b}{N} \left[E(\bar{X}_N - \mu)^2\right]^{1/2} \sum_{|j| \le q} \left[E(S_{1,N-|j|} + S_{|j|+1,N})^2\right]^{1/2}.$$

By Lemma 16.2,

$$E(\bar{X}_N - \mu)^2 = \frac{1}{N} \sum_{|j| \le N} \left(1 - \frac{|j|}{N}\right) \gamma_j = O(N^{-1}).$$

Similarly $E(S_{1,N-|j|} + S_{|j|+1,N})^2 = O(N)$. Therefore,

$$E|\hat{\sigma}_2^2| = O(qN^{-1} + N^{-1}N^{-1/2}qN^{1/2}) = O(q/N).$$

We now turn to the verification of (16.47). We will show that $E\hat{\sigma}_1^2 \to \sum_j \gamma_j$ and $\mathrm{Var}[\hat{\sigma}_1^2] \to 0$.

By (16.16),

$$E\hat{\sigma}_1^2 = \sum_{|j| \le q} \omega_q(j) \frac{N - |j|}{N} \gamma_j \to \sum_{j=-\infty}^{\infty} \gamma_j.$$

By (16.15), it remains to show that

$$\sum_{|k|,|\ell| \le q} |\mathrm{Cov}(\tilde{\gamma}_k, \tilde{\gamma}_\ell)| \to 0. \tag{16.49}$$

To lighten the notation, without any loss of generality, *we assume from now on that* $\mu = 0$, so that

$$\mathrm{Cov}(\tilde{\gamma}_k, \tilde{\gamma}_\ell) = \frac{1}{N^2} \mathrm{Cov}\left(\sum_{i=1}^{N-|k|} X_i X_{i+|k|}, \sum_{j=1}^{N-|\ell|} X_j X_{j+|\ell|} \right).$$

Therefore, by stationarity,

$$\begin{aligned} |\mathrm{Cov}(\tilde{\gamma}_k, \tilde{\gamma}_\ell)| &\le \frac{1}{N^2} \sum_{i,j=1}^{N} \left|\mathrm{Cov}\left(X_i X_{i+|k|}, X_j X_{j+|\ell|}\right)\right| \\ &= \frac{1}{N} \sum_{|r|<N} \left(1 - \frac{|r|}{N}\right) \left|\mathrm{Cov}\left(X_0 X_{|k|}, X_r X_{r+|\ell|}\right)\right|. \end{aligned}$$

The last sum can be split into three terms corresponding to $r = 0$, $r < 0$ and $r > 0$.

The contribution to the left–hand side of (16.49) of the term corresponding to $r = 0$ is

$$N^{-1} \sum_{|k|,|\ell| \le q} \left|\mathrm{Cov}\left(X_0 X_{|k|}, X_0 X_{|\ell|}\right)\right| = O(q^2/N).$$

The terms corresponding to $r < 0$ and $r > 0$ are handled in the same way, so we focus on the contribution of the summands with $r > 0$ which is

$$N^{-1} \sum_{|k|,|\ell| \le q} \sum_{r=1}^{N-1} \left(1 - \frac{r}{N}\right) \left|\operatorname{Cov}\left(X_0 X_{|k|}, X_r X_{r+|\ell|}\right)\right|.$$

We now use the decompositions

$$\begin{aligned}
&\operatorname{Cov}\left(X_0 X_{|k|}, X_r X_{r+|\ell|}\right) \\
&\quad = \operatorname{Cov}\left(X_0 X_{|k|}, X_r^{(r)} X_{r+|\ell|}^{(r+|\ell|)}\right) + \operatorname{Cov}\left(X_0 X_{|k|}, X_r X_{r+|\ell|} - X_r^{(r)} X_{r+|\ell|}^{(r+|\ell|)}\right)
\end{aligned}$$

and

$$\begin{aligned}
&\operatorname{Cov}\left(X_0 X_{|k|}, X_r^{(r)} X_{r+|\ell|}^{(r+|\ell|)}\right) \\
&\quad = \operatorname{Cov}\left(X_0 X_{|k|}^{(|k|)}, X_r^{(r)} X_{r+|\ell|}^{(r+|\ell|)}\right) + \operatorname{Cov}\left(X_0 (X_{|k|} - X_{|k|}^{(|k|)}), X_r^{(r)} X_{r+|\ell|}^{(r+|\ell|)}\right).
\end{aligned}$$

By Definition 16.1, X_0 depends on $\varepsilon_0, \varepsilon_{-1}, \ldots,$ while the random variables $X_{|k|}^{(k)}, X_r^{(r)}$ and $X_{r+|\ell|}^{(r+|\ell|)}$ depend on $\varepsilon_1, \varepsilon_2, \ldots, \varepsilon_{k \vee (r+|\ell|)}$ and errors independent of the ε_i. Therefore $\operatorname{Cov}\left(X_0 X_{|k|}^{(|k|)}, X_r^{(r)} X_{r+|\ell|}^{(r+|\ell|)}\right)$ is equal to

$$\begin{aligned}
&E\left[X_0 X_{|k|}^{(|k|)} X_r^{(r)} X_{r+|\ell|}^{(r+|\ell|)}\right] - E\left[X_0 X_{|k|}\right] E\left[X_r^{(r)} X_{r+|\ell|}^{(r+|\ell|)}\right] \\
&\quad = E[X_0] E\left[X_{|k|}^{(|k|)} X_r^{(r)} X_{r+|\ell|}^{(r+|\ell|)}\right] - E[X_0] E\left[X_{|k|}^{(|k|)}\right]\left[X_r^{(r)} X_{r+|\ell|}^{(r+|\ell|)}\right] = 0.
\end{aligned}$$

We thus obtain

$$\begin{aligned}
\operatorname{Cov}\left(X_0 X_{|k|}, X_r X_{r+|\ell|}\right) &= \operatorname{Cov}\left(X_0 (X_{|k|} - X_{|k|}^{(|k|)}), X_r^{(r)} X_{r+|\ell|}^{(r+|\ell|)}\right) \\
&\quad + \operatorname{Cov}\left(X_0 X_{|k|}, X_r X_{r+|\ell|} - X_r^{(r)} X_{r+|\ell|}^{(r+|\ell|)}\right).
\end{aligned}$$

By assumption (16.19), it remains to verify that

$$N^{-1} \sum_{|k|,|\ell| \le q} \sum_{r=1}^{N-1} \left|\operatorname{Cov}\left(X_0 X_{|k|}, X_r X_{r+|\ell|} - X_r^{(r)} X_{r+|\ell|}^{(r+|\ell|)}\right)\right| \to 0.$$

This is done using the technique introduced in the proof of Theorem 16.1. By the Cauchy-Schwarz inequality, the problem reduces to showing that

$$N^{-1} \sum_{|k|,|\ell| \le q} \sum_{r=1}^{N-1} \left\{E[X_0^2 X_{|k|}^2]\right\}^{1/2} \left\{E\left[\left(X_r X_{r+|\ell|} - X_r^{(r)} X_{r+|\ell|}^{(r+|\ell|)}\right)^2\right]\right\}^{1/2} \to 0.$$

Using (16.65), this in turn is bounded by constant times

$$N^{-1} \sum_{|k|,|\ell|\le q} \sum_{r=1}^{\infty} \left\{ E\left[X_r - X_r^{(r)}\right]^4 \right\}^{1/4},$$

which tends to zero by L^4–m–approximability and the condition $q^2/N \to 0$. □

Proof of Proposition 16.3. We only prove the left relation in (16.27). The element in the k-th row and ℓ-th column of $\hat{\boldsymbol{\Sigma}}(\boldsymbol{\beta}) - \hat{\boldsymbol{\Sigma}}(\hat{\mathbf{C}}\hat{\boldsymbol{\beta}})$ is given by

$$\begin{aligned} &\sum_{h=0}^{q} \frac{w_q(h)}{N} \sum_{1\le n\le N-h} \left(\beta_{kn}\beta_{\ell,n+h} - \hat{c}_k\hat{\beta}_{kn}\hat{c}_\ell\hat{\beta}_{\ell,n+h}\right) \\ &\quad + \sum_{h=1}^{q} \frac{w_q(h)}{N} \sum_{1\le n\le N-h} \left(\beta_{k,n+h}\beta_{\ell,n} - \hat{c}_k\hat{\beta}_{k,n+h}\hat{c}_\ell\hat{\beta}_{\ell,n}\right). \end{aligned} \tag{16.50}$$

For reasons of symmetry it suffices to study (16.50), which can be decomposed into

$$\begin{aligned} &\sum_{h=0}^{q} \frac{w_q(h)}{N} \sum_{1\le n\le N-h} \beta_{kn}\left(\beta_{\ell,n+h} - \hat{c}_\ell\hat{\beta}_{\ell,n+h}\right) \\ &\quad + \sum_{h=0}^{q} \frac{w_q(h)}{N} \sum_{1\le n\le N-h} \hat{c}_\ell\hat{\beta}_{\ell,n+h}\left(\beta_{kn} - \hat{c}_k\hat{\beta}_{kn}\right). \end{aligned} \tag{16.51}$$

As both summands above can be treated similarly, we will only treat (16.51). For any $\varepsilon > 0$ we have

$$\begin{aligned} &P\left(\left|\sum_{h=0}^{q} \frac{w_q(h)}{N} \sum_{1\le n\le N-h} \beta_{kn}\left(\beta_{\ell,n+h} - \hat{c}_\ell\hat{\beta}_{\ell,n+h}\right)\right| > \varepsilon\kappa\right) \\ &\quad \le P\left(\left|\sum_{h=0}^{q} \frac{w_q(h)}{N} \sum_{1\le n\le N-h} \beta_{kn}\left(\beta_{\ell,n+h} - \hat{c}_\ell\hat{\beta}_{\ell,n+h}\right)\right| > \frac{\varepsilon}{q}\sum_{h=0}^{q} w_q(h)\right) \\ &\quad \le \sum_{h=0}^{q} P\left(\frac{1}{N}\left|\sum_{1\le n\le N-h} \beta_{kn}\left(\beta_{\ell,n+h} - \hat{c}_\ell\hat{\beta}_{\ell,n+h}\right)\right| > \frac{\varepsilon}{q}\right). \end{aligned} \tag{16.52}$$

In order to show that (16.52) tends to 0 as $N \to \infty$, we introduce a slowly increasing sequence $\alpha_N \to \infty$ such that $q^4\alpha_N/N \to 0$ and we let C_0 such that $N \max_{1\le\ell\le d} E\|v_\ell - \hat{c}_\ell\hat{v}_\ell\|^2 \le C_0$. By Cauchy-Schwarz and Markov inequality we have

$$
\begin{aligned}
&P\left(\left|\sum_{1\le n\le N-h}\beta_{kn}\left(\beta_{\ell,n+h}-\hat{c}_\ell\hat{\beta}_{\ell,n+h}\right)\right|>\frac{\varepsilon N}{q}\right)\\
&\le P\left(\sum_{n=1}^{N}\beta_{kn}^2\sum_{n=1}^{N}\left(\beta_{\ell n}-\hat{c}_\ell\hat{\beta}_{\ell n}\right)^2>\frac{\varepsilon^2 N^2}{q^2}\right)\\
&\le P\left(\frac{1}{N}\sum_{n=1}^{N}\beta_{kn}^2>q\alpha_N\right)+P\left(\frac{1}{N}\sum_{n=1}^{N}\left(\beta_{\ell n}-\hat{c}_\ell\hat{\beta}_{\ell n}\right)^2>\frac{\varepsilon^2}{q^3\alpha_N}\right)\\
&\le \frac{E\beta_{k1}^2}{q\alpha_N}+P\left(\frac{1}{N}\sum_{n=1}^{N}\|Y_n\|^2\|v_\ell-\hat{c}_\ell\hat{v}_\ell\|^2>\frac{\varepsilon^2}{q^3\alpha_N}\right)\\
&\le \frac{E\|Y_1\|^2}{q\alpha_N}+P\left(\frac{1}{N}\sum_{n=1}^{N}\|Y_n\|^2>2E\|Y_1\|^2\right)\\
&\quad+P\left(\|v_\ell-\hat{c}_\ell\hat{v}_\ell\|^2>\frac{\varepsilon^2}{2E\|Y_1\|^2q^3\alpha_N}\right)\\
&\le \frac{E\|Y_1\|^2}{q\alpha_N}+\frac{\operatorname{Var}\left(\frac{1}{N}\sum_{n=1}^{N}\|Y_n\|^2\right)}{E^2\|Y_1\|^2}+\frac{2C_0\,E\|Y_1\|^2q^3\alpha_N}{N\varepsilon^2}.
\end{aligned}
$$

It can be easily shown that for U, V in L_H^4

$$
\nu_2\left(\|U\|^2-\|V\|^2\right)\le \nu_4^2(U-V)+2\left\{\nu_4(U)+\nu_4(V)\right\}\nu_4(U-V).
$$

An immediate consequence is that L^4–m–approximability of $\{Y_n\}$ implies L^2–m–approximability of the scalar sequence $\{\|Y_n\|^2\}$. A basic result for stationary sequences gives

$$
\operatorname{Var}\left(\frac{1}{N}\sum_{n=1}^{N}\|Y_n\|^2\right)\le\frac{1}{N}\sum_{h\in\mathbb{Z}}\left|\operatorname{Cov}\left(\|Y_0\|^2,\|Y_h\|^2\right)\right|,
$$

where the by Lemma 16.2 the autocovariances are absolutely summable. Hence the summands in (16.52) are bounded by

$$
C_1\left\{\frac{1}{q\alpha_N}+\frac{1}{N}+\frac{q^3\alpha_N}{N\varepsilon^2}\right\},
$$

where the constant C_1 depends only on the law of $\{Y_n\}$. The proof of the proposition follows immediately from our assumptions on q and α_N. □

16.9 Proof of Theorem 16.6

The proof of this result needs some preliminary lemmas, which we establish first. We can assume without loss of generality that $K(u) = 0$ if $|u| > 1$. Let m be a positive integer and recall that $\mathbf{X}_\ell^{(m)}$ is defined in Assumption 16.4. The long term covariance matrix associated with the stationary sequence $\{\mathbf{X}_\ell^{(m)}, 1 \le \ell < \infty\}$ is given by

$$\boldsymbol{\Sigma}^{(m)} = E\mathbf{X}_1^{(m)}(\mathbf{X}_1^{(m)})^T + \sum_{\ell=1}^{\infty} E\mathbf{X}_1^{(m)}(\mathbf{X}_{\ell+1}^{(m)})^T + \sum_{\ell=1}^{\infty} E\mathbf{X}_{\ell+1}^{(m)}(\mathbf{X}_1^{(m)})^T.$$

The corresponding Bartlett estimator is defined as

$$\boldsymbol{\Sigma}_N^{(m)} = \sum_{k=-(N-1)}^{N-1} K(k/B_N)\hat{\boldsymbol{\Gamma}}_k^{(m)},$$

where

$$\hat{\boldsymbol{\Gamma}}_k^{(m)} = \frac{1}{N} \sum_{\ell=\max(1,1-k)}^{\min(N,N-k)} \mathbf{X}_\ell^{(m)}(\mathbf{X}_{\ell+k}^{(m)})^T$$

are the sample covariances of lag k. Since K is symmetric, $K(0) = 1$ and $K(u) = 0$ outside $[-1, 1]$, we have that, for all sufficiently large N,

$$\boldsymbol{\Sigma}_N^{(m)} = \hat{\boldsymbol{\Gamma}}_0^{(m)} + \sum_{k=1}^{B_N} K(k/B_N)\hat{\boldsymbol{\Gamma}}_k^{(m)} + \sum_{k=1}^{B_N} K(k/B_N)(\hat{\boldsymbol{\Gamma}}_k^{(m)})^T.$$

We start with the consistency of $\boldsymbol{\Sigma}_N^{(m)}$.

Lemma 16.3. *If Assumptions 16.2-16.6 are satisfied, then we have for every m,*

$$\boldsymbol{\Sigma}_N^{(m)} \xrightarrow{P} \boldsymbol{\Sigma}^{(m)},$$

as $N \to \infty$.

Proof. Since the sequence $\mathbf{X}_\ell^{(m)}$ is m-dependent we have that

$$\boldsymbol{\Sigma}^{(m)} = E\mathbf{X}_1\mathbf{X}_1^T + \sum_{\ell=1}^{m} E\mathbf{X}_1\mathbf{X}_{\ell+1}^T + \sum_{\ell=1}^{m} E\mathbf{X}_{\ell+1}\mathbf{X}_1^T.$$

It follows from the ergodic theorem that for any fixed k and m

$$\hat{\boldsymbol{\Gamma}}_k^{(m)} \xrightarrow{P} E\mathbf{X}_1^{(m)}(\mathbf{X}_{1+k}^{(m)})^T.$$

So using Assumptions 16.5(i), 16.5(ii) and 16.6 we get that

$$\hat{\boldsymbol{\Gamma}}_0^{(m)} + \sum_{k=1}^{m} K(k/B_N)\hat{\boldsymbol{\Gamma}}_k^{(m)} + \sum_{k=1}^{m} K(k/B_N)(\hat{\boldsymbol{\Gamma}}_k^{(m)})^T$$

$$\xrightarrow{P} E\mathbf{X}_1\mathbf{X}_1^T + \sum_{\ell=1}^{m} E\mathbf{X}_1\mathbf{X}_{\ell+1}^T + \sum_{\ell=1}^{m} E\mathbf{X}_{\ell+1}\mathbf{X}_1^T.$$

Lemma 16.3 is proven if we show that

$$\sum_{k=m+1}^{B_N} K(k/B_N)\hat{\boldsymbol{\Gamma}}_k^{(m)} \xrightarrow{P} 0 \tag{16.53}$$

and

$$\sum_{k=m+1}^{B_N} K(k/B_N)(\hat{\boldsymbol{\Gamma}}_k^{(m)})^T \xrightarrow{P} 0. \tag{16.54}$$

Clearly, it is enough to prove (16.53).

Let

$$\mathbf{E}_N^{(m)} = \sum_{k=m+1}^{B_N} K(k/B_N)\hat{\boldsymbol{\Gamma}}_k^{(m)}.$$

Elementary arguments show that

$$\begin{aligned}
\mathbf{E}_N^{(m)} &= \sum_{k=m+1}^{B_N} K(k/B_N)\hat{\boldsymbol{\Gamma}}_k^{(m)} \\
&= \sum_{k=m+1}^{B_N} K(k/B_N)\frac{1}{N}\sum_{\ell=1}^{N-k} \mathbf{X}_\ell^{(m)}\left(\mathbf{X}_{\ell+k}^{(m)}\right)^T \\
&= \sum_{\ell=1}^{N-(m+1)} \mathbf{X}_\ell^{(m)}\mathbf{H}_{\ell,N}^{(m)},
\end{aligned}$$

where

$$\mathbf{H}_{\ell,N}^{(m)} = \sum_{k=m+1}^{\min(N-\ell,B_N)} \frac{K(k/B_N)}{N}\left(\mathbf{X}_{\ell+k}^{(m)}\right)^T.$$

Let

$$E_N^{(m)}(i,j) = \sum_{\ell=1}^{N-(m+1)} X_{i\ell}^{(m)}H_{\ell,N}^{(m)}(j), \quad 1 \le i,j \le pq,$$

where $X_{i\ell}^{(m)}$ and $H_{\ell,N}^{(m)}(j)$ are the ith and the jth coordinates of the vectors $\mathbf{X}_{\ell,N}^{(m)}$ and $\mathbf{H}_{\ell,N}^{(m)}$, respectively. Next we write

$$\begin{aligned}
E\left(E_N^{(m)}(i,j)\right)^2 &= E\left(\sum_{\ell=1}^{N-(m+1)} X_{i\ell}^{(m)}H_{\ell,N}^{(m)}(j)\right)^2 \\
&= \sum\sum_{\substack{1\le r\le N-(m+1)\\ 1\le \ell\le N-(m+1)}} E\left(H_{\ell,N}^{(m)}(j)X_{i\ell}^{(m)}X_{ir}^{(m)}H_{r,N}^{(m)}(j)\right) \\
&= E_{1,N}^{(m)}(i,j) + E_{2,N}^{(m)}(i,j),
\end{aligned}$$

where

$$E_{1,N}^{(m)}(i,j) = \sum\sum_{\substack{1\le r\le N-(m+1)\\ 1\le \ell\le N-(m+1)\\ |r-\ell|\le m}} E\left(H_{\ell,N}^{(m)}(j)X_{i\ell}^{(m)}X_{ir}^{(m)}H_{r,N}^{(m)}(j)\right),$$

and

$$E_{2,N}^{(m)}(i,j) = \sum\sum_{\substack{1\le r\le N-(m+1)\\ 1\le \ell\le N-(m+1)\\ |r-\ell|> m}} E\left(H_{\ell,N}^{(m)}(j)X_{i\ell}^{(m)}X_{ir}^{(m)}H_{r,N}^{(m)}(j)\right).$$

Notice that $\mathbf{X}_\ell^{(m)}$ is independent of $\mathbf{H}_{\ell,N}^{(m)}$, $\mathbf{H}_{r,N}^{(m)}$ and $\mathbf{X}_r^{(m)}$, if $r > m+\ell$. Hence

$$\begin{aligned}
&E\left(H_{\ell,N}^{(m)}(j)X_{i\ell}^{(m)}X_{ir}^{(m)}H_{r,N}^{(m)}(j)\right)\\
&= \begin{cases} EX_{i\ell}^{(m)}E\left(H_{\ell,N}^{(m)}(j)X_{ir}^{(m)}H_{r,N}^{(m)}(j)\right) & r > m+\ell,\\ EX_{ir}^{(m)}E\left(H_{\ell,N}^{(m)}(j)X_{i\ell}^{(m)}H_{r,N}^{(m)}(j)\right) & \ell > m+r,\\ E\left(H_{\ell,N}^{(m)}(j)X_{i\ell}^{(m)}X_{ir}^{(m)}H_{r,N}^{(m)}(j)\right) & |\ell-r|\le m, \end{cases}\\
&= \begin{cases} 0 & |\ell-r| > m,\\ E\left(H_{\ell,N}^{(m)}(j)X_{i\ell}^{(m)}X_{ir}^{(m)}H_{r,N}^{(m)}(j)\right) & |\ell-r|\le m. \end{cases}
\end{aligned}$$

Thus we have

$$EE_{2,N}^{(m)}(i,j) = 0.$$

Let M be an upper bound on $|K(t)|$. Using the fact that $\mathbf{X}_\ell^{(m)}$ is an m-dependent sequence, we now obtain the following:

$$E(H_{\ell,N}^{(m)}(j))^2 = \sum_{k=m+1}^{\min(N-\ell,B_N)}\sum_{v=m+1}^{\min(N-\ell,B_N)} \frac{K(k/B_N)}{N}\frac{K(v/B_N)}{N}E\left(X_{j,\ell+k}^{(m)}X_{j,\ell+v}^{(m)}\right) \tag{16.55}$$

$$\begin{aligned}
&\le \frac{M^2}{N^2}\sum_{k=m+1}^{\min(N-\ell,B_N)}\sum_{v=m+1}^{\min(N-\ell,B_N)} E\left(X_{j,\ell+k}^{(m)}X_{j,\ell+v}^{(m)}\right)\\
&\le \frac{M^2}{N^2}B_N\sum_{r=-m}^{m} E\left|X_{j0}^{(m)}X_{jr}^{(m)}\right|\\
&= O\left(\frac{B_N}{N^2}\right).
\end{aligned}$$

In the next step we will first use the Cauchy-Schwarz inequality, then the independence of $H_{\ell,N}^{(m)}(j)$ and $X_{i\ell}^{(m)}$ and the independence of $H_{r,N}^{(m)}(j)$ and $X_{ir}^{(m)}$ to get

$$\begin{aligned}
\left|E_{2,N}^{(m)}(i,j)\right| &\le \sum_{\substack{1\le r\le N-(m+1)\\ 1\le\ell\le N-(m+1)\\ |r-\ell|\le m}}\sum E\left|H_{\ell,N}^{(m)}(j)X_{i\ell}^{(m)}X_{ir}^{(m)}H_{r,N}^{(m)}(j)\right| \\
&\le \sum_{\substack{1\le r\le N-(m+1)\\ 1\le\ell\le N-(m+1)\\ |r-\ell|\le m}}\sum \left(E\left(H_{\ell,N}^{(m)}(j)X_{i\ell}^{(m)}\right)^2\right)^{1/2}\left(E\left(X_{ir}^{(m)}H_{r,N}^{(m)}(j)\right)^2\right)^{1/2} \\
&\le \sum_{\substack{1\le r\le N-(m+1)\\ 1\le\ell\le N-(m+1)\\ |r-\ell|\le m}}\sum \left(E\left(H_{\ell,N}^{(m)}(j)\right)^2\right)^{1/2}\left(E\left(X_{i\ell}^{(m)}\right)^2\right)^{1/2}\left(E\left(X_{ir}^{(m)}\right)^2\right)^{1/2} \\
&\qquad\times\left(E\left(H_{r,N}^{(m)}(j)\right)^2\right)^{1/2} \\
&\le 2mNO\left(\frac{B_N^{1/2}}{N}\right)O(1)O(1)O\left(\frac{B_N^{1/2}}{N}\right) \\
&= O\left(\frac{B_N}{N}\right) \\
&= o(1),
\end{aligned}$$

where we also used (16.55) and Assumption 16.6. This completes the proof of Lemma 16.3. □

In the following, i denotes the imaginary unit, i.e. $i^2 = -1$.

Lemma 16.4. *If Assumptions 16.2-16.6 are satisfied, then for all* $1 \le j \le d$ *we have*

$$\limsup_{N\to\infty}\limsup_{m\to\infty}\sup_{-\infty<t<\infty} E\left(\frac{1}{N^{1/2}}\sum_{k=1}^{N}(X_{jk}-X_{jk}^{(m)})e^{ikt}\right)^2 = 0, \qquad (16.56)$$

$$\limsup_{N\to\infty}\limsup_{m\to\infty}\sup_{-\infty<t<\infty} E\left(\frac{1}{N^{1/2}}\sum_{k=1}^{N}X_{jk}e^{ikt}\right)^2 < \infty \qquad (16.57)$$

and

$$\limsup_{N\to\infty}\limsup_{m\to\infty}\sup_{-\infty<t<\infty} E\left(\frac{1}{N^{1/2}}\sum_{k=1}^{N}X_{jk}^{(m)}e^{ikt}\right)^2 < \infty. \qquad (16.58)$$

Proof. First we note that

$$E\left(\sum_{k=1}^{N}(X_{jk}-X_{jk}^{(m)})e^{ikt}\right)^2 = \sum_{1\le k\le N} E((X_{jk}-X_{jk}^{(m)})e^{ikt})^2$$
$$+2\sum_{1\le k<\ell\le N} E\left[(X_{jk}-\gamma_{jk}^{(m)})(X_{j\ell}-X_{j\ell}^{(m)})\right]e^{i(k+\ell)t}.$$

It follows from Assumption 16.4 that there is a sequence $c_1(m)\to 0$ such that

$$\left|\sum_{1\le k\le N} E(X_{jk}-X_{jk}^{(m)})^2 e^{i2kt}\right| \le Nc_1(m).$$

Next we write

$$\sum_{1\le k<\ell\le N} E\left[(X_{jk}-X_{jk}^{(m)})X_{j\ell}\right]e^{i(k+\ell)t}$$
$$=\sum_{1\le k<\ell\le N} E\left[(X_{jk}-X_{jk}^{(m)})(X_{j\ell}-X_{j\ell}^{\ell-k})\right]e^{i(k+\ell)t},$$

since $(\mathbf{X}_k, \mathbf{X}_k^{(m)})$ and $\mathbf{X}_\ell^{\ell-k}$ are independent. Using the Cauchy-Schwarz inequality first, then Assumption 16.4 again, we get that

$$\sum_{1\le k<\ell\le N}\left|E\left[(X_{jk}-X_{jk}^{(m)})(X_{j\ell}-X_{j\ell}^{(\ell-k)})\right]e^{i(k+\ell)t}\right|$$
$$\le \sum_{1\le k<\ell\le N}\left[E(X_{jk}-X_{jk}^{(m)})^2\right]^{1/2}\left[E(X_{j\ell}-X_{j\ell}^{(\ell-k)})^2\right]^{1/2}$$
$$\le N\left[E(X_{j1}-X_{j1}^{(m)})^2\right]^{1/2}\sum_{1\le k<\infty}\left[E(X_{j1}-X_{j1}^{(k)})^2\right]^{1/2}$$
$$= Nc_2(m)$$

with some sequence $c_2(m)\to 0$. Similar arguments show that

$$\sum_{1\le k<\ell\le N}\left|E\left[(X_{jk}-X_{jk}^{(m)})X_{j\ell}^{(m)}\right]e^{i(k+\ell)t}\right| = Nc_3(m),$$

with some sequence $c_3(m)\to 0$, completing the proof of (16.56).

Similarly to the proof of (16.56), we write

$$E\left(\sum_{k=1}^{N} X_{jk}e^{ikt}\right)^2 = \sum_{k=1}^{N}\sum_{\ell=1}^{N} EX_{jk}X_{j\ell}e^{i(k+\ell)t}$$
$$= \sum_{k=1}^{N} EX_{jk}^2 e^{2ikt} + 2\sum_{1\le k<\ell\le N} EX_{jk}X_{j\ell}e^{i(k+\ell)t}$$
$$= EX_{j1}^2 \sum_{k=1}^{N} e^{2ikt} + 2\sum_{1\le k<\ell\le N} EX_{jk}(X_{j\ell} - X_{j\ell}^{(\ell-k)})e^{i(k+\ell)t},$$

since by the independence of X_{jk} and $X_{j\ell}^{(\ell-k)}$ we have that $EX_{jk}X_{j\ell}^{(\ell-k)} = 0$. Using the Cauchy-Schwarz inequality with Assumption 16.4 we get that

$$\left|\sum_{1\le k<\ell\le N} EX_{jk}\left(X_{j\ell} - X_{j\ell}^{(\ell-k)}\right)e^{i(k+\ell)t}\right| \le cN$$

with some constant c, completing the proof of (16.57). The same arguments can be used to prove (16.58). □

Next we define $\mathbf{S}_N(t) = \sum_{k=1}^N \mathbf{X}_k e^{ikt}$ and $\mathbf{S}_N^{(m)}(t) = \sum_{k=1}^N \mathbf{X}_k^{(m)} e^{ikt}$. Let $\mathbf{S}_N^*(t)$ be the conjugate transpose of $\mathbf{S}_N(t)$ and introduce

$$\mathbf{I}_N(t) = \frac{1}{N}\mathbf{S}_N(t)\mathbf{S}_N^*(t)$$
$$= \frac{1}{N}\sum_{k=1}^{N}\mathbf{X}_k e^{ikt}\sum_{\ell=1}^{N}\mathbf{X}_\ell^T e^{-i\ell t}$$
$$= \frac{1}{N}\sum_{\ell=1}^{N}\sum_{k=1}^{N} e^{it(k-\ell)}\mathbf{X}_k\mathbf{X}_\ell^T$$
$$= \sum_{k=1-N}^{N-1} e^{-itk}\frac{1}{N}\sum_{\ell=\max(1,1-k)}^{\min(N,N-k)}\mathbf{X}_k\mathbf{X}_{\ell+k}^T$$
$$= \sum_{k=1-N}^{N-1} e^{-itk}\hat{\boldsymbol{\Gamma}}_k.$$

Similarly we define

$$\mathbf{I}_N^{(m)}(t) = \frac{1}{N}\mathbf{S}_N^{(m)}(t)\left(\mathbf{S}_N^{(m)}(t)\right)^* = \sum_{k=1-N}^{N-1} e^{-itk}\hat{\boldsymbol{\Gamma}}_k^{(m)}.$$

Lemma 16.5. *If Assumptions 16.2-16.6 are satisfied, then we have*

$$\limsup_{N\to\infty}\limsup_{m\to\infty}\sup_{-\infty<t<\infty} E\left|\mathbf{I}_N(t) - \mathbf{I}_N^{(m)}(t)\right| = 0.$$

Proof. By the triangle inequality we have

$$\left|\mathbf{I}_N(t) - \mathbf{I}_N^{(m)}(t)\right| = \left|\frac{1}{N}\mathbf{S}_N(t)\mathbf{S}_N^*(t) - \frac{1}{N}\mathbf{S}_N^{(m)}(t)\left(\mathbf{S}_N^{(m)}(t)\right)^*\right|$$
$$\leq \frac{1}{N}\left|\mathbf{S}_N(t)(\mathbf{S}_N^*(t) - (\mathbf{S}_N^{(m)}(t))^*)\right|$$
$$+ \frac{1}{N}\left|(\mathbf{S}_N(t) - \mathbf{S}_N^{(m)}(t))(\mathbf{S}_N^{(m)}(t))^*\right|.$$

Now the result follows from Lemma 16.4 via the Cauchy-Schwartz inequality. □

Proof of Theorem 16.6. Recall that the Fourier transform, $\hat{K}(u)$, of K is $\hat{K}(u) = \{2\pi\}^{-1}\int_{-\infty}^{\infty} K(s)e^{-isu}ds$. Since K and $\hat{K}$ are in L^1 and both are Lipschitz functions, the inversion formula gives $K(s) = \int_{-\infty}^{\infty}\hat{K}(u)e^{isu}du$. From the relationship between K and $\hat{K}$ and from the fact that K is supported on the interval $[-1, 1]$, we obtain:

$$\begin{aligned}
\boldsymbol{\Sigma}_N &= \sum_{k=-B_N}^{B_N} K(k/B_N)\hat{\boldsymbol{\Gamma}}_k \\
&= \sum_{k=1-N}^{N-1} K(k/B_N)\hat{\boldsymbol{\Gamma}}_k \\
&= \sum_{k=1-N}^{N-1}\left(\int_{-\infty}^{\infty}\hat{K}(u)e^{i(k/B_N)u}du\right)\hat{\boldsymbol{\Gamma}}_k \\
&= \int_{-\infty}^{\infty}\hat{K}(u)\sum_{k=1-N}^{N-1} e^{-i(-u/B_N)k}\hat{\boldsymbol{\Gamma}}_k du \\
&= \int_{-\infty}^{\infty}\hat{K}(u)\mathbf{I}_N(-u/B_N)du.
\end{aligned}$$

Similarly,

$$\boldsymbol{\Sigma}_N^{(m)} = \int_{-\infty}^{\infty}\hat{K}(u)\mathbf{I}_N^{(m)}(-u/B_N)du.$$

Hence we have

$$\begin{aligned}
E\left|\boldsymbol{\Sigma}_N - \boldsymbol{\Sigma}_N^{(m)}\right| &= E\left|\int_{-\infty}^{\infty}\hat{K}(u)\left(\mathbf{I}_N(u/B_N) - \mathbf{I}_N^{(m)}(u/B_N)\right)du\right| \\
&\leq \int_{-\infty}^{\infty}\left|\hat{K}(u)\right| E\left|\left(\mathbf{I}_N(u/B_N) - \mathbf{I}_N^{(m)}(u/B_N)\right)\right| du \\
&\leq \sup_{-\infty<t<\infty}\left\|\mathbf{I}_N(t) - \mathbf{I}_N^{(m)}(t)\right\|_1\int_{-\infty}^{\infty}\left|\hat{K}(u)\right| du.
\end{aligned}$$

Applying Lemma 16.5 we conclude that

$$\left|\boldsymbol{\Sigma}_N - \boldsymbol{\Sigma}_N^{(m)}\right| \xrightarrow{P} 0,$$

as $\min(N, m) \to \infty$. On the other hand, by Lemma 16.3, for every fixed m

$$\boldsymbol{\Sigma}_N^{(m)} \xrightarrow{P} \boldsymbol{\Sigma}^{(m)}.$$

Since

$$\boldsymbol{\Sigma}^{(m)} \to \boldsymbol{\Sigma},$$

as $m \to \infty$, the proof of the theorem is complete. □

16.10 Proofs of Theorems 16.7 and 16.8

The proof of Theorem 16.7 relies on Theorem A.1 of Aue *et al.* (2009), which we state here for ease of reference.

Theorem 16.10. *Suppose* $\{\boldsymbol{\xi}_n\}$ *is a* d*–dimensional* L^2*–*m*–approximable mean zero sequence. Then*

$$N^{-1/2}\mathbf{S}_N(\cdot, \boldsymbol{\xi}) \xrightarrow{d} \mathbf{W}(\boldsymbol{\xi})(\cdot), \tag{16.59}$$

where $\{\mathbf{W}(\boldsymbol{\xi})(x),\ x \in [0,1]\}$ *is a mean zero Gaussian process with covariances*

$$\mathrm{Cov}(\mathbf{W}(\boldsymbol{\xi})(x), \mathbf{W}(\boldsymbol{\xi})(y)) = \min(x, y)\boldsymbol{\Sigma}(\boldsymbol{\xi}).$$

The convergence in (16.59) *is in the* d*–dimensional Skorokhod space* $D_d([0,1])$.

Proof of Theorem 16.7. Let

$$G_N(x, \boldsymbol{\xi}) = \frac{1}{N}\mathbf{L}_n(x, \boldsymbol{\xi})^T \hat{\boldsymbol{\Sigma}}(\boldsymbol{\xi})^{-1}\mathbf{L}_n(x, \boldsymbol{\xi})^T.$$

We notice that replacing the $\mathbf{L}_N(x, \hat{\boldsymbol{\eta}})$ with $\mathbf{L}_N(x, \hat{\boldsymbol{\beta}})$ does not change the test statistic in (16.31). Furthermore, since by the second part of Proposition 16.3 $|\hat{\boldsymbol{\Sigma}}(\hat{\boldsymbol{\eta}}) - \hat{\boldsymbol{\Sigma}}(\hat{\boldsymbol{\beta}})| = o_P(1)$, it is enough to study the limiting behavior of the sequence $G_N(x, \hat{\boldsymbol{\beta}})$. This is done by first deriving the asymptotics of $G_N(x, \boldsymbol{\beta})$ and then analyzing the effect of replacing $\boldsymbol{\beta}$ with $\hat{\boldsymbol{\beta}}$.

Let $\boldsymbol{\beta}_i^{(m)}$ be the m-dependent approximations for $\boldsymbol{\beta}_i$ which are obtained by replacing $Y_i(t)$ in (16.25) by $Y_i^{(m)}(t)$. For a vector $\mathbf{v}$ in R^d we let $|\mathbf{v}|$ be its Euclidian norm. Then

$$\begin{aligned}
E|\boldsymbol{\beta}_1 - \boldsymbol{\beta}_1^{(m)}|^2 &= E\sum_{\ell=1}^{d}\left(\beta_{\ell 1} - \beta_{\ell 1}^{(m)}\right)^2 \\
&= \sum_{\ell=1}^{d} E\Big(\int (Y_1(t) - Y_1^{(m)}(t))v_\ell(t)dt\Big)^2 \\
&\le \sum_{\ell=1}^{d} E\int (Y_1(t) - Y_1^{(m)}(t))^2 dt \int v_\ell^2(t)dt \\
&= d\ \nu_2^2(Y_1 - Y_1^{(m)}).
\end{aligned}$$

Since by Lyapunov's inequality we have $\nu_2(Y_1 - Y_1^{(m)}) \le \nu_4(Y_1 - Y_1^{(m)})$, (16.4) yields that $\sum_{m\ge 1}(E|\boldsymbol{\beta}_1 - \boldsymbol{\beta}_1^{(m)}|^2)^{1/2} < \infty$. Thus Theorem 16.10 implies that

$$\frac{1}{\sqrt{N}}\mathbf{S}_N(x,\boldsymbol{\beta}) \stackrel{D^d[0,1]}{\longrightarrow} \mathbf{W}(\boldsymbol{\beta})(x).$$

The coordinatewise absolute convergence of the series $\boldsymbol{\Sigma}(\boldsymbol{\beta})$ follows from part (a) of Theorem 16.5. By assumption the estimator $\hat{\boldsymbol{\Sigma}}(\boldsymbol{\beta})$ is consistent and consequently

$$\int G_N(x,\boldsymbol{\beta})dx \stackrel{D[0,1]}{\longrightarrow} \sum_{\ell=1}^{d}\int B_\ell^2(x)dx$$

follows from the continuous mapping theorem.

We turn now to the effect of changing $G_N(x,\boldsymbol{\beta})$ to $G_N(x,\hat{\boldsymbol{\beta}})$. Due to the quadratic structure of $G_N(x,\boldsymbol{\xi})$, we have $G_N(x,\hat{\boldsymbol{\beta}}) = G_N(x,\hat{\mathbf{C}}\hat{\boldsymbol{\beta}})$ when $\hat{\mathbf{C}} = \mathrm{diag}(\hat{c}_1,\hat{c}_2,\ldots,\hat{c}_d)$. To finish the proof it is thus sufficient to show that

$$\sup_{x\in[0,1]}\frac{1}{\sqrt{N}}|\mathbf{S}_N(x,\boldsymbol{\beta}) - \mathbf{S}_N(x,\hat{\mathbf{C}}\hat{\boldsymbol{\beta}})| = o_P(1) \tag{16.60}$$

and

$$|\hat{\boldsymbol{\Sigma}}(\boldsymbol{\beta}) - \hat{\boldsymbol{\Sigma}}(\hat{\mathbf{C}}\hat{\boldsymbol{\beta}})| = o_P(1). \tag{16.61}$$

Relation (16.61) follows from Proposition 16.3. To show (16.60) we observe that by the Cauchy-Schwarz inequality and Theorem 16.2

$$\begin{aligned}
&\sup_{x\in[0,1]} \frac{1}{N}|\mathbf{S}_N(x,\boldsymbol{\beta}) - \mathbf{S}_N(x,\hat{\mathbf{C}}\hat{\boldsymbol{\beta}})|^2 \\
&= \sup_{x\in[0,1]} \frac{1}{N}\sum_{\ell=1}^{d}\left|\int \sum_{n=1}^{\lfloor Nx\rfloor} Y_n(t)(v_\ell(t) - \hat{c}_\ell\hat{v}_\ell(t))dt\right|^2 \\
&\le \frac{1}{N}\sup_{x\in[0,1]}\int\left(\sum_{n=1}^{\lfloor Nx\rfloor} Y_n(t)\right)^2 dt \times \sum_{\ell=1}^{d}\int (v_\ell(t) - \hat{c}_\ell\hat{v}_\ell(t))^2 dt \\
&\le \frac{1}{N}\int \max_{1\le k\le N}\left(\sum_{n=1}^{k} Y_n(t)\right)^2 dt \times O_P(N^{-1}).
\end{aligned}$$

Define

$$g(t) = E|Y_1(t)|^2 + 2\big(E|Y_1(t)|^2\big)^{1/2}\sum_{r\ge1}\big(E|Y_{1+r}(t) - Y_{1+r}^{(r)}(t)|^2\big)^{1/2}.$$

Then by similar arguments as in Section 16.8 we have

$$E\left(\sum_{n=1}^{N} Y_n(t)\right)^2 \le N\, g(t).$$

Hence by Menshov's inequality (see e.g. Section 10 of Billingsley (1999)) we infer that

$$E\max_{1\le k\le N}\left(\sum_{n=1}^{k} Y_n(t)\right)^2 \le (\log_2 4N)^2 N\, g(t).$$

Notice that (16.4) implies $\int g(t)dt < \infty$. In turn we obtain that

$$\frac{1}{N}\int \max_{1\le k\le N}\left(\sum_{n=1}^{k} Y_n(t)\right)^2 dt = O_P\big((\log N)^2\big),$$

which proves (16.60). □

Proof of Theorem 16.8. Notice that if the mean function changes from $\mu_1(t)$ to $\mu_2(t)$ at time $k^* = \lfloor N\theta\rfloor$, then $\mathbf{L}_N(x,\hat{\boldsymbol{\eta}})$ can be written as

$$\mathbf{L}_N(x,\hat{\boldsymbol{\beta}}) + N\begin{cases} x(1-\theta)[\hat{\boldsymbol{\mu}}_1 - \hat{\boldsymbol{\mu}}_2] & \text{if } x\le\theta;\\ \theta(1-x)[\hat{\boldsymbol{\mu}}_1 - \hat{\boldsymbol{\mu}}_2] & \text{if } x>\theta,\end{cases} \tag{16.62}$$

where

$$\hat{\boldsymbol{\mu}}_1 = \left[\int \mu_1(t)\hat{v}_1(t)dt, \int \mu_1(t)\hat{v}_2(t)dt, \ldots, \int \mu_1(t)\hat{v}_d(t)dt\right]^T,$$

and $\hat{\boldsymbol{\mu}}_2$ is defined analogously.

It follows from (16.62) that $T_N(d)$ can be expressed as the sum of three terms:

$$T_N(d) = T_{1,N}(d) + T_{2,N}(d) + T_{3,N}(d),$$

where

$$T_{1,N}(d) = \frac{1}{N}\int_0^1 \mathbf{L}_N(x,\hat{\boldsymbol{\beta}})^T \hat{\boldsymbol{\Sigma}}(\hat{\eta})^{-1}\mathbf{L}_N(x,\hat{\boldsymbol{\beta}})dx;$$

$$T_{2,N}(d) = \frac{N}{2}\theta(1-\theta)[\hat{\boldsymbol{\mu}}_1 - \hat{\boldsymbol{\mu}}_2]^T \hat{\boldsymbol{\Sigma}}(\hat{\eta})^{-1}[\hat{\boldsymbol{\mu}}_1 - \hat{\boldsymbol{\mu}}_2];$$

$$T_{3,N}(d) = \int_0^1 g(x,\theta)\mathbf{L}_N(x,\hat{\boldsymbol{\beta}})^T \hat{\boldsymbol{\Sigma}}(\hat{\eta})^{-1}[\hat{\boldsymbol{\mu}}_1 - \hat{\boldsymbol{\mu}}_2]dx,$$

with $g(x,\theta) = 2\{x(1-\theta)I_{\{x\le\theta\}} + \theta(1-x)I_{\{x>\theta\}}\}$.

Since $\boldsymbol{\Omega}$ in (16.33) is positive definite (p.d.), $\hat{\boldsymbol{\Sigma}}(\hat{\eta})$ is almost surely p.d. for large enough N (N is random). Hence for large enough N the term $T_{1,N}(d)$ is nonnegative. We will show that $N^{-1}T_{2,N}(d) \ge \kappa_1 + o_P(1)$, for a positive constant κ_1, and $N^{-1}T_{3,N}(d) = o_P(1)$. To this end we notice the following. Ultimately all eigenvalues of $\hat{\boldsymbol{\Sigma}}(\hat{\eta})$ are positive. Let $\lambda^*(N)$ and $\lambda_*(N)$ denote the largest, respectively, the smallest eigenvalue. By Lemma 2.2, $\lambda^*(N) \to \lambda^*$ a.s. and $\lambda_*(N) \to \lambda_*$ a.s., where λ^* and λ_* are the largest and smallest eigenvalue of $\boldsymbol{\Omega}$. Next we claim that

$$|\hat{\boldsymbol{\mu}}_1 - \hat{\boldsymbol{\mu}}_2| = |\boldsymbol{\mu}_1 - \boldsymbol{\mu}_2| + o_P(1).$$

To obtain this, we use the relation $\|\hat{v}_i - \hat{c}_j v_j\| = o_P(1)$ which can be proven similarly as Lemma A.1 of Berkes *et al.* (2009), but the law of large numbers in a Hilbert space must be replaced by the ergodic theorem. The ergodicity of $\{Y_n\}$ follows from the representation $Y_n = f(\varepsilon_n, \varepsilon_{n-1}, \ldots)$. Notice that because of the presence of a change point it cannot be claimed that $\|\hat{v}_i - \hat{c}_j v_j\| = O_P(N^{-1/2})$.

It follows that if N is large enough, then

$$[\hat{\boldsymbol{\mu}}_1 - \hat{\boldsymbol{\mu}}_2]^T \hat{\boldsymbol{\Sigma}}(\hat{\eta})^{-1}[\hat{\boldsymbol{\mu}}_1 - \hat{\boldsymbol{\mu}}_2] > \frac{1}{2\lambda^*}|\hat{\boldsymbol{\mu}}_1 - \hat{\boldsymbol{\mu}}_2|^2 = \frac{1}{2\lambda^*}|\boldsymbol{\mu}_1 - \boldsymbol{\mu}_2|^2 + o_P(1).$$

To verify $N^{-1}T_{3,N}(d) = o_P(1)$, observe that

$$\begin{aligned}
&\sup_{x\in[0,1]} \left|\mathbf{L}_N(x,\hat{\boldsymbol{\beta}})^T \hat{\boldsymbol{\Sigma}}(\hat{\eta})^{-1}[\hat{\boldsymbol{\mu}}_1 - \hat{\boldsymbol{\mu}}_2]\right| \\
&\quad\le \sup_{x\in[0,1]} |\mathbf{L}_N(x,\hat{\boldsymbol{\beta}})||\hat{\boldsymbol{\Sigma}}(\hat{\eta})^{-1}||\hat{\boldsymbol{\mu}}_1 - \hat{\boldsymbol{\mu}}_2| \\
&\quad= o_P(N)|\boldsymbol{\mu}_1 - \boldsymbol{\mu}_2|.
\end{aligned}$$

We used the matrix norm $|A| = \sup_{|x|\le 1}|Ax|$ and $|\hat{\boldsymbol{\Sigma}}(\hat{\eta})^{-1}| \xrightarrow{\text{a.s}} |\boldsymbol{\Omega}^{-1}| < \infty$. □

16.11 Proof of Theorem 16.9

We first establish a technical bound which implies the consistency of the estimator $\hat{\sigma}_{\ell k}$ given in (16.40). Let $\hat{c}_\ell = \text{sign}(\langle v_\ell, \hat{v}_\ell\rangle)$ and $\hat{d}_k = \text{sign}(\langle u_k, \hat{u}_k\rangle)$.

Lemma 16.6. *Under the assumptions of Theorem 16.9 we have*

$$\limsup_{N\to\infty} NE\,|\sigma_{\ell k} - \hat{c}_\ell\, \hat{d}_k\, \hat{\sigma}_{\ell k}|^2 \le \kappa_1\left(\frac{1}{\alpha_k^2} + \frac{1}{(\alpha'_\ell)^2}\right),$$

where κ_1 is a constant independent of k and ℓ.

Proof. It follow from elementary inequalities that

$$|\sigma_{\ell k} - \hat{c}_\ell\, \hat{d}_k\, \hat{\sigma}_{\ell k}|^2 \le 2T_1^2 + 2T_2^2,$$

where

$$T_1 = \frac{1}{N}\iint\left(\sum_{i=1}^{N}(X_i(s)Y_i(t) - E[X_i(s)Y_i(t)])\right)u_k(s)v_\ell(t)dt\,ds; \qquad (16.63)$$

$$T_2 = \frac{1}{N}\sum_{i=1}^{N}\iint E[X_i(s)Y_i(t)]\big[u_k(t)v_\ell(s) - \hat{d}_k\hat{u}_k(t)\hat{c}_\ell\hat{v}_\ell(s)\big]dt\,ds. \qquad (16.64)$$

By the Cauchy-Schwarz inequality and the inequality

$$|ab - cd|^2 \le 2a^2(b-d)^2 + 2d^2(a-c)^2, \qquad (16.65)$$

we obtain

$$T_1^2 \le \frac{1}{N^2}\iint\left(\sum_{i=1}^{N} X_i(s)Y_i(t) - E[X_i(s)Y_i(t)]\right)^2 dt\,ds; \qquad (16.66)$$

$$T_2^2 = 2v_2^2(X)v_2^2(Y)\big(\|u_k - \hat{d}_k\hat{u}_k\|^2 + \|v_\ell - \hat{c}_\ell\hat{v}_\ell\|^2\big). \qquad (16.67)$$

Hence by similar arguments as we used for the proof of Theorem 16.1 we get $NET_1^2 = O(1)$. The proof follows now immediately from Lemma 2.3 and Theorem 16.1.

Now we are ready to verify (16.42). We have

$$\hat{\psi}_{KL}(t,s) = \sum_{k=1}^{K}\sum_{\ell=1}^{L}\hat{\lambda}_\ell^{-1}\hat{\sigma}_{\ell k}\hat{u}_k(t)\hat{v}_\ell(s).$$

The orthogonality of the sequences $\{u_k\}$ and $\{v_\ell\}$ and (16.41) imply that

$$\iint\left(\sum_{k>K}\sum_{\ell>L}\lambda_\ell^{-1}\sigma_{\ell k}u_k(t)v_\ell(s)\right)^2 dt\,ds$$
$$= \sum_{k>K}\sum_{\ell>L}\iint \lambda_\ell^{-2}\sigma_{\ell k}^2 u_k^2(t)v_\ell^2(s)dt\,ds$$
$$= \sum_{k>K}\sum_{\ell>L}\lambda_\ell^{-2}\sigma_{\ell k}^2 \to 0 \quad (L, K\to\infty).$$

Therefore, letting

$$\psi_{KL}(t,s) = \sum_{k=1}^{K}\sum_{\ell=1}^{L}\lambda_\ell^{-1}\sigma_{\ell k}u_k(t)v_\ell(s),$$

(16.42) will follow once we show that

$$\iint\left[\psi_{KL}(t,s) - \hat{\psi}_{KL}(t,s)\right]^2 dt\,ds \xrightarrow{P} 0 \quad (N\to\infty).$$

Notice that by the Cauchy-Schwarz inequality the latter relation is implied by

$$KL\sum_{k=1}^{K}\sum_{\ell=1}^{L}\iint\left[\lambda_\ell^{-1}\sigma_{\ell k}u_k(t)v_\ell(s) - \hat{\lambda}_\ell^{-1}\hat{\sigma}_{\ell k}\hat{u}_k(t)\hat{v}_\ell(s)\right]^2 dt\,ds \xrightarrow{P} 0 \quad (N\to\infty). \tag{16.68}$$

A repeated application of (16.65) and some basic algebra yield

$$\begin{aligned}\frac{1}{4}&\left[\lambda_\ell^{-1}\sigma_{\ell k}u_k(t)v_\ell(s) - \hat{\lambda}_\ell^{-1}\hat{\sigma}_{\ell k}\hat{u}_k(t)\hat{v}_\ell(s)\right]^2\\ &\le \lambda_\ell^{-2}|\sigma_{\ell k} - \hat{c}_\ell\,\hat{d}_k\,\hat{\sigma}_{\ell k}|^2\hat{u}_k^2(t)\hat{v}_\ell^2(s) + \hat{\sigma}_{\ell k}^2\left|\lambda_\ell^{-1} - \hat{\lambda}_\ell^{-1}\right|^2\hat{u}_k^2(t)\hat{v}_\ell^2(s)\\ &\quad + \sigma_{\ell k}^2\lambda_\ell^{-2}\left|u_k(t) - \hat{d}_k\hat{u}_k(t)\right|^2 v_\ell^2(s) + \sigma_{\ell k}^2\lambda_\ell^{-2}\left|v_\ell(s) - \hat{c}_\ell\hat{v}_\ell(s)\right|^2\hat{u}_k^2(t).\end{aligned}$$

Hence

$$\begin{aligned}\frac{1}{4}&\iint\left[\lambda_\ell^{-1}\sigma_{\ell k}u_k(t)v_\ell(s) - \hat{\lambda}_\ell^{-1}\hat{\sigma}_{\ell k}\hat{u}_k(t)\hat{v}_\ell(s)\right]^2 dt\,ds\\ &\le \lambda_\ell^{-2}|\sigma_{\ell k} - \hat{c}_\ell\,\hat{d}_k\,\hat{\sigma}_{\ell k}|^2 + \hat{\sigma}_{\ell k}^2|\lambda_\ell^{-1} - \hat{\lambda}_\ell^{-1}|^2\\ &\quad + \sigma_{\ell k}^2\lambda_\ell^{-2}\left(\|u_k - \hat{d}_k\hat{u}_k\|^2 + \|v_\ell - \hat{c}_\ell\hat{v}_\ell\|^2\right).\end{aligned}$$

Thus in order to get (16.68) we will show that

$$KL\sum_{k=1}^{K}\sum_{\ell=1}^{L}\lambda_\ell^{-2}|\sigma_{\ell k} - \hat{c}_\ell\,\hat{d}_k\hat{\sigma}_{\ell k}|^2 \xrightarrow{P} 0; \tag{16.69}$$

$$KL\sum_{k=1}^{K}\sum_{\ell=1}^{L}\hat{\sigma}_{\ell k}^2|\lambda_\ell^{-1} - \hat{\lambda}_\ell^{-1}|^2 \xrightarrow{P} 0; \tag{16.70}$$

$$KL\sum_{k=1}^{K}\sum_{\ell=1}^{L}\sigma_{\ell k}^2\lambda_\ell^{-2}\left(\|u_k - \hat{d}_k\hat{u}_k\|^2 + \|v_\ell - \hat{c}_\ell\hat{v}_\ell\|^2\right) \xrightarrow{P} 0. \tag{16.71}$$

We start with (16.69). By Lemma 16.6 and Assumption 16.9 we have

$$E\left(KL\sum_{k=1}^{K}\sum_{\ell=1}^{L}\lambda_\ell^{-2}|\sigma_{\ell k} - \hat{c}_\ell\,\hat{d}_k\hat{\sigma}_{\ell k}|^2\right) \to 0 \quad (N\to\infty).$$

Next we prove relation (16.70). In order to shorten the proof we replace $\hat{\sigma}_{\ell k}$ by $\sigma_{\ell k}$. Otherwise we would need a further intermediate step, requiring similar arguments which follow. Now for any $0 < \varepsilon < 1$ we have

$$
\begin{aligned}
P&\Big(KL\sum_{k=1}^{K}\sum_{\ell=1}^{L}\sigma_{\ell k}^2|\lambda_\ell^{-1} - \hat{\lambda}_\ell^{-1}|^2 > \varepsilon\Big)\\
&= P\Big(KL\sum_{k=1}^{K}\sum_{\ell=1}^{L}\sigma_{\ell k}^2\lambda_\ell^{-2}\Big|\frac{\hat{\lambda}_\ell - \lambda_\ell}{\hat{\lambda}_\ell}\Big|^2 > \varepsilon\Big)\\
&\le P\Big(\max_{1\le \ell\le L}\Big|\frac{\hat{\lambda}_\ell - \lambda_\ell}{\hat{\lambda}_\ell}\Big|^2 > \frac{\varepsilon}{\Psi KL}\Big)\\
&\le \sum_{\ell=1}^{L} P\Big(\Big|\frac{\hat{\lambda}_\ell - \lambda_\ell}{\hat{\lambda}_\ell}\Big|^2 > \frac{\varepsilon}{\Psi KL} \cap |\lambda_\ell - \hat{\lambda}_\ell| < \varepsilon\lambda_\ell\Big)\\
&\quad + \sum_{\ell=1}^{L} P\Big(\Big|\frac{\hat{\lambda}_\ell - \lambda_\ell}{\hat{\lambda}_\ell}\Big|^2 > \frac{\varepsilon}{\Psi KL} \cap |\lambda_\ell - \hat{\lambda}_\ell| \ge \varepsilon\lambda_\ell\Big)\\
&\le \sum_{\ell=1}^{L}\Big[P\Big(|\hat{\lambda}_\ell - \lambda_\ell|^2 > \frac{\varepsilon}{\Psi KL}\lambda_\ell(1-\varepsilon)\Big) + P\Big(|\lambda_\ell - \hat{\lambda}_\ell|^2 \ge \varepsilon^2\lambda_\ell^2\Big)\Big]\\
&\le \kappa_2\Big(\frac{KL^2}{\epsilon N\lambda_L} + \frac{1}{\varepsilon N\lambda_L^2}\Big),
\end{aligned}
$$

by an application of the Markov inequality and Theorem 16.2. According to our Assumption 16.9 this also goes to zero for $N \to \infty$.

Finally we prove (16.71). By Lemma 2.3 and Theorem 16.1 we infer that

$$
\begin{aligned}
E&\Big(KL\sum_{k=1}^{K}\sum_{\ell=1}^{L}\sigma_{\ell k}^2\lambda_\ell^{-2}\Big(\|u_k - \hat{d}_k\hat{u}_k\|^2 + \|v_\ell - \hat{c}_\ell\hat{v}_\ell\|^2\Big)\Big)\\
&\le \kappa_3\frac{KL}{N}\sum_{k=1}^{K}\sum_{\ell=1}^{L}\sigma_{\ell k}^2\lambda_\ell^{-2}\Big(\frac{1}{\alpha_k^2} + \frac{1}{\alpha_\ell'^2}\Big)\\
&\le 2\kappa_3\Psi\frac{KL}{N\min\{h_L, h_K'\}^2}.
\end{aligned}
$$

Assumption 16.9 (ii) assures that the last term goes to zero. This completes the proof.

Chapter 17
Spatially distributed functional data

Chapters 13, 14 and 16 focused on functional time series. The present chapter and Chapter 18 deal with curves observed at spatial locations. The data consist of curves $X(\mathbf{s}_k; t),\ t \in [0, 1]$, observed at spatial locations $\mathbf{s}_1, \mathbf{s}_2, \ldots, \mathbf{s}_N$. We propose methods for the estimation of the mean function and the FPC's for such data. We also develop a significance test for the correlation of two such functional spatial fields. The test we consider in this section is an extension of the test of Chapter 9 in which the pairs of curves were assumed to be independent. The main feature of spatially distributed curves is that the curves at neighboring locations look similar, so the dependence cannot be neglected, and, together with the spatial distribution of the locations $\mathbf{s}_1, \mathbf{s}_2, \ldots, \mathbf{s}_N$, is the main feature of the data. After validating the finite sample performance of the test by means of a simulation study, we apply it to determine if there is correlation between long term trends in the so called critical ionospheric frequency and changes in the direction of the internal magnetic field of the Earth. The test provides conclusive evidence for correlation thus solving a long standing space physics conjecture. This conclusion is not apparent if the spatial dependence of the curves is neglected. This chapter focuses on methodological and computational issues. Chapter 18 investigates the asymptotic properties of the sample mean and of the EFPC's for spatially distributed functions.

This chapter is organized as follows. Section 17.1 introduces spatially distributed functional data in greater detail, and provides the motivation for the research presented in this Chapter. In Section 17.2, we briefly describe the fundamental concepts of spatial statistics required to understand the remaining sections. Sections 17.3 and 17.4 focus, respectively, on the estimation of the mean function and the FPC's in the spatial setting. Section 17.5 demonstrates by means of a simulation study that the methods we propose improve on the standard approach, and discusses their relative performance and computational cost. In Section 17.6, we develop a test for the correlation of two functional spatial fields. This test requires estimation of a covariance tensor. After addressing this issue in Section 17.7, we study in Section 17.8 the finite sample properties of several implementations of the test. Finally, in Section 17.9, we apply the methodology developed in the previous section to test for the correlation between the ionospheric critical frequency and magnetic curves.

17.1 Introduction

We consider data consisting of curves $X(\mathbf{s}_k; t)$, $t \in [0, T]$, observed at spatial locations $\mathbf{s}_1, \mathbf{s}_2, \ldots, \mathbf{s}_N$. Such functional data structures are quite common. A well–known example is the Canadian temperature and precipitation data used in several chapters of Ramsay and Silverman (2005). The annual curves are available at 35 locations, some of which are quite close, and so the curves look very similar, others are very remote with notably different curves. Figure 17.1 shows the temperature curves together with the simple average and the average estimated by one of the methods described in this chapter. Conceptually, the average temperature in Canada should be computed as the average over a fine regular grid spanning the whole country. In reality, there are only several dozen locations mostly in the inhabited southern strip. Computing an average over these locations will bias the estimate. Data at close–by locations contribute similar information, and should get smaller weights than data at sparse locations. This is the fundamental principle of spatial statistics which however received only limited attention in the framework of functional data analysis. The above example shows a simple application of our methodology as a descriptive tool. The estimation and testing problems we solve are related to long term trends, so the data we study do not look like the annual averages shown in Figure 17.1, but rather consist of curves which exhibit temporal evolution. Evaluation of the significance of trends has been an important issue in statistical geophysics, so, in addition to its value in exploratory analysis, the methodology we develop is useful in inferential procedures. In addition to the mean function, we also study the estimation of the FPC's.

Many environmental and geophysical data sets that fall into the framework considered in this chapter. For example, the Australian rainfall data set, studied by Delaigle and Hall (2010) among others, consists of daily rainfall measurements

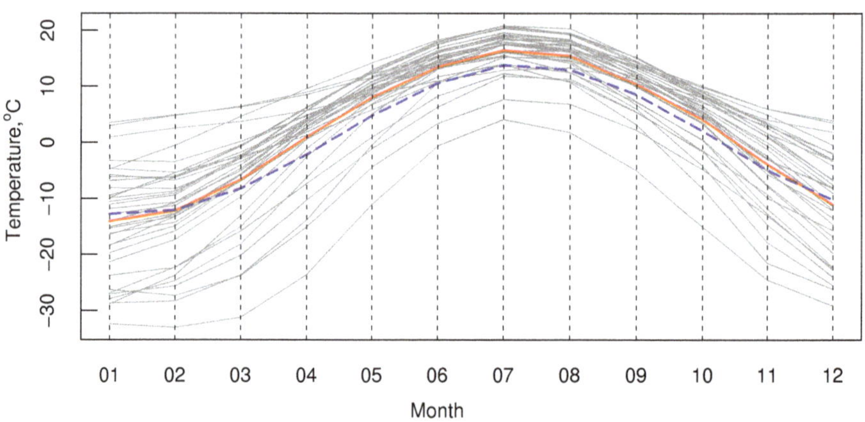

Fig. 17.1 Average annual temperature curves at 35 locations in Canada. The continuous thick line is the simple average, the dashed line is the average that takes into account spatial locations and dependence.

from 1840 to 1990 at 191 Australian weather stations. Snow water curves measured at several dozen locations in every state over many decades have been studied in the purely spatial framework, e.g. Carroll *et al.* (1995) and Carroll and Cressie (1996). Useful insights can potentially be gained by studying the whole curves reflecting the temporal dynamics, rather than just temporal averages. Another important example are pollution curves: $X(\mathbf{s}_k;t)$ is the concentration of a pollutant at time t at location $\mathbf{s}_k$. Data of this type were studied by Kaiser *et al.* (2002). A functional framework might be convenient because such data are typically available only at sparsely distributed time points t_j, which can be different at different locations. In many studies, $X(\mathbf{s}_k;t)$ is the count at time t of an infectious disease cases, where $\mathbf{s}_k$ is a representative location, e.g. a "middle point" of a county. Delicado *et al.* (2010) review other examples and contributions to the methodology for spatially distributed functional data. The work with geostatistical functional data has focused on kriging, see Delicado *et al.* (2010), Nerini *et al.* (2010) and Giraldo *et al.* (2010, 2011).

The data set that most directly motivated the research described in this chapter consists of the curves of the so–called F2–layer critical frequency, foF2. Three such curves are shown in Figure 17.2. In principle, foF2 curves are available at over 200 locations throughout the globe, but sufficiently complete data are available at only 30-40 locations which are very unevenly spread; for example, there is a dense network of observatories over Europe and practically no data over the oceans. The study of this data set has been motivated by the hypothesis of Roble and Dickinson (1989) who suggested that the increasing amounts of (radiative) greenhouse gases

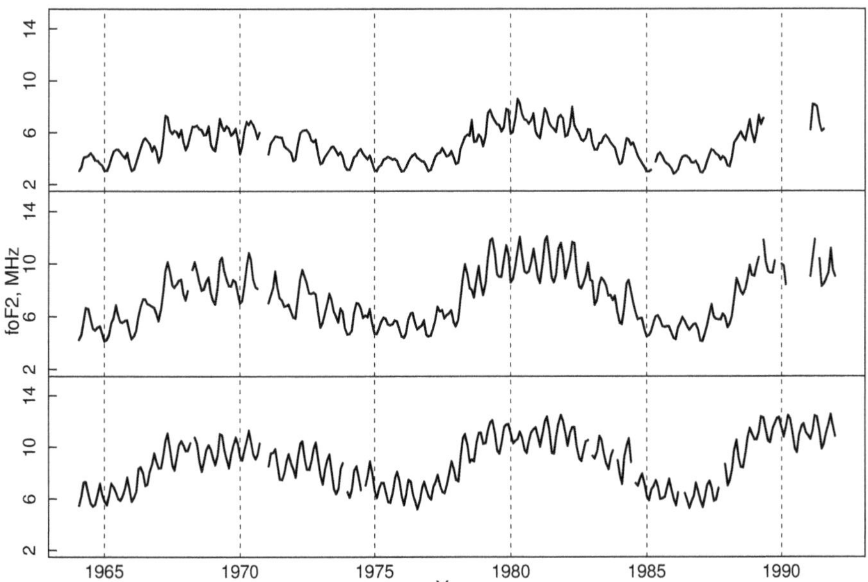

Fig. 17.2 F2-layer critical frequency curves at three locations. Top to bottom (latitude in parentheses): Yakutsk (62.0), Yamagawa (31.2), Manila (14.7). The functions exhibit a latidudal trend in amplitude.

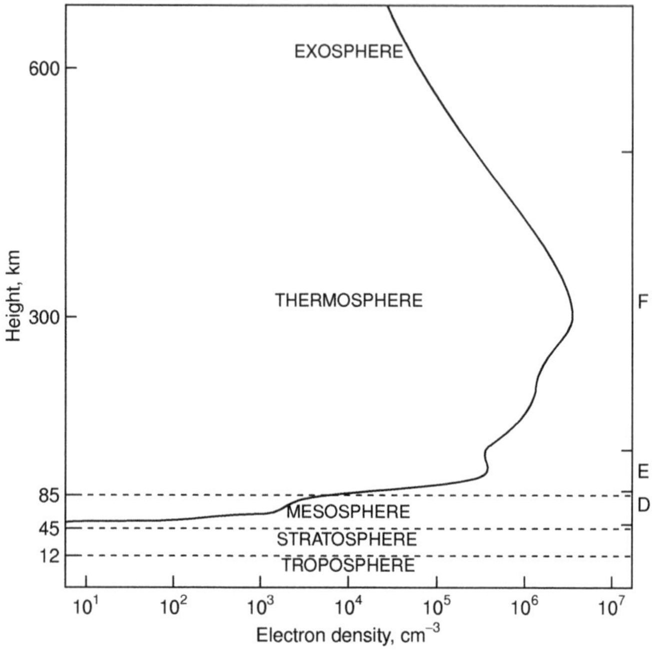

Fig. 17.3 Typical profile of day time ionosphere. The curve shows electron density as a function of height. The right vertical axix indicates the D, E and F regions of the ionosphere.

should lead to global cooling in mesosphere and thermosphere, as opposed to the global warming in lower troposphere, cf. Figure 17.3. Rishbeth (1990) pointed out that such cooling would result in a thermal contraction and the global lowering of the ionospheric peak electron densities. The height of the peak density of the F region of the ionosphere, see Figure 17.3, can be computed from the critical frequency foF2. The last twenty years have seen very extensive research in this area, see Lastovicka *et al.* (2008) for a partial overview. One of the difficulties in determining a global trend is that the foF2 curves appear to exhibit trends in opposing directions over various regions. A possible explanation suggests that these trends are caused by long term trends in the magnetic field of the Earth. There has however been no agreement in the space physics community if this is indeed the case. The results of Section 17.9 confirm that there is a strong connection between the magnetic field and ionospheric trends.

Throughout this chapter, $\{X(\mathbf{s})\}$ denotes a random field defined on a spatial domain and taking values in the Hilbert space $L^2 = L^2([0,1])$. The value of the function $X(\mathbf{s})$ at time $t \in [0,1]$ is denoted by $X(\mathbf{s};t)$. For inferential procedures, we assume that the random functions $X(\mathbf{s})$ are identically distributed. If this is the case, then

$$X(\mathbf{s};t) = \mu(t) + \sum_{i=1}^{\infty} \xi_i(\mathbf{s}) v_i(t), \tag{17.1}$$

where $\mu(t) = X(\mathbf{s}; t)$, the v_i are the eigenfunctions of the covariance operator

$$C = E[\langle X(\mathbf{s}) - \mu, \cdot\rangle (X(\mathbf{s}) - \mu)],$$

and $\xi_i(\mathbf{s}) = \langle X(\mathbf{s}) - \mu, v_i\rangle$. Note that the mean function μ and the FPC's v_i do not depend on $\mathbf{s}$. Even if model (17.1) does not hold, our estimates of the mean function and the FPC's provide useful descriptive statistics, as illustrated in Figure 17.1. For the applications we have in mind, it is enough to assume that the spatial domain is a subset of the plane or a two–dimensional sphere.

Recall that for functions, $X_1, X_2, \ldots, X_N$, the sample mean is defined as

$$\bar{X}_N = N^{-1} \sum_{n=1}^{N} X_n,$$

and the sample covariance operator as

$$\widehat{C}(x) = N^{-1} \sum_{n=1}^{N} \left[\left\langle (X_n - \bar{X}_N), x \right\rangle (X_n - \bar{X}_N) \right], \quad x \in L^2.$$

The EFPC's are computed as the eigenvalues of $\widehat{C}$. These are the estimates produced by several software packages, including the popular `R` package `fda`. For sparse data measured with error, nontrivial modifications are needed, see Section 1.5. In either case, the consistency of the sample mean and the EFPC's relies on the assumption that the functional observations form a simple random sample. In Chapter 16, we showed that the the consistency holds with the same rates for weakly dependent functional time series. However, if the functions $X_k = X(\mathbf{s}_k)$ are spatially distributed, the sample mean and the EFPC's need not be consistent, see Chapter 18. This happens if the spatial dependence is strong or if there are clusters of the points $\mathbf{s}_k$. For moderately dependent spatially separated curves, these estimators are consistent. We will demonstrate that in finite samples better estimators are available though. We will then use these improved estimators as part of the procedure for testing the independence of two functional fields $\{X(\mathbf{s}), \mathbf{s} \in \mathbf{S}\}$ and $\{Y(\mathbf{s}), \mathbf{s} \in \mathbf{S}\}$. First we review some essential concepts of spatial statistics.

17.2 Essentials of spatial statistics

In order to make this chapter self–contained, we discuss in this section some relevant concepts and methods of spatial statistics. We focus only on geostatistical data, i.e. observations available at irregularly distributed points of a spatial domain. The book of Schabenberger and Gotway (2005) offers an accessible and comprehensive introduction to spatial statistics, a reader interested in a quick introduction to basic ideas of geostatistics, which goes beyond the information presented in this section, is referred to Chapters 2 and 3 of Gelfand *et al.* (2010). In this section, we assume that all data are scalars.

A sample of spatial data is

$$\{X(\mathbf{s}_k), \ \mathbf{s}_k \in \mathbf{S}, \ k = 1, 2, \ldots, N\}.$$

The spatial domain $\mathbf{S}$ is typically a subset of the two–dimensional plane or sphere. The observed value $X(\mathbf{s}_k)$ is viewed as a realization of a random variable, so $\{X(\mathbf{s}), \ \mathbf{s} \in \mathbf{S}\}$ is a scalar random field. Just as in time series analysis, stationary random fields play a fundamental role in modeling spatial data. To define arbitrary shifts, we must assume that $\mathbf{S}$ is either the whole Euclidean space $\mathbb{R}^d$, or the whole sphere. The random field $\{X(\mathbf{s}), \ \mathbf{s} \in \mathbf{S}\}$ is then strictly stationary if

$$\{X(\mathbf{s}_1 + \mathbf{h}), X(\mathbf{s}_2 + \mathbf{h}), \ldots, X(\mathbf{s}_m + \mathbf{h})\} \stackrel{d}{=} \{X(\mathbf{s}_1), X(\mathbf{s}_2), \ldots, X(\mathbf{s}_m)\}$$

for any points $\mathbf{s}_1, \mathbf{s}_2, \ldots, \mathbf{s}_m \in \mathbf{S}$ and any shift $\mathbf{h}$. If we assume only that the mean $EX(\mathbf{s})$ and the covariances $\text{Cov}(X(\mathbf{s}), X(\mathbf{s}+\mathbf{h}))$ do not depend on $\mathbf{s}$, then the field is called second–order stationary. For such a field, we define the covariance function

$$C(\mathbf{h}) = \text{Cov}(X(\mathbf{s}), X(\mathbf{s} + \mathbf{h})).$$

If $C(\mathbf{h})$ depends only on the length h of $\mathbf{h}$, we say that the random field is isotropic. The covariance function of an isotropic random field is typically parametrized as

$$C(h) = \sigma^2 \phi(h), \quad h \geq 0, \ \phi(0) = 1.$$

The function ϕ is then called the correlation function. The function $\phi(\cdot)$ quantifies the strength of linear dependence between observations distance h apart and the smoothness of the field. The following correlation functions are frequently used. The powered exponential correlation function is defined by

$$\phi(h) = \exp\left\{-\left(\frac{h}{\rho}\right)^p\right\}, \quad \rho > 0, \ 0 < p \leq 2.$$

If $p = 1$, this correlation function is called exponential, if $p = 2$, it is called Gaussian. A very general family of correlation functions is the so–called Matérn class. The Matérn class correlation functions are defined as

$$\phi(h) = \frac{2^{1-\nu}}{\Gamma(\nu)} \left(\frac{h}{\rho}\right)^\nu K_\nu(h/\rho), \quad \rho > 0, \ \nu > 0,$$

where K_ν is the modified Bessel function, see Stein (1999) for the details. The function K_ν decays monotonically and approximately exponentially fast; numerical calculations show that $K_\nu(s)$ practically vanishes if $s > \nu$.

The correlation function of an isotropic random field is positive definite, and every positive definite function ϕ is a correlation function of a (Gaussian) random field. This follows from Kolmogorov's consistency theorem. A positive definite function ϕ is called a *valid* correlation function. There exist examples of correlation functions which are valid in one dimension, but are no longer valid in a higher dimension, or on a manifold. For this reason, when working with globally distributed data, we use the chordal distance defined as the Euclidean distance in the three–dimensional space. Denoting the latitude by L and the longitude by l, the chordal distance, $0 \leq d_{k,\ell} \leq 2$, between two points, $\mathbf{s}_k, \mathbf{s}_\ell$, on the unit sphere is given by

$$d_{k,\ell} = 2\left[\sin^2\left(\frac{L_k - L_\ell}{2}\right) + \cos L_k \cos L_\ell \sin^2\left(\frac{l_k - l_\ell}{2}\right)\right]^{1/2}. \tag{17.2}$$

If we work with distance (17.2), we can use any correlation function which is valid is the three–dimensional Euclidean space.

In spatial statistics, the concept of *intrinsic* stationarity is very useful. The field $\{X(\mathbf{s}),\ \mathbf{s} \in \mathbf{S}\}$ is said to be intrinsically stationary if $\mathrm{Var}[X(\mathbf{s}+\mathbf{h}) - X(\mathbf{s})]$ does not depend on $\mathbf{s}$. Notice that a second–order stationary field is intrinsically stationary. The converse is not true. The Brownian motion is intrinsically stationary (has stationary increments), but it is not a stationary process. If $\{X(\mathbf{s}),\ \mathbf{s} \in \mathbf{S}\}$ is intrinsically stationary, we define the *semivariogram* by

$$\gamma(\mathbf{h}) = \frac{1}{2}\mathrm{Var}[X(\mathbf{s}+\mathbf{h}) - X(\mathbf{s})].$$

(The variogram is defined as $2\gamma(\cdot)$.) The semivariogram of a second order stationary field with the covariance function $C(\cdot)$ is given by

$$\gamma(\mathbf{h}) = C(\mathbf{0}) - C(\mathbf{h}). \tag{17.3}$$

Even for second order stationary fields, the estimation of the covariance function proceeds through the estimation of the semivariogram. One advantage of this approach is that the semivariogram is less sensitive to the misspecification or biased estimation of the mean. Typically isotropy is assumed. First, an empirical variogram is computed at several available lags $h > 0$. Then a parametric model (derived from a valid covariance function via (17.3)) is fitted. There are several versions of the empirical variogram. The classical estimator proposed by Matheron is given by

$$\hat{\gamma}(d) = \frac{1}{|N(d)|}\sum_{N(d)}(X(\mathbf{s}_k) - X(\mathbf{s}_\ell))^2, \tag{17.4}$$

where $N(d)$ is the set of pairs $(\mathbf{s}_k, \mathbf{s}_\ell)$ approximately distance d apart, and $|N(d)|$ is the count of pairs in $N(d)$. A robust estimator proposed by Cressie and Hawkins is defined as

$$\hat{\gamma}(d) = \left(0.457 + \frac{0.494}{|N(d)|}\right)^{-1}\left(\frac{1}{|N(d)|}\sum_{N(d)}|X(\mathbf{s}_k) - X(\mathbf{s}_\ell)|^{1/2}\right)^4. \tag{17.5}$$

The precise definition of the summations in (17.4) and (17.5) requires the introduction of a binning parameter which allows to treat pairs of points as being approximately distance d apart. For details, we refer to Section 4.4 of Schabenberger and Gotway (2005), where other ways of variogram estimation are also discussed. Examples of empirical variograms and fitted parametric models are given in Figure 17.11.

17.3 Estimation of the mean function

We represent the observed functions as

$$X(\mathbf{s}_k; t) = \mu(t) + \varepsilon(\mathbf{s}_k; t), \quad k = 1, 2, \ldots, N, \tag{17.6}$$

where ε is an unobservable field with $E\varepsilon(\mathbf{s}; t) = 0$. We propose three methods of estimating the mean function μ, which we call M1, M2, M3. As will become apparent in this section, several further variants, not discussed here, are conceivable. But the results of Section 17.5 show that while all these methods offer an improvement over the simple sample mean, their performance is comparable. All methods assume that the expected inner products

$$C_{k\ell} = E[\langle \varepsilon(\mathbf{s}_k), \varepsilon(\mathbf{s}_\ell) \rangle] \tag{17.7}$$

depend only on the distance $d(\mathbf{s}_k, \mathbf{s}_\ell)$, this defines a "functional" second order stationarity. The estimation of the $C_{k\ell}$ is the central issue, and will be discussed as we introduce methods M1 and M2. Method M3 does not require the estimation of the $C_{k\ell}$, but it requires the estimation of the corresponding covariances for the projections of the functions $X(\mathbf{s}_k)$ onto several basis functions.

Methods M1 and M2 estimate μ by the weighted average

$$\hat{\mu}_N = \sum_{n=1}^{N} w_n X(\mathbf{s}_n). \tag{17.8}$$

The optimal weights w_k are defined to minimize $E \left\| \sum_{n=1}^{N} w_n X(\mathbf{s}_n) - \mu \right\|^2$ subject to the condition $\sum_{n=1}^{N} w_n = 1$. Using the method of the Lagrange multiplier, we seek to minimize the objective function

$$\begin{aligned} \varphi(w_1, w_2, \ldots, w_N, r) &= E \left\| \sum_{n=1}^{N} w_n X(\mathbf{s}_n) - \mu \right\|^2 - 2r \left(\sum_{n=1}^{N} w_n - 1 \right) \\ &= \sum_{k,\ell=1}^{N} w_k w_\ell C_{k\ell} - 2r \left(\sum_{n=1}^{N} w_n - 1 \right). \end{aligned} \tag{17.9}$$

To compute the optimal weights, observe that

$$\frac{\partial \varphi}{\partial w_n} = 2 \sum_{k=1}^{N} w_k C_{kn} - 2r, \quad n = 1, 2, \ldots, N;$$

$$\frac{\partial \varphi}{\partial r} = -2 \left(\sum_{n=1}^{N} w_n - 1 \right).$$

The unknowns $w_1, w_2, \ldots, w_N, r$ are solutions to the system of $N + 1$ equations

$$\sum_{n=1}^{N} w_n = 1, \quad \sum_{k=1}^{N} w_k C_{kn} - r = 0, \qquad n = 1, 2, \ldots, N. \tag{17.10}$$

Set $\mathbf{w} = (w_1, \ldots, w_N)^T$. An easy way to solve equations (17.10) is to compute $\mathbf{v} = \mathbf{C}^{-1}\mathbf{1}$, where $\mathbf{C} = [C_{k\ell}, 1 \le k, \ell \le N]$, and then set $\mathbf{w} = a\mathbf{v}$, where a is a constant such that $\mathbf{1}^T\mathbf{w} = 1$.

Method M1. At each time point t_j, we fit a parametric spatial model to the scalar field $X(\mathbf{s}; t_j)$. To focus attention, we provide formulas for the exponential model

$$\operatorname{Cov}(X(\mathbf{s}_k; t_j), X(\mathbf{s}_\ell; t_j)) = \sigma^2(t_j) \exp\left\{-\frac{d(\mathbf{s}_k, \mathbf{s}_\ell)}{\rho(t_j)}\right\}. \tag{17.11}$$

It is clear how they can be modified for other popular models. Observe that under model (17.11),

$$\begin{aligned}
C_{k\ell} &= E \int (X(\mathbf{s}_k; t) - \mu(t))\,(X(\mathbf{s}_\ell; t) - \mu(t))\,dt \\
&= \int \operatorname{Cov}(X(\mathbf{s}_k; t_j), X(\mathbf{s}_\ell; t_j))dt \\
&= \int \sigma^2(t) \exp\left\{-\frac{d(\mathbf{s}_k, \mathbf{s}_\ell)}{\rho(t)}\right\} dt.
\end{aligned}$$

One way to estimate $C_{k\ell}$ is to set

$$\widehat{C}_{k\ell} = \int \hat{\sigma}^2(t) \exp\left\{-\frac{d(\mathbf{s}_k, \mathbf{s}_\ell)}{\hat{\rho}(t)}\right\} dt, \tag{17.12}$$

with the estimates $\hat{\sigma}^2(t_j)$ and $\hat{\rho}(t_j)$ obtained using some version of empirical variogram, for example (17.4) or (17.5).

If the sample size N is small, the ordinary nonlinear least squares method needed to obtain $\hat{\sigma}^2(t_j)$ and $\hat{\rho}(t_j)$ may fail to converge for some t_j. An example based on the critical frequency data is given in Figure 17.4. The convergence does however take place for most t_j, so the integral in (17.12) can be approximated using a Riemann sum.

Another way to proceed, is to replace the $\hat{\rho}(t_j)$ by their average $\hat{\rho} = m^{-1}\sum_{j=1}^{m} \hat{\rho}(t_j)$, where m is the count of the t_j at which the variogram is estimated successfully. Then, the $C_{k\ell}$ are approximated by

$$\widehat{C}_{k\ell} = \left(\int \hat{\sigma}^2(t)dt\right) \exp\left\{-\frac{d(\mathbf{s}_k, \mathbf{s}_\ell)}{\hat{\rho}}\right\}.$$

As explained above, in order to compute the weights w_j in (17.10), it is enough to know the matrix $\mathbf{C}$ only up to a multiplicative constant. Thus we may set

$$\widehat{C}_{k\ell} = \exp\left\{-\frac{d(\mathbf{s}_k, \mathbf{s}_\ell)}{\hat{\rho}}\right\}. \tag{17.13}$$

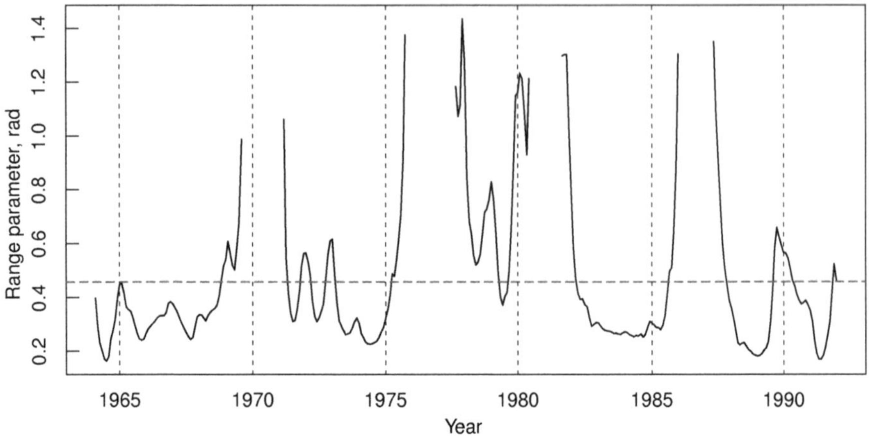

Fig. 17.4 The range parameter $\rho(t_j)$ of the scaled foF2 curves, determined using method $M1$, as a function of time. The horizontal line is its average value, $\bar{\rho} = 0.474$. The gaps indicate the times t_j where the method failed to converge.

Once the matrix $\mathbf{C}$ has been estimated, we compute the weights w_j, and estimate the mean via (17.8).

If (17.12) is used, we refer to this method as M1a, if (17.13) is used, we call it M1b.

Method M1 relies on the estimation of the variograms

$$2\gamma(\mathbf{s}_k, \mathbf{s}_\ell; t_j) = E\left(X(\mathbf{s}_k; t_j) - X(\mathbf{s}_\ell; t_j)\right)^2 \qquad (17.14)$$
$$= 2\mathrm{Var}(X(\mathbf{s}_k; t_j)) - 2\mathrm{Cov}(X(\mathbf{s}_k; t_j), X(\mathbf{s}_\ell; t_j)),$$

which lead to the estimates in a parametric model. The model is the same for every t_j, but the estimates $(\hat{\sigma}^2(t_j), \hat{\rho}(t_j)$, for the exponential model) depend on t_j. An advantage of this approach is that even if, for small N, parameter estimates may not converge at some t_j, it is still possible to obtain estimates (17.12) and (17.13). Method M2, described below, requires only one optimization, so it is much faster than M1, but this optimization may fail to converge for small N. (This has not happened though for our real and simulated data.)

Method M2. We define the *functional* variogram

$$2\gamma(\mathbf{s}_k, \mathbf{s}_\ell) = E\|X(\mathbf{s}_k) - X(\mathbf{s}_\ell)\|^2 \qquad (17.15)$$
$$= 2E\|X(\mathbf{s}_k) - \mu\|^2 - 2E\left[\langle X(\mathbf{s}_k) - \mu, X(\mathbf{s}_\ell) - \mu\rangle\right]$$
$$= 2E\|X(\mathbf{s}) - \mu\|^2 - 2C_{k\ell}.$$

The variogram (17.15) can be estimated by its empirical counterparts, like (17.4) or (17.5), with the $|X(\mathbf{s}_k) - X(\mathbf{s}_\ell)|$ replaced by

$$\|X(\mathbf{s}_k) - X(\mathbf{s}_\ell)\| = \left\{\int (X(\mathbf{s}_k; t) - X(\mathbf{s}_\ell; t))^2\, dt\right\}^{1/2}.$$

Next, we fit a parametric model, for example we postulate that

$$\gamma(\mathbf{s}_k, \mathbf{s}_\ell) = \sigma_f^2 \left(1 - \exp\left\{-\frac{d(\mathbf{s}_k, \mathbf{s}_\ell)}{\rho_f}\right\}\right). \tag{17.16}$$

The subscript f is used to emphasize the *functional* variogram. Denoting by $\hat{\rho}_f$ the resulting estimate, we estimate the C_{kl} by (17.13) with $\hat{\rho}$ replaced by $\hat{\rho}_f$.

Method M3. This method uses a basis expansion of the functional data, it does not use the weighted sum (17.8). Suppose $B_j, 1 \le j \le K$, are elements of a functional basis with K so large that for each k

$$X(\mathbf{s}_k) \approx \sum_{j \le K} \langle B_j, X(\mathbf{s}_k) \rangle B_j \tag{17.17}$$

to a good approximation. By (17.6), we obtain for every j

$$\langle B_j, X(\mathbf{s}_k) \rangle = \langle B_j, \mu \rangle + \langle B_j, \varepsilon(\mathbf{s}_k) \rangle, \quad k = 1, 2, \ldots, N. \tag{17.18}$$

For every fixed j, the $\langle B_j, X(\mathbf{s}_k) \rangle$ are observations of a second order stationary and isotropic scalar spatial field with a constant unknown mean $\langle B_j, \mu \rangle$. This mean can be estimated by postulating a covariance structure for each $\langle B_j, X(\mathbf{s}_k) \rangle$, for example

$$\mathrm{Cov}\left(\langle B_j, X(\mathbf{s}_k) \rangle, \langle B_j, X(\mathbf{s}_\ell) \rangle\right) = \sigma_j^2 \exp\left\{-\frac{d(\mathbf{s}_k, \mathbf{s}_\ell)}{\rho_j}\right\}.$$

The mean $\langle B_j, \mu \rangle$ is estimated by a weighted average of the $\langle B_j, X(\mathbf{s}_k) \rangle$. The weights depend on j and are computed using (17.10) with the C_{kn} replaced by $\mathrm{Cov}(\langle B_j, X(\mathbf{s}_k) \rangle, \langle B_j, X(\mathbf{s}_n) \rangle$. Denote the resulting estimate by $\hat{\mu}_j$. The mean function μ is then estimated by

$$\hat{\mu}(t) = \sum_{j \le K} \hat{\mu}_j B_j(t).$$

Choosing an appropriate K in (17.17) is a complex theoretical problem, but in practice, at least for data sets that motivate this research, it is easy to find K such that the approximation (17.17) is visually satisfactory. In fact, the functional objects in `R` are created using approximation (17.17), so in practice it can be treated as an equality.

Remark 17.1. Methods M1 and M2 produce an estimated mean function which is a linear combination of the observed curves. Such estimates belong to the same class of functions as the original data, in particular they inherit their smoothness (or roughness). Method M3 produces estimates that are linear combinations of the basis functions. Such estimates will typically be smoother than the real data.

17.4 Estimation of the principal components

Assume now that the mean function μ has been estimated, and this estimate is subtracted from the data. To simplify the formulas, in the following we thus assume that $EX(\mathbf{s}) = 0$.

We consider analogs of methods M2 and M3. Extending Method M1 is possible, but presents a computational challenge because a parametric spatial model would need to be estimated for every pair (t_i, t_j). For the ionosonde data studied in Section 17.9, there are 336 points t_j. Estimation on a single data set would be feasible, but not a simulation study based on thousands of replications. In both approaches, which we term CM2 and CM3, the FPC's are estimated by expansions of the form

$$v_j(t) = \sum_{\alpha=1}^{K} x_\alpha^{(j)} B_\alpha(t), \tag{17.19}$$

where the B_α are elements of an *orthonormal* basis. We first describe an analog of method M3, which is conceptually and computationally simpler.

Method CM3. The starting point is the expansion

$$X(\mathbf{s}; t) = \sum_{j=1}^{\infty} b_j(\mathbf{s}) B_j(t),$$

where, by the orthonormality of the B_j, the $b_j(\mathbf{s})$ form an observable field $b_j(\mathbf{s}_k) = \langle B_j, X(\mathbf{s}_k)\rangle$. Using the orthonormality of the B_j again, we obtain

$$\begin{aligned} C(B_j) &= E\left[\left\langle \sum_{\alpha=1}^{\infty} b_\alpha(\mathbf{s}) B_\alpha, B_j \right\rangle \sum_{i=1}^{\infty} b_i(\mathbf{s}) B_i\right] \\ &= E\left[b_j(\mathbf{s}) \sum_{i=1}^{\infty} b_i(\mathbf{s}) B_i\right] \\ &= \sum_{i=1}^{\infty} E[b_i(\mathbf{s}) b_j(\mathbf{s})] B_i. \end{aligned} \tag{17.20}$$

Thus, to estimate C, we must estimate the means $E[b_i(\mathbf{s}) b_j(\mathbf{s})]$.

Fix i and j, and define the scalar field z by $z(\mathbf{s}) = b_i(\mathbf{s}) b_j(\mathbf{s})$. We can postulate a parametric model for the covariance structure of the field $z(\cdot)$, and use an empirical variogram to estimate $\mu_z = Ez(\mathbf{s})$ as a weighted average of the $z(\mathbf{s}_k)$. Denote the resulting estimate by $\hat{r}_{ij}$. The empirical version of (17.20) is then

$$\widehat{C}(B_j) = \sum_{i=1}^{K} \hat{r}_{ij} B_i. \tag{17.21}$$

Relation (17.21) defines the estimator $\widehat{C}$ which acts on the span of $B_j, 1 \le j \le K$. Its eigenfunctions are of the form $x = \sum_{1 \le \alpha \le K} x_\alpha B_\alpha$. Observe that

$$\widehat{C}(x) = \sum_\alpha x_\alpha \sum_i \hat{r}_{i\alpha} B_i = \sum_i \left(\sum_\alpha \hat{r}_{i\alpha} x_\alpha \right) B_i.$$

On the other hand,

$$\lambda x = \sum_i \lambda x_i B_i.$$

Since the B_i form an orthonormal basis, we obtain

$$\sum_\alpha \hat{r}_{i\alpha} x_\alpha = \lambda x_i.$$

Setting

$$\mathbf{x} = [x_1, x_2, \ldots, x_K]^T, \quad \widehat{\mathbf{R}} = [\hat{r}_{ij}, \ 1 \le i, j \le K],$$

we can write the above as a matrix equation

$$\widehat{\mathbf{R}}\mathbf{x} = \lambda \mathbf{x}. \tag{17.22}$$

Denote the solutions to (17.22) by

$$\hat{\mathbf{x}}^{(j)} = [\hat{x}_1^{(j)}, \hat{x}_2^{(j)}, \ldots, \hat{x}_k^{(j)}]^T, \ \hat{\lambda}_j, \quad 1 \le j \le K. \tag{17.23}$$

The $\hat{\mathbf{x}}^{(j)}$ satisfy $\sum_{\alpha=1}^K \hat{x}_\alpha^{(j)} \hat{x}_\alpha^{(i)} = \delta_{ij}$. Therefore the $\hat{v}_j$ defined by

$$\hat{v}_j = \sum_{\alpha=1}^K \hat{x}_\alpha^{(j)} B_\alpha \tag{17.24}$$

are also orthonormal (because the B_j are orthonormal). The $\hat{v}_j$ given by (17.24) are the estimators of the FPC's, and the $\hat{\lambda}_j$ in (17.23) of the corresponding eigenvalues.

As in method M3, the value of K can be taken to the number of basis functions used to create the functional objects in R, so it can be a relatively large number, e.g. $K = 49$. Even though the range of j in (17.23) and (17.24) runs up to K, only the first few estimated FPC's $\hat{v}_j$ would be used in further work.

Method CM2. Recall that under the assumption of zero mean function, the covariance operator is defined by $C(x) = E[\langle X(\mathbf{s}), x\rangle X(\mathbf{s})]$. For independent data it is estimated by the simple average

$$\frac{1}{N} \sum_{n=1}^N \langle X(\mathbf{s}_n), \cdot \rangle X(\mathbf{s}_n) = \frac{1}{N} \sum_{n=1}^N C_k, \tag{17.25}$$

where C_k is the operator defined by

$$C_k(x) = \langle X(\mathbf{s}_k), x\rangle X(\mathbf{s}_k).$$

As for the mean, more precise estimates can be obtained by using the weighted average

$$\widehat{C} = \sum_{k=1}^{N} w_k C_k. \tag{17.26}$$

Before discussing the estimation of the weights w_k, we explain how the FPC's v_j and their eigenvalues λ_j can be estimated using (17.26) and the representation (17.19). As in method CM3, set $x = \sum_{1 \le \alpha \le K} x_\alpha B_\alpha$, and observe that

$$\widehat{C}(x) = \sum_{j=1}^{K} \left(\sum_{\alpha=1}^{K} s_{j\alpha} x_\alpha \right) B_j,$$

where

$$s_{j\alpha} = \sum_{k=1}^{N} w_k \left\langle X_k, B_j \right\rangle \left\langle X_k, B_\alpha \right\rangle.$$

Thus, analogously to (17.22), we obtain a matrix equation $\mathbf{Sx} = \lambda \mathbf{x}$, from which the estimates of the v_j, λ_j can be found as in (17.23) and (17.24).

We now return to the estimation of the weights w_k in (17.26). One way to define the optimal weights is to require that they minimize the expected Hilbert–Schmidt norm of $\widehat{C} - C$. Recall that the Hilbert–Schmidt norm of an operator K is defined by

$$\|K\|_{\mathcal{S}}^2 = \sum_{i=1}^{\infty} \|K(e_i)\|^2 = \sum_{i=1}^{\infty} \int |K(e_i)(t)|^2 dt,$$

where $\{e_i, i \ge 1\}$ is any orthonormal basis in L^2. Since $\| \cdot \|_{\mathcal{S}}$ is a norm in the the Hilbert space $\mathcal{S}$ of the Hilbert–Schmidt operators with the inner product

$$\langle K_1, K_2 \rangle_{\mathcal{S}} = \sum_{i=1}^{\infty} \langle K_1(e_i), K_2(e_i) \rangle,$$

we can repeat all algebraic manipulations needed to obtain the weight w_i in (17.8). The optimal weights in (17.26) thus satisfy

$$\sum_{n=1}^{N} w_n = 1, \qquad \sum_{k=1}^{N} w_k \kappa_{kn} - r = 0, \quad n = 1, 2, \ldots, N, \tag{17.27}$$

where

$$\kappa_{k\ell} = E[\langle C_k - C, C_\ell - C \rangle_{\mathcal{S}}].$$

Finding the weights thus reduces to estimating the expected inner products $\kappa_{k\ell}$.

Since method M2 of Section 17.3 relies only on estimating inner product in the Hilbert space L^2, it can be extended to the Hilbert space $\mathcal{S}$. First observe that, analogously to (17.15),

$$E\|C_k - C_\ell\|_{\mathcal{S}}^2 = 2E\|C_k - C\|_{\mathcal{S}}^2 - 2\kappa_{k\ell}.$$

We can estimate the variogram

$$\gamma_C(d) = E\| \langle X(\mathbf{s}), \cdot \rangle X(\mathbf{s}) - \langle X(\mathbf{s}+\mathbf{d}), \cdot \rangle X(\mathbf{s}+\mathbf{d})\|_{\mathcal{S}}^2, \quad d = \|\mathbf{d}\|$$

by fitting a parametric model. In formulas (17.4) and (17.5), the squared distances $(X(\mathbf{s}_k) - X(\mathbf{s}_\ell))^2$ must be replaced by the squared norms $\|C_k - C_\ell\|_{\mathcal{S}}^2$. These norms are equal to

$$\|C_k - C_\ell\|_{\mathcal{S}}^2 = \sum_{i=1}^{\infty} \int (f_{ik} X_k(t) - f_{i\ell} X_\ell(t))^2 \, dt,$$

where

$$f_{ik} = \int X_k(t) e_i(t) dt.$$

The inner products f_{ik} can be computed using the `R` package `fda`.

17.5 Finite sample performance of the estimators

In this section, we report the results of a simulation study designed to compare the performance of the methods proposed in Sections 17.3 and 17.4 in a realistic setting motivated by the ionosonde data. It is difficult to design an exhaustive simulation study due to the number of possible combinations of the point distributions, dependence structures, shapes of mean functions and the FPC's and ways of implementing the methods (choice of spatial models, variogram estimation etc.). We do however think that our study provides useful information and guidance for practical application of the proposed methodology.

Data generating processes. We generate functional data at location $\mathbf{s}_k$ as

$$X(\mathbf{s}_k; t) = \mu(t) + \sum_{i=1}^{p} \xi_i(\mathbf{s}_k) v_i(t), \tag{17.28}$$

where the v_i are orthonormal functions, cf. model (17.1), and the scalar fields ξ_i are independent.

To evaluate the estimators of the mean, we use $p = 2$ and

$$v_1(t) = \sqrt{2}\sin(2\pi t \cdot 6), \quad v_2(t) = \sqrt{2}\sin(2\pi t/2). \tag{17.29}$$

We use two mean functions

$$\mu(t) = 2\sqrt{2}\sin(6\pi t) \tag{17.30}$$

and

$$\mu(t) = \sqrt{t}\sin(6\pi t). \tag{17.31}$$

The mean function (17.30) resembles the mean shape for the ionosonde data. It is however a member of the Fourier basis, and can be isolated using only one basis function, what could possibly artificially enhance the performance of method M3. We therefore also consider the mean function (17.31). Combining the mean function (17.30) and the FPC's (17.29), we obtain functions which very closely resemble the shapes of the ionosonde curves. In the above formulas, time is rescaled so that $t \in [0, 1]$.

To evaluate the estimators of the FPC's, we set

$$X(\mathbf{s}_k; t) = \xi_1(\mathbf{s}_k)\frac{e_1(t) + e_2(t)}{\sqrt{2}} + \xi_2(\mathbf{s}_k)e_3(t), \tag{17.32}$$

where $e_1(t) = \sqrt{2}\sin(2\pi t \cdot 7)$, $e_2(t) = \sqrt{2}\sin(2\pi t \cdot 2)$, $e_3(t) = \sqrt{2}\sin(3\pi t \cdot 3)$. Direct verification, which uses the independence of the fields ξ_1 and ξ_2, shows that the FPC's are $v_1 = 2^{-1/2}(e_1 + e_2)$ and $v_2 = e_3$.

To complete the description of the data generating processes, we must specify the dependence structure of the scalar spatial fields ξ_1 and ξ_2. We use the Gaussian and exponential models:

$$\begin{aligned} \text{Gaussian:} \quad & c(\mathbf{s}_k, \mathbf{s}_\ell) = c_0 + \sigma^2 \exp\{-d^2(k,\ell)/\rho^2\}, \\ \text{Exponential:} \quad & c(\mathbf{s}_k, \mathbf{s}_\ell) = c_0 + \sigma^2 \exp\{-d(k,\ell)/\rho\}. \end{aligned} \tag{17.33}$$

The distances are the chordal distances (17.2) between the locations described below. To make simulated data look similar to the real foF2 data we set $\sigma_1 = 1$, $\rho_1 = \pi/6$ for the field $\xi_1(\mathbf{s})$ and $\sigma_2 = 0.1$, $\rho_2 = \pi/4$ for $\xi_2(\mathbf{s})$. For the simulated data we set $c_0 = 0$. These parameters are the same for both the Gaussian and exponential models. They result in effective ranges that differ by about 20%.

The locations $\mathbf{s}_k$ are selected to match the locations of the real ionosonde stations. For the sample size 218 we use all available locations, shown in Figure 17.5. The selected 32 locations correspond to the ionosondes with the longest record history. The 100 stations were selected randomly out of the 218 stations.

Details of implementation. All methods require the specification of parametric spatial model for various variograms. Even though for some methods the variograms are defined for L^2– or $\mathcal{S}$–valued objects, only *scalar* models are required. In this simulation study, we employ the exponential model. Methods M3, CM2 and CM3 require the specification of a basis $\{B_j\}$ and the number K of the basis functions. We use the Fourier basis and $K = 1 + 4[\sqrt{\#\{t_j\}}]$, where $\#\{t_j\}$ is the count of the points at which the curves are observed. For our real and simulated data $K = 1 + 4[\sqrt{336}] = 73$, a number that falls between the recommended values of 49 and 99 for the number of basis functions. Specifically, the basis functions B_j are

$$\{1, \sqrt{2}\sin(2\pi i t), \sqrt{2}\cos(2\pi i t); \quad i = 1, 2, \ldots, 36\}. \tag{17.34}$$

All methods require the estimation of a parametric model on an empirical variogram. In our study we use only estimators (17.4) and (17.5), and refer to them, respectively, as MT and CH.

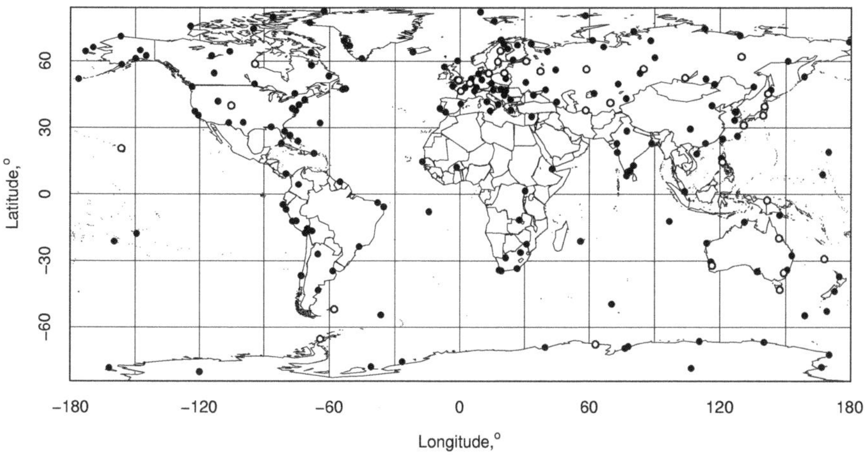

Fig. 17.5 Locations of 218 ionosonde stations. Circles represent the 32 stations with the longest complete records.

Results of the simulation study. For comparison of different methods we introduce the quantity L which is the average of the integrated absolute differences between real and estimated mean functions or FPC's. For the mean function, L is defined by

$$L = \frac{1}{R}\sum_{r=1}^{R}\int |\hat{\mu}_r(t) - \mu(t)|dt, \tag{17.35}$$

where R is the number of replications, we use $R = 10^3$. For the FPC's, the definition is fully analogous. We also compute the standard deviation for L, based on the normal approximation for R independent runs. We use the L^1 distance rather than the L^2 distance, so as not to favor a priori methods which minimiza the L^2 distance.

The results of the simulations for the mean function (17.31) are shown in Figure 17.6. The DGP's have exponential covariance functions. If the ξ_i in (17.28) have Gaussian covariances, the results are not visually distinguishable. The errors values for mean (17.30) are slightly different, but the relative position of the box plots practically does not change. All methods M1, M2 and M3 are significantly better than the sample average. Method M2 strikes the best balance between the computational cost and the precision of estimation.

Errors in the estimation of the FPC's in model (17.32) are shown in Figure 17.7. The displayed errors are those for the ξ_i with exponential covariances and the CH variogram. The results for Gaussian covariances and the MT variogram are practically the same. The performance of methods CM2 and CM3 is comparable, and they are both much better than using the eigenfunctions of the empirical covariance operator (17.25), which is the standard method implemented in the fda package. The computational complexity of methods CM2 and CM3 is the same.

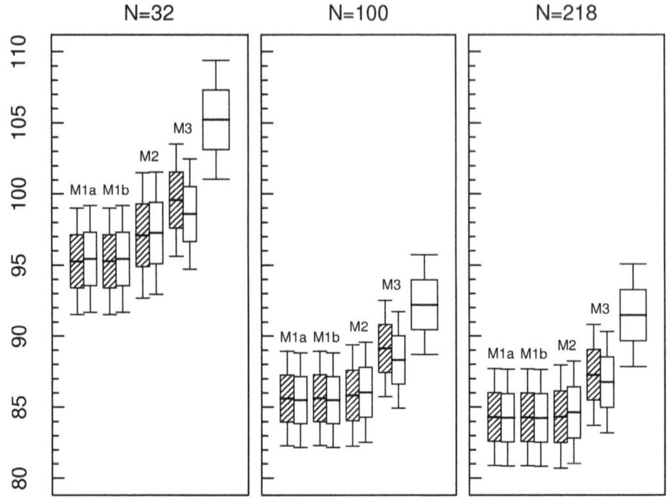

Fig. 17.6 Errors in the estimation of the mean function for sample sizes: 32, 100, 218. The dashed boxes are estimates using the CH variogram, empty are for the MT variogram. The right–most box for each N corresponds to the simple average. The bold line inside each box plot represents the average value of L (17.35). The upper and lover sides of rectangles shows one standard deviation, and horizontal lines show two standard deviations.

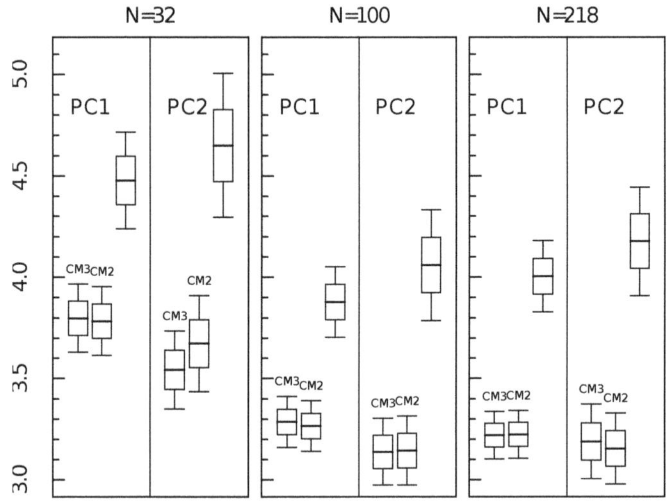

Fig. 17.7 Errors in the estimation of the FPC's for sample sizes: 32, 100, 218 . The bold line inside each box plot represents the average value of L. The upper and lover sides of rectangles shows one standard deviation, and horizontal lines show two standard deviations.

Conclusions. For simulated data generated to resemble the ionosonde data, all methods introduced in Sections 17.3 and 17.4 have integrated absolute deviations (away from a true curve) statistically significantly smaller than the standard methods designed for iid curves. Methods M2 and CM2, based on weighted averages estimated using functional variograms, offer a computationally efficient and unified approach to the estimation of the mean function and of the FPC's in this spatial setting.

17.6 Testing for correlation of two spatial fields

Motivated by the problem of testing for correlation between foF2 and magnetic curves, described in detail in Section 17.9, we now propose a relevant statistical significance test.

The data are observed at N spatial locations: $\mathbf{s}_1, \mathbf{s}_2, \ldots, \mathbf{s}_N$. At location $\mathbf{s}_k$, we observe two curves:

$$X_k = X(\mathbf{s}_k) = X(\mathbf{s}_k; t), \quad t \in [0, 1],$$

and

$$Y_k = Y(\mathbf{s}_k) = Y(\mathbf{s}_k; t), \quad t \in [0, 1].$$

We assume that the sample $\{X_k\}$ is a realization of a random field $\{X(\mathbf{s}), s \in \mathbf{S}\}$, and the sample $\{Y_k\}$ is a realization of a random field $\{Y(\mathbf{s}), s \in \mathbf{S}\}$. We want to test the null hypothesis:

H_0: for each $s \in \mathbf{S}$, the random functions $X(\mathbf{s})$ and $Y(\mathbf{s})$ are independent

against the alternative that H_0 does not hold. The test statistic will detect departures from H_0 that manifest themselves in the lack of the correlation between the projections $\langle x, X(\mathbf{s})\rangle$ and $\langle y, Y(\mathbf{s})\rangle$, for any $x, y \in L^2$. In fact, the lack of correlation will be tested only for x and y from sufficiently large subspaces, those spanned by the first p FPC's of $X(\mathbf{s})$, and the first q FPC's of $Y(\mathbf{s})$. The idea of the test, thus requires that the pairs $(X(\mathbf{s}), Y(\mathbf{s}))$ have the same distribution for every $s \in \mathbf{S}$. The construction of the test assumes that both fields, $\{X(\mathbf{s}), s \in \mathbf{S}\}$ and $\{Y(\mathbf{s}), s \in \mathbf{S}\}$ are strictly stationary, even though this assumption could be weakened to the stationarity of some fourth order moments. Since we provide only a heuristic derivation of the test, we are not concerned here with optimal assumptions. To lighten the notation, assume that

$$EX_k(t) = 0 \quad \text{and} \quad EY_n(t) = 0.$$

The mean functions will be estimated and subtracted using one of the methods of Section 17.3.

We now explain the idea of the test. We approximate the curves X_n and Y_n by the expansions

$$X_n(t) \approx \sum_{i=1}^{p} \langle X_n, v_i\rangle v_i(t), \quad Y_n(t) \approx \sum_{j=1}^{q} \langle Y_n, u_j\rangle u_j(t),$$

where the v_i and the u_j are the corresponding FPC's. At this point, the functions v_i, $1 \le i \le p$, and u_j, $1 \le j \le q$, are deterministic, so the independence of the curves X_n of the curves Y_n implies the independence of the vectors

$$[\langle X_n, v_1\rangle, \langle X_n, v_2\rangle, \ldots, \langle X_n, v_p\rangle]^T, \quad 1 \le n \le N$$

and

$$[\langle Y_n, u_1\rangle, \langle Y_n, u_2\rangle, \ldots, \langle Y_n, u_q\rangle]^T, \quad 1 \le n \le N.$$

Then, under H_0, the expected value of the sample covariances

$$A_N(i,j) = \frac{1}{N}\sum_{n=1}^{N} \langle X_n, v_i\rangle \langle Y_n, u_j\rangle \tag{17.36}$$

If some of the estimated $A_N(i,j)$ are too large, we reject the null hypothesis.

To construct a test statistic, we introduce the quantities

$$V_{k\ell}(i,i') = E[\langle v_i, X_k\rangle \langle v'_i, X_\ell\rangle], \quad U_{k\ell}(j,j') = E[\langle u_j, Y_k\rangle \langle u'_j, Y_\ell\rangle].$$

Note that $V_{k\ell}(i,i') = 0$ and $U_{k\ell}(j,j') = 0$, if the observations in each sample are independent (and have mean zero). Thus, the $V_{k\ell}(i,i')$ and the $U_{k\ell}(j,j')$ are specific to dependent data, they do not occur in the testing procedure developed in Chapter 9 for independent curves. Setting $X_{ik} = \langle v_i, X_k\rangle$, $Y_{jk} = \langle u_j, Y_k\rangle$, observe that if the X_{ik} are uncorrelated with the Y_{jk}, then

$$E[\sqrt{N}A_N(i,j)\sqrt{N}A_N(i',j')] = \frac{1}{N}E\left[\sum_{k=1}^{N} X_{ik}Y_{jk}\sum_{\ell=1}^{N} X_{i'\ell}Y_{j'\ell}\right]$$
$$= \frac{1}{N}\sum_{k=1}^{N}\sum_{\ell=1}^{N} E[X_{ik}X_{i'\ell}]E[Y_{jk}Y_{j'\ell}] = \frac{1}{N}\sum_{k=1}^{N}\sum_{\ell=1}^{N} V_{k\ell}(i,i')U_{k\ell}(j,j').$$

The covariance tensor of the $\sqrt{N}A_N(i,j)$ thus has the entries

$$\sigma_N(i,j;\, i',j') = \frac{1}{N}\sum_{k,\ell=1}^{N} V_{k\ell}(i,i')U_{k\ell}(j,j'). \tag{17.37}$$

The idea of the test, is to approximate the distribution of the matrix

$$\mathbf{A}_N = [A_N(i,j),\ 1 \le i \le p,\ 1 \le j \le q]$$

via $\sqrt{N}\mathbf{A}_N \approx \mathbf{Z}$, where $\mathbf{Z}$ is a $p \times q$ Gaussian matrix whose elements have covariances $E[Z(i,j)Z(i',j')] = \sigma_N(i,j;\, i',j')$.

We now explain how to implement this idea. Denote by $\hat{\lambda}_i, \hat{\gamma}_j$ and $\hat{v}_i, \hat{u}_j$ the eigenvalues and the eigenfunctions estimated either by method CM2 or CM3. The covariances $A_N(i,j)$ are then estimated by

$$\hat{A}_N(i,j) = \frac{1}{N}\sum_{n=1}^{N} \langle X_n, \hat{v}_i\rangle \langle Y_n, \hat{u}_j\rangle.$$

If the observations within each sample are independent, the test statistic introduced in Chapter 9 is

$$N \sum_{i=1}^{p} \sum_{j=1}^{q} \hat{\lambda}_i^{-1} \hat{\gamma}_j^{-1} \hat{A}_N^2(i, j).$$

Since $\lambda_i = E[\langle v_i, X \rangle^2]$, the sum above is essentially the sum of all correlations, and it usually tends to a chi–squared distribution with pq degrees of freedom, as shown in Chapter 9. This is however not necessarily true for dependent data. To explain, set $\mathbf{a}_N = \mathrm{vec}(\mathbf{A}_N)$, i.e. $\mathbf{a}_N$ is a column vector of length pq consisting of the columns of $\mathbf{A}_N$ stacked on top of each other, starting with the first column. Then $\sqrt{N}\mathbf{a}_N$ is approximated by a Gaussian vector $\mathbf{z}$ with covariance matrix $\boldsymbol{\Sigma}$ constructed from the entries (17.37). It follows that

$$\hat{S}_N = N \hat{\mathbf{a}}_N^T \hat{\boldsymbol{\Sigma}}^{-1} \hat{\mathbf{a}}_N \approx \chi^2_{pq}, \tag{17.38}$$

where $\hat{\mathbf{a}}_N = \mathrm{vec}(\hat{\mathbf{A}}_N)$, see e.g. Theorem 2.9 of Seber and Lee (2003), which states that for a zero mean normal vector $\mathbf{N}$ with covariance matrix $\boldsymbol{\Sigma}$, the quadratic form $\mathbf{N}^T \boldsymbol{\Sigma}^{-1} \mathbf{N}$ has chi–square distribution. The entries of the matrix $\hat{\boldsymbol{\Sigma}}$ are

$$\hat{\sigma}_N(i, j;\, i', j') = \frac{1}{N} \sum_{k,\ell=1}^{N} \hat{V}_{k\ell}(i, i') \hat{U}_{k\ell}(j, j'), \tag{17.39}$$

where $\hat{V}_{k\ell}(i, i')$ and $\hat{U}_{k\ell}(j, j')$ are estimators of $V_{k\ell}(i, i')$ and $U_{k\ell}(j, j')$, respectively. This estimation is discussed in Section 17.7. The test rejects H_0 if $\hat{S}_N > \chi^2_{pq}(1-\alpha)$, where $\chi^2_{pq}(1-\alpha)$ is the $100(1-\alpha)$th percentile of the chi–squared distribution with pq degrees of freedom. One can use Monte Carlo versions of the above test, for example the test based on the approximation

$$\hat{T}_N := N \hat{\mathbf{a}}_N^T \hat{\mathbf{a}}_N \approx \mathbf{w}^T \hat{\boldsymbol{\Sigma}} \mathbf{w}, \tag{17.40}$$

where the components of $\mathbf{w}$ are are iid standard normal.

The test procedure can be summarized as follows:

1. Subtract the mean functions, estimated by one of the methods of section 17.3, from both samples.
2. Estimate the FPC's by method CM2 or CM3.
3. Using a model for the covariance tensor (17.39), see Section 17.7, compute the test statistic $\hat{S}_N$. (This tensor is not needed to compute $\hat{T}_N$, but it is needed to find its Monte Carlo distribution.)
4. Find the P–value using either a Monte-Carlo distribution or the χ^2 approximation.

We now turn to the important issue of modeling and estimation of the matrix $\boldsymbol{\Sigma}$.

17.7 Modeling and estimation of the covariance tensor

The estimation of the $V_{k\ell}(i,i')$ involves only the X_n, and the estimation of the $U_{k\ell}(j,j')$ only the Y_n, so we describe only the procedure for the $V_{k\ell}(i,i')$. We assume that the mean has been estimated and subtracted, so we define

$$C_h(x) = E[\langle X(\mathbf{s}), x\rangle\, X(\mathbf{s}+\mathbf{h})], \quad h = \|\mathbf{h}\|.$$

The estimation of the $V_{k\ell}(i,i')$ relies on the identity

$$V_{k\ell}(i,i') = \langle C_h(v_i), v_{i'}\rangle\,, \quad h = d(\mathbf{s}_k,\mathbf{s}_\ell),$$

and an extension of the multivariate intrinsic model, see e.g. Chapter 22 of Wackernagel (2003). A most direct extension is to assume that

$$C_h = C r(h), \tag{17.41}$$

where C is a covariance operator, i.e. a symmetric positive definite operator with summable eigenvalues, and $r(h)$ is a correlation function of a scalar random field. Since $r(0) = 1$, we have $C = C_0$, so C in (17.41) must be the the covariance operator of each $X(\mathbf{s})$. If we assume the intrinsic model (17.41), then

$$V_{k\ell}(i,j) = \big\langle r(h)C(v_i), v_j\big\rangle = \lambda_i\delta_{ij} r(d(\mathbf{s}_k,\mathbf{s}_\ell)). \tag{17.42}$$

To allow more modeling flexibility, we postulate that

$$V_{k\ell}(i,j) = \lambda_i\delta_{ij} r_i(d(\mathbf{s}_k,\mathbf{s}_\ell)). \tag{17.43}$$

Under (17.42) (equivalently, under (17.41)), each scalar field $\langle X(\mathbf{s}), v_i\rangle$ has the same correlation function, only their variances are different. Under (17.43), the fields $\langle X(\mathbf{s}), v_i\rangle$ can have different correlation functions. As will be seen below, model (17.43) also leads to a valid covariance matrix. The correlations $r_i(d(\mathbf{s}_k,\mathbf{s}_\ell))$ and the variances λ_i can be estimated using a parametric model for the scalar field $\xi_i(\mathbf{s}) = \langle X(\mathbf{s}), v_i\rangle$. The resulting estimates $\hat{r}_i(d(\mathbf{s}_k,\mathbf{s}_\ell))$ and $\hat{\lambda}_i$ lead to the estimates $\hat{V}_{k\ell}(i,j)$ via (17.43). Analogous estimates of the functional field Y are $\hat{\gamma}_j(d(\mathbf{s}_k,\mathbf{s}_\ell))$, $\hat{\tau}_j$ and $\hat{U}_{k\ell}(i,j)$. For ease of reference, we note that under model (17.43) and H_0, the covariance tensor,

$$\left[\frac{1}{N}\sum_{k=1}^{N}\sum_{\ell=1}^{N}\hat{V}_{k\ell}(i,i')\hat{U}_{k\ell}(j,j'),\ 1\le i,i'\le p,\ 1\le j,j'\le q\right],$$

has the following matrix representation

$$\hat{\boldsymbol{\Sigma}} = \operatorname{diag}\left(\sum_{k=1}^{N}\sum_{\ell=1}^{N}\hat{\boldsymbol{\Sigma}}_{\xi_1}(k,\ell)\hat{\boldsymbol{\Sigma}}_{\eta_1}(k,\ell),\ldots,\sum_{k=1}^{N}\sum_{\ell=1}^{N}\hat{\boldsymbol{\Sigma}}_{\xi_p}(k,\ell)\hat{\boldsymbol{\Sigma}}_{\eta_q}(k,\ell)\right), \tag{17.44}$$

where

$$\hat{\boldsymbol{\Sigma}}_{\xi_i}(k,\ell) = \frac{1}{\sqrt{N}}\hat{\lambda}_i\hat{r}_i(d(\mathbf{s}_k,\mathbf{s}_\ell))$$

and

$$\hat{\boldsymbol{\Sigma}}_{\eta_j}(k,\ell) = \frac{1}{\sqrt{N}}\hat{\gamma}_j\hat{\tau}_j(d(\mathbf{s}_k,\mathbf{s}_\ell)).$$

This form is used to construct the Monte Carlo tests discussed in Section 17.8.

The matrix $\widehat{\boldsymbol{\Sigma}}$ with the estimates just specified is positive definite, as the following verification shows. (The matrix $\boldsymbol{\Sigma}$ is also positive definite by the same argument.) To verify that the matrix $\widehat{\boldsymbol{\Sigma}}$ is positive definite, we must show that

$$\sum_{i,j}\sum_{i',j'}\hat{\sigma}(i,j;i',j')b_{ij}b_{i'j'} \ge 0, \tag{17.45}$$

where $[b_{ij},\ 1 \le i \le p, 1 \le j \le q]$ is an arbitrary $p \times q$ matrix. Observe that

$$\begin{aligned}
&\sum_{i,j}\sum_{i',j'}\hat{\sigma}(i,j;i',j')b_{ij}b_{i'j'} \\
&= \frac{1}{N}\sum_{i,j}\sum_{i',j'}\sum_{k,\ell}\hat{V}_{k\ell}(i,i')\hat{U}_{k\ell}(j,j')b_{ij}b_{i'j'} \\
&= \frac{1}{N}\sum_{i,j}\sum_{i',j'}\sum_{k,\ell}\hat{\lambda}_i\delta_{ii'}\hat{r}_i(d(\mathbf{s}_k,\mathbf{s}_\ell))\hat{\gamma}_j\delta_{jj'}\hat{\tau}_j(d(\mathbf{s}_k,\mathbf{s}_\ell))b_{ij}b_{i'j'} \\
&= \frac{1}{N}\sum_{i,j}b_{ij}^2\sum_{k,\ell}\hat{\lambda}_i\hat{r}_i(d(\mathbf{s}_k,\mathbf{s}_\ell))\hat{\gamma}_j\hat{\tau}_j(d(\mathbf{s}_k,\mathbf{s}_\ell)).
\end{aligned}$$

Thus, (17.45) will follow once we have shown that for any i, j,

$$\sum_{k,\ell}\hat{\lambda}_i\hat{r}_i(d(\mathbf{s}_k,\mathbf{s}_\ell))\hat{\gamma}_j\hat{\tau}_j(d(\mathbf{s}_k,\mathbf{s}_\ell)) \ge 0.$$

Since $\hat{\lambda}_i\hat{r}_i$ is a covariance function, there are mean zero random variables

$$\varepsilon_1(i), \varepsilon_2(i), \ldots, \varepsilon_N(i) \tag{17.46}$$

such that $\hat{\lambda}_i\hat{r}_i(d(\mathbf{s}_k,\mathbf{s}_\ell)) = E[\varepsilon_k(i)\varepsilon_\ell(i)]$. Similarly, there are random variables

$$\eta_1(j), \eta_2(j), \ldots, \eta_N(j) \tag{17.47}$$

such that $\hat{\gamma}_j\hat{\tau}_j(d(\mathbf{s}_k,\mathbf{s}_\ell)) = E[\eta_k(j)\eta_\ell(j)]$. The families (17.46) and (17.47) can be assumed independent. Using the above construction, we obtain

$$\begin{aligned}
&\sum_{k,\ell}\hat{\lambda}_i\hat{r}_i(d(\mathbf{s}_k,\mathbf{s}_\ell))\hat{\gamma}_j\hat{\tau}_j(d(\mathbf{s}_k,\mathbf{s}_\ell)) = \sum_{k,\ell}E[\varepsilon_k(i)\varepsilon_\ell(i)]E[\eta_k(j)\eta_\ell(j)] \\
&= \sum_{k,\ell}E[\{\varepsilon_k(i)\eta_k(j)\}\{\varepsilon_\ell(i)\eta_\ell(j)\}] = E\left[\sum_{k=1}^{N}\varepsilon_k(i)\eta_k(j)\right]^2 \ge 0.
\end{aligned}$$

17.8 Size and power of the correlation test

As in Section 17.5, our objective is to evaluate the finite sample performance of the test introduced in Section 17.6 in a realistic setting geared toward the application presented in Section 17.9.

Data generating processes. We generate samples of zero mean Gaussian processes

$$X(\mathbf{s};t) = \sum_{i=1}^{p} \xi_i(\mathbf{s})v_i(t); \quad Y(\mathbf{s};t) = \sum_{j=1}^{q} \eta_j(\mathbf{s})u_j(t). \tag{17.48}$$

The process X is designed to resemble in distribution appropriately transformed and centered foF2 curves; the process Y the centered magnetic curves. Following the derivation presented in Section 17.9, we use $p = 7$ and $q = 1$. The curves v_i and u_1 are the estimated FPC's of the real data. The scalar Gaussian spatial fields ξ_i and η_1 follow parametric models estimated for real data, details of the models are presented in Table 17.1. The ξ_i are independent. Under H_0, the ξ_i are independent of η_1. The dependence under H_A can be generated in many ways. We considered the following scenarios: ξ_1 and η_1 are dependent, ξ_i and η_1 are independent for $i \neq 1$, then ξ_2 and η_1 are dependent, ξ_i and η_1 are independent for $i \neq 2$, etc. To produce two dependent spatial fields ξ_i and η, we generated N iid pairs $\mathbf{x}_i = [x_{1i}, x_{2i}]^T$, $1 \leq i \leq N$, where

$$\mathbf{x}_i \sim N\left(\mathbf{0}, \begin{pmatrix} 1 & \rho \\ \rho & 1 \end{pmatrix}\right).$$

Then we merged all x_{1i} into vector $\mathbf{y}_1 = [x_{11}, \ldots, x_{1N}]^T$ and all x_{2i} into vector $\mathbf{y}_2 = [x_{21}, \ldots, x_{2N}]^T$. Performing the Cholesky rotation, we obtain correlated spatial vectors:

$$\boldsymbol{\xi}_i = \mathbf{V}\mathbf{y}_1, \quad (\boldsymbol{\Sigma}_{\xi_i} = \mathbf{V}\mathbf{V}^T), \qquad \boldsymbol{\eta} = \mathbf{U}\mathbf{y}_2, \quad (\boldsymbol{\Sigma}_{\eta} = \mathbf{U}\mathbf{U}^T).$$

We used sample sizes $N = 32$ and $N = 100$ corresponding to the locations determined as in Section 17.5.

Testing procedures. We studied the finite sample behavior of three methods, which we call S, SM and T. Method S rejects H_0 if the statistic $\hat{S}_N$ (17.38) exceeds a chi–square critical value. Method SM uses a Monte Carlo distribution of the statistic $\hat{S}_N$: after estimating all parameters from the data and assuming the Gaussian distribution of the fields ξ_i and η_1, we can replicate the values of the statistic $\hat{S}_N$ under H_0 using the covariance matrix (17.44). Method T uses the statistics $\hat{T}_N$ (9.3), and approximates its distribution by the Monte Carlo distribution of $\mathbf{w}^T \hat{\boldsymbol{\Sigma}} \mathbf{w}$, as explained in Section 17.6. For determining the critical values in methods SM and T, we used 10^7 Monte Carlo replications. The empirical size and power are based on 10^5 independent runs.

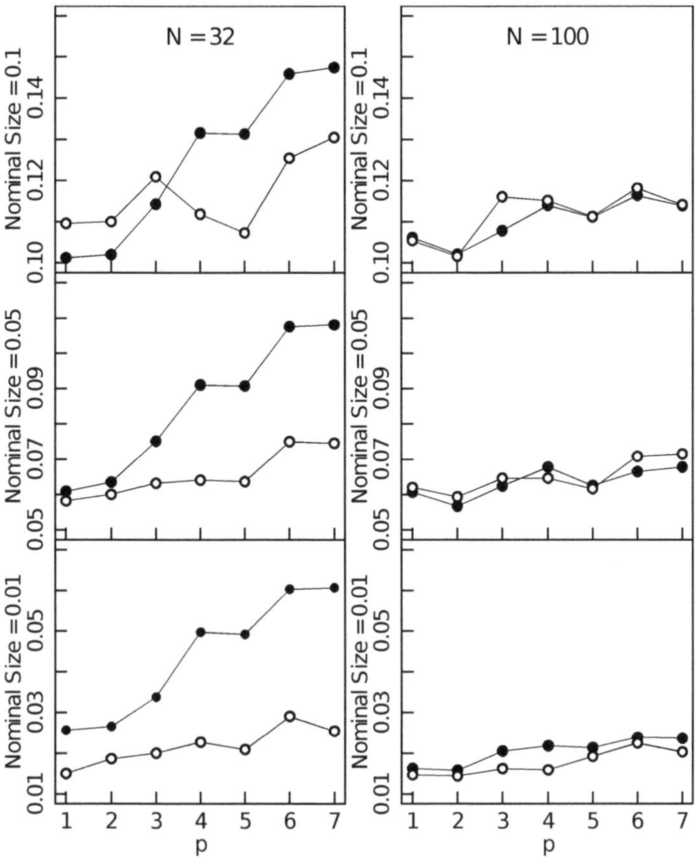

Fig. 17.8 Size of the correlation test as a function of p. Solid disks represent method S (based on χ^2 distribution). Circles represent method SM (based on the Monte-Carlo distribution).

Conclusions. As Figure 17.8 shows, the empirical size is higher than the nominal size, and it tends to increase with the number p of principal components used to construct the test, especially for $N = 32$. The usuall recommendation is to use p which explains about 85% of the variance. For the foF2 data with $N = 32$, this corresponds to $p = 4$. Tests of independence typically have larger than nominal size because real or simulated data may have some spurious dependencies; to put it simply, one cannot get "more independent data". Applied to real data in Section 17.9, all tests (S, SM and T) lead to extremely strong rejections, so the inflated empirical size is not a problem. Figure 17.8 also shows that the Monte Carlo approximation is useful for $N = 32$, this is the sample size we must use in Section 17.9. The size of test T is practically indistinguishable for that of test SM. Figure 17.9 shows the power of method SM; power curves for method T are practically the same, method S has higher power. The simulation study shows that a strong rejection when the test is applied to real data can be viewed as a reliable evidence of dependence.

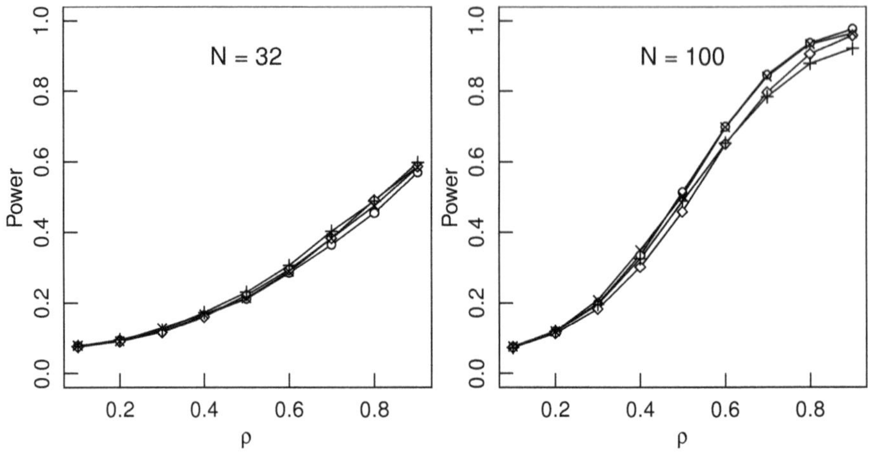

Fig. 17.9 Power of the correlation test SM as a function of the population correlation ρ. Each line represents one of the four possible correlated spatial field $\boldsymbol{\xi}_1 - \boldsymbol{\eta}, \boldsymbol{\xi}_2 - \boldsymbol{\eta}, \boldsymbol{\xi}_3 - \boldsymbol{\eta}, \boldsymbol{\xi}_4 - \boldsymbol{\eta}$. The test was performed using $p = 4$, which explains about 85% of variance of the foF2 curves. Since all curves in the graphs are practically the same, we do not specify which curve represents a particular dependent pair $\boldsymbol{\xi}_i - \boldsymbol{\eta}$.

17.9 Application to critical ionospheric frequency and magnetic curves

In this section, we apply the correlation test to foF2 and magnetic curves.

Description of the data. The F2 layer of the ionosphere is the upper part of the F layer shown in Figure 17.3. The F2 layer electron critical frequency, foF2, is measured using an instrument called the ionosonde, a type of radar. The foF2 frequency is used to estimate the location of the peak electron density, so an foF2 trend corresponds to a trend in the average height of the ionosphere over a spatial location. The foF2 data have therefore been used to test the hypothesis of ionospheric global cooling discussed in Section 17.1. Hourly values of foF2 are available from the SPIDR database `http://spidr.ngdc.noaa.gov/spidr/` for more than 200 ionosondes. We use monthly averages for 32 selected ionosondes, with sufficiently complete records, for the period 1964 – 1992. Their locations are shown in Figure 17.5. Three typical foF2 curves are shown in Figure 17.2. We omit the details of the procedure for obtaining curves like those shown in Figure 17.2, but we emphasize that it requires a great deal of work. In particular, the SPIDR data suffer from two problems. First, for some data, the amplitude is artificially magnified ten times, and needs to be converted into standard units (MHz). Second, in many cases, missing observations are not replaced by the standard notation 9999, but rather just skipped. Thus if one wants to use equally-spaced time series, skipped data must be found and replaced by missing values. For filling in missing values, we perform linear interpolation. We developed a customized C++ code to handle these issues. We emphasize that one of the reasons why this global data set has not been analyzed prior to the work of Gromenko *et al.* (2011) is that useable data had been derived only over relatively small regions, like Western Europe, see e.g. Bremer (1998), and more often only for a single location, see e.g. Lastovicka *et al.* (2006).

We use the foF2 data to test a hypotheses on long term ionospheric trends extending over several decades. We thus removed annual and higher frequency variations using 16 month averaging with MODWT filter, see Chapter 5 of Percival and Walden (2000). This leads to 32 time series at different locations, each containing 336 equally–spaced temporal observations. The amplitude of the foF2 curves exhibits a nonlinear latitudal trend; it decreases as the latitude increases, see Figure 17.2. To remove this trend, which may potentially bias the test, we assume that the foF2 signal, $F(\mathbf{s};t)$, at location $\mathbf{s}$ follows the model

$$F(\mathbf{s};t) = G(L(\mathbf{s}))X(\mathbf{s};t), \tag{17.49}$$

where $X(\mathbf{s};t)$ is a constant amplitude field, and $G(\cdot)$ is a scaling function which depends only on the *magnetic* latitude L (in radians). Since the trend in the amplitude of $F(\mathbf{s};t)$ is caused by the solar radiation which is nonlinearly proportional to the zenith angle, we postulate that the function $G(\cdot)$ has the form

$$G(L) = a + b\cos^c(L). \tag{17.50}$$

The parameters a, b, c are estimated as follows. Let $\mathbf{s}_0$ be the position of the ionosonde closest to the magnetic equator. For identifiability , we set $G(L(\mathbf{s}_0)) = 1$. For the remaining locations $\mathbf{s}_k$, we compute $\hat{G}(L(\mathbf{s}_k))$ as the average, over all 336 time points t_j of the ratio $F(\mathbf{s}_k;t_j)/F(\mathbf{s}_0;t_j)$. Figure 17.10 shows these ratios as a function of the magnetic and geographic latitude. The ratios in the magnetic latitude show much less spread, and this is another reason why we work with the magnetic latitude. The curve $G(L)$ (17.50) is fitted to the $\hat{G}(L(\mathbf{s}_k))$ in magnetic latitude by nonlinear least squares. The fitted values are $a = 0.5495, b = 0.4488, c = 4.2631$.

We now describe how we construct the curves that reflect the relevant long term changes in the internal magnetic field of the earth. The height of the F2 layer (and so the foF2 frequency) can be affected by a vertical plasma drift which responds to the magnetic field. The vertical plasma drift is due to the wind effect, and is given by (we use the same notations as in Mikhailov and Marin (2001))

$$W = (V_{nx}\cos D - V_{ny}\sin D)\sin I\cos I + V_{nz}\sin^2 I.$$

In the above formula, V_{nx} , V_{ny} and V_{nz} are, respectively, meridional (parallel to constant longitude lines), zonal (parallel to constant latitude lines) and vertical components of the thermospheric neutral wind; I and D are inclination and declination of the earth magnetic field, see Figure 13.2 in Kivelson and Russell (1997). Usually $V_{nz} \ll V_{nx}, V_{ny}$, and assuming that the difference between magnetic and geographic coordinates, D, is small (at least for low- and mid-latitude regions) we can simplify the above formula to $W = V_{nx}\sin I\cos I$. Thus, only the meridional thermospheric wind is significant. Measuring neutral wind components (V_{nx}, V_{ny}, V_{nz})

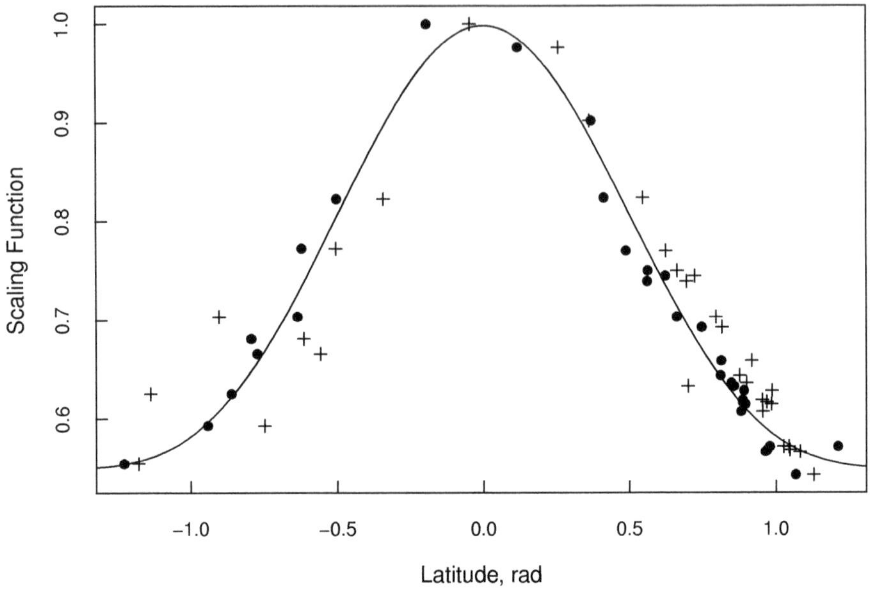

Fig. 17.10 Dots represent the scaling function $G_L(\mathbf{s}_i)$ in the magnetic coordinate system and crosses are same in the geographic coordinate system. Line is the best fit for G_L in the magnetic coordinate system.

is difficult, and long term wind records are not available. We therefore replace V_{nx} by its average. For our test, which uses correlations, the specific value of this average plays no role, so we define the magnetic curves as

$$Y(\mathbf{s};t) = \sin I(\mathbf{s};t) \cos I(\mathbf{s};t). \tag{17.51}$$

The curves $I(\mathbf{s};t)$ are computed using the international geomagnetic reference field (IGRF); the software is available at http://www.ngdc.noaa.gov/IAGA/vmod/.

The test is applied to the curves $X(\mathbf{s}_k;t)$ defined by (17.49) and (17.50), and to the curves $Y(\mathbf{s}_k;t)$ defined by (17.51).

Application of the correlation test. We first estimate and subtract the mean functions of the fields $X(\mathbf{s}_k)$ and $Y(\mathbf{s}_k)$ using method M2 (the other spatial methods give practically the same estimates). The principal components v_i and u_i are estimated using method CM2 (method CM3 gives practically the same curves).

We apply the test, for all $1 \le p \le 7$ and $q = 1$. The first seven eigenvalues of the field X (computed per (17.22) or its analog for method CM2) explain about 95% of the variance. The first eigenvalue of the field Y explains about 99% of the variance. The eigenfunction u_1 is approximately equal to the linear function: $u_1(t) \sim t$. This means that at any location, after removing the average, the magnetic field either linearly increases or decreases, with slopes depending on the location, see Figure 17.12. To lighten the notation, we drop the "hats" from the estimated scores and denote the zero mean vector $[\xi_i(\mathbf{s}_1), \ldots, \xi_i(\mathbf{s}_N)]^T$ by $\boldsymbol{\xi}_i$, and

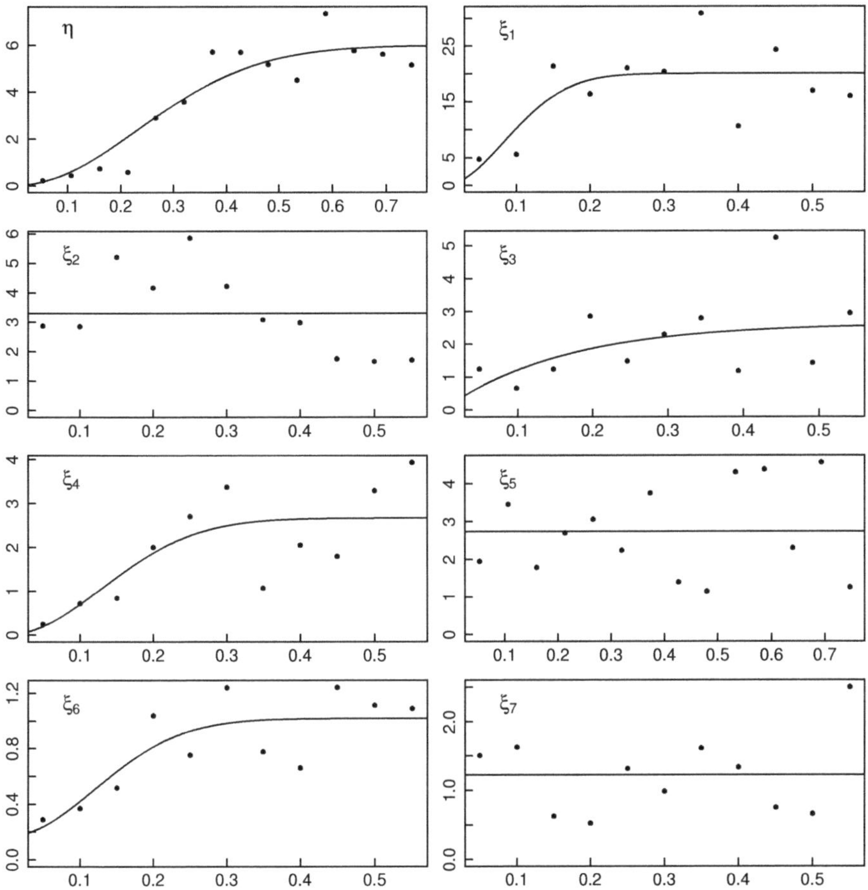

Fig. 17.11 Fitted empirical variograms for different spatial fields. The horizontal axes show normalized lag distance. The dots correspond to estimated variograms.

$[\eta_1(\mathbf{s}_1), \ldots, \eta_1(\mathbf{s}_N)]^T$ by $\boldsymbol{\eta}$ The covariances $\boldsymbol{\Sigma}_{\xi_i}$ and $\boldsymbol{\Sigma}_{\eta}$ are estimated using parametric spatial models determined by the inspection of the empirical variograms. In this application, it is sufficient to use only two covariance models, the Gaussian and the exponential models define in (17.33). When the scores do not have a spatial structure, we use the sample variance (flat variogram). The fitted variograms are shown in Figure 17.11, The estimated models and their parameters are listed in Table 17.1.

The P–values for different number of FPC's $1 \le p \le 7$ are summarized in Table 17.2. Independent of p and a specific implementation of the test, all P–values are very small, and so the rejection of the null hypothesis is conclusive; we conclude that there is a statistically significant correlation between the foF2 curves $X(\mathbf{s}_k)$ and the magnetic curves $Y(\mathbf{s}_k)$. We also applied the test of Chapter 9, which neglects any spatial dependence. The P-values for that test hover around the 5% level, but still

Table 17.1 Models and estimated covariance parameters for the transformed foF2 curves and the magnetic curves.

Spatial field	Model	Parameters		
		c_0	σ^2	ρ
η	Gaussian	–	5.99 ± 0.48	0.32 ± 0.04
ξ_1	Gaussian	–	20.05 ± 2.20	0.12 ± 0.03
ξ_2	–	–	3.30 ± 0.43	–
ξ_3	Exponential	–	2.63 ± 0.52	0.16 ± 0.07
ξ_4	Gaussian	–	2.66 ± 0.39	0.18 ± 0.05
ξ_5	–	–	2.74 ± 0.32	–
ξ_6	Gaussian	0.16 ± 0.02	0.85 ± 0.24	0.17 ± 0.06
ξ_7	–	–	1.22 ± 0.18	–

Table 17.2 P–values of the correlation tests applied to the transformed foF2 data. The first column shows the number of FPC's, the second column shows cumulative variances computed as the ratios of the eigenvalues estimated using method CM2. Testing procedures S, SM and T are defined in Section 17.8. The "simple" procedure neglects the spatial dependence of the curves.

p	CV, %	Spatial			Simple
		S	SM	T	
1	47.88	$6.22 \cdot 10^{-5}$	$3.05 \cdot 10^{-4}$	$3.05 \cdot 10^{-4}$	0.035
2	62.59	$3.26 \cdot 10^{-6}$	$2.91 \cdot 10^{-4}$	$2.99 \cdot 10^{-4}$	0.095
3	73.67	$4.53 \cdot 10^{-8}$	$2.43 \cdot 10^{-4}$	$2.32 \cdot 10^{-4}$	0.043
4	84.40	$1.47 \cdot 10^{-26}$	$1.6 \cdot 10^{-7}$	$2.24 \cdot 10^{-5}$	0.039
5	88.70	$4.95 \cdot 10^{-26}$	$2.6 \cdot 10^{-7}$	$2.27 \cdot 10^{-5}$	0.046
6	92.21	$6.73 \cdot 10^{-27}$	$5.9 \cdot 10^{-7}$	$2.21 \cdot 10^{-5}$	0.060
7	94.57	$2.12 \cdot 10^{-32}$	$1.6 \cdot 10^{-7}$	$1.92 \cdot 10^{-5}$	0.030

point toward rejection. The evidence is however much less clear cut. This may partially explain why this issue has been a matter of much debate in the space physics community. The correlation between the foF2 and magnetic curves is far from obvious. Figure 17.12 shows these pairs at all 32 locations. It is hard to conclude by eye the the direction of the magnetic field change impacts the foF2 curves.

Discussion. A very important role in our analysis is played by the transformation (17.50). Applying the test to the original foF2 curves $F(\mathbf{s}_k;t)$, gives the P–values 0.209 ($p = 1$) and 0.011 ($p = 2$) for the spatial S test, and 0.707 ($p = 1$), 0.185 ($p = 2$), 0.139 ($p = 3$) for the "simple" test. These values of p explain over 90% of the variance. As explained above, the amplitude of the field $F(\mathbf{s}_k;t)$ evolves with the latitude. This invalidates the assumption of a mean function which is independent of the spatial location. Thus even for the spatial test, the mean function confounds the first FPC. However, the spatial estimation of the mean function and of the FPC's "quickly corrects" for the violation of assumptions, and the null hypothesis is rejected for $p \geq 2$. When the spatial structure is neglected (and no latitudal transformation is applied) no correlation between the foF2 curves and magnetic curves is found.

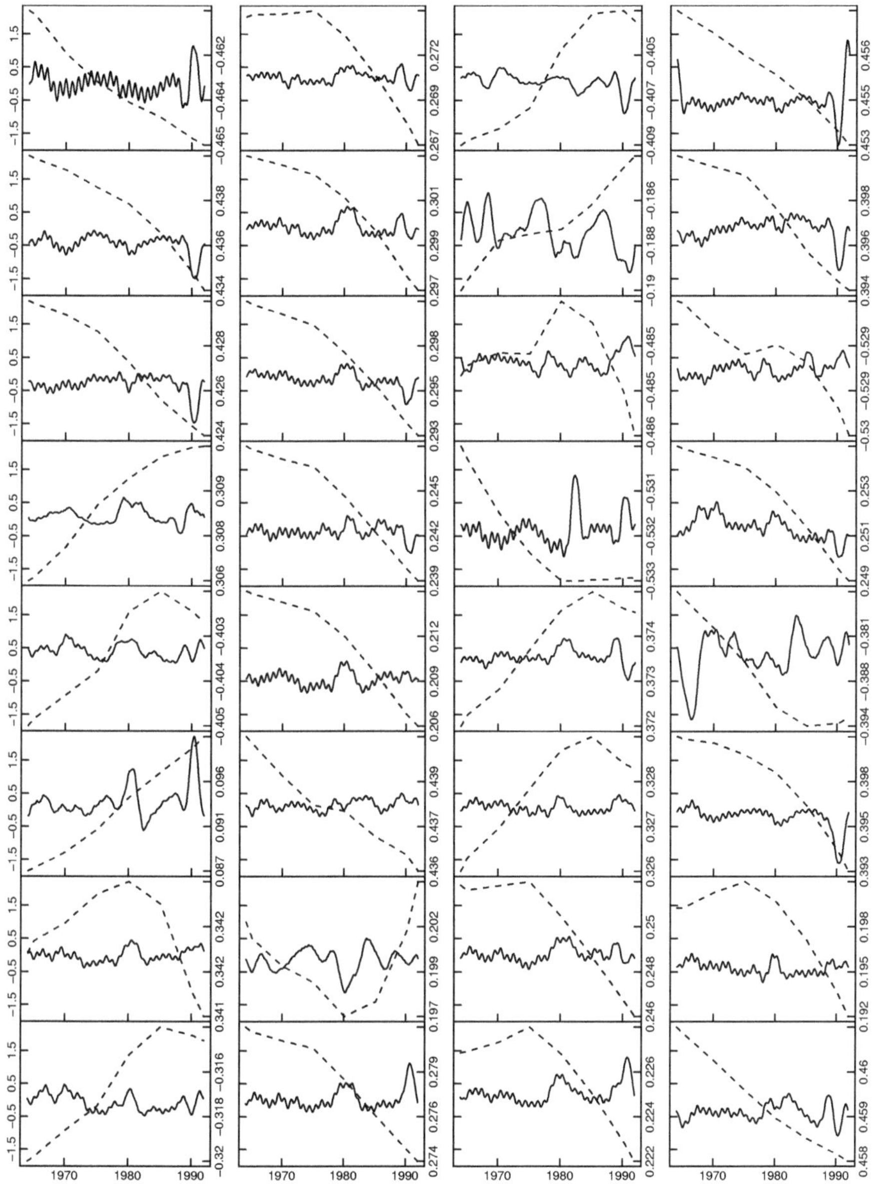

Fig. 17.12 Transformed and centered foF2 curves (continuous) and centered magnetic curves (dashed) at 32 locations denoted with circles in Figure 17.5. The scales for the two families of curves are different: the foF2 curves have the same scale, the scale of the magnetic curves changes in each box, to show the direction of the long term trend.

The rejection of the null hypothesis means that after adjusting the foF2 curves for the latitude and the global mean, their regional variability is correlated with the regional changes in the in the magnetic field. This means that long term magnetic trends must be considered as additional covariates in testing for long term trends in the foF2 curves. (The main covariate is the solar activity which drives the shape of the mean function.)

A broader conclusion of the work presented in this chapter is that methods of functional data analysis must be applied with care to curves obtained at spatial locations. Neglecting the spatial dependence can lead to incorrect conclusions and biased estimates. The same applies to space physics research. If trends or models are estimated separately at each spatial location, one should not rely on results obtained by some form of a simple averaging. Interestingly, the results related to global ionospheric trends are often on the borderline of statistical significance if the spatial dependence structure is neglected. Standard t-tests lead either to rejection or acceptance, depending on a specific method used (a similar phenomenon is observed in the last column of Table 17.2). It is hoped that the methodology presented in this Chapter will be useful in addressing such issues.

Chapter 18
Consistency of the simple mean and the empirical functional principal components for spatially distributed curves

In this chapter, we continue to study functional data that consist of curves $X(\mathbf{s}_k;t)$, $t \in [0,1]$, observed at spatial points $\mathbf{s}_1, \mathbf{s}_2, \ldots, \mathbf{s}_N$. In Chapter 17, we have seen that in this context the simple sample average and the EFPC's are not the optimal estimators of their population counterparts, and that better estimators can be constructed by using weighted averages. The simple sample average and the EFPC's are however often default estimators, and it is important to understand when they are consistent. In this chapter, we establish conditions for them to be consistent. These conditions involve an interplay of the assumptions on an appropriately defined dependence between the functions $X(\mathbf{s}_k)$ and the assumptions on the spatial distribution of the points $\mathbf{s}_k$. The rates of convergence may be the same as for iid functional samples, but generally depend on the strength of dependence and appropriately quantified distances between the points $\mathbf{s}_k$. We also formulate conditions for the lack of consistency. The general results are established using an approach based on the estimation of the expected moments of an appropriate norm of the difference between the estimator and the estimand which splits the norm into terms that reflect the assumptions on the spatial dependence and the distribution of the points. This technique is broadly applicable to all statistics obtained by simple averaging of functional data. We specialize the general rates of consistency to spatial functional models of practical interest.

A general theoretical framework has to address several problems. The first issue is the dimensionality of the index space. While in time series analysis, the process is indexed by an equispaced scalar parameter, we need here a d-dimensional index space. For model building this makes a big difference since the dynamics and dependence of the process have to be described in all directions, and the typical recurrence equations used in time series cannot be employed. The model building is further complicated by the fact that the index space is often continuous (geostatistical data). Rather than defining a random field $\{\xi(\mathbf{s}); \mathbf{s} \in \mathbb{R}^d\}$ via a specific model equations, dependence conditions are imposed, in terms of the decay of the covariances or using mixing conditions. Another feature peculiar to random field theory is the design of the sampling points; the distances between them play a fundamental role. Different asymptotics hold in the presence of clusters and for sparsely distributed points. At least three types of point distributions have been considered, see Cressie (1993): When the region R_N where the points $\{\mathbf{s}_{i,N}; 1 \le i \le N\}$ are sampled remains bounded, then we are in the so-called *infill domain sampling* case. Classical asymptotic results, like the law of large numbers or the central limit theorem will usually fail, see Lahiri (1996). The other extreme situation is described by the *increasing domain sampling*. Here a minimum separation between the sampling points $\{\mathbf{s}_{i,N}\} \in R_N$ for all i and N is required. This is of course only possible if $\mathrm{diam}(R_N) \to \infty$. We shall also explore the *nearly infill* situation studied by Lahiri (2003) and Park *et al.* (2009). In this case, the domain of the sampling region becomes unbounded ($\mathrm{diam}(R_N) \to \infty$), but at the same time the number of sites in any given subregion tends to infinity, i.e. the points become more dense. These issues are also studied by Zhang (2004), Loh (2005), Lahiri and Zhu (2006), Du *et al.* (2009). We formalize these concepts in Sections 18.2 and 18.3. Finally, the interplay of the geostatistical spatial structure and the functional temporal structure must be cast into a workable framework.

For the reasons explained above, the framework advocated in Chapter 16, designed for functional time series, is inappropriate for functional spatial fields. The starting point for the theory of Chapter 16 is the representation $X_k = f(\varepsilon_k, \varepsilon_{k-1}, \ldots)$ of a function X_k in terms of iid error functions ε_k. While all time series models used in practice admit such a representation, no analog representations exist for geostatistical spatial data. (Even though not widely used, spatial autoregressive processes have been proposed, but no Volterra (nonlinear moving average) type expansions have been developed for them.)

This chapter is organized as follows. Section 18.1 describes in greater detail the objectives of this research by developing several examples which show how spatially distributed functional data differ from functional random samples and from functional time series. In simple settings, it illustrates what kind of consistency or inconsistency results can be expected, and what kind of difficulties must be overcome. After the stage has been set, we formulate the asymptotic assumptions in Section 18.2. A crucial part of these assumptions consists of conditions on the spatial distribution of the points $\mathbf{s}_k$. Section 18.3 compares our conditions to those typically assumed for scalar spatial processes. In Sections 18.4 and 18.5, we establish consistency results, respectively, for the functional mean and the covariance operator. These sections also contain examples specializing the general results to more specific settings. Section 18.6 explains, by means of general theorems and examples, when the sample principal components are not consistent. The proofs are collected in Section 18.7.

18.1 Motivating examples

We have shown in this book that the FPC's play a fundamental role in functional data analysis, much greater than the usual multivariate principal components. This is mostly due to the fact that the Karhunen-Loève expansion allows to represent functional data in a concise way, a property we have used in various settings. Depending on the structure of the data, various aspects of the estimation of the FPC's are emphasized. This section illustrates the role of the spatial dependence and the distribution of the curves. As in Chapter 17, we assume that the observed curves are identically distributed, so that the mean function and the FPC's are well–defined.

In Section 2.5, we saw that if the functional observations X_k are independent, then

$$\limsup_{N\to\infty} NE\|\widehat{C}_N - C\|_{\mathcal{S}}^2 < \infty \tag{18.1}$$

and that, consequently,

$$\max_{1\le k\le d} E\|\hat{c}_k\hat{v}_k - v_k\|^2 = O\left(N^{-1}\right).$$

In Section 16.2, we showed that (18.1) continues to hold for weakly dependent time series, in particular for m–dependent X_k. Our first example shows why m–dependence does not necessarily imply (18.1) for spatially distributed curves.

Example 18.1. Suppose $X_k = X(\mathbf{s}_k)$, where $\mathbf{s}_1, \mathbf{s}_2, \ldots, \mathbf{s}_N$ are points in an arbitrary metric space, and the random field $X(\cdot)$ is such that $X(\mathbf{s})$ is independent of $X(\mathbf{s}')$ if the distance between $\mathbf{s}$ and $\mathbf{s}'$, $d(\mathbf{s},\mathbf{s}')$, is greater than m. Set

$$B_N(m) = \{(k,\ell):\ 1\le k,\ell\le N \quad \text{and} \quad d(\mathbf{s}_k,\mathbf{s}_\ell)\le m\},$$

and denote by $|B_N(m)|$ the count of pairs in $B_N(m)$. A brief calculation which uses the identity

$$NE\|\widehat{C}_N - C\|_{\mathcal{S}}^2 = N\iint \operatorname{Var}\left\{N^{-1}\sum_{k=1}^{N}(X_k(t)X_k(s) - E[X_k(t)X_k(s)])\right\} dt\, ds \tag{18.2}$$

and the Cauchy inequality, leads to the bound

$$NE\|\widehat{C}_N - C\|_{\mathcal{S}}^2 \le N^{-1}\,|B_N(m)|\,E\|X(\mathbf{s})\|^4.$$

If the $\mathbf{s}_k$ are the points in $\mathbb{R}^d$ with integer coordinates, then $|B_N(m)|$ is asymptotically proportional to mN, implying $\limsup_{N\to\infty} N^{-1}\,|B_N(m)| < \infty$, and the standard rate (18.1). But if there are too many pairs in $B_N(m)$ this rate will no longer hold.

Example 18.1 shows that if the points $\mathbf{s}_k$ are not equispaced and too densely distributed, then the standard rate (18.1) need not hold. The next example shows that in such cases the EFPC's $\hat{v}_k$ may not converge at all.

Example 18.2. This example presents only an intuitive idea. A more precise argument, with a numerical example, is developed in Example 18.6.

Consider a functional random field

$$X(\mathbf{s};t) = \sum_{j=1}^{\infty} \xi_j(\mathbf{s}) e_j(t), \quad \mathbf{s} \in \mathbb{R}^d, \ t \in [0,1], \tag{18.3}$$

where $\{e_j, j \geq 1\}$ is a complete orthonormal system and the $\xi_j(\mathbf{s})$ are mean zero random variables with $E[\xi_j(\mathbf{s})\xi_j(\mathbf{s}+\mathbf{h})] = \lambda_j \rho_j(h), h = \|\mathbf{h}\|$, where $\sum_{j=1}^{\infty} \lambda_j < \infty$ and each $\rho_j(\cdot)$ is a positive correlation function. Direct verification shows that $C(x) = \sum_{j=1}^{\infty} \lambda_j \langle e_j, x \rangle e_j$, so the λ_j are the eigenvalues of C, and the e_j the corresponding eigenfunctions.

Now consider a sequence $\mathbf{s}_n \to \mathbf{0}$. Because of the positive dependence, $X(\mathbf{s}_n)$ is close to $X(\mathbf{0})$, so $\widehat{C}_N$, as an arithmetic average, is close to the random operator $X^\star = \langle X(\mathbf{0}), \cdot \rangle X(\mathbf{0})$. Observe that $X^\star(X(\mathbf{0})) = \|X(\mathbf{0})\|^2 X(\mathbf{0})$. Thus $\|X(\mathbf{0})\|^2 = \sum_{j=1}^{\infty} \xi_j^2(\mathbf{0})$ is an eigenvalue of $X^\star$. Since it is random, it cannot be close to any of the λ_j. The eigenfunctions of $\widehat{C}_N$ are also close to random functions in L^2, and do not converge to the FPC's e_j.

The above examples show that if the points $\mathbf{s}_n$ are too close to each other, then the empirical functional principal components are not consistent estimates of the population principal components. Other examples of the lack of consistency are known, see Johnstone and Lu (2009) and references therein. They fall into the "small n large p" framework, and the lack of consistency is due to noisy data which are not sparsely represented. A solution is to perform the principal component analysis on transformed data which admits a sparse representation. (A different asymptotic approach is taken by Jung and Marron (2009).) The spatial functional data introduced in Chapter 17 admit natural sparse representations; the lack of consistency may be due to dependence and densely distributed locations of the observations. It is actually not crucial that the $\mathbf{s}_n$ be close to each other. What matters is the interplay of the spatial distances between these points and the strength of dependence between the curves. To illustrate, suppose in Example 18.2, the covariance between $X(\mathbf{s}_n)$ and $X(\mathbf{0})$ is

$$E[\langle X(\mathbf{s}_n), x \rangle \langle X(\mathbf{0}), y \rangle] = \sum_{j=0}^{\infty} \lambda_j \exp\left\{-\frac{\|\mathbf{s}_n\|}{\rho_j}\right\} \langle e_j, x \rangle \langle e_j, y \rangle.$$

In a finite sample, small $\|\mathbf{s}_n\|$ have the same effect as large ρ_j, i.e. as stronger dependence.

These considerations show that it is useful to have general criteria for functional spatial data, which combine the spatial distribution of the points and the strength of dependence, and which ensure that the functional principal components can be consistently estimated, and, consequently, that further statistical inference for spatial functional data can be carried out. Such criteria should hold for practically useful models for functional spatial data. The next example discusses such models, with a rigorous formulation presented in Section 18.2.

Example 18.3. Suppose $\{e_j, j \geq 1\}$ is an arbitrary *fixed* orthonormal basis in L^2. Under very mild assumptions, every constant mean functional random field admits the representation

$$X(\mathbf{s}) = \mu + \sum_{j \geq 1} \xi_j(\mathbf{s}) e_j, \tag{18.4}$$

where the $\xi_j(\mathbf{s})$ are zero mean random variables. In principle, all properties of X, including the spatial dependence structure, can be equivalently stated as properties of the family of the scalar fields ξ_j. Representation (18.4) is thus the most natural and convenient model for spatially distributed functional data.

Assume that $\mu = 0$ and the field X is strictly stationary (in space). (Strict stationarity can be replaced by weaker moment conditions formulated in Section 18.2.) Suppose we want to predict $X(\mathbf{s}_0)$ using a linear combination of the curves $X(\mathbf{s}_1), X(\mathbf{s}_2), \ldots, X(\mathbf{s}_N)$, i.e. we want to minimize

$$\begin{aligned} E\left\| X(\mathbf{s}_0) - \sum_{n=1}^{N} a_n X(\mathbf{s}_n) \right\|^2 \\ = E\left\langle X(\mathbf{s}_0), X(\mathbf{s}_0) \right\rangle - 2 \sum_{n=1}^{N} a_n E\left\langle X(\mathbf{s}_n), X(\mathbf{s}_0) \right\rangle \\ + \sum_{k,\ell=1}^{N} a_k a_\ell E\left\langle X(\mathbf{s}_k), X(\mathbf{s}_\ell) \right\rangle . \end{aligned} \tag{18.5}$$

Thus for the problem of the least squares linear prediction of a mean zero spatial process (kriging), we need to know only

$$K(\mathbf{s}, \mathbf{s}') = E\left[\left\langle X(\mathbf{s}), X(\mathbf{s}') \right\rangle\right]. \tag{18.6}$$

By the orthonormality of the e_j in (18.18),

$$\begin{aligned} E\left[\left\langle X(\mathbf{s}), X(\mathbf{s}') \right\rangle\right] &= E\left[\left\langle \sum_{j=1}^{\infty} \xi_j(\mathbf{s}) e_j, \sum_{i=1}^{\infty} \xi_i(\mathbf{s}') e_i \right\rangle\right] \\ &= \sum_{j=1}^{\infty} \sum_{i=1}^{\infty} E[\xi_j(\mathbf{s}) \xi_i(\mathbf{s}')] \left\langle e_j, e_i \right\rangle = \sum_{j=1}^{\infty} E[\xi_j(\mathbf{s}) \xi_j(\mathbf{s}')]. \end{aligned}$$

Thus, the functional covariances (18.6) are fully determined by the covariances

$$K_j(\mathbf{s}, \mathbf{s}') = E[\xi_j(\mathbf{s}) \xi_j(\mathbf{s}')]. \tag{18.7}$$

Notice that we do not need to know the cross covariances $E[\xi_j(\mathbf{s}) \xi_i(\mathbf{s}')]$ for $i \neq j$. Thus, if we are interested in kriging, we can assume that the spatial processes $\xi_j(\cdot)$ in (18.18) are independent. Such an assumption simplifies the verification of some fourth order properties discussed in the following sections. This observation remains true if the spatial field does not have zero mean, i.e. if we observe realizations of $Z(\mathbf{s}) = \mu(\mathbf{s}) + X(\mathbf{s})$. A brief calculation shows that for kriging, it is enough to know $\mu(\cdot)$ and the covariances (18.7). Stein (1999) and Cressie (1993) provide rigorous accounts of kriging for scalar spatial data.

Our next example shows how representation (18.4) and the independence of the ξ_j allow to derive the standard rate (18.1), if the points $\mathbf{s}_k$ are equispaced on the line and the covariances decay exponentially. In the following sections, we construct a theory that allows us to obtain the standard and nonstandard rates of consistency in much more general settings. We will use the following well–known Lemma, which follows from a direct verification using the bivariate normal density.

Lemma 18.1. *Suppose X and Y are jointly normal mean zero random variables such that $EX^2 = \sigma^2,\ EY^2 = \nu^2,\ E[XY] = \rho\sigma\nu$. Then*

$$\mathrm{Cov}(X^2, Y^2) = 2\rho^2\sigma^2\nu^2.$$

Example 18.4. Suppose $X(\mathbf{s};t)$ is an arbitrary functional random field observed at locations $\mathbf{s}_1, \mathbf{s}_2, \ldots, \mathbf{s}_N$. Then, by (18.2),

$$\begin{aligned} &NE\|\widehat{C} - C\|_{\mathcal{S}}^2 \\ &= N^{-1} \sum_{k,\ell=1}^{\infty} \iint \mathrm{Cov}(X(\mathbf{s}_k;t)X(\mathbf{s}_k;u),\ X(\mathbf{s}_\ell;t)X(\mathbf{s}_\ell;u))dt\,du. \end{aligned} \tag{18.8}$$

Without any further assumptions, a sufficient condition for the EFPC's to be consistent with the rate $N^{-1/2}$ is that the right–hand side of (18.8) is bounded from above by a constant. Under additional assumptions, more precise sufficient conditions are possible.

Suppose first that representation (18.18) holds with independent strictly stationary scalar fields $\xi_j(\cdot)$. Define the covariances

$$E[\xi_j(\mathbf{s}_k)\xi_j(\mathbf{s}_\ell)] = \gamma_j(\mathbf{s}_k - \mathbf{s}_\ell), \quad \mathrm{Cov}(\xi_j^2(\mathbf{s}_k), \xi_j^2(\mathbf{s}_\ell)) = \tau_j(\mathbf{s}_k - \mathbf{s}_\ell).$$

Using (18.8), we see that under these assumptions,

$$NE\|\widehat{C} - C\|_{\mathcal{S}}^2 = N^{-1} \sum_{k,\ell=1}^{\infty} \left\{ \sum_{i \neq j} \gamma_i(\mathbf{s}_k - \mathbf{s}_\ell)\gamma_j(\mathbf{s}_k - \mathbf{s}_\ell) + \sum_{j=1}^{\infty} \tau_j(\mathbf{s}_k - \mathbf{s}_\ell) \right\}.$$

Thus (18.1) holds, if

$$\limsup_{N\to\infty} N^{-1} \sum_{k,\ell=1}^{N} \left\{ \sum_{j=1}^{\infty} \gamma_j(\mathbf{s}_k - \mathbf{s}_\ell) \right\}^2 < \infty \tag{18.9}$$

and

$$\limsup_{N\to\infty} N^{-1} \sum_{k,\ell=1}^{N} \sum_{j=1}^{\infty} \left|\tau_j(\mathbf{s}_k - \mathbf{s}_\ell)\right| < \infty. \tag{18.10}$$

Suppose now, in addition, that X is Gaussian with

$$E[\xi_j(\mathbf{s}_k)\xi_j(\mathbf{s}_\ell)] = \sigma_j^2 \exp\left\{-\rho_j^{-1} d(\mathbf{s}_k, \mathbf{s}_\ell)\right\} \tag{18.11}$$

so that

$$\mathrm{Cov}(\xi_j^2(\mathbf{s}_k), \xi_j^2(\mathbf{s}_\ell)) = 2\sigma_j^4 \exp\left\{-2\rho_j^{-1} d(\mathbf{s}_k, \mathbf{s}_\ell)\right\}. \tag{18.12}$$

Suppose the points $\mathbf{s}_k$ are equispaced on the line. Denoting the smallest distance between the points by d, we see that

$$\begin{aligned} N^{-1} \sum_{k,\ell=1}^{N} & \left\{ \sum_{j=1}^{\infty} \gamma_j(\mathbf{s}_k - \mathbf{s}_\ell) \right\}^2 \\ &= \sum_{j=1}^{\infty} \sigma_j^2 + 2N^{-1} \sum_{m=1}^{N-1} (N-m) \left\{ \sum_{j=1}^{\infty} \sigma_j^2 \exp\left(-\rho_j^{-1} md\right) \right\}^2. \end{aligned}$$

If we assume that

$$\sum_{j \ge 1} \sigma_j^2 < \infty \quad \text{and} \quad \sup_{j \ge 1} \rho_j < \rho < \infty, \tag{18.13}$$

then Conditions (18.9) and (18.10), and so the standard rate (18.1), hold. Condition (18.13) means that the correlation functions of all processes $\xi_j(\cdot)$ must decay uniformly sufficiently fast.

To verify (18.9), observe that

$$\begin{aligned} N^{-1} \sum_{m=1}^{N-1} & (N-m) \left\{ \sum_{j=1}^{\infty} \sigma_j^2 \exp\left(-\rho_j^{-1} md\right) \right\}^2 \\ &\le \sum_{m=1}^{N-1} \left\{ \sum_{j=1}^{\infty} \sigma_j^2 \exp\left(-\rho_j^{-1} md\right) \right\}^2 \\ &\le \sum_{m=1}^{N-1} \left\{ \sum_{j=1}^{\infty} \sigma_j^2 \exp\left(-\rho^{-1} md\right) \right\}^2 \\ &= O(1) \left(\sum_{j=1}^{\infty} \sigma_j^2 \right)^2 = O(1). \end{aligned}$$

The verification of (18.10) is analogous because (18.13) implies $\sum_{j=1}^{\infty} \sigma_j^4 < \infty$.

We will see that Condition (18.13) (formulated analogously for several classes of models) is applicable in much more general settings than equispaced points on the line.

18.2 Models and Assumptions

We assume that $\{X(\mathbf{s}), \mathbf{s} \in \mathbb{R}^d\}$ is a random field taking values in $L^2 = L^2([0,1])$, i.e. each $X(\mathbf{s})$ is a square integrable function defined on $[0,1]$. The value of this function at $t \in [0,1]$ is denoted by $X(\mathbf{s};t)$. With the usual inner product in L^2, the norm of $X(\mathbf{s})$ is

$$\|X(\mathbf{s})\| = \left\{\int X^2(\mathbf{s};t)dt\right\}^{1/2}.$$

The mean function $\mu(\mathbf{s}) = \{EX(\mathbf{s};t),\ t \in [0,1]\}$ and the covariance operator is then defined for $x \in L^2$ by

$$C_{\mathbf{s},\mathbf{s}}(x) = E[\langle X(\mathbf{s}) - \mu(\mathbf{s}),\, x\rangle\,(X(\mathbf{s}) - \mu(\mathbf{s}))].$$

More generally we define the cross–covariance operators

$$C_{\mathbf{s}_1,\mathbf{s}_2}(x) = E[\langle X(\mathbf{s}_1) - \mu(\mathbf{s}_1),\, x\rangle\,(X(\mathbf{s}_2) - \mu(\mathbf{s}_2))].$$

For the existence of $C_{\mathbf{s}_1,\mathbf{s}_2}$, a minimal assumption is that the variables have finite second moments in the sense that

$$E\|X(\mathbf{s})\|^2 < \infty, \quad \forall\, \mathbf{s}. \tag{18.14}$$

To think of our observations $X(\mathbf{s})$ as curves in L^2 is convenient and motivated this work, but our results require only the general assumption that $\{X(\mathbf{s}), \mathbf{s} \in \mathbb{R}^d\}$ is a field taking values in some separable Hilbert space. In particular, our results hold when L^2 is replaced by $\mathbb{R}^p$.

Our goal is to estimate the mean functions and the FPC's. The FPC's are eigenfunctions of the covariance operator, as we will describe in some detail in the next section, and likewise estimation of the FPC's is based on estimation of covariance operators. A minimal requirement for these population parameters to exist is that all locations share a common mean curve and that the covariance operator is the same for all locations, respectively:

$$\mu(\mathbf{s}) = \mu \quad \text{and} \quad C_{\mathbf{s},\mathbf{s}} = C. \tag{18.15}$$

To develop an estimation framework, we impose conditions on the decay of the cross–covariances $E[\langle X(\mathbf{s}_1) - \mu, X(\mathbf{s}_2) - \mu\rangle]$, as the distance between $\mathbf{s}_1$ and $\mathbf{s}_2$ increases. We shall use the distance function defined by the Euclidian norm in $\mathbb{R}^d$, denoted $\|\mathbf{s}_1 - \mathbf{s}_2\|_2$, but other distance functions can be used as well.

Assumption 18.1. *The spatial process* $\{X(\mathbf{s}), \mathbf{s} \in \mathbb{R}^d\}$ *satisfies* (18.14) *and* (18.15). *In addition,*

$$|E\langle X(\mathbf{s}_1) - \mu, X(\mathbf{s}_2) - \mu\rangle| \le h(\|\mathbf{s}_1 - \mathbf{s}_2\|_2), \tag{18.16}$$

where $h : [0,\infty) \to [0,\infty)$ *with* $h(x) \searrow 0$*, as* $x \to \infty$.

If $\{e_j\}$ is an orthonormal basis in L^2, then it can be easily seen that (18.16) is equivalent to

$$\left|\sum_{j\geq 1}\langle C_{\mathbf{s}_1,\mathbf{s}_2}(e_j), e_j\rangle\right| \leq h\big(\|\mathbf{s}_1-\mathbf{s}_2\|_2\big). \tag{18.17}$$

For any such orthonormal basis, an expansion of $X(\mathbf{s})$ yields

$$X(\mathbf{s};t) = \mu + \sum_{j=1}^{\infty}\xi_j(\mathbf{s})e_j(t), \quad \mathbf{s}\in\mathbb{R}^d,\ t\in[0,1], \tag{18.18}$$

where $\xi_j(\mathbf{s}) = \langle X(\mathbf{s})-\mu, e_j\rangle$. Using the relation

$$\langle C_{\mathbf{s}_1,\mathbf{s}_2}(e_j), e_j\rangle = E\big[\xi_j(\mathbf{s}_1)\xi_j(\mathbf{s}_2)\big],$$

the more specifical assumption

$$\big|E\big[\xi_j(\mathbf{s}_1)\xi_j(\mathbf{s}_2)\big]\big| \leq \phi_j(\|\mathbf{s}_1-\mathbf{s}_2\|_2), \tag{18.19}$$

on the scalar fields, gives (18.16), if

$$\sum_{j\geq 1}\phi_j\big(\|\mathbf{s}_1-\mathbf{s}_2\|_2\big) \leq h\big(\|\mathbf{s}_1-\mathbf{s}_2\|_2\big). \tag{18.20}$$

Examples 18.5 and 18.6 consider typical spatial covariance functions, and show when condition (18.20) holds with a function h as in Assumption 18.1.

Example 18.5. Suppose that the fields $\{\xi_j(\mathbf{s}),\ \mathbf{s}\in\mathbb{R}^d\}$, $j\geq 1$, are zero mean, strictly stationary and α-*mixing*. That is

$$\sup_{(A,B)\in\sigma(\xi_j(\mathbf{s}))\times\sigma(\xi_j(\mathbf{s}+\mathbf{h}))}|P(A)P(B)-P(A\cap B)| \leq \alpha_j(\mathbf{h}),$$

with $\alpha_j(\mathbf{h})\to 0$ if $\|\mathbf{h}\|_2\to\infty$. Let $\alpha_j'(h) = \sup\{\alpha_j(\mathbf{h}) : \|\mathbf{h}\|_2 = h\}$. Then $\alpha_j^*(h) = \sup\{\alpha_j'(x) : x\geq h\}\searrow 0$ as $h\to\infty$. Using stationarity and the main result in Rio (1993) it follows that

$$\begin{aligned}|E[\xi_j(\mathbf{s}_1)\xi_j(\mathbf{s}_2)]| &= |E[\xi_j(\mathbf{0})\xi_j(\mathbf{s}_2-\mathbf{s}_1)]|\\ &\leq 2\int_0^{2\alpha_j(\mathbf{s}_2-\mathbf{s}_1)} Q_j^2(u)du\\ &\leq 2\int_0^{2\alpha_j^*(\|\mathbf{s}_2-\mathbf{s}_1\|_2)} Q_j^2(u)du\\ &=: \phi_j(\|\mathbf{s}_2-\mathbf{s}_1\|_2),\end{aligned}$$

where $Q_j(u) = \inf\{t : P(|\xi_j(\mathbf{0})|>t)\leq u\}$ is the quantile function of $|\xi_j(\mathbf{0})|$. Note that $\alpha_h(\mathbf{h})\leq 1/4$ for any $\mathbf{h}$, and thus $\phi_j(x)\leq 2\int_0^1 Q_j^2(u)du = 2E[\xi_j^2(\mathbf{0})]$. If $\sum_{j\geq 1}E\xi_j^2(\mathbf{0})<\infty$, then (18.16) holds with $h(x) = \sum_{j\geq 1}\phi_j(x)$. (Note that $|h(x)|\searrow 0$ follows from $\alpha_j^*(x)\searrow 0$ and the monotone convergence theorem.)

We note that α-mixing is one of the classical assumptions in random field literature to establish limit theorems. It is in fact a much stronger assumption than ours and it is suitable if one needs more delicate results, like a central limit theorem (see e.g. Bolthausen (1982)) or uniform laws of large numbers, see Jenish and Prucha (2009). Besides the restriction to scalar observations, many papers restrict to the so-called "purely increasing domain sampling", an assumption that we are going to relax in the following.

Example 18.6. Suppose (18.19) holds, and set $h(x) = \sum_{j\geq 1} \phi_j(x)$. If each ϕ_j is a powered exponential covariance function defined by

$$\phi_j(x) = \sigma_j^2 \exp\left\{-\left(\frac{x}{\rho_j}\right)^p\right\}.$$

then h satisfies the conditions of Assumption 18.1 if

$$\sum_{j\geq 1} \sigma_j^2 < \infty \quad \text{and} \quad \sup_{j\geq 1} \rho_j < \infty. \tag{18.21}$$

Condition (18.21) is also sufficient if all ϕ_j are in the Matérn class, see Stein (1999), with the same ν, i.e.

$$\phi_j(x) = \sigma_j^2 x^\nu K_\nu(x/\rho_j),$$

because the modified Bessel function K_ν decays monotonically and approximately exponentially fast; numerical calculations show that $K_\nu(s)$ practically vanishes if $s > \nu$. Condition (18.21) is clearly sufficient for spherical ϕ_j defined (for $d = 3$) by

$$\phi_j(x) = \begin{cases} \sigma_j^2\left(1 - \dfrac{3x}{2\rho_j} + \dfrac{x^3}{2\rho_j^3}\right), & x \leq \rho_j \\ 0, & x > \rho_j \end{cases}$$

because ϕ_j is decreasing on $[0, \rho_j]$.

Assumption 18.1 is appropriate when studying the estimation of the mean function. For the estimation of the covariance operator, we need to impose a different assumption. If z and y are elements of a Hilbert space, the operator $z \otimes y$, is defined by

$$x \mapsto z \otimes y(x) = \langle z, x\rangle y.$$

In the following assumption, we suppose that the mean of the functional field is zero. This is justified by notational convenience and because we deal with the consistent estimation of the mean function separately.

Assumption 18.2. *The spatial process* $\{X(\mathbf{s}), \mathbf{s} \in \mathbb{R}^d\}$ *satisfies* (18.15) *with* $\mu \equiv 0$ *and has 4 moments, i.e.* $E\langle X(\mathbf{s}), x\rangle = 0$, $\forall x \in L^2$, *and* $E\|X(\mathbf{s})\|^4 < \infty$. *In addition,*

$$\left|E\langle X(\mathbf{s}_1) \otimes X(\mathbf{s}_1) - C\,,\, X(\mathbf{s}_2) \otimes X(\mathbf{s}_2) - C\rangle_{\mathcal{S}}\right| \leq H\big(\|\mathbf{s}_1 - \mathbf{s}_2\|_2\big), \tag{18.22}$$

where $H : [0, \infty) \to [0, \infty)$ *with* $H(x) \searrow 0$, *as* $x \to \infty$.

Assumption 18.2 cannot be verified using only conditions on the covariances of the scalar fields ξ_j in (18.4) because these covariances do not specify the 4th order structure of the model. This can be done if the random field is Gaussian, as illustrated in Example 18.12, or if additional structure is imposed. If the scalar fields $\xi_i(\cdot)$ are independent, the following lemma can be used to verify (18.22).

Lemma 18.2. *Let* $X(\mathbf{s})$ *have representation* (18.4) *with zero mean and* $E\|X(\mathbf{s})\|^4 < \infty$. *Assume further that* $\xi_i(\cdot)$ *and* $\xi_j(\cdot)$ *are independent if* $i \neq j$. *Then*

$$\begin{aligned}
&\left|E\langle X(\mathbf{s}_1)\otimes X(\mathbf{s}_1)-C\,,\ X(\mathbf{s}_2)\otimes X(\mathbf{s}_2)-C\rangle_{\mathcal{S}}\right| \\
&\quad\leq \left|\sum_{j\geq 1}\mathrm{Cov}\big(\xi_j^2(\mathbf{s}_1),\xi_j^2(\mathbf{s}_2)\big)\right| + \left|\sum_{j\geq 1}E\big[\xi_j(\mathbf{s}_1)\xi_j(\mathbf{s}_2)\big]\right|^2.
\end{aligned}$$

Proof. If $\xi_i(\cdot)$ and $\xi_j(\cdot)$ are independent for $i \neq j$, then the e_j are the eigenvalues of C, and the $\xi_j(\mathbf{s})$ are the principal component scores with $E\xi_j^2(\mathbf{s}) = \lambda_j$. Using continuity of the inner product and dominated convergence we obtain

$$\begin{aligned}
&\left|E\langle X(\mathbf{s}_1)\otimes X(\mathbf{s}_1)-C\,,\ X(\mathbf{s}_2)\otimes X(\mathbf{s}_2)-C\rangle_{\mathcal{S}}\right| \\
&= \left|E\sum_{j\geq 1}\Big\langle \langle X(\mathbf{s}_1),e_j\rangle X(\mathbf{s}_1)-C(e_j)\,,\ \langle X(\mathbf{s}_2),e_j\rangle X(\mathbf{s}_2)-C(e_j)\Big\rangle\right| \\
&= \left|E\sum_{j\geq 1}\Big\langle \xi_j(\mathbf{s}_1)\sum_{\ell\geq 1}\xi_\ell(\mathbf{s}_1)e_\ell-\lambda_j e_j\,,\ \xi_j(\mathbf{s}_2)\sum_{k\geq 1}\xi_k(\mathbf{s}_2)e_k-\lambda_j e_j\Big\rangle\right| \\
&= \left|E\sum_{j\geq 1}\Big\{\xi_j(\mathbf{s}_1)\xi_j(\mathbf{s}_2)\sum_{\ell\geq 1}\xi_\ell(\mathbf{s}_1)\xi_\ell(\mathbf{s}_2)+\lambda_j^2-\lambda_j\xi_j^2(\mathbf{s}_1)-\lambda_j\xi_j^2(\mathbf{s}_2)\Big\}\right| \\
&\leq \left|\sum_{j\geq 1}\mathrm{Cov}\big(\xi_j^2(\mathbf{s}_1),\xi_j^2(\mathbf{s}_2)\big)\right| + \left|\sum_{j\geq 1}\sum_{\ell\neq j}E\big[\xi_j(\mathbf{s}_1)\xi_j(\mathbf{s}_2)\big]\times E\big[\xi_\ell(\mathbf{s}_1)\xi_\ell(\mathbf{s}_2)\big]\right| \\
&\leq \left|\sum_{j\geq 1}\mathrm{Cov}\big(\xi_j^2(\mathbf{s}_1),\xi_j^2(\mathbf{s}_2)\big)\right| + \left|\sum_{j\geq 1}E\big[\xi_j(\mathbf{s}_1)\xi_j(\mathbf{s}_2)\big]\right|^2. \qquad \square
\end{aligned}$$

As already noted, for spatial processes assumptions on the distribution of the sampling points are as important as those on the covariance structure. To formalize the different sampling schemes, we use the following measure of "minimal dispersion" of some point cloud $\mathfrak{S}$:

$$I_\rho(\mathbf{s},\mathfrak{S}) = |\{\mathbf{y}\in\mathfrak{S} : \|\mathbf{s}-\mathbf{y}\|_2 \leq \rho\}|/|\mathfrak{S}| \quad \text{and} \quad I_\rho(\mathfrak{S}) = \sup\{I_\rho(\mathbf{s},\mathfrak{S}),\ \mathbf{s}\in\mathfrak{S}\},$$

where $|\mathfrak{S}|$ denotes the number of elements of $\mathfrak{S}$. The quantity $I_\rho(\mathfrak{S})$ is the maximal fraction of $\mathfrak{S}$–points in a ball of radius ρ centered at an element of $\mathfrak{S}$. Notice that $1/|\mathfrak{S}| \leq I_\rho(\mathfrak{S}) \leq 1$. We call $\rho \mapsto I_\rho(\mathfrak{S})$ the *intensity function* of $\mathfrak{S}$.

Definition 18.1. For a sampling scheme $\mathfrak{S}_N = \{\mathbf{s}_{i,N}; 1 \le i \le S_N\}$, $S_N \to \infty$, we consider the following conditions:

(i) there is a $\rho > 0$ such that $\limsup_{N\to\infty} I_\rho(\mathfrak{S}_N) > 0$;
(ii) for some sequence $\rho_N \to \infty$ we have $I_{\rho_N}(\mathfrak{S}_N) \to 0$;
(iii) for any fixed $\rho > 0$ we have $S_N I_\rho(\mathfrak{S}_N) \to \infty$.

We call a deterministic sampling scheme $\mathfrak{S}_N = \{\mathbf{s}_{i,N}; 1 \le i \le S_N\}$

Type A if (i) holds;
Type B if (ii) and (iii) hold;
Type C if (ii) holds, but there is a $\rho > 0$ such that $\limsup_{N\to\infty} S_N I_\rho(\mathfrak{S}_N) < \infty$.

If the sampling scheme is stochastic we call it *Type A, B* or *C* if relations (i), (ii) and (iii) hold with $I_\rho(\mathfrak{S}_N)$ replaced by $EI_\rho(\mathfrak{S}_N)$.

Type A sampling is related to purely infill domain sampling which corresponds to $I_\rho(\mathfrak{S}_N) = 1$ for all $N \ge 1$, provided ρ is large enough. However, in contrast to the purely infill domain sampling, it still allows for a non-degenerate asymptotic theory for sparse enough subsamples (in the sense of Type B or C).

Example 18.7. Assume that $\mathfrak{S}_N$ are sampling points on the line with $s_{2k} = 1/k$ and $s_{2k+1} = k$, $1 \le k \le N$. Then, for $\rho = 1$, $\lim_{N\to\infty} I_\rho(\mathfrak{S}_N) = 1/2$, so this sampling scheme is of Type A. But the subsample corresponding to odd indices is of Type C.

A brief reflection shows that assumptions (i) and (ii) are mutually exclusive. Combining (ii) and (iii) implies that the points intensify (at least at certain spots) excluding the purely increasing domain sampling. Hence the Type B sampling corresponds to the nearly infill domain sampling. If only (ii) holds, but (iii) does not (Type C sampling) then the sampling scheme corresponds to purely increasing domain sampling.

Our conditions are more general than those proposed so far. Their relation to more specific sampling designs previously used is discussed in Section 18.3.

18.3 Regular spatial designs

We continue to assume a spatial design $\mathfrak{S}_N = \{\mathbf{s}_{k,N}, 1 \le k \le S_N\}$. The two special cases we discuss are closely related to those considered by Lahiri (2003). The points are assumed to be on a grid of an increasing size, or to have a density. The results of this section show how our more general assumptions look in these special cases, and provide additional intuition behind the sampling designs formulated in Definition 18.1. They also set a framework for some results of Sections 18.4 and 18.5.

Non-random regular design. Let $\mathcal{Z}(\boldsymbol{\delta})$ be a lattice in $\mathbb{R}^d$ with increments δ_i in the i-th direction. Let $\delta_0 = \min\{\delta_1, \dots, \delta_d\}$, $\Delta^d = \prod_{i=1}^d \delta_i$ and let $R_N = \alpha_N R_0$, where R_0 is some bounded Riemann measurable Borel-set in $\mathbb{R}^d$ containing the origin. A set is Riemann measurable if its indicator function is Riemann integrable. This condition excludes highly irregular sets R_0. The scaling parameters $\alpha_N > 0$ are assumed to be non-decreasing and will be specified below in Lemma 18.4. We assume without loss of generality that $\text{Vol}(R_0) = 1$, hence $\text{Vol}(R_N) = \alpha_N^d$. Typical examples are $R_0 = \{x \in \mathbb{R}^d : \|x\| \le z_{1,d}\}$, with $z_{1,d}$ equal to the radius of the d-dimensional sphere with volume 1, or $R_0 = [-1/2, 1/2]^d$. The sampling points $\mathfrak{S}_N$ are defined as $\{\mathbf{s}_{k,N}, 1 \le k \le S_N\} = \mathcal{Z}(\eta_N \boldsymbol{\delta}) \cap R_N$, where η_N is chosen such that the sample size $S_N \sim N$. It is intuitively clear that $\text{Vol}(R_N) \approx \eta_N^d \Delta^d S_N$, suggesting

$$\eta_N = \frac{\alpha_N}{\Delta N^{1/d}}. \tag{18.23}$$

A formal proof that η_N in (18.23) assures $S_N \sim N$ is immediate from the following

Lemma 18.3. *Let K be a bounded set in $\mathbb{R}^d$, and assume that K is Riemann measurable with* $\text{Vol}(K) = 1$. *If $\beta_N \to 0$, then*

$$|K \cap \mathcal{Z}(\beta_N \boldsymbol{\delta})| \sim \frac{1}{\Delta^d \beta_N^d}.$$

Proof. Let $K \subset M_1 \subset M_2$ where M_1 and M_2 are rectangles in $\mathbb{R}^d$ having no intersecting margin (M_1 is an inner subset of M_2). The points $\{x_{i,N}\} = \mathcal{Z}(\beta_N \boldsymbol{\delta}) \cap M_2$ can be seen as the vertices of rectangles $J_{i,N} = x_{i,N} + \{\mathbf{t} \circ \beta_N \boldsymbol{\delta}, \mathbf{t} \in [0,1]^d\}$, where $\circ$ denotes the Hadamard (entrywise) product. For large enough N, the sets $L_{i,N} = J_{i,N} \cap M_1$ define a partition of M_1. Then, by the assumed Riemann measurability,

$$\begin{aligned}
\int_{M_1} I_K(x)dx &= \liminf_{N\to\infty} \beta_N^d \Delta^d \sum_i \inf\{I_K(x) : x \in L_{i,N}\} \\
&\le \liminf_{N\to\infty} \beta_N^d \Delta^d \sum_i I_K(x_{i,N}) \\
&\le \limsup_{N\to\infty} \beta_N^d \Delta^d \sum_i I_K(x_{i,N}) \\
&\le \limsup_{N\to\infty} \beta_N^d \Delta^d \sum_i \sup\{I_K(x) : x \in L_{i,N}\} \\
&= \int_{M_1} I_K(x)dx.
\end{aligned}$$
□

The following Lemma relates the non-random regular design to Definition 18.1. We write $a_N \gg b_N$ if $\limsup b_N/a_N < \infty$.

Lemma 18.4. *In the above described design the following pairs of statements are equivalent:*

(i) α_N remains bounded $\Leftrightarrow$ Type A sampling;
(ii) $\alpha_N \to \infty$ and $\alpha_N = o(N^{1/d})$ $\Leftrightarrow$ Type B sampling;
(iii) $\alpha_N \gg N^{1/d}$ $\Leftrightarrow$ Type C sampling.

Proof. Let $U_\varepsilon(x)$ be the sphere in $\mathbb{R}^d$ with center x and radius ε. Assume first that $\alpha_N = o(N^{1/d})$, which covers (i) and (ii). In this case the volume of the rectangles $L_{i,n}$ as described in the proof of Lemma 18.3 satisfies

$$\mathrm{Vol}(L_{i,n}) = \Delta^d \eta_N^d = \frac{\alpha_N^d}{N} \to 0. \tag{18.24}$$

Hence $|U_\rho(x) \cap \mathcal{Z}(\eta_N \boldsymbol{\delta})|$ is asymptotically proportional to

$$\mathrm{Vol}(U_\rho(x))/\mathrm{Vol}(L_{i,n}) = V_d \left(\frac{\rho}{\alpha_N}\right)^d N,$$

where V_d is the volume of the d-dimensional unit sphere. Now if we fix an arbitrary $\rho_0 > 0$ then there are constants $0 < C_L < C_U < \infty$, such that for any $\rho \geq \rho_0$ and $N \geq N_0$ and $x \in \mathbb{R}^d$

$$C_L \left(\frac{\rho}{\alpha_N}\right)^d \leq \frac{|U_\rho(x) \cap \mathcal{Z}(\eta_N \boldsymbol{\delta})|}{N} \leq C_U \left(\frac{\rho}{\alpha_N}\right)^d.$$

By the required Riemann measurability we can find an $x \in R_0$ such that for some small enough ε we have $U_{2\varepsilon}(x) \subset R_0$. Then $U_{2\varepsilon\alpha_N}(\alpha_N x) \subset R_N$. Hence for any $2\rho_0 \leq \rho \leq \varepsilon\alpha_N$,

$$\begin{aligned} C_L\left(\frac{\rho}{\alpha_N}\right)^d &\leq \frac{|U_{\rho/2}(\alpha_N x) \cap \mathfrak{S}_N|}{N} \leq I_\rho(\mathfrak{S}_N) \leq \frac{|U_{2\rho}(\alpha_N x) \cap \mathfrak{S}_N|}{N} \\ &\leq C_U\left(\frac{\rho}{\alpha_N}\right)^d. \end{aligned}$$

With the help of the above inequalities (i) and (ii) are easily checked.

Now we prove (iii). We notice that by (18.24) $\alpha_N \gg N^{1/d}$ is equivalent to $\mathrm{Vol}(L_{i,n})$ does not converge to 0. Assume first that we have Type C sampling. Then by the arguments above we find an x and a $\rho > 0$ such that $U_\rho(\alpha_N x) \subset R_N$. Thus

$$|U_\rho(\alpha_N x) \cap \mathcal{Z}(\eta_N \boldsymbol{\delta})| \leq S_N I_\rho(\mathfrak{S}_N).$$

As this quantity remains bounded, $\mathrm{Vol}(L_{i,n})$ does not converge to 0.

On the other hand, if $\mathrm{Vol}(L_{i,n})$ does not converge to 0, then for any $\rho > 0$ and any $x \in \mathbb{R}^d$ we have $\limsup_{N\to\infty} |U_\rho(x) \cap \mathcal{Z}(\eta_N \boldsymbol{\delta})| < \infty$, and thus for arbitrarily large ρ

$$I_\rho(\mathfrak{S}_N) \leq \sup_x \frac{|U_\rho(x) \cap \mathcal{Z}(\eta_N \boldsymbol{\delta})|}{S_N} \to 0.$$

The claim follows immediately. □

Randomized design. Let $\{\mathbf{s}_k, 1 \leq k \leq N\}$ be iid random vectors with a density $f(\mathbf{s})$ which has support on a Borel set $R_0 \subset \mathbb{R}^d$ containing the origin and satisfying $\mathrm{Vol}(R_0) = 1$. Again we assume Riemann measurability for R_0 to exclude highly irregular sets. For the sake of simplicity we shall assume that on R_0 the density is bounded away from zero, so that we have $0 < f_L \le \inf_{x\in R_0} f(x)$. The point set $\{\mathbf{s}_{k,N},\ 1 \le k \le N\}$ is defined by $\mathbf{s}_{k,N} = \alpha_N \mathbf{s}_k$ for $k = 1, \ldots, N$. For fixed N, this is equivalent to: $\{\mathbf{s}_{k,N},\ 1 \le k \le N\}$ is an iid sequence on $R_N = \alpha_N R_0$ with density $\alpha_N^{-d} f(\alpha_N^{-1}\mathbf{s})$.

We cannot expect to obtain a full analogue of Lemma 18.4 in the randomized setup. For Type C sampling, the problem is much more delicate, and a closer study shows that it is related to the oscillation behavior of multivariate empirical processes. While Stute (1984) gives almost sure upper bounds, we would need here sharp results on the moments of the modulus of continuity of multivariate empirical process. Such results exist, see Einmahl and Ruymgaart (1987), but are connected to technical assumptions on the bandwidth for the modulus (here determined by α_N) which are not satisfied in our setup. Since a detailed treatment would be very difficult, we only state the following lemma.

Lemma 18.5. *In the above described sampling scheme the following statements hold:*

(i) α_N remains bounded $\Rightarrow$ Type A sampling;
(ii) $\alpha_N \to \infty$ and $\alpha_N = o(N^{1/d}) \Rightarrow$ Type B sampling;

Proof. By Jensen's inequality we infer that

$$\begin{aligned} E I_\rho(\mathfrak{S}_N) &= E \sup_{x\in R_N} \frac{1}{N} \sum_{k=1}^{N} I\{\mathbf{s}_{k,N} \in U_\rho(x) \cap R_N\} \\ &\ge \sup_{x\in R_N} P\big(\mathbf{s}_{1,N} \in U_\rho(x) \cap R_N\big) \\ &= \sup_{x\in R_0} P\big(\mathbf{s}_1 \in U_{\rho/\alpha_N}(x) \cap R_0\big) \\ &= \sup_{x\in R_0} \int_{U_{\rho/\alpha_N}(x)\cap R_0} f(s)ds. \end{aligned}$$

We have two scenarios. First, α_N remains bounded. Then we can choose ρ big enough such that $U_{\rho/\alpha_N}(0)$ covers R_0 for all N. It follows that $\limsup_{N\to\infty} E I_\rho(\mathfrak{S}_N) = 1$ and (i) follows.

Second, $\alpha_N \to \infty$. Then for large enough N, R_0 contains a ball with radius ρ/α_N. It follows that

$$E I_\rho(\mathfrak{S}_N) \ge f_L\, V_d \left(\frac{\rho}{\alpha_N}\right)^d. \tag{18.25}$$

Now statement (ii) follows easily. □

18.4 Consistency of the sample mean function

Our goal is to establish the consistency of the sample mean for functional spatial data. We consider Type B or Type C sampling and obtain rates of convergence. We start with a general setup, and show that the rates can be improved in special cases. The general results are applied to functional random fields with specific covariance structures. The proofs of the main results, Propositions 18.1, 18.2, 18.3, are collected in Section 18.7.

For independent or weakly dependent functional observations X_k,

$$E\left\|\frac{1}{N}\sum_{k=1}^{N} X_k - \mu\right\|^2 = O\left(N^{-1}\right). \tag{18.26}$$

Proposition 18.1 shows that for general functional spatial processes, the rate of consistency may be much slower than $O\left(N^{-1}\right)$; it is the maximum of $h(\rho_N)$ and $I_{\rho_N}(\mathfrak{S}_N)$ with ρ_N from (ii) of Definition 18.1. Intuitively, the sample mean is consistent if there is a sequence of increasing balls which contain a fraction of points which tends to zero, and the decay of the correlations compensates for the increasing radius of these balls.

Proposition 18.1. *Let Assumption 18.1 hold, and assume that* $\mathfrak{S}_N$ *defines a* non-random *design of Type A, B or C. Then for any* $\rho_N > 0$,

$$E\left\|\frac{1}{N}\sum_{k=1}^{N} X(\mathbf{s}_{k,N}) - \mu\right\|^2 \leq h(\rho_N) + h(0) I_{\rho_N}(\mathfrak{S}_N). \tag{18.27}$$

Hence, under the Type B or Type C non-random sampling, with ρ_N *as in (ii) of Definition 18.1, the sample mean is consistent.*

Example 18.8. Assume that N points $\{\mathbf{s}_{k,N},\ 1 \leq k \leq N\}$ are on a regular grid in $\alpha_N[-1/2, 1/2]^d$. Then $I_\rho(\mathfrak{S}_N)$ is proportional to $(\rho/\alpha_N)^d$. For example, if $h(x) = 1/(1+x)^2$, then choosing $\rho_N = \alpha_N^{d/(d+2)}$ we obtain that

$$h(\rho_N) + h(0) I_{\rho_N}(\mathfrak{S}_N) \ll \alpha_N^{-2d/(d+2)} \vee N^{-1}.$$

(Recall that $I_{\rho_N}(\mathfrak{S}_N) \geq N^{-1}$.) A stronger result is obtained in Proposition 18.2 below.

We now consider the special case, where we have a regular sampling design. Here we are able to obtain the strongest results.

Proposition 18.2. *Assume the nonrandom sampling design. Let Assumption 18.1 hold with h such that* $x^{d-1}h(x)$ *is monotone on* $[b, \infty)$, $b > 0$. *Then under Type B sampling*

$$E\left\|\frac{1}{S_N}\sum_{k=1}^{S_N} X(\mathbf{s}_{k,N}) - \mu\right\|^2$$
$$\leq \frac{1}{\alpha_N^d}\left\{d(3\Delta)^d \int_0^{K\alpha_N} x^{d-1} h(x)\,dx + o(1) \sup_{x\in[0,K\alpha_N]} x^{d-1}h(x)\right\}, \tag{18.28}$$

for some large enough constant K which is independent of N. Under Type C sampling $1/\alpha_N^d$ *in (18.28) is replaced by* $O\left(N^{-1}\right)$.

The technical assumptions on h pose no practical problem, they are satisfied for all important examples, see Example 18.6. A common situation is that $x^{d-1}h(x)$ is increasing on $[0, b]$ and decreasing thereafter.

Our first example shows that for most typical covariance functions, under nearly infill domain sampling, the rate of consistency may be much slower than for the iid case, if the size of the domain does not increase fast enough.

Example 18.9. Suppose the functional spatial process has representation (18.4), and (18.19) holds with with the covariance functions ϕ_j as in Example 18.6 (powered exponential, Matérn or spherical). Define $h(x) = \sum_{j\geq 1}\phi_j(x)$, and assume that condition (18.21) holds. Assumption 18.1 is then satisfied and

$$\int_0^\infty x^{d-1}h(x)dx < \infty \quad \text{and} \quad \sup_{x\in\mathbb{R}} x^{d-1}h(x) < \infty. \tag{18.29}$$

Therefore, for the nonrandom sampling,

$$E\left\|\frac{1}{S_N}\sum_{k=1}^{S_N} X(\mathbf{s}_{k,N}) - \mu\right\|^2 = \begin{cases} O\left(\alpha_N^{-d} \vee N^{-1}\right), & \text{under Type B sampling} \\ O\left(N^{-1}\right), & \text{under Type C sampling} \end{cases} \tag{18.30}$$

The next example shows that formula (18.30) is far from universal, and that the rate of consistency may be even slower if the covariances decay slower than exponential.

Example 18.10. Consider the general setting of Example 18.9, but assume that each covariance function ϕ_j has the quadratic rational form

$$\phi_j(x) = \sigma_j^2\left\{1 + \left(\frac{x}{\rho_j}\right)^2\right\}^{-1}.$$

Condition (18.21) implies that $h(x) = \sum_{j\geq 1}\phi_j(x)$ satisfies Assumption 18.1, but now $h(x) \sim x^{-2}$, as $x \to \infty$. Because of this rate, condition (18.29) holds only for $d = 1$ (and so for this dimension (18.30) also holds). If $d \geq 2$, (18.29) fails, and to find the rate of the consistency, we must use (18.28) directly. We focus only on Type B sampling, and assume implicitly that the rate is slower than N^{-1}. We assume (18.21) throughout this example.

If $d = 2$,

$$\begin{aligned}\int_0^{K\alpha_N} x^{d-1}h(x)dx &= \sum_j \sigma_j^2\int_0^{K\alpha_N} x\left\{1 + \left(\frac{x}{\rho_j}\right)^2\right\}^{-1}dx \\ &= \sum_j \sigma_j^2\rho_j^2\, O\left(\int_1^{K\alpha_N} x^{-1}dx\right) = O(\ln\alpha_N)\end{aligned}$$

and similarly $\sup_{x\in[0,K\alpha_N]} x^{d-1}h(x) = O(1)$.

If $d \geq 3$, the leading term is

$$\int_0^{K\alpha_N} x^{d-1} h(x)dx = O\left(\alpha_N^{d-3}\right).$$

We summarize these calculations as

$$E\left\|\frac{1}{S_N}\sum_{k=1}^{S_N} X(\mathbf{s}_{k,N}) - \mu\right\|^2 = \begin{cases} O\left(\alpha_N^{-1}\right), & \text{if } d = 1 \\ O\left(\alpha_N^{-2}\ln(\alpha_N)\right), & \text{if } d = 2 \\ O\left(\alpha_N^{-2}\right), & \text{if } d \geq 3, \end{cases}$$

for Type B sampling scheme (provided the rate is slower than N^{-1}).

The last example shows that for very persistent spatial dependence, the rate of consistency can be essentially arbitrarily slow.

Example 18.11. Assume that $h(x)$ decays only at a logarithmic rate, $h(x) = \{\log(x \vee e)\}^{-1}$. Then, for any $d \geq 1$, the left hand side in (18.28) is $\ll (\log \alpha_N)^{-1}$.

We now turn to the case of the random design.

Proposition 18.3. *Assume the random sampling design of Section 18.3. If the sequence $\{\mathbf{s}_{k,N}\}$ is independent of the process X, and if Assumption 18.1 holds, then we have for any $\varepsilon_N > 0$*

$$E\left\|\frac{1}{N}\sum_{k=1}^{N} X(\mathbf{s}_{k,N}) - \mu\right\|^2 \leq 6\,h(0) \sup_{\mathbf{s}\in R_0} f^2(\mathbf{s})\,\varepsilon_N^d + h(\alpha_N\varepsilon_N) + \frac{h(0)}{N}.$$

Choosing ε_N such that $\varepsilon_N \to 0$ and $\alpha_N\varepsilon_N \to \infty$, it follows that under Type B or Type C sampling, the sample mean is consistent.

The bound in Proposition 18.3 can be easily applied to any specific random sampling design and any model for the functions ϕ_j in (18.18). It nicely shows that what matters for the rate of consistency is the interplay between the rate of growth of the sampling domain and the rate of decay of dependence.

Let us explain in slightly more detail a Type C sampling situation. Here typically we have $\alpha_N = N^{1/d}$. Then taking $\varepsilon_N = aN^{-1/d}\log N$, $a > 0$, we see that the rate of consistency is $h(a\log N) \vee N^{-1}$. For typical covariance functions ϕ_j, like powered exponential, Matérn or spherical, $h(a\log N)$ decays faster than N^{-1}. In such cases, the rate of consistency is, up to some logarithmic factor, the same as for an iid sample. For ease of reference, we formulate the following corollary, which can be used in practical applications.

Corollary 18.1. *Assume the random sampling design with the sequence $\{\mathbf{s}_{k,N}\}$ independent the process X. Suppose that $X(s)$ has representation* (18.18) *and that* (18.19) *holds with the ϕ_j in one of the families specified in Example 18.6. If Condition* (18.21) *holds, and $\alpha_N \geq N^{1/d}$ then* (18.26) *holds up to some multiplicative logarithmic factor.*

18.5 Consistency of the empirical covariance operator

In Section 18.4 we found the rates of consistency for the functional sample mean. We now turn to the rates for the sample covariance operator. Assuming the functional observations have mean zero, the natural estimator of the covariance operator C is the sample covariance operator given by

$$\widehat{C}_N = \frac{1}{N}\sum_{k=1}^{N} X(\mathbf{s}_k) \otimes X(\mathbf{s}_k).$$

In general, the sample covariance operator is defined by

$$\hat{\Gamma}_N = \frac{1}{N}\sum_{k=1}^{N} \left(X(\mathbf{s}_k) - \bar{X}_N\right) \otimes \left(X(\mathbf{s}_k) - \bar{X}_N\right),$$

where

$$\bar{X}_N = \frac{1}{N}\sum_{k=1}^{N} X(\mathbf{s}_k).$$

Both operators are implemented in statistical software packages, for example in the popular `R` package `fda` and in a similar MATLAB package, see Ramsay *et al.* (2009), The operator $\hat{\Gamma}_N$ is used to compute the EFPC's for centered data, while $\widehat{C}_N$ for data without centering.

We first derive the rates of consistency for $\widehat{C}_N$ assuming $EX(\mathbf{s}) = 0$. Then we turn to the operator $\hat{\Gamma}_N$. The proofs are obtained by applying the technique developed for the estimation of the functional mean. It is a general approach based on the estimation of the second moments of an appropriate norm (between estimator and estimand) so that the conditions in Definition 18.1 can come into play. It is broadly applicable to all statistics obtained by simple averaging. The proofs are thus similar to those presented in the simplest case in Section 18.7, but the notation becomes more cumbersome because of the increased complexity of the objects to be averaged. To conserve space these proofs are not included.

We begin by observing that

$$\begin{aligned} E\left\|\widehat{C}_N - C\right\|_{\mathcal{S}}^2 &= \langle \widehat{C}_N - C\,,\, \widehat{C}_N - C\rangle_{\mathcal{S}} \\ &= \frac{1}{N^2}\sum_{k=1}^{N}\sum_{\ell=1}^{N}\langle X(\mathbf{s}_k)\otimes X(\mathbf{s}_k) - C\,,\, X(\mathbf{s}_\ell)\otimes X(\mathbf{s}_\ell) - C\rangle_{\mathcal{S}}. \end{aligned}$$

It follows that under Assumption 18.2

$$E\left\|\widehat{C}_N - C\right\|_{\mathcal{S}}^2 \le \frac{1}{N^2}\sum_{k=1}^{N}\sum_{\ell=1}^{N} H\left(\|\mathbf{s}_k - \mathbf{s}_\ell\|_2\right). \tag{18.31}$$

Relation (18.31) is used as the starting point of all proofs, cf. the proof of Proposition 18.1 in Section 18.4. Modifying the proofs of Section 18.4, we arrive at the following results.

Proposition 18.4. *Let Assumption 18.2 hold, and assume that $\mathfrak{S}_N$ defines a* non-random *design of Type A, B or C. Then for any $\rho_N > 0$*

$$E\,\big\|\widehat{C}_N - C\,\big\|_{\mathcal{S}}^2 \le H(\rho_N) + H(0)I_{\rho_N}(\mathfrak{S}_N).$$

Hence under the Type B or Type C non-random sampling, with ρ_N as in (ii) of Definition 18.1, the empirical covariance operator is consistent.

Proposition 18.5. *Assume the nonrandom sampling design. Let Assumption 18.2 hold, with some function H such that $x^{d-1}H(x)$ is monotone on $[b,\infty)$, $b > 0$. Then under Type B sampling*

$$E\,\big\|\widehat{C}_N - C\,\big\|_{\mathcal{S}}^2 \le \frac{1}{\alpha_N^d}\left\{d(3\Delta)^d \int_0^{K\alpha_N} x^{d-1}H\big(x\big)dx + o(1) \sup_{x\in[0,K\alpha_N]} x^{d-1}H(x)\right\},$$

for some large enough constant K which is independent of N. Under Type C sampling, the factor $1/\alpha_N^d$ is replaced by $O\big(N^{-1}\big)$.

Proposition 18.6. *Assume the random sampling design of Section 18.3. If the sequence $\{\mathbf{s}_{k,N}\}$ is independent of the process X and if Assumption 18.2 holds, then we have for any $\varepsilon_N > 0$,*

$$E\,\big\|\widehat{C}_N - C\,\big\|^2 \le 6\,H(0) \sup_{\mathbf{s}\in R_0} f^2(\mathbf{s})\,\varepsilon_N^d + H(\alpha_N\varepsilon_N) + \frac{H(0)}{N}.$$

It follows that under Type B or Type C sampling the sample covariance operator is consistent.

Example 18.12. Let X have representation (18.4), in which the scalar fields $\xi_j(\cdot)$ are independent and Gaussian, and (18.11) (18.12) and (18.13) hold.

It follows that for some large enough constant A,

$$\Big|\sum_{j\ge 1}\mathrm{Cov}\big(\xi_j^2(\mathbf{s}_1),\xi_j^2(\mathbf{s}_2)\big)\Big| + \Big|\sum_{j\ge 1}E\big[\xi_j(\mathbf{s}_1)\xi_j(\mathbf{s}_2)\big]\Big|^2 \le A\exp\big(-2\rho^{-1}\|\mathbf{s}_1-\mathbf{s}_2\|_2\big).$$

Hence by Lemma 18.2, Assumption 18.2 holds with $H(x) = A\exp\big(-2\rho^{-1}\|\mathbf{s}_1 - \mathbf{s}_2\|_2\big)$. Proposition 18.4 yields consistency of the estimator under Type B or Type C sampling, as

$$E\,\|\widehat{C}_N - C\,\|_{\mathcal{S}}^2 \le A\Big(\exp(-2\rho^{-1}\rho_N) + I_{\rho_N}(\mathfrak{S}_N)\Big).$$

If we assume a regular sampling design, then by Proposition 18.5

$$E\,\|\widehat{C}_N - C\,\|_{\mathcal{S}}^2 \le A\left(\frac{1}{\alpha_N^d} + \frac{1}{N}\right).$$

Introducing the (unobservable) operator

$$\tilde{\Gamma}_N = \frac{1}{N}\sum_{k=1}^{N}(X(\mathbf{s}_k)-\mu)\otimes(X(\mathbf{s}_k)-\mu),$$

we see that

$$\tilde{\Gamma}_N - \hat{\Gamma}_N = (\bar{X}_N-\mu)\otimes(\bar{X}_N-\mu).$$

Therefore

$$E\|\hat{\Gamma}_N - C\|_{\mathcal{S}}^2 \le 2E\|\tilde{\Gamma}_N - C\|_{\mathcal{S}}^2 + 2E\|(\bar{X}_N-\mu)\otimes(\bar{X}_N-\mu)\|_{\mathcal{S}}^2.$$

The bounds in Propositions 18.4, 18.5 and 18.6 apply to $E\|\tilde{\Gamma}_N - C\|_{\mathcal{S}}^2$. Observe that

$$E\|(\bar{X}_N-\mu)\otimes(\bar{X}_N-\mu)\|_{\mathcal{S}}^2 = E\|\bar{X}_N-\mu\|^4.$$

If $X(\mathbf{s})$ are bounded variables, i.e. $\sup_{t\in[0,1]}|X(\mathbf{s};t)| \le B < \infty$ a.s., then $\|\bar{X}_N-\mu\|^4 \le 4B^2\|\bar{X}_N-\mu\|^2$. It follows that under Assumption 18.1 we obtain the same order of magnitude for the bounds of $E\|\bar{X}_N-\mu\|^4$ as we have obtained in Propositions 18.1, 18.2 and 18.3 for $E\|\bar{X}_N-\mu\|^2$. In general $E\|\bar{X}_N-\mu\|^4$ can neither be bounded in terms of $E\|\bar{X}_N-\mu\|^2$ nor with $E\|\hat{C}_N - C\|_{\mathcal{S}}^2$. To bound fourth order moments, conditions on the covariance between the variables $Z_{k,\ell} := \langle X(\mathbf{s}_{k,N})-\mu\,,\, X(\mathbf{s}_{\ell,N})-\mu\rangle$ and $Z_{i,j}$ for all $1\le i,j,k,\ell\le N$ are unavoidable. However, a simpler general approach is to require higher order moments of $\|X(\mathbf{s})\|$. More precisely, we notice that for any $p>1$, by the Hölder inequality,

$$E\|\bar{X}_N-\mu\|^4 \le \left(E\|\bar{X}_N-\mu\|^2\right)^{1/p}\left(E\|\bar{X}_N-\mu\|^{\frac{4p-2}{p-1}}\right)^{(p-1)/p}.$$

Thus as long as $E\|X(\mathbf{s})\|^{\frac{4p-2}{p-1}} < \infty$, we conclude that, by stationarity,

$$E\|\bar{X}_N-\mu\|^4 \le M(p)\left(E\|\bar{X}_N-\mu\|^2\right)^{1/p},$$

where $M(p)$ depends on the distribution of $X(\mathbf{s})$ and on p, but not on N. It is now evident how the results of Section 18.4 can be used to obtain bounds for $E\|\hat{\Gamma}_N - C\|_{\mathcal{S}}^2$. We state in Proposition 18.7 the version for the general non-random design. The special cases follow, and the random designs are treated analogously. It follows that if Assumptions 18.1 and 18.2 hold, then $E\|\hat{\Gamma}_N - C\|_{\mathcal{S}}^2 \to 0$, under Type B or C sampling, provided $E\|X(\mathbf{s})\|^{4+\delta} < \infty$.

Proposition 18.7. *Let Assumptions 18.1 and 18.2 hold and assume that for some $\delta > 0$ we have $E\|X(\mathbf{s})\|^{4+\delta} < \infty$. Assume further that $\mathfrak{S}_N$ defines a non-random design of Type A, B or C. Then for any $\rho_N > 0$ we have*

$$E\|\hat{\Gamma}_N - C\|_{\mathcal{S}}^2 \le 2\left\{H(\rho_N)+H(0)I_{\rho_N}(\mathfrak{S}_N)\right\} + 2C(\delta)\left\{h(\rho_N)+h(0)I_{\rho_N}(\mathfrak{S}_N)\right\}^{\frac{\delta}{2+\delta}}. \tag{18.32}$$

If $X(\mathbf{s}_1)$ is a.s. bounded by some finite constant B, then we can formally let δ in (18.32) go to ∞, with $C(\infty) = 4B^2$.

18.6 Inconsistent empirical functional principal components

We begin by formalizing the intuition behind Example 18.2. By Lemma 2.3, the claims in that example follow from Proposition 18.8. Recall that $X^\star = X(\mathbf{0}) \otimes X(\mathbf{0})$, and observe that for $x \in L^2$,

$$X^\star(x)(t) = \left(\int X(\mathbf{0}; u) x(u) du \right) X(\mathbf{0}; t) = \int c^\star(t, u) x(u) du,$$

where

$$c^\star(t, u) = X(\mathbf{0}; t) X(\mathbf{0}; u).$$

Since

$$E \iint \left(c^\star(t, u)\right)^2 dt\, du = E \|X(\mathbf{0})\|^4 < \infty,$$

the operator $X^\star$ is Hilbert–Schmidt almost surely.

Proposition 18.8. *Suppose representation* (18.18) *holds with stationary mean zero Gaussian processes* ξ_j *such that*

$$E[\xi_j(\mathbf{s})\xi_j(\mathbf{s} + \mathbf{h})] = \lambda_j \rho_j(h), \quad h = \|\mathbf{h}\|,$$

where each ρ_j *is a continuous correlation function, and* $\sum_j \lambda_j < \infty$. *Assume the processes* ξ_j *and* ξ_i *are independent if* $i \neq j$. *If* $\mathfrak{S}_N = \{\mathbf{s}_1, \mathbf{s}_2, \ldots, \mathbf{s}_n\} \subset \mathbb{R}^d$ *with* $\mathbf{s}_n \to \mathbf{0}$, *then*

$$\lim_{N \to \infty} E \|\widehat{C}_N - X^\star\|_{\mathcal{S}}^2 = 0. \tag{18.33}$$

Proposition 18.8 is proven in Section 18.7.

We now present a very specific example that illustrates Proposition 18.8.

Example 18.13. Suppose

$$X(s; t) = \zeta_1(s) e_1(t) + \sqrt{\lambda} \zeta_2(s) e_2(t), \tag{18.34}$$

where the ζ_1 and ζ_2 are iid processes on the line, and $0 < \lambda < 1$. Assume that the processes ζ_1 and ζ_2 are Gaussian with mean zero and covariances $E[\zeta_j(s)\zeta_j(s + h)] = \exp\{-h^2\}$, $j = 1, 2$. Thus, each $Z_j := \zeta_j(0)$ is standard normal. Rearranging the terms, we obtain

$$\begin{aligned} X^\star(x) = &\left(Z_1^2 \langle x, e_1 \rangle + \sqrt{\lambda} Z_1 Z_2 \langle x, e_2 \rangle \right) e_1 \\ &+ \left(\sqrt{\lambda} Z_1 Z_2 \langle x, e_1 \rangle + \lambda Z_2^2 \langle x, e_2 \rangle \right) e_2. \end{aligned}$$

The matrix

$$\begin{bmatrix} Z_1^2 & \sqrt{\lambda} Z_1 Z_2 \\ \sqrt{\lambda} Z_1 Z_2 & \lambda Z_2^2 \end{bmatrix}$$

has only one positive eigenvalue $Z_1^2+\lambda Z_2^2 = \|X(0)\|^2$. A normalized eigenfunction associated with it is

$$f := \frac{X(0)}{\|X(0)\|} = \left[Z_1^2 + \lambda Z_2^2\right]^{-1/2}\left(Z_1 e_1 + \sqrt{\lambda} Z_2 e_2\right). \tag{18.35}$$

Denote by $\hat{v}_1$ a normalized eigenfunction corresponding to the largest eigenvalue of $\widehat{C}_N$. By Lemma 2.3, $\hat{v}_1$ is close in probability to $\text{sign}(\langle \hat{v}_1, f\rangle) f$. It is thus not close to $\text{sign}(\langle \hat{v}_1, e_1\rangle) e_1$.

Ten simulated $\hat{v}_1$, with $e_1(t) = \sqrt{2}\sin(2\pi t)$, $e_2(t) = \sqrt{2}\cos(2\pi t)$, $\lambda = 0.5$, are shown in Figure 18.1. The EFPC $\hat{v}_1$ is a linear combination of e_1 and e_2 with random weights. As formula (18.35) suggests, the function e_1 is likely to receive a larger weight. The weights, and so the simulated $\hat{v}_1$, cluster because both Z_1 and Z_2 are standard normal.

We now state a general result showing that Type A sampling generally leads to inconsistent estimators if the spatial dependence does not vanish.

Proposition 18.9. *Assume that* $E\langle X(\mathbf{s}_1) - \mu\,,\, X(\mathbf{s}_2) - \mu\rangle \geq b(\|\mathbf{s}_1 - \mathbf{s}_2\|_2) > 0$, *where* $b(x)$ *is non-increasing. Then under Type A sampling the sample mean* $\bar{X}_N$ *is*

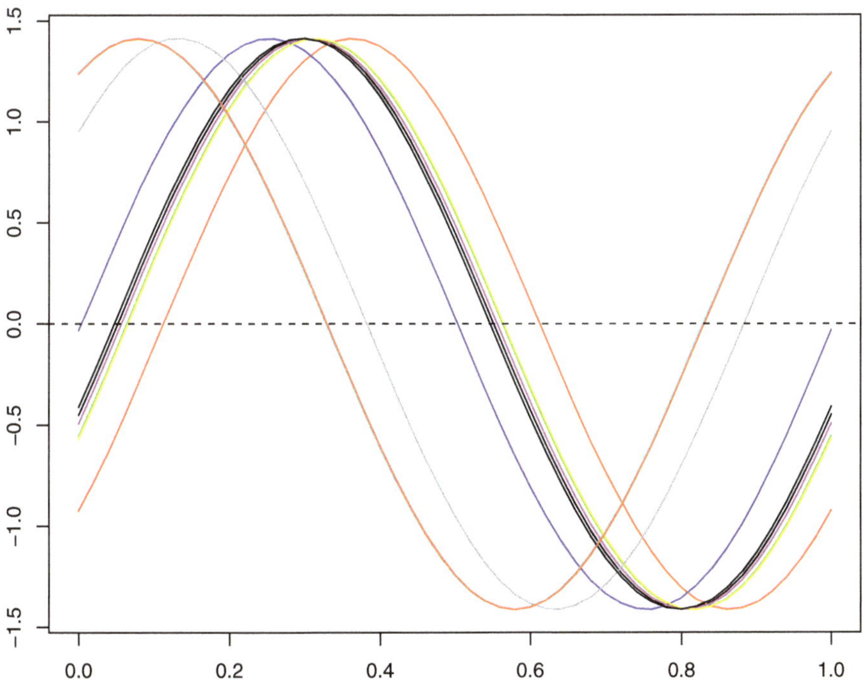

Fig. 18.1 Ten simulated EFPC's $\hat{v}_1$ for process (18.34) with $\lambda = 0.5$ and $e_1(t) = \sqrt{2}\sin(2\pi t)$, $e_2(t) = \sqrt{2}\cos(2\pi t)$ $(N = 100)$.

not a consistent estimator of μ. *Similarly, if* $EX(\mathbf{s}) = 0$ *and*

$$E\langle X(\mathbf{s}_1) \otimes X(\mathbf{s}_1) - C\ ,\ X(\mathbf{s}_2) \otimes X(\mathbf{s}_2) - C\rangle_{\mathcal{S}} \geq B(\|\mathbf{s}_1 - \mathbf{s}_2\|_2) > 0, \quad (18.36)$$

where $B(x)$ *is non-increasing, then under Type A sampling the sample covariance* $\widehat{C}_N$ *is not a consistent estimator of* C.

We illustrate Proposition 18.9 with an example that complements Example 18.2 and Proposition 18.8 in a sense that in Proposition 18.8 the functional model was complex, but the spatial distribution of the $\mathbf{s}_k$ simple. In Example 18.14, we allow a general Type A distribution, but consider the simple model (18.34).

Example 18.14. We focus on condition (18.36) for the FPC's. For the general model (18.18), the left–hand side of (18.36) is equal to

$$\kappa(\mathbf{s}_1, \mathbf{s}_2) = \sum_{i,j\geq 1} \mathrm{Cov}(\xi_i(\mathbf{s}_1)\xi_j(\mathbf{s}_1), \xi_i(\mathbf{s}_2)\xi_j(\mathbf{s}_2)).$$

If the processes ξ_j satisfy the assumptions of Proposition 18.8, then, by Lemma 18.1,

$$\begin{aligned}&\mathrm{Cov}(\xi_i(\mathbf{s}_1)\xi_j(\mathbf{s}_1), \xi_i(\mathbf{s}_2)\xi_j(\mathbf{s}_2))\\ &\quad = \lambda_i^2 r_i + \lambda_j^2 r_j + \lambda_i\lambda_j \frac{r_i + r_j}{2} - \left(\lambda_i^{3/2} r_i + \lambda_j^{3/2} r_j\right)\sqrt{\lambda_i + \lambda_j},\end{aligned}$$

where $r_i = \rho_i(\|\mathbf{s}_1 - \mathbf{s}_2\|)$.

To calculate $\kappa(\mathbf{s}_1, \mathbf{s}_2)$ in a simple case, corresponding to (18.34), suppose

$$\lambda_1 = 1, \lambda_2 = \lambda,\ 0 < \lambda < 1,\ \ \lambda_i = 0,\ i > 2,\ \text{and}\ \rho_1 = \rho_2 = \rho. \quad (18.37)$$

Then,

$$\kappa(\mathbf{s}_1, \mathbf{s}_2) = f(\lambda)\rho(\|\mathbf{s}_1 - \mathbf{s}_2\|),$$

where

$$f(\lambda) = (3 - 2\sqrt{2})(1 + \lambda^2) + 2\left[1 + \lambda + \lambda^2 - (1 + \lambda^{3/2})(1 + \lambda)^{1/2}\right].$$

The function f increases from about 0.17 at $\lambda = 0$ to about 0.69 at $\lambda = 1$.

We have verified that if the functional random field (18.18) satisfies the assumptions of Proposition 18.8 and (18.37), then $\widehat{C}$ is an inconsistent estimator of C under Type A sampling, whenever $\rho(h)$ is a nonincreasing function of h.

18.7 Proofs of the results of Sections 18.4, 18.5 and 18.6

Proof of Proposition 18.1. By Assumption 18.1 we have

$$\begin{aligned}
E\left\|\frac{1}{N}\sum_{k=1}^{N}X(\mathbf{s}_{k,N})-\mu\right\|^2 &= \frac{1}{N^2}\sum_{k=1}^{N}\sum_{\ell=1}^{N}E\langle X(\mathbf{s}_{k,N})-\mu\,,\,X(\mathbf{s}_{\ell,N})-\mu\rangle \\
&\leq \frac{1}{N^2}\sum_{k=1}^{N}\sum_{\ell=1}^{N}h\big(\|\mathbf{s}_{k,N}-\mathbf{s}_{\ell,N}\|_2\big) \\
&\leq \frac{1}{N^2}\sum_{k=1}^{N}\sum_{\ell=1}^{N}\big(h(\rho_N)I\{\|\mathbf{s}_{k,N}-\mathbf{s}_{\ell,N}\|_2\geq\rho_N\} \\
&\qquad\qquad + h(0)I\{\|\mathbf{s}_{k,N}-\mathbf{s}_{\ell,N}\|_2\leq\rho_N\}\big) \\
&\leq h(\rho_N)+h(0)\,I_{\rho_N}(\mathfrak{S}_N).
\end{aligned}$$

□

The following Lemma is a simple calculus problem and will be used in the proof of Proposition 18.2.

Lemma 18.6. *Assume that f is a non-negative function which is monotone on $[0,b]$ and on $[b,\infty)$. Then*

$$\sum_{k=0}^{L}f\left(\frac{k}{N}\right)\frac{1}{N}\leq\int_0^{L/N}f(x)dx+\frac{2}{N}\sup_{x\in[0,L/N]}|f(x)|.$$

Proof of Proposition 18.2. By Assumption 18.1,

$$\begin{aligned}
E\left\|\frac{1}{S_N}\sum_{k=1}^{S_N}X(\mathbf{s}_{k,N})-\mu\right\|^2 &= \frac{1}{S_N^2}\sum_{k=1}^{S_N}\sum_{\ell=1}^{S_N}E\langle X(\mathbf{s}_{k,N})-\mu\,,\,X(\mathbf{s}_{\ell,n})-\mu\rangle \\
&\leq \frac{1}{S_N^2}\sum_{k=1}^{S_N}\sum_{\ell=1}^{S_N}h\big(\|\mathbf{s}_{k,N}-\mathbf{s}_{\ell,n}\|_2\big).
\end{aligned}$$

Let $\mathbf{a}=(a_1,\ldots,a_d)$ and $\mathbf{b}=(b_1,\ldots,b_d)$ be two elements on $\mathcal{Z}(\boldsymbol{\delta})$. We define $d(\mathbf{a},\mathbf{b})=\min_{1\leq i\leq d}v_i(\mathbf{a},\mathbf{b})$, where $v_i(\mathbf{a},\mathbf{b})$ is the number of edges between a_i and b_i. For any two points $\mathbf{s}_{k,N}$ and $\mathbf{s}_{\ell,N}$ we have

$$d(\mathbf{s}_{k,N},\mathbf{s}_{\ell,N})=m\quad\text{from some } m\in\{0,\ldots,KN^{1/d}\},\tag{18.38}$$

where K depends on diam(R_0). It is easy to see that the number of points on the grid having distance m from a given point is less than $2d(2m+1)^d$, $m\geq 0$. Hence the number of pairs for which (18.38) holds is less than $2d(2m+1)^{d-1}N$. On the other hand, if $d(\mathbf{s}_{k,N},\mathbf{s}_{\ell,N})=m$, then $\|\mathbf{s}_{k,N}-\mathbf{s}_{\ell,N}\|_2\geq m\delta_0\eta_N$. Let us assume without loss of generality that $\delta_0=1$. Noting that there is no loss of generality if we assume that $x^{\delta-1}h(x)$ is also monotone on $[0,b]$, we obtain by Lemma 18.6 for large enough N and $K < K' < K''$

$$
\begin{aligned}
&\frac{1}{S_N^2}\sum_{k=1}^{S_N}\sum_{\ell=1}^{S_N} h\big(\|\mathbf{s}_{k,N}-\mathbf{s}_{\ell,n}\|_2\big)\\
&\quad\le 2d\sum_{m=1}^{K'N^{1/d}}\frac{(2m+1)^{d-1}}{N}h\big(m\eta_N\big)+\frac{2h(0)}{N}\\
&\quad\le 2d\left(\frac{3}{\eta_N}\right)^{d-1}\sum_{m=0}^{K'N^{1/d}+1}\left(\frac{m}{N}N\eta_N\right)^{d-1}h\left(\frac{m}{N}N\eta_N\right)\frac{1}{N}+\frac{2h(0)}{N}\\
&\quad\le 2d\left(\frac{3}{\eta_N}\right)^{d-1}\left(\int_0^{K''N^{1/d-1}}(N\eta_N x)^{d-1}h\big(N\eta_N x\big)dx\right.\\
&\qquad\qquad\left.+\frac{2}{N}\sup_{x\in[0,K''\alpha_N/\Delta]}x^{d-1}h(x)\right)+\frac{2h(0)}{N}\\
&\quad=\frac{(3\Delta)^d d}{\alpha_N^d}\int_0^{K''\alpha_N/\Delta}x^{d-1}h\big(x\big)dx\\
&\qquad+\frac{4d(3\Delta)^{d-1}}{\alpha_N^{d-1}N^{1/d}}\sup_{x\in[0,K''\alpha_N/\Delta]}x^{d-1}h(x)+\frac{2h(0)}{N}.
\end{aligned}
$$

By Lemma 18.4, Type B sampling implies $\alpha_N\to\infty$ and $\alpha_N = o\big(N^{1/d}\big)$. This shows (18.28). Under Type C sampling $1/\alpha_N^d \ll 1/N$. The proof is finished. □

Proof of Proposition 18.3. This time we have

$$
\begin{aligned}
E\left\|\frac{1}{N}\sum_{k=1}^{N}X(\mathbf{s}_{k,N})-\mu\right\|^2 &\le \frac{1}{N^2}\sum_{k=1}^{N}\sum_{\ell=1}^{N}Eh\big(\|\mathbf{s}_{k,N}-\mathbf{s}_{\ell,N}\|_2\big)\\
&\le \alpha_N^{-2d}\int_{R_N}\int_{R_N}h\big(\|\mathbf{s}-\mathbf{r}\|_2\big)f(\alpha_N^{-1}\mathbf{s})f(\alpha_N^{-1}\mathbf{r})\,d\mathbf{s}d\mathbf{r}+\frac{h(0)}{N}\\
&=\int_{R_0}\int_{R_0}h\big(\alpha_N\|\mathbf{s}-\mathbf{r}\|_2\big)f(\mathbf{s})f(\mathbf{r})\,d\mathbf{s}d\mathbf{r}+\frac{h(0)}{N}.
\end{aligned}
$$

Furthermore, for any $\varepsilon_N > 0$,

$$
\begin{aligned}
&\int_{R_0}\int_{R_0}h\big(\alpha_N\|\mathbf{s}-\mathbf{r}\|_2\big)f(\mathbf{s})f(\mathbf{r})\,d\mathbf{s}d\mathbf{r}\\
&\quad\le h(0)\int_{R_0}\int_{R_0}f(\mathbf{s})f(\mathbf{r})I\big\{\|\mathbf{s}-\mathbf{r}\|_2\le\varepsilon_N\big\}\,d\mathbf{s}d\mathbf{r}+h(\alpha_N\varepsilon_N)\\
&\quad\le h(0)\sup_{\mathbf{s}\in R_0}f^2(\mathbf{s})\times\int_{R_0}\int_{R_0}I\big\{\|\mathbf{s}-\mathbf{r}\|_2\le\varepsilon_N\big\}d\mathbf{s}d\mathbf{r}+h(\alpha_N\varepsilon_N).
\end{aligned}
$$

Now for fixed $\mathbf{r}$ it is not difficult to show that $\int_{R_0} I\big\{\|\mathbf{s}-\mathbf{r}\|_2\le\varepsilon_N\big\}d\mathbf{s}\le 6\varepsilon_N^d$. (The constant 6 could be replaced with $\pi^{d/2}/\Gamma(d/2+1)$).

Proof of Proposition 18.8. Observe that

$$\|\widehat{C}_N - X^\star\|_{\mathcal{S}}^2 = \iint \left\{ \frac{1}{N} \sum_{n=1}^{N} [X(\mathbf{s}_n;t)X(\mathbf{s}_n;u) - X(\mathbf{0};t)X(\mathbf{0};u)] \right\}^2 dt\,du.$$

Therefore,

$$\|\widehat{C}_N - X^\star\|_{\mathcal{S}}^2 \le 2I_1(N) + 2I_2(N),$$

where

$$I_1(N) = \iint \left\{ \frac{1}{N} \sum_{n=1}^{N} X(\mathbf{s}_n;t)(X(\mathbf{s}_n;u) - X(\mathbf{0};u)) \right\}^2 dt\,du$$

and

$$I_2(N) = \iint \left\{ \frac{1}{N} \sum_{n=1}^{N} X(\mathbf{0};u)(X(\mathbf{s}_n;t) - X(\mathbf{0};t)) \right\}^2 dt\,du.$$

We will show that $EI_1(N) \to 0$. The argument for $I_2(N)$ is the same. Observe that

$$\begin{aligned}
&I_1(N)\\
&= \frac{1}{N^2} \sum_{k,\ell=1}^{N} \iint X(\mathbf{s}_k;t)(X(\mathbf{s}_k;u) - X(\mathbf{0};u))X(\mathbf{s}_\ell;t)(X(\mathbf{s}_\ell;u) - X(\mathbf{0};u))dt\,du\\
&= \frac{1}{N^2} \sum_{k,\ell=1}^{N} \int X(\mathbf{s}_k;t)X(\mathbf{s}_\ell;t)dt \int (X(\mathbf{s}_k;u) - X(\mathbf{0};u))(X(\mathbf{s}_\ell;u) - X(\mathbf{0};u))du.
\end{aligned}$$

Thus,

$$\begin{aligned}
&EI_1(N)\\
&\le \frac{1}{N^2} \sum_{k,\ell=1}^{N} \left\{ E\left(\int X(\mathbf{s}_k;t)X(\mathbf{s}_\ell;t)dt \right)^2 \right\}^{1/2} \left\{ E\left(\int Y_k(u)Y_\ell(u)du \right)^2 \right\}^{1/2},
\end{aligned}$$

where

$$Y_k(u) = X(\mathbf{s}_k;u) - X(\mathbf{0};u).$$

We first deal with the integration over t:

$$\begin{aligned}
&E\left(\int X(\mathbf{s}_k;t)X(\mathbf{s}_\ell;t)dt \right)^2 \le E \int X^2(\mathbf{s}_k;t)dt \int X^2(\mathbf{s}_\ell;t)dt\\
&\quad = E\left[\|X(\mathbf{s}_k)\|^2 \|X(\mathbf{s}_\ell)\|^2 \right] \le \left\{ E\|X(\mathbf{s}_k)\|^4 \right\}^{1/2} \left\{ E\|X(\mathbf{s}_\ell)\|^4 \right\}^{1/2} = E\|X(\mathbf{0})\|^4.
\end{aligned}$$

We thus see that

$$
\begin{aligned}
&E I_1(N) \\
&\le \left\{E\|X(\mathbf{0})\|^4\right\}^{1/2} \frac{1}{N^2} \sum_{k,\ell=1}^{N} \left\{E\left(\int Y_k(u)Y_\ell(u)du\right)^2\right\}^{1/2} \\
&\le \left\{E\|X(\mathbf{0})\|^4\right\}^{1/2} \frac{1}{N^2} \sum_{k,\ell=1}^{N} \left\{E\left(\int Y_k^2(u)du\right)^2\right\}^{1/4} \left\{E\left(\int Y_\ell^2(u)du\right)^2\right\}^{1/4} \\
&= \left\{E\|X(\mathbf{0})\|^4\right\}^{1/2} \left[\frac{1}{N} \sum_{k=1}^{N} \left\{E\left(\int Y_k^2(u)du\right)^2\right\}^{1/4}\right]^2.
\end{aligned}
$$

Consequently, to complete the verification of (18.33), it suffices to show that

$$
\lim_{N\to\infty} \frac{1}{N} \sum_{k=1}^{N} \left\{E\left(\int Y_k^2(u)du\right)^2\right\}^{1/4} = 0.
$$

The above relation will follow from

$$
\lim_{k\to\infty} E\left(\int Y_k^2(u)du\right)^2 = 0. \tag{18.39}
$$

To verify (18.39), first notice that, by the orthonormality of the e_j,

$$
\int Y_k^2(u)du = \sum_{j=1}^{\infty} \left(\xi_j(\mathbf{s}_k) - \xi_j(\mathbf{0})\right)^2.
$$

Therefore, by the independence of the processes ξ_j,

$$
\begin{aligned}
E\left(\int Y_k^2(u)du\right)^2 &= \sum_{j=1}^{\infty} E\left(\xi_j(\mathbf{s}_k) - \xi_j(\mathbf{0})\right)^4 \\
&\quad + \sum_{i\ne j} E\left(\xi_i(\mathbf{s}_k) - \xi_i(\mathbf{0})\right)^2 E\left(\xi_j(\mathbf{s}_k) - \xi_j(\mathbf{0})\right)^2.
\end{aligned}
$$

The covariance structure was specified so that

$$
E\left(\xi_j(\mathbf{s}_k) - \xi_j(\mathbf{0})\right)^2 = 2\lambda_j(1 - \rho_j(\|\mathbf{s}_k\|)),
$$

so the normality yields

$$
\begin{aligned}
E\left(\int Y_k^2(u)du\right)^2 &\le 12 \sum_{j=1}^{\infty} \lambda_j^2 (1 - \rho_j(\|\mathbf{s}_k\|))^2 \\
&\quad + 4\left\{\sum_{j=1}^{\infty} \lambda_j (1 - \rho_j(\|\mathbf{s}_k\|))\right\}^2.
\end{aligned}
$$

The right hand side tends to zero by the Dominated Convergence Theorem. This establishes (18.39), and completes the proof of (18.33). □

Proof of Proposition 18.9. We only check inconsistency of the sample mean. In view of the proof of Proposition 18.1 we have now the lower bound

$$E\left\|\frac{1}{N}\sum_{k=1}^{N} X(\mathbf{s}_{k,N}) - \mu\right\|^2 \geq \frac{1}{N^2}\sum_{k=1}^{N}\sum_{\ell=1}^{N} b(\|\mathbf{s}_{k,N} - \mathbf{s}_{\ell,N}\|_2)$$
$$\geq b(\rho) I_\rho^2(\mathfrak{S}_N),$$

which is by assumption bounded away from zero for $N \to \infty$. □

References

Akaike, H. (1978). A Bayesian analysis of the minimum AIC procedure. *Ann. Inst. Statist. Math.*, **30A,** 9–14.

Akhiezier, N. I. and Glazman, I. M. (1993). *Theory of Linear Operators in Hilbert Space.* Dover, New York.

Andersen, T. G. and Bollerslev, T. (1997a). Heterogeneous information arrivals and return volatility dynamics: uncovering the long run in high frequency data. *Journal of Finance*, **52,** 975–1005.

Andersen, T. G. and Bollerslev, T. (1997b). Intraday periodicity and volatility persistence in financial markets. *Journal of Empirical Finance*, **2–3,** 115–158.

Anderson, T. W. (1984). *An Introduction to Multivariate Statistical Analysis.* Wiley, New York.

Anderson, T. W. (1994). *The Statistical Analysis of Time Series.* Wiley and Sons.

Andrews, D. W. K. (1991). Heteroskedasticity and autocorrelation consistent covariance matrix estimation. *Econometrica*, **59,** 817–858.

Andrews, D. W. K. and Monahan, J. C. (1992). An improved heteroskedasticity and autocorrelation consistent covariance matrix estimator. *Econometrica*, **60,** 953–966.

Antoch, J., Husková, M. and Prásková, Z. (1997). Effect of dependence on statistics for determination of change. *Journal of Statistical Planning and Inference*, **60,** 291–310.

Antoniadis, A., Paparoditis, E. and Sapatinas, T. (2006). A functional wavelet–kernel approach for time series prediction. *Journal of the Royal Statistical Society, Series* B, **68,** 837–857.

Aston, J. A. D. and Kirch, C. (2011a). Detecting and estimating epidemic changes in dependent functional data. CRiSM Research Report 11–07. University of Warwick.

Aston, J. A. D. and Kirch, C. (2011b). Estimation of the distribution of change-points with application to fMRI data. CRiSM Research Reports. University of Warwick.

Aue, A., Gabrys, R., Horváth, L. and Kokoszka, P. (2009). Estimation of a change–point in the mean function of functional data. *Journal of Multivariate Analysis*, **100,** 2254–2269.

Aue, A., Hörmann, S., Horváth, L. and Hušková, M. (2010). Sequential stability test for functional linear models. Technical Report. University of California Davis.

Aue, A., Hörmann, S., Horváth, L. and Reimherr, M. (2009). Break detection in the covariance structure of multivariate time series models. *The Annals of Statistics*, **37,** 4046–4087.

Benko, M., Härdle, W. and Kneip, A. (2009). Common functional principal components. *The Annals of Statistics*, **37,** 1–34.

Berkes, I., Gabrys, R., Horváth, L. and Kokoszka, P. (2009). Detecting changes in the mean of functional observations. *Journal of the Royal Statistical Society (B)*, **71,** 927–946.

Berkes, I., Hörmann, S. and Horváth, L. (2008). The functional central limit theorem for a family of GARCH observations with applications. *Statistics and Probability Letters*, **78,** 2725–2730.

Berkes, I., Hörmann, S. and Schauer, J. (2009). Asymptotic results for the empirical process of stationary sequences. *Stochastic Processes and their Applications*, **119,** 1298–1324.

Berkes, I. and Horváth, L. (2001). Strong approximation for the empirical process of a GARCH sequence. *The Annals of Applied Probability*, **11,** 789–809.

Berkes, I. and Horváth, L. (2003a). Asymptotic results for long memory LARCH sequences. *The Annals of Applied Probability*, **13,** 641–668.

Berkes, I. and Horváth, L. (2003b). Limit results for the empirical process of squared residuals in GARCH models. *Stochastic Processes and their Applications*, **105,** 279–298.

Berkes, I., Horváth, L. and Kokoszka, P. (2005). Near integrated GARCH sequences. *Annals of Applied Probability*, **15,** 890–913.

Berkes, I., Horváth, L., Kokoszka, P. and Shao, Q-M. (2005). Almost sure convergence of the Bartlett estimator. *Periodica Mathematica Hungarica*, **51,** 11–25.

Berkes, I., Horváth, L., Kokoszka, P. and Shao, Q-M. (2006). On discriminating between long-range dependence and changes in mean. *The Annals of Statistics*, **34,** 1140–1165.

Berkes, I., Horváth, L. and Kokoszka, P. S. (2003). GARCH processes: structure and estimation. *Bernoulli*, **9,** 201–227.

Besse, P., Cardot, H. and Stephenson, D. (2000). Autoregressive forecasting of some functional climatic variations. *Scandinavian Journal of Statistics*, **27,** 673–687.

Bhansali, R. J. (1993). Order selection for linear time series models: a review. In *Developments in Time Series Analysis, London* (ed. T. Subba Rao), pp. 50–6. Chapman and Hall.

Billingsley, P. (1968). *Convergence of Probability Measures*. Wiley, New York.

Billingsley, P. (1995). *Probability and Measure*, 3rd edn. Wiley, New York.

Billingsley, P. (1999). *Convergence of Probability Measures; Second Edition*. Wiley, New York.

Boente, G. and Fraiman, R. (2000). Kernel–based functional principal components. *Statistics and Probability Letters*, **48,** 335–345.

Boente, G., Rodriguez, D. and Sued, M. (2011). Testing the equality of covariance operators. In *Recent Advances in Functional Data Analysis and Related Topics* (ed. F. Ferraty). Physica–Verlag.

Bolthausen, E. (1982). On the central limit theorem for stationary mixing random fields. *The Annals of Probability*, **10**, 1047–1050.

Borggaard, C. and Thodberg, H. (1992). Optimal minimal neural interpretation of spectra. *The Annals of Chemistry*, **64**, 545–551.

Bosq, D. (2000). *Linear Processes in Function Spaces*. Springer, New York.

Bosq, D. and Blanke, D. (2007). *Inference and Prediction in Large Dimensions*. Wiley.

Box, G. E. P., Jenkins, G. M. and Reinsel, G. C. (1994). *Time Series Analysis: Forecasting and Control*, Third edn. Prentice Hall, Englewood Cliffs.

Bradley, R. C. (2007). *Introduction to Strong Mixing Conditions*, volume 1,2,3. Kendrick Press.

Bremer, J. (1998). Trends in the ionospheric E and F regions over Europe. *Annales Geophysicae*, **16**, 986–996.

Brockwell, P. J. and Davis, R. A. (1991). *Time Series: Theory and Methods*. Springer, New York.

Brodsky, B. E. and Darkhovsky, B. S. (1993). *Nonparametric Methods in Change–Point Problems*. Kluwer.

Cai, T. and Hall, P. (2006). Prediction in functional linear regression. *The Annals of Statistics*, **34**, 2159–2179.

Campbell, J. Y., Lo, A. W. and MacKinlay, A. C. (1997). *The Econometrics of Financial Markets*. Princeton University Press, New Jersey.

Cardot, H., Faivre, R. and Goulard, M. (2003c). Functional approaches for predicting land use with the temproal evolution of coarse resolution remote sensing data. *Journal of Applied Statistics*, **30**, 1185–1199.

Cardot, H., Ferraty, F., Mas, A. and Sarda, P. (2003). Testing hypothesis in the functional linear model. *Scandinavian Journal of Statistics*, **30**, 241–255.

Cardot, H., Ferraty, F. and Sarda, P. (2003b). Spline estimators for the functional linear model. *Statistica Sinica*, **13**, 571–591.

Carey, J. R., Liedo, P., Harshman, L., Müller, H. G., Partridge, L. and Wang, J. L. (2002). Life history responce of mediterranean fruit flies to dietary restrictions. *Aging Cell*, **1**, 140–148.

Carroll, S. S. and Cressie, N. (1996). A comparison of geostatistical methodologies used to estimate snow water equivalent. *Water Resources Bulletin*, **32**, 267–278.

Carroll, S. S., Day, G. N., Cressie, N. and Carroll, T. R. (1995). Spatial modeling of snow water equivalent using airborne and ground–based snow data. *Environmetrics*, **6**, 127–139.

Cattell, R. B. (1966). The scree test for the number of factors. *Journal of Multivariate Behavioral Research*, **1**, 245–276.

Chatfield, C. (1998). Durbin–Watson test. In *Encyclopedia of Biostatistics* (eds P. Armitage and T. Colton), volume 2, pp. 1252–1253. Wiley.

Chiou, J-M. and Müller, H-G. (1998). Quasi–likelihood regression with unknown link and variance functions. *Journal of the American Statistical Association*, **92,** 72–83.

Chiou, J-M. and Müller, H-G. (2007). Diagnostics for functional regression via residual processes. *Computational Statistics and Data Analysis*, **15,** 4849–4863.

Chiou, J-M., Müller, H-G. and Wang, J-L. (2004). Functional response models. *Statistica Sinica*, **14,** 675–693.

Chiou, J-M., Müller, H-G., Wang, J-L. and Carey, J. R. (2003). A functional multiplicative effects model for longitudal data, with application to reproductive histories of female medflies. *Statistica Sinica*, **13,** 1119–1133.

Chitturi, R. V. (1976). Distribution of multivariate white noise autocorrelation. *Journal of the American Statistical Association*, **71,** number 353, 223–226.

Clarkson, D. B., Fraley, C., Gu, C. and Ramsay, J. O. (2005). *S+ Functional Data Analysis*. Springer.

Cochrane, D. and Orcutt, G. H. (1949). Application of least squares regression to relationships containing auto-correlated error terms. *Journal of the American Statistical Association*, **44,** 32–61.

Cook, D. R. (1977). Detection of influential observations in linear regression. *Technometrics*, **19,** 15–18.

Cook, R. D. (1994). On interpretation of regression plots. *Journal of the American Statistical Association*, **89,** 177–189.

Cook, R. D. and Weisberg, S. (1982). *Residuals and Inference in Regression*. Chapman and Hall.

Cressie, N. A. C. (1993). *Statistics for Spatial Data*. Wiley.

Csörgő, M. and Horváth, L. (1993). *Weighted Approximations in Probability and Statistics*. Wiley, New York.

Csörgő, M. and Horváth, L. (1997). *Limit Theorems in Change-Point Analysis*. Wiley, New York.

Cuevas, A., Febrero, M. and Fraiman, R. (2002). Linear functional regression: the case of fixed design and functional response. *The Canadian Journal of Statistics*, **30,** 285–300.

Cuevas, A., Febrero, M. and Fraiman, R. (2004). An ANOVA test for functional data. *Computational Statistics and Data Analysis*, **47,** 111–122.

Cupidon, J., Gilliam, D. S., Eubank, R. and Ruymgaart, F. (2007). The delta method for analytic functions of random operators with application to functional data. *Bernoulli*, **13,** 1179–1194.

Daglis, I. A., Kozyra, J. U., Kamide, Y., Vassiliadis, D., Sharma, A. S., Liemohn, M.W., Gonzalez, W. D., Tsurutani, B. T. and Lu, G. (2003). Intense space storms: Critical issues and open disputes. *Journal of Geophysical Research*, **108,** doi:10.1029/2002JA009722.

DasGupta, A. (2008). *Asymptotic Theory of Statistics and Probability*. Springer.

Dauxois, J., Pousse, A. and Romain, Y. (1982). Asymptotic theory for principal component analysis of a vector random function. *Journal of Multivariate Analysis*, **12,** 136–154.

Debnath, L. and Mikusinski, P. (2005). *Introduction to Hilbert Spaces with Applications*. Elsevier.

Delaigle, A. and Hall, P. (2010). Defining probability density function for a distribution of random functions. *The Annals of Statistics*, **38,** 1171–1193.

Delicado, P. (2007). Functional k–sample problem when data are density functions. *Computational Statistics*, **22,** 391–410.

Delicado, P., Giraldo, R., Comas, C. and Mateu, J. (2010). Statistics for spatial functional data: some recent contributions. *Environmetrics*, **21,** 224–239.

Didericksen, D., Kokoszka, P. and Zhang, X. (2011). Empirical properties of forecasts with the functional autoregressive model. *Computational Statistics*, Forthcoming.

Doukhan, P. (1994). *Mixing: Properties and Examples.* Lecture Notes in Statistics. Springer.

Doukhan, P. and Louhichi, S. (1999). A new weak dependence and applications to moment inequalities. *Stochastic Processes and their Applications*, **84,** 313–343.

Du, J., Zhang, H. and Mandrekar, V. S. (2009). Fixed-domain asymptotic properties of tapered maximum likelihood estimators. *Annals of Statistics*, **37,** 3330–3361.

Durbin, J. and Watson, G. S. (1950). Testing for serial correlation in least squares regression. I. *Biometrika*, **37,** 409–428.

Durbin, J. and Watson, G. S. (1951). Testing for serial correlation in least squares regression. II. *Biometrika*, **38,** 159–178.

Durbin, J. and Watson, G. S. (1971). Testing for serial correlation in least squares regression. III. *Biometrika*, **58,** 1–19.

Einmahl, J. H. J and Ruymgaart, F. G. (1987). The order of magnitude of the moments of the modulus of continuity of the multiparameter Poisson and empirical processes. *Journal of Multivariate Analysis*, **21,** 263–273.

Engle, R. F. (1982). Autoregressive conditional heteroskedasticity with estimates of the variance of United Kingdom inflation. *Econometrica*, **50,** 987–1007.

Eubank, R. L. and Hsing, T. (2008). Canonical correlation for stochastic processes. *Stochastic Processes and their Applications*, **118,** 1634–1661.

Febrero, M., Galeano, P. and González-Manteiga, W. (2008). Outlier detection in functional data by depth measures with application to identify abnormal NO_x levels. *Environmetrics*, **19,** 331–345; DOI: 10.1002/env.878.

Febrero-Bande, M., Galeano, P. and González-Manteigna, W. (2010). Measures of influence for the functional linear model with scalar response. *Journal of Multivariate Analysis*, **101,** 327–339.

Ferraty, F. (2011) (ed.). *Recent Advances in Functional Data Analysis and Related Topic*. Physica–Verlag.

Ferraty, F. and Romain, Y. (2011) (eds). *The Oxford Handbook of Functional Data Analysis.* Oxford University Press.

Ferraty, F. and Vieu, P. (2006). *Nonparametric Functional Data Analysis: Theory and Practice*. Springer.

Ferraty, F., Vieu, P. and Viguier-Pla, S. (2007). Factor–based comparison of groups of curves. *Computaional Statistics & Data Analysis*, **51,** 4903–4910.

Fraiman, R. and Muniz, G. (2001). Trimmed means for functional data. *Test*, **10,** 419–440.

Fremdt, S., Horváth, L., Kokoszka, P. and Steinebach, J. (2011). Testing the equality of covariance operators in functional samples. Technical report. Universität zu Köln.

Gabrys, R., Horváth, L. and Kokoszka, P. (2010). Tests for error correlation in the functional linear model. *Journal of the American Statistical Association*, **105,** 1113–1125.

Gabrys, R. and Kokoszka, P. (2007). Portmanteau test of independence for functional observations. *Journal of the American Statistical Association*, **102,** 1338–1348.

Gelfand, A. E., Diggle, P. J., Fuentes, M. and Guttorp, P. (2010) (eds). *Handbook of Spatial Statistics*. CRC Press.

Gervini, D. (2008). Robust functional estimation using the spatial median and spherical principal components. *Biometrika*, **95,** 587–600.

Giraitis, L., Kokoszka, P. S. and Leipus, R. (2000). Stationary ARCH models: dependence structure and Central Limit Theorem. *Econometric Theory*, **16,** 3–22.

Giraitis, L., Kokoszka, P. S., Leipus, R. and Teyssière, G. (2003). Rescaled variance and related tests for long memory in volatility and levels. *Journal of Econometrics*, **112,** 265–294.

Giraldo, R., Delicado, P. and Mateu, J. (2010). Ordinary kriging for function–valued spatial data. *Environmental and Ecological Statistics*, **18,** 411–426.

Giraldo, R., Delicado, P. and Mateu, J. (2011). A generalization of cokriging and multivariable spatial prediction for functional data. Technical report. Universitat Politécnica de Catalunya, Barcelona.

Gohberg, I., Golberg, S. and Kaashoek, M. A. (1990). *Classes of Linear Operators*. Operator Theory: Advances and Applications, volume 49. Birkhaüser.

Gohberg, I. C. and Krein, M. C. (1969). *Introduction to the Theory of Linear Nonselfadjoint Operators in Hilbert Space*. Translations of Mathematical Monographs. AMS.

Graham, A. (1981). *Kronecker Products and Matrix Calculus with Applications*. John Wiley and Sons.

Griswold, C., Gomulkiewicz, R. and Heckman, N. (2008). Hypothesis testing in comparative and experimental studies of function-valued traits. *Evolution*, **62,** 1229–42.

Gromenko, O. and Kokoszka, P. (2011). Testing the equality of mean functions of ionospheric critical frequency curves. Technical Report. Utah State University.

Gromenko, O., Kokoszka, P., Zhu, L. and Sojka, J. (2011). Estimation and testing for spatially indexed curves with application to ionospheric and magnetic field trends. *The Annals of Applied Statistics*, Forthcoming.

Gu, C. (2002). *Smoothing Spline ANOVA Models*. Springer.

Guillaume, D. M., Dacorogna, M. M., Dave, R. D., Müller, U. A., Olsen, R. B. and Pictet, O. V. (1997). From the bird's eye to the microscope: a survey of new

stylized facts of the intra-daily foreign exchange markets. *Finance and Stochastics*, **1,** 95–129.
Hall, P. and Heyde, C. C. (1980). *Martingale Limit Theory and its Applications*. Academic Press, New York.
Hall, P. and Hosseini-Nasab, M. (2006). On properties of functional principal components. *Journal of the Royal Statistical Society (B)*, **68,** 109–126.
Hall, P. and Hosseini-Nasab, M. (2007). Theory for high–order bounds in functional principal components analysis. Technical Report. The University of Melbourne.
Hall, P. and Keilegom, I. Van (2007). Two–sample tests in functional data analysis starting from discrete data. *Statistica Sinica*, **17,** 1511–1531.
Hamilton, J. D. (1994). *Time Series Analysis*. Princeton University Press, Princeton, NJ.
Hannan, E. J. (1980). The estimation of the order of an ARMA process. *The Annals of Statistics*, **8,** 1071–1081.
Hannan, E. J. and Quinn, B. G. (1979). The determination of the order of an autoregression. *J. Royal Statist. Soc.* B, **41,** 190–195.
Hannan, E. J. and Rissannen, J (1982). Recursive estimation of mixed autoregressive-moving average order. *Biometrika*, **69,** 81–94; Correction (1983) **70,** 303.
Hansen, B. E. (1995). Rethinking the univariate approach to unit root testing: using covariates to increase power. *Econometric Theory*, **11,** 1148–1171; Code available at `http://www.ssc.wisc.edu/~bhansen`.
He, G., Müller, H-G. and Wang, J-L. (2003). Functional canonical analysis for square integrable stochastic processes. *Journal of Multivariate Analisis*, **85,** 54–77.
He, G., Müller, H-G. and Wang, J-L. (2004). Methods of canonical analysis for functional data. *Journal of Statistical Planning and Inference*, **122,** 141–159.
Hörmann, S. (2008). Augmented GARCH sequences: Dependence structure and asymptotics. *Bernoulli*, **14,** 543–561.
Hörmann, S., Horváth, L. and Reeder, R. (2010). A functional version of the ARCH model. Technical Report. University of Utah.
Hörmann, S. and Kokoszka, P. (2010). Weakly dependent functional data. *The Annals of Statistics*, **38,** 1845–1884.
Hörmann, S. and Kokoszka, P. (2011). Consistency of the mean and the principal components of spatially indexed functional data. *Bernoulli*, Forthcoming.
Horn, R. A. and Johnson, C. R. (1985). *Matrix Analysis*. Cambridge University Press.
Horn, R. A. and Johnson, C. R. (1991). *Topics in Matrix Analysis*. Cambridge University Press.
Horváth, L., Horváth, Z. and Hušková, M. (2008). Ratio tests for change point detection. In *Beyond Parametrics in Interdisciplinary Research: Festschrift in Honor of Professor Pranab K. Sen*, IMS Collections, pp. 293–304. IMS.
Horváth, L., Hušková, M. and Kokoszka, P. (2010). Testing the stability of the functional autoregressive process. *Journal of Multivariate Analysis*, **101,** 352–367.

Horváth, L., Kokoszka, P. and Reeder, R. (2011). Estimation of the mean of functional time series and a two sample problem. *Journal of the Royal Statistical Society (B)*, Forthcoming.

Horváth, L., Kokoszka, P. and Reimherr, M. (2009). Two sample inference in functional linear models. *Canadian Journal of Statistics*, **37,** 571–591.

Horváth, L., Kokoszka, P. S. and Steinebach, J. (1999). Testing for changes in multivariate dependent observations with applications to temperature changes. *Journal of Multivariate Analysis*, **68,** 96–119.

Horváth, L. and Reeder, R. (2011). Detecting changes in functional linear models. Technical Report. University of Utah.

Horváth, L. and Reeder, R. (2011b). A test of significance in functional quadratic regression. Technical Report. University of Utah. preprint available at `http://arxiv.org/abs/1105.0014`.

Hosking, J. R. M. (1980). The multivariate portmanteau statistic. *Journal of the American Statistical Association*, **75,** 602–608.

Hosking, J. R. M. (1981). Equivalent forms of the multivariate portmanteau statistics. *Journal of the Royal Statistical Society (B)*, **43,** 261–262.

Hosking, J. R. M. (1984). Modeling persistence in hydrological time series using fractional differencing. *Water Resources Research*, **20,** number 12, 1898–1908.

Hosking, J. R. M. (1989). Corrigendum: Equivalent forms of the multivariate portmanteau statistics. *Journal of the Royal Statistical Society (B)*, **51,** 303–303.

Hotelling, H. (1933). Analysis of a complex of statistical variables into principal components. *Journal of Educational Psychology*, **24,** 498–520.

Hušková, M., Prášková, Z. and Steinebach, J. (2007). On the detection of changes in autoregressive time series I. Asymptotics. *Journal of Statistical Planning and Inference*, **137,** 1243–1259.

Izem, R. and Marron, J. S. (2007). Functional data analysis of nonlinear modes of variation. *Electronic Journal of Statistics*, **1,** 641–676.

Jach, A. and Kokoszka, P. (2008). Wavelet domain test for long–range dependence in the presence of a trend. *Statistics*, **42,** 101–113.

Jach, A., Kokoszka, P., Sojka, J. and Zhu, L. (2006). Wavelet–based index of magnetic storm activity. *Journal of Geophysical Research*, **111,** A09215.

Jenish, N. and Prucha, I. R. (2009). Central limit theorems and uniform laws of large numbers for arrays or random fields. *Journal of Econometrics*, **150,** 86–98.

Jiofack, J. G. A. and Nkiet, G. M. (2010). Testing for lack of dependence between functional variables. *Statistics and Probability Letters*, **80,** 1210–1217.

Johnson, R. A. and Wichern, D. W. (2002). *Applied Multivariate Statistical Analysis*. Prentice Hall.

Johnstone, I. M. and Lu, A. Y. (2009). On consistency and sparcity for principal components analysis in high dimensions. *Journal of the Americal Statistical Association*, **104,** 682–693.

Jolliffe, I. T. (2002). *Principal Component Analysis*. Springer.

Jones, M. C. and Rice, J. A. (1992). Displaying the important features of a large collection of similar curves. *The American Statistician*, **46,** 140–145.

Jung, S. and Marron, J. S. (2009). PCA consistency in high dimension and low sample size. *The Annals of Statistics*, **37,** 4104–4130.

Kaiser, M. S., Daniels, M. J., Furakawa, K. and Dixon, P. (2002). Analysis of particulate matter air pollution using Markov random field models of spatial dependence. *Environmetrics*, **13,** 615–628.

Kamide, Y., Baumjohann, W., a nd W. D. Gonzalez, I. A. Daglis, Grande, M., Joselyn, J. A., McPherron, R. L., Phillips, J. L., Reeves, E. G. D., Rostoker, G., Sharma, A. S., Singer, H. J., Tsurutani, B. T. and Vasyliunas, V. M. (1998). Current understanding of magnetic storms: Storm–substorm relationships. *Journal of Geophysical Research*, **103,** 17705–17728.

Kargin, V. and Onatski, A. (2008). Curve forecasting by functional autoregression. *Journal of Multivariate Analysis*, **99,** 2508–2526.

Kiefer, J. (1959). K-sample analogues of the Kolmogorov-Smirnov and Cramér-v.Mises tests. *Ann. Math. Statist.*, **30,** 420–447.

Kirkpatrick, M. and Heckman, N. (1989). A quantitiative genetic model for growth, shape, reaction norms and other infinite–dimensional characters. *Journal of Mathematical Biology*, **27,** 429–450.

Kivelson, M. G. and Russell, C. T. (1997) (eds). *Introduction to Space Physics*. Cambridge University Press.

Kokoszka, P., Maslova, I., Sojka, J. and Zhu, L. (2008). Testing for lack of dependence in the functional linear model. *Canadian Journal of Statistics*, **36,** 207–222.

Kokoszka, P. and Reimherr, M. (2011). Determining the order of the functional autoregressive model. Technical Report. University of Chicago.

Kokoszka, P. and Zhang, X. (2010). Improved estimation of the kernel of the functional autoregressive process. Technical Report. Utah State University.

Kokoszka, P. and Zhang, X. (2011). Functional prediction of cumulative intraday returns. Technical Report. Utah State University.

Koul, H. L. (2002). *Weighted Empirical Processes in Dynamic Nonlinear Models*. Springer.

Kraus, D. and Panaretos, V. M. (2011). Statistical inference on the second–order structure of functional data in the presence of influential observations. Technical report. École Polytechnique Fédérale de Lausanne.

Kuelbs, J. (1973). The invariance principle for Banach space valued random variables. *Journal of Multivariate Analysis*, **3,** 161–172.

Lahiri, S. N. (1996). On inconsistency of estimators based on spatial data under infill asymptotics. *Sankhya Series* A, **58,** 403–417.

Lahiri, S. N. (2003). Central limit theorems for weighted sums of a spatial process under a class of stochastic and fixed designs. *Sankhya, Series* A, **65,** 356–388.

Lahiri, S. N. and Zhu, J. (2006). Resampling methods for spatial regression models under a class of stochastic designs. *Annals of Statistics*, **34,** 1774–1813.

Lastovicka, J., A, V. Mikhailov, Ulich, T., Bremer, J., Elias, A., Ortiz de Adler, N., Jara, V., Abbarca del Rio, R., Foppiano, A., Ovalle, E. and Danilov, A. (2006). long term trends in foF2: a comparison of various methods. *Journal of Atmospheric and Solar-Terrestrial Physics*, **68,** 1854–1870.

Lastovicka, J., Akmaev, R. A., Beig, G., Bremer, J., Emmert, J. T., Jacobi, C., Jarvis, J. M., Nedoluha, G., Portnyagin, Yu. I. and Ulich, T. (2008). Emerging pattern of global change in the upper atmosphere and ionosphere. *Annales Geophysicae*, **26,** 1255–1268.

Laukaitis, A. and Račkauskas, A. (2002). Functional data analysis of payment systems. *Nonlinear Analysis: Modeling and Control*, **7,** 53–68.

Laukaitis, A. and Račkauskas, A. (2005). Functional data analysis for clients segmentation tasks. *European Journal of Operational Research*, **163,** 210–216.

Lavielle, M. and Teyssiére, G. (2006). Detection of multiple change-points in multivariate time series. *Lithuanian Mathematical Journal*, **46,** 287–306.

Lehmann, E. L. (1999). *Elements of Large Sample Theory*. Springer.

Leng, X. and Müller, H-G. (2006). Classification using functional data analysis for temporal gene expression data. *Bioinformatics*, **22,** 68–76.

Leon, S. (2006). *Linear Algebra with Applications*. Pearson.

Leurgans, S. E., Moyeed, R. A. and Silverman, B. W. (1993). Canonical correlation analysis when the data are curves. *Journal of the Royal Statistical Society (B)*, **55,** 752–740.

Li, W. K. and McLeod, A. I. (1981). Distribution of the residual autocorrelations in multivariate ARMA time series models. *Journal of the Royal Statistical Society (B)*, **43,** 231–239.

Li, Y. and Hsing, T. (2007). On rates of convergence in functional linear regression. *Journal of Multivariate Analysis*, **98,** 1782–1804.

Li, Y. and Hsing, T. (2010). Deciding the dimension of effective dimension reduction space for functional and high-dimensional data. *The Annals of Statistics*, **38,** 3028–3062.

Li, Y., Wang, N. and Carroll, R. J. (2010). Generalized functional linear models with semiparametric single–index interactions. *Journal of the American Statistical Association*, **105,** 621–633.

Liu, W. and Wu, W. B. (2010). Asymptotics of spectral density estimates. *Econometric Theory*, **26,** 1218–1245.

Ljung, G. and Box, G. (1978). On a measure of lack of fit in time series models. *Biometrika*, **66,** 67–72.

Loh, W.-L. (2005). Fixed-domain asymptotics for a subclass of Matern-type Gaussian random fields. *Annals of Statistics*, **33,** 2344–2394.

López-Pintado, S. and Romo, J. (2009). On the concept of depth for functional data. *Journal of the American Statistical Association*, **104,** 718–734.

Lütkepohl, H. (2005). *New Introduction to Multiple Time Series Analysis*. Springer.

Ma, P. and Zhong, W. (2008). Penalized clustering of large scale functional data with multiple covariates. *Journal of the American Statistical Association*, **103,** 625–636.

Malfait, N. and Ramsay, J. O. (2003). The historical functional model. *Canadian Journal of Statistics*, **31,** 115–128.

Mas, A. (2002). Weak convergence for the covariance operators of a Hilbertian linear process. *Stochastic Processes and their Applications*, **99,** 117–135.

Maslova, I., Kokoszka, P., Sojka, J. and Zhu, L. (2009). Removal of nonconstant daily variation by means of wavelet and functional data analysis. *Journal of Geophysical Research*, **114,** A03202.

Maslova, I., Kokoszka, P., Sojka, J. and Zhu, L. (2010a). Estimation of Sq variation by means of multiresolution and principal component analyses. *Journal of Atmospheric and Solar–Terrestial Physics*, **72,** 625–632.

Maslova, I., Kokoszka, P., Sojka, J. and Zhu, L. (2010b). Statistical significance testing for the association of magnetometer records at high–, mid– and low latitudes during substorm days. *Planetary and Space Science*, **58,** 437–445.

McKeague, I. and Sen, B. (2010). Fractals with point impacts in functional linear regression. *The Annals of Statistics*, **38,** 2559–2586.

McMurry, T. and Politis, D. N. (2010). Resampling methods for functional data. In *Oxford Handbook on Statistics and FDA* (eds F. Ferraty and Y. Romain). Oxford University Press.

Mikhailov, A. V. and Marin, D. (2001). An interpretation of the f0F2 and hmF2 long-term trends in the framework of the geomagnetic control concept. *Annales Geophysicae*, **19,** 733–748.

Móricz, F. (1976). Moment inequalities and the strong law of large numbers. *Zeitschrift für Wahrscheinlichkeitstheorie und verwandte Gebiete*, **35,** 299–314.

Müller, H-G., and Yao, F. (2008). Functional additive models. *Journal of the American Statistical Association*, **103,** 1534–1544.

Müller, H.-G. (2009). Functional modeling of longitudinal data. In *Longitudinal Data Analysis* (eds G. Fitzmaurice, M. Davidian, G. Verbeke and G. Molenberghs), pp. 223–252. Wiley, New York.

Müller, H-G. and Stadtmüller, U. (2005). Generalized functional linear models. *The Annals of Statistics*, **33,** 774–805.

Nerini, D., Monestiez, P. and Mantéa, C. (2010). Cokriging for spatial functional data. *Journal of Multivariate Analysis*, **101,** 409–418.

Newey, W. K. and West, K. D. (1987). A simple, positive semi-definite, heteroskedasticity and autocorrelation consistent covariance matrix. *Econometrica*, **55,** 703–08.

Noble, B. (1969). *Applied Linear Algebra.* Prentice Hall, Englewood Cliffs, NJ.

Opsomer, J., Wand, Y. and Yang, Y. (2001). Nonparametric regression with correlated errors. *Statistical Science*, **16,** 134–153.

Panaretos, V. M., Kraus, D. and Maddocks, J. H. (2010). Second-order comparison of Gaussian random functions and the geometry of DNA minicircles. *Journal of the American Statistical Association*, **105,** 670–682.

Park, B. U., Kim, T. Y., Park, J-S. and Hwang, S. Y. (2009). Practically applicable central limit theorem for spatial statistics. *Mathematical Geosciences*, **41,** 555–569.

Pearson, K. (1901). On lines and planes of closest fit to systems of points in space. *Philosophical Magazine*, **2,** 559–572.

Percival, D. B. and Walden, A. T. (2000). *Wavelet Methods for Time Series Analysis.* Cambridge University Press, Cambridge.

Pötscher, B. and Prucha, I. (1997). *Dynamic Non–linear Econonometric Models. Asymptotic Theory*. Springer.

Qu, Z. and Perron, P. (2007). Estimating and testing structural changes in multivariate regressions. *Econometrica*, **75,** 459–502.

Rackauskas, A. and Suquet, C. (2006). Testing epidemic changes of infinite dimensional parameters. *Statistical Inference for Stochastic Processes*, **9,** 111–134.

Ramsay, J., Hooker, G. and Graves, S. (2009). *Functional Data Analysis with R and MATLAB*. Springer.

Ramsay, J. O. and Silverman, B. W. (2002). *Applied Functional Data Analysis*. Springer.

Ramsay, J. O. and Silverman, B. W. (2005). *Functional Data Analysis*. Springer.

Rao, P. and Griliches, Z. (1969). Small-sample properties of several two-stage regression methods in the context of auto-correlated errors. *Journal of the American Statistical Association*, **64,** 253–272.

Reed, M. and Simon, B. (1972). *Methods of Modern Mathematical Physics I: Functional Analysis*. Academic Press.

Reeder, R. (2011). Limit theorems in functional data analysis with applications. Ph.D. Thesis. University of Utah.

Reiss, P. T. and Ogden, R. T. (2007). Functional principal component regression and functional partial least squares. *Journal of the American Statistical Association*, **102,** 984–996.

Reiss, P. T. and Ogden, R. T. (2009a). Smoothing parameter selection for a class of semiparametric linear models. *Journal of the Royal Statistical Society (B)*, **71,** 505–523.

Reiss, P. T. and Ogden, R. T. (2010). Functional generalized linear models with images as predictors. *Biometrics*, **66,** 61–69.

Riesz, F. and Sz.-Nagy, B. (1990). *Functional Analysis*. Dover.

Rio, E. (1993). Covariance inequalities for strongly mixing processes. *Ann. Inst. H. Poincaré Probab. Statist.*, **29,** 587–597.

Rishbeth, H. (1990). A greenhouse effect in the ionosphere? *Planet. Space Sci.*, **38,** 945–948.

Robinson, P. M. (1998). Inference without smoothing in the presence of nonparametric autocorrelation. *Econometrica*, **66,** 1163–1182.

Roble, R. G. and Dickinson, R. E. (1989). How will changes in carbon dioxide and methane modify the mean structure of the mesosphere and thermosphere? *Geophys. Res. Lett.*, **16,** 1441–1444.

Rostoker, G. (2000). Effects of substorms on the stormtime ring current index Dst. *Annales Geophysicae*, **18,** 1390–1398.

Ruppert, D. (2011). *Statistics and Data Analysis for Financial Ingineering*. Springer.

Sacks, J. and Ylvisaker, D. (1966). Designs for regression problems with correlated errors. *The Annals of Mathematical Statistics*, **37,** 66–89.

Schabenberger, O. and Gotway, C. A. (2005). *Statistical Methods for Spatial Data Analysis*. Chapman & Hall/CRC.

Schmidt, F. (1907). Zur Theorie der linearen und nichtlinearen Integralgleichungen. Teil I. Entwicklung willkurlicher Funktionen nach Systemen vorgeschriebener. *Mathematische Annalen*, **63,** 433–476.

Schmoyer, R. L. (1994). Permutations test for correlation in regression errors. *Journal of the American Statistical Association*, **89,** 1507–1516.

Seber, G. A. F. (1984). *Multivariate Observations*. Wiley, New York.

Seber, G. A. F. and Lee, A. J. (2003). *Linear Regression Analysis*. Wiley, New York.

Shao, X. (2010). A self–normalized approach to confidence interval construction in time series. *Journal of the Royal Statistical Society (B)*, **72,** 343–366.

Shao, X. and Zhang, X. (2010). Testing for change points in time series. *Journal of the American Statistical Association*, **105,** 1228–1240.

Shi, J. Q. and Choi, T. (2011). *Gaussian Process Regression Analysis for Functional Data*. CRC Press.

Shibata, R. (1980). Asymptotically efficient selection of the order of the model for estimating parameters of a linear process. *The Annals of Statistics*, **8,** 147–164.

Shumway, R. H. and Stoffer, D. S. (1991). Dynamic linear models with switching. *Journal of the American Statistical Association*, **86,** 763–769; Correction V87 p. 913.

Shumway, R. H. and Stoffer, D. S. (2006). *Time Series Analysis and Its Applications with R Examples*. Springer.

Srivastava, M. S. and Worsley, K. J. (1986). Likelihood ratio tests for a change in the multivariate normal mean. *Journal of the American Statistical Association*, **81,** 199–204.

Stadtlober, E., Hörmann, S. and Pfeiler, B. (2008). Qualiy and performance of a PM10 daily forecasting model. *Athmospheric Environment*, **42,** 1098–1109.

Stein, M. L. (1999). *Interpolation of Spatial Data: Some Theory for Krigging*. Springer.

Stout, W. F. (1974). *Almost Sure Convergence*. Academic Press, New York.

Stute, W. (1984). The oscillation behavior of empirical processes: the multivariate case. *The Annals of Probability*, **12,** 361–379.

Sugiura, M. (1964). Hourly values of equatorial Dst for the IGY. *Ann. Int. Geophysical Year*, **35,** number 9; Pergamon Press, Oxford.

Székely, G. J. and Rizzo, M. L. (2009). Brownian distance covariances. *The Annals of Applied Statistics*, **3,** 1236–1265.

Székely, G. J., Rizzo, M. L. and Bakirov, N. K. (2007). Measuring and testing dependence by correlation of distances. *The Annals of Statistics*, **35,** 2769–2794.

Taniguchi, M. and Kakizawa, Y. (2000). *Asumptotic Theory of Statistical Inference for Time Series*. Springer.

Thiel, H. (1965). The analysis of disturbances in regression analysis. *Journal of the American Statistical Association*, **60,** 1067–1079.

Thiel, H. and Nagar, A. L. (1961). Testing the independence of regression disturbances. *Journal of the American Statistical Association*, **57,** 793–806.

Tsay, R. S. (2005). *Analysis of Financial Time Series*. Wiley.

Wackernagel, H. (2003). *Multivariate Geostatistics*. Springer.

Wu, W. (2005). *Nonlinear System Theory: Another Look at Dependence*. Proceedings of The National Academy of Sciences of the United States, volume 102. National Academy of Sciences.

Wu, W. (2007). Strong invariance principles for dependent random variables. *The Annals of Probability*, **35,** 2294–2320.

Xiao, Z., Linton, O. B., Carroll, R. J. and Mammen, E. (2003). More efficient local polynomial estimation in nonparametric regression with autocorrelated errors. *Journal of the American Statistical Association*, **98,** 980–992.

Xu, W-Y. and Kamide, Y. (2004). Decomposition of daily geomagnetic variations by using method of natural orthogonal component. *Journal of Geophysical Research*, **109,** A05218; DOI:10.1029/203JA010216.

Yao, F. and Müller, H.-G. (2010). Functional quadratic regression. *Biometrika*, **97,** 49–64.

Yao, F., Müller, H-G. and Wang, J-L. (2005a). Functional data analysis for sparse longitudinal data. *Journal of the American Statistical Association*, **100,** 577–590.

Yao, F., Müller, H-G. and Wang, J-L. (2005b). Functional linear regression analysis for longitudinal data. *The Annals of Statistics*, **33,** 2873–2903.

Zamba, K. D. and Hawkins, D. M (2006). A multivariate change-point model for statistical process control. *Technometrics*, **48,** 539–549.

Zhang, H. (2004). Inconsistent estimation and asymptotically equivalent interpolations in model-based geostatistics. *Journal of the American Statistical Association*, **99,** 250–261.

Zhang, J-T. and Chen, J. (2007). Statistical inference for functional data. *The Annals of Statistics*, **35,** 1052–1079.

Zhang, X., Shao, X., Hayhoe, K. and Wuebbles, D. (2011). Testing the structural stability of temporally dependent functional observations and application to climate projections. Technical Report. University of Illinois at Urbana–Champaign.

Zuo, Y. and Serfling, R. (2000). General notions of statistical depth function. *The Annals of Statistics*, **28,** 461–482.

Index

GPSR Compliance
The European Union's (EU) General Product Safety Regulation (GPSR) is a set of rules that requires consumer products to be safe and our obligations to ensure this.

If you have any concerns about our products, you can contact us on

ProductSafety@springernature.com

In case Publisher is established outside the EU, the EU authorized representative is:

Springer Nature Customer Service Center GmbH
Europaplatz 3
69115 Heidelberg, Germany

www.ingramcontent.com/pod-product-compliance
Ingram Content Group UK Ltd.
Pitfield, Milton Keynes, MK11 3LW, UK
UKHW022216051026
15624UKWH00001B/189
* 9 7 8 1 4 6 1 4 3 6 5 4 6 *